W9-COE-286

STATE UNIVERSITY COLLEGE
1981
BROCKPORT, N. Y.

Current Topics in Membranes and Transport

Volume 15
Molecular Mechanisms of Photoreceptor Transduction

Contributors

G. Aguirre
Bruce L. Bastian
D. A. Baylor
M. W. Bitensky
M. Deric Bownds
Joel E. Brown
Roger Calhoon
G. J. Chader
Adolph I. Cohen
John E. Dowling
Thomas G. Ebrey
Gordon L. Fain
Debòra B. Farber
Darrell Fleischman
R. T. Fletcher
Bernard K.-K. Fung
Geoffrey H. Gold
James B. Hurley
Paul Kilbride
Juan I. Korenbrot
Hermann Kühn
T. D. Lamb
P. A. Liebman
Stuart A. Lipton
Y. P. Liu
P. A. McNaughton
G. Matthews
Edward P. Meyertholen
William H. Miller
Grant D. Nicol
Burks Oakley II
Sanford E. Ostroy
W. Geoffrey Owen
Lawrence H. Pinto
E. N. Pugh, Jr.
M. M. Rasenick
R. Santos-Anderson
Hitoshi Shichi
Robert T. Sorbi
Peter J. Stein
Lubert Stryer
Roberta A. Svoboda
Ete Z. Szuts
M. T'so
Vincent Torre
Motoyuki Tsuda
Geraldine Waloga
G. L. Wheeler
Meegan J. Wilson
A. Yamazaki
K.-W. Yau

Current Topics in Membranes and Transport

Edited by

Felix Bronner
Department of Oral Biology
University of Connecticut Health Center
Farmington, Connecticut

Arnost Kleinzeller
Department of Physiology
University of Pennsylvania School of Medicine
Philadelphia, Pennsylvania

VOLUME 15

Molecular Mechanisms of Photoreceptor Transduction

Guest Editor

William H. Miller
Department of Ophthalmology and Visual Science
and Department of Physiology
Yale University School of Medicine
New Haven, Connecticut

Volume 15 is part of the series (p. xxi) from the Yale Department of Physiology under the editorial supervision of:

Joseph F. Hoffman
Department of Physiology
Yale University School of Medicine
New Haven, Connecticut

Gerhard Giebisch
Department of Physiology
Yale University School of Medicine
New Haven, Connecticut

1981

Academic Press
A Subsidiary of Harcourt Brace Jovanovich, Publishers

New York London Toronto Sydney San Francisco

ACADEMIC PRESS, INC.
111 Fifth Avenue, New York, New York 10003

United Kingdom Edition published by
ACADEMIC PRESS, INC. (LONDON) LTD.
24/28 Oval Road, London NW1 7DX

LIBRARY OF CONGRESS CATALOG CARD NUMBER: 70–117091

ISBN 0–12–153315–8

PRINTED IN THE UNITED STATES OF AMERICA

81 82 83 84 9 8 7 6 5 4 3 2 1

Opinionis enim commenta delet dies, naturae judicia confirmat.
Cicero, *De Natura Deorum II.2.5*

Time obliterates the fictions of opinion and confirms the decisions of nature.
JOHNSON

Contents

CHAPTER 7. **Light Control of Cyclic-Nucleotide Concentration in the Retina**

THOMAS G. EBREY, PAUL KILBRIDE, JAMES B. HURLEY, ROGER CALHOON, AND MOTOYUKI TSUDA

CHAPTER 8. **Cyclic-GMP Phosphodiesterase and Calmodulin in Early-Onset Inherited Retinal Degenerations**

G. J. CHADER, Y. P. LIU, R. T. FLETCHER, G. AGUIRRE, R. SANTOS-ANDERSON, AND M. T'SO

CHAPTER 9. **Control of Rod Disk Membrane Phosphodiesterase and a Model for Visual Transduction**

P. A. LIEBMAN AND E. N. PUGH, JR.

CHAPTER 10. **Interactions of Rod Cell Proteins with the Disk Membrane: Influence of Light, Ionic Strength, and Nucleotides**

HERMANN KÜHN

CHAPTER 14. **Cyclic-Nucleotide Metabolism in Vertebrate Photoreceptors: A Remarkable Analogy and an Unraveling Enigma**

M. W. BITENSKY, G. L. WHEELER, A. YAMAZAKI, M. M. RASENICK, AND P. J. STEIN

CHAPTER 15. **Guanosine Nucleotide Metabolism in the Bovine Rod Outer Segment: Distribution of Enzymes and a Role of GTP**

HITOSHI SHICHI

CHAPTER 16. **Calcium Tracer Exchange in the Rods of Excised Retinas**

ETE Z. SZUTS

List of Contributors

Numbers in parentheses indicate the pages on which the authors' contributions begin.

G. Aguirre, Section Ophthalmology, School of Veterinary Medicine, University of Pennsylvania, Philadelphia, Pennsylvania 19104 (133)

Bruce L. Bastian, Department of Ophthalmology, Jules Stein Eye Institute, UCLA School of Medicine, Los Angeles, California 90024 (341)

D. A. Baylor, Department of Neurobiology, Stanford University School of Medicine, Stanford, California 94305 (3)

M. W. Bitensky,[1] Department of Pathology, Yale University School of Medicine, New Haven, Connecticut 06510 (237)

M. Deric Bownds, Laboratory of Molecular Biology, Department of Zoology, and Neurosciences Training Program, University of Wisconsin-Madison, Madison, Wisconsin 53706 (203)

Joel E. Brown, Department of Physiology and Biophysics, State University of New York at Stony Brook, Stony Brook, New York 11794 (369)

Roger Calhoon, Department of Physiology and Biophysics, University of Illinois, Urbana, Illinois 61801 (121)

G. J. Chader, Laboratory of Vision Research, National Eye Institute, National Institutes of Health, U.S. Department of Health and Human Services, Bethesda, Maryland 20205 (133)

Adolph I. Cohen, Departments of Anatomy-Neurobiology, and Ophthalmology, Washington University School of Medicine, St. Louis, Missouri 63110 (215)

John E. Dowling, Department of Biology, Harvard University, Cambridge, Massachusetts 02138 (381)

Thomas G. Ebrey, Department of Physiology and Biophysics, University of Illinois, Urbana, Illinois 61801 (121)

Gordon L. Fain, Department of Ophthalmology, Jules Stein Eye Institute, UCLA School of Medicine, Los Angeles, California 90024 (341)

Debora B. Farber, Jules Stein Eye Institute, UCLA School of Medicine, Los Angeles, California 90024, and Developmental Neurology Laboratory, Veterans Administration Medical Center, Sepulveda, California 91343 (231)

Darrell Fleischman, Charles F. Kettering Research Laboratory, Yellow Springs, Ohio 45387 (109)

R. T. Fletcher, Laboratory of Vision Research, National Eye Institute, National Institutes of Health, U.S. Department of Health and Human Services, Bethesda, Maryland 20205 (133)

[1]Present address: Division of Life Sciences, Los Alamos Scientific Laboratory, Los Alamos, New Mexico 87544.

Bernard K.-K. Fung,[2] Department of Structural Biology, Sherman Fairchild Center, Stanford University School of Medicine, Stanford, California 94305 (93)

Geoffrey H. Gold,[3] Department of Physiology, University of California School of Medicine, San Francisco, California 94143 (59, 307)

James B. Hurley, Department of Structural Biology, Sherman Fairchild Center, Stanford University School of Medicine, Stanford, California 94305 (93, 121)

Paul Kilbride, Pharmacology Department, Washington University Medical School, St. Louis, Missouri 63110 (121)

Juan I. Korenbrot, Departments of Physiology and Biochemistry, University of California School of Medicine, San Francisco, California 94143 (307)

Hermann Kühn, Institut für Neurobiologie der Kernforschungsanlage Jülich, D-5170 Jülich, Federal Republic of Germany (171)

T. D. Lamb, Physiological Laboratory, Cambridge University, Cambridge CB2 3EG, England (19)

P. A. Liebman, Department of Anatomy, School of Medicine, University of Pennsylvania, Philadelphia, Pennsylvania 19104 (157)

Stuart A. Lipton, Department of Neurology and Neurobiology, Harvard Medical School, Boston, Massachusetts 02115 (381)

Y. P. Liu, Laboratory of Vision Research, National Eye Institute, National Institutes of Health, U.S. Department of Health and Human Services, Bethesda, Maryland 20205 (133)

G. Matthews,[4] Department of Neurobiology, Stanford University School of Medicine, Stanford, California 94305 (3)

P. A. McNaughton, Physiological Laboratory, Cambridge University, Cambridge CB2 3EG, England (19)

Edward P. Meyertholen, Department of Biological Sciences, Purdue University, West Lafayette, Indiana 47907 (393)

William H. Miller, Department of Ophthalmology and Visual Science, and Department of Physiology, Yale University School of Medicine, New Haven, Connecticut 06510 (417, 441)

Grant D. Nicol,[5] Yale University School of Medicine, New Haven, Connecticut 06510 (417)

Burks Oakley II,[6] Department of Biological Sciences, Purdue University, West Lafayette, Indiana 47907 (405)

[2]Present address: Department of Radiation Biology and Biophysics, University of Rochester Medical Center, Rochester, New York 14627.

[3]Present address: Department of Physiology, Yale University School of Medicine, New Haven, Connecticut 06510.

[4]Present address: Department of Neurobiology and Behavior, State University of New York at Stony Brook, Stony Brook, New York 11794.

[5]Present address: Department of Electrical Engineering and Computer Science, University of California, Berkeley, California 94720.

[6]Present address: Department of Electrical Engineering, University of Illinois at Urbana-Champaign, Urbana, Illinois 61801.

Sanford E. Ostroy, Department of Biological Sciences, Purdue University, West Lafayette, Indiana 47907 (393)

W. Geoffrey Owen,[7] Physiological Laboratory, Cambridge University, Cambridge CB2 3EG, England (33)

Lawrence H. Pinto, Department of Biological Sciences, Purdue University, West Lafayette, Indiana 47907 (405)

E. N. Pugh, Jr., Department of Psychology, University of Pennsylvania, Philadelphia, Pennsylvania 19104 (157)

M. M. Rasenick, Department of Pathology, Yale University School of Medicine, New Haven, Connecticut 06510 (237)

R. Santos-Anderson, Department of Ophthalmology, University of Illinois Medical School, Chicago, Illinois 60612 (133)

Hitoshi Shichi, Laboratory of Vision Research, National Eye Institute, National Institutes of Health, U.S. Department of Health and Human Services, Bethesda, Maryland 20205 (273)

Robert T. Sorbi, Istituto di Fisologia Umana, University of Parma, Parma 43100, Italy (331)

Peter J. Stein,[8] Department of Biological Sciences, Purdue University, West Lafayette, Indiana 47907 (237, 393)

Lubert Stryer, Department of Structural Biology, Sherman Fairchild Center, Stanford University School of Medicine, Stanford, California 94305 (93)

Roberta A. Svoboda, Department of Biological Sciences, Purdue University, West Lafayette, Indiana 47907 (393)

Ete Z. Szuts, Laboratory of Sensory Physiology, Marine Biological Laboratory, Woods Hole, Massachusetts 02543 (291)

M. T'so, Department of Ophthalmology, University of Illinois Medical School, Chicago, Illinois 60612 (133)

Vincent Torre, Physiological Laboratory, Cambridge University, Cambridge CB2 3EG, England (33)

Motoyuki Tsuda, Department of Physics, Sapporo Medical College, Sapporo 060, Japan (121)

Geraldine Waloga,[9] Department of Physiology and Biophysics, State University of New York at Stony Brook, Stony Brook, New York 11794 (369)

G. L. Wheeler, Department of Chemistry, University of New Haven, West Haven, Connecticut 06516 (237)

[7]Present address: Department of Biophysics and Medical Physics, University of California, Berkeley, California 94720.

[8]Present address: Department of Pathology, Yale University School of Medicine, New Haven, Connecticut 06510.

[9]Present address: Department of Physiology, Boston University School of Medicine, Boston, Massachusetts 02215.

Meegan J. Wilson, Department of Biological Sciences, Purdue University, West Lafayette, Indiana 47907 (393)

A. Yamazaki, Department of Pathology, Yale University School of Medicine, New Haven, Connecticut 06510 (237)

K.-W. Yau,[10] Physiological Laboratory, Cambridge University, Cambridge CB2 3EG, England (19)

[10]Present address: Department of Physiology and Biophysics, The University of Texas Medical Branch, Galveston, Texas 77550.

Preface

The contributions to this book are designed to review current knowledge of the molecular transmitter systems inside the vertebrate rod outer segment. These molecular systems link the absorption of light by rhodopsin with the decreased sodium ion permeability of the plasma membrane (reviewed in the "Handbook of Sensory Physiology," Vol. VII/2, Springer-Verlag). Calcium and guanosine 3′:5′-monophosphate (cGMP), both of which are plentiful in the rod outer segment, are thought to be key elements of these molecular transmitter systems. Both are also suspected of having their cytosol-free concentrations controlled by the state of illumination. The problem is to delineate the roles of calcium, cGMP, and associated molecules acting as possible internal (within the rod outer segment) intermediary transmitter substances.

The concept of the internal transmitter in photoreception originated with the observation by M. G. F. Fuortes (1959) that the nonphotoreceptor eccentric cell of the *Limulus* ommatidium decreases in resistance when the ommatidium is illuminated. "It was suggested," according to Borsellino *et al.* (1965), "that the results could be interpreted most easily by assuming that the [photoreceptor] rhabdome produces [externally] a [transmitter] substance that diffuses to the eccentric cell's membrane and changes its resistance. As an alternative interpretation, Dr. Hartline (personal communication) [H. K. Hartline, then at Rockefeller University] suggested that light might change the resistance of the retinula cells without [directly] affecting the eccentric cell. . . . This view had also been proposed by Tomita, Kikuchi, and Tanaka (1960) based on comparison of responses recorded with intracellular and extracellular electrodes."

That light does change the resistance of both invertebrate and vertebrate photoreceptors is now established (reviewed in the "Handbook of Sensory Physiology," Vol. VII/2, Springer-Verlag). We are concerned here only with the vertebrate which responds to illumination with a decreased conductance to sodium ion (Sillman *et al.,* 1969; Penn and Hagins, 1969). Baylor and Fuortes (1970) analyzed this decreased conductance caused by light in turtle cones and were led in accordance with the *Limulus* photoreceptor interpretation to assume "that absorption of photons leads to the production of an intermediary substance (see Fuortes and Hodgkin, 1964; Borsellino, Fuortes and Smith, 1965) and that this substance decreases membrane conductance by interacting with the [plasma membrane sodium] channels." While an internal intermediary substance may be "released" from rod disks by light to transmit to the separate plasma membrane, the internal transmitter concept is nevertheless independent of photoreceptor organelle morphology. The concept applies equally to vertebrate cones in which the disks are continuous infoldings of the plasma membrane, arthropod rhabdomal microvilli of the plasma membrane, molluskan *Pecten* distal sense cell appendages which are ciliary leaflets of the plasma membrane, and vertebrate rods in which the disks are for the most part separate from the plasma membrane. The common denominator of transduction which is independent of the morphology is that the absorption of light by 11-*cis*-retinal causes a change in photoreceptor permeability by unknown intermediary molecular events. Although the internal transmitter concept is general, detailed knowledge is most complete for the rod outer segment. Therefore the plan of this volume is to review for the rod outer segment (1) its electrical response, (2) biochemical data on the control of calcium and cGMP, and (3) the physiological effects of these molecules.

WILLIAM H. MILLER

REFERENCES

Baylor, D. A., and Fuortes, M. G. F. (1970). Electrical responses of single cones in the retina of the turtle. *J. Physiol. (London)* **207,** 77–92.

Borsellino, A., Fuortes, M. G. F., and Smith, T. G. (1965). Visual responses in *Limulus*. *Cold Spring Harbor Symp. Quant. Biol.* **30,** 429–443.

Fuortes, M. G. F. (1959). Initiation of impulses in visual cells of *Limulus*. *J. Physiol. (London)* **148,** 14–28.

Fuortes, M. G. F., and Hodgkin, A. L. (1964). Changes in time scale and sensitivity in the ommatidia of *Limulus*. *J. Physiol. (London)* **172,** 239–263.

Penn, R. D., and Hagins, W. A. (1969). Signal transmission along retinal rods and the origin of the electroretinographic a-wave. *Nature (London)* **223,** 201–205.

Sillman, A. J., Ito, H., and Tomita, T. (1969). Studies on the mass receptor potential of the isolated frog retina. II. On the basis of the ionic mechanism. *Vision Res.* **9,** 1443–1451.

Tomita, T., Kikuchi, R., and Tanaka, I. (1960). Excitation and inhibition in lateral eye of horsehoe crab. *In* "Electrical Activity of Single Cells" (Y. Katsuki, ed.), pp. 11–23. Igakushoin, Tokyo.

Yale Membrane Transport Processes Volumes

Joseph F. Hoffman (ed.). (1978). "Membrane Transport Processes," Vol. 1. Raven, New York.

Daniel C. Tosteson, Yu. A. Ovchinnikov, and Ramon Latorre (eds.). (1978). "Membrane Transport Processes," Vol. 2. Raven, New York.

Charles F. Stevens and Richard W. Tsien (eds.). (1979). "Membrane Transport Processes," Vol. 3: Ion Permeation through Membrane Channels. Raven, New York.

Emile L. Boulpaep (ed.). (1980). "Cellular Mechanisms of Renal Tubular Ion Transport": Volume 13 of *Current Topics in Membranes and Transport* (F. Bronner and A. Kleinzeller, eds.). Academic Press, New York.

William H. Miller (ed.). (1981). "Molecular Mechanisms of Photoreceptor Transduction": Volume 15 of *Current Topics in Membranes and Transport* (F. Bronner and A. Kleinzeller, eds.). Academic Press, New York.

In preparation

Clifford L. Slayman (ed.). "Electrogenic Ion Pumps": Volume 16 of *Current Topics in Membranes and Transport* (A. Kleinzeller and F. Bronner, eds.). Academic Press, New York.

Contents of Previous Volumes

Volume 4

Volume 5

Volume 6

Volume 7

Volume 8

Volume 9

Volume 10

Volume 11

Volume 12

Volume 13

Cellular Mechanisms of Renal Tubular Ion Transport

Volume 14

Carriers and Membrane Transport Proteins

Part I
The Rod Physiological Response

Properties of the physiological response are what must be mediated by internal intermediary transmitter substances

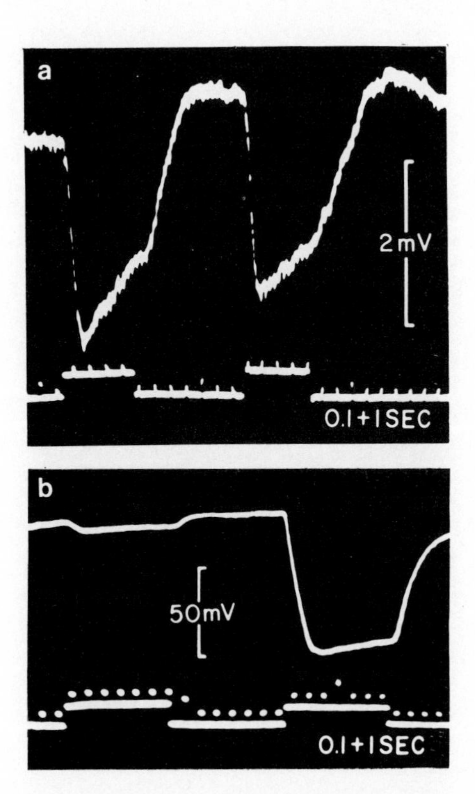

First documented intracellular recording from a vertebrate photoreceptor (a) contrasted with the response from a horizontal cell (b). First response in each record is to a 0.2-mm-diameter light spot; second response is to diffuse illumination. In both cases the response is hyperpolarizing, but only the horizontal response shows a substantial area effect. (From T. Tomita, Mechanisms subserving color coding in the vertebrate retina; reproduced from Fig. 2, Abstracts II, cIII.1, IOPAB; Int. Biophysics Meeting, Paris-Orsay, June 1964; with permission.)

CURRENT TOPICS IN MEMBRANES AND TRANSPORT, VOLUME 15

Chapter 1

The Photocurrent and Dark Current of Retinal Rods

G. MATTHEWS[1] *AND D. A. BAYLOR*

Department of Neurobiology
Stanford University School of Medicine
Stanford, California

Intracellular recordings in retinas from a variety of vertebrates have shown that rod and cone photoreceptors give graded hyperpolarizations upon absorbing light

[1]Present address: Department of Neurobiology and Behavior, SUNY at Stony Brook, Stony Brook, New York.

ISBN 0-12-153315-8

(Bortoff, 1964; Tomita, 1965, 1970; Toyoda *et al.*, 1969; Werblin and Dowling, 1969; Baylor and Fuortes, 1970). Light is thought to hyperpolarize by reducing an inward current of sodium ions across the photoreceptor membrane; evidence for this ionic mechanism is discussed by Owen and Torre in this volume. The decrease in sodium conductance appears to result from changes in the cytoplasmic concentration of an internal transmitter substance, the identity of which has not yet been established. In rods the involvement of an internal transmitter is indicated by the finding that most of the visual pigment, rhodopsin, is located in membranous intracellular disks which are not in continuity with the external plasma membrane of the outer segment (Cohen, 1970; Yoshikami *et al.*, 1974) where the sodium conductance change occurs (Penn and Hagins, 1972). In cones, a similar mechanism is suggested by the hyperbolic relation between conductance and light intensity (Baylor and Fuortes, 1970) and by the kinetics of the light response (Baylor *et al.*, 1974). In each kind of receptor, it is assumed that the transmitter is free to diffuse within the cell, exists at a concentration dependent on light intensity, and acts to regulate the membrane conductance, perhaps by direct combination with ionic channels. Attempts to identify the internal transmitter substance are described in several later chapters in this volume.

In this chapter we review some recent electrophysiological experiments on the conductance-controlling processes in the rod outer segment. The strategy was to examine a single outer segment's membrane current, a variable more directly related to the primary transduction process than the intracellular potential, which has been widely studied in previous work. A more detailed account of the work described here can be found in Baylor *et al.* (1979a,b, 1980) and Yau *et al.* (1979).

I. MEASUREMENT OF THE ROD OUTER SEGMENT'S MEMBRANE CURRENT

A. Rationale

Intracellular recording reveals the electrical signals of rods in their full complexity. The variable observed, membrane potential, is important because it regulates the transfer of information to other cells, and such recordings have given important insights into the transduction mechanism. Nevertheless, membrane potential reflects not only the primary channel-blocking process in the outer segment but also several secondary signal-shaping processes. These mechanisms complicate analysis of the blocking process by intracellular recording. One such signal-shaping mechanism is the electrical coupling interaction between rods (Fain *et al.*, 1975; Schwartz, 1975, 1976; Copenhagen and Owen, 1976a,b; see Gold, Chapter 4, this volume). Here current flows between receptor interiors over a network of specialized gap junctions, causing the internal potential of a given receptor to represent an average over a group of neighboring cells (Lamb and Simon, 1976). The averaging operation reduces the amplitude of fluctuations such as single photon effects so much that they cannot be individually resolved. Another group of signal-shaping processes are the conductance changes that occur in response to the light-induced hyperpolarization (see Owen and Torre, Chapter 3, this volume, and Bastian and Fain, Chapter 19, this volume). These processes modify the hyperpolarization so that it does not accurately reflect the kinetics of the change in the light-sensitive conductance.

B. Method

To obtain a clearer picture of the primary events, a technique was developed for recording transmembrane current from a single rod outer segment (Yau *et al.*, 1977; Baylor *et al.*, 1979a). The method is based on that of Neher and Sakmann (1976) for recording single channel currents from muscle fibers and relies on the fact that light suppresses a net inward membrane current present along the length of the outer segment in darkness (Penn and Hagins, 1972). To observe the current an outer segment projecting from a piece of retina is drawn by gentle suction into the fire-polished tip of a glass micropipet. If the fit is snug and the leakage resistance between the electrode wall and the outer segment membrane sufficiently high, the current can be accurately measured by a current-to-voltage transducer connected to the inside of the suction electrode. The arrangement is diagrammed in Fig. 1.

Experiments were performed on the retina of the toad *Bufo marinus*, chosen because of the large size of its red rod outer segments and the availability of physiological information from previous intracellular recordings (Brown and Pinto, 1974; Fain *et al.*, 1975; Cervetto *et al.*, 1977) and microspectrophotometry (Harosi, 1975). A retina was isolated from a thoroughly dark-adapted eye under infrared light using a dissecting microscope fitted with an infrared–visible image converter. The retina was then chopped into small pieces about 200 μm on a side and placed in a saline-filled chamber on the stage of a compound inverted microscope equipped with an infrared viewing system. Light stimuli were steps or flashes in a narrow spectral band centered at 500 nm wavelength, near the absorption maximum for the rhodopsin present in the red rods (Harosi, 1975). The stimuli could be applied in the form of narrow slits of variable dimensions, position, and orientation, as well as diffusely over the entire piece of retina. Usually the stimuli were plane-polarized with the electric vector transverse to the long axis of the rod (i.e., in the preferred orientation).

C. Independence of Current Signals from Secondary Shaping Processes

Current signals recorded in this way show little if any contribution from the excitation of neighboring rods electrically coupled to the central cell (Baylor *et*

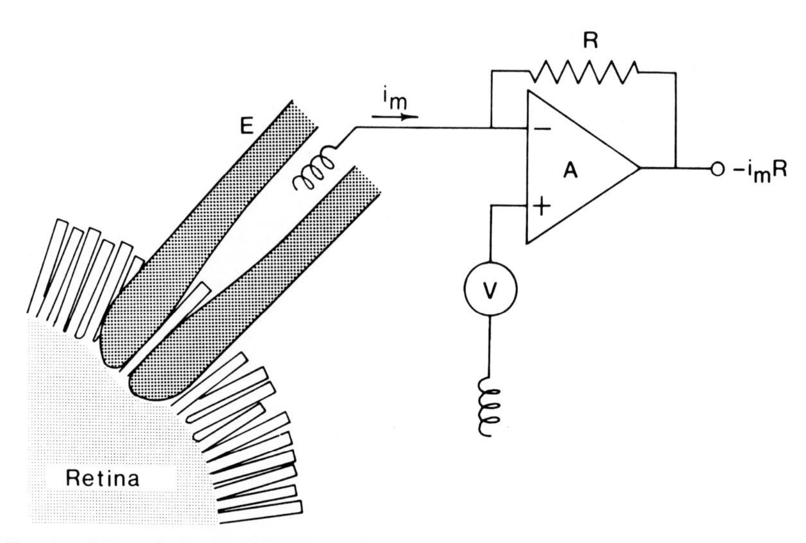

FIG. 1. Schematic diagram of an arrangement for recording the membrane current of a rod outer segment in a piece of retina. A suitably oriented outer segment is drawn by gentle suction into the tip of a fire-polished glass electrode (E). The inside of the electrode is connected to the current-to-voltage transducer amplifier (A). The voltage output of this amplifier is proportional to the membrane current. A variable voltage source (V) is used to compensate for electrode potentials and to measure electrode resistance. Figure reproduced with permission from Yau *et al.* (1977).

al., 1979a,b). Single photons absorbed in adjacent cells give no detectable response, presumably because they change the driving force on the dark current of the recorded cell by less than 1%. For larger signals the photocurrent is found to be very similar for local stimulation of only the recorded outer segment and diffuse illumination of the entire piece of retina. This is consistent with the finding (Bader *et al.*, 1979) that current in the light-sensitive conductance depends only weakly on the driving force in the physiological range of potentials. We assume provisionally that the recorded currents are very similar to those that would be obtained if the internal potential were clamped at the dark level.

II. GENERAL FEATURES OF THE ROD PHOTOCURRENT

A. Dependence on Flash Intensity: Relation to Dark Current

Brief flashes gave transient reductions in the inward dark current of the outer segment. A family of superimposed responses to flashes of increasing intensity is shown in Fig. 2. After a dim flash the photocurrent rose with an S-shaped delay to a peak at 1–2 sec after the flash and then declined. As found previously with

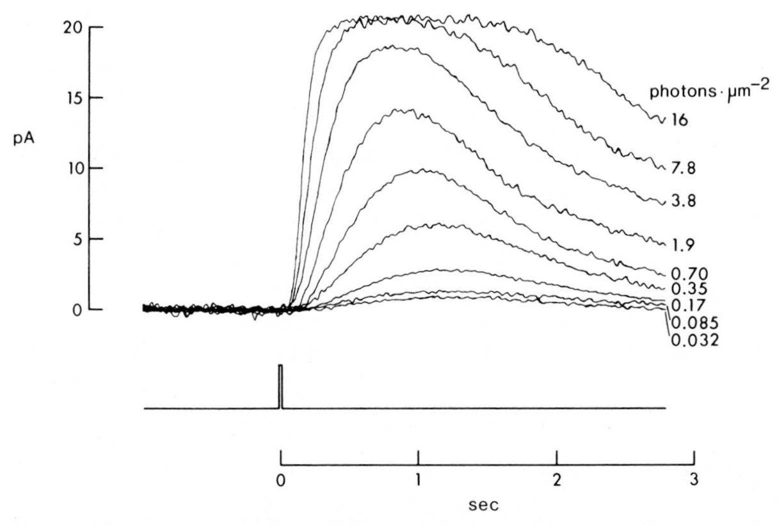

FIG. 2. Superimposed responses to flashes of increasing intensity. The change in membrane current from its dark level is plotted against time, with outward currents upward. The pulse beneath the traces indicates the timing of the 20-msec 500-nm flash of light delivered locally to the outer segment of the recorded cell. The numbers to the right of each trace give the light intensity in photons per square micrometer per flash. The lowest three traces are averages of 10, 8, and 22, responses respectively; other traces are based on 1–4 responses. Currents were low-pass-filtered at 20 Hz. Figure reproduced with permission from Baylor *et al.* (1979a).

other recording methods (Penn and Hagins, 1972; Fain *et al.,* 1975), the peak response amplitude grew with increasing flash strength approximately according to a rectangular hyperbola; the half-saturating photon density was near one photon per square micrometer for polarized 500-nm light. Based on an effective collecting area of about 35 μm^2, a half-saturating flash gave roughly 35 photoisomerizations. Although in Fig. 2 the ordinate is the change in membrane current from the level in darkness, it may be shown that the plateau of the saturating response corresponds to zero membrane current (Penn and Hagins, 1972; Baylor *et al.,* 1979a); thus for the cell in Fig. 2 there was an inward dark current of about 20 pA which was abolished by bright light.

B. Difference in Form of Voltage and Current Responses to Bright Flashes: Interpretation

The response to the brightest flash in Fig. 2 shows no sign of the initial spikelike transient or "nose" characteristic of the intracellular voltage response of rods (Schwartz, 1973; Brown and Pinto, 1974; Fain *et al.,* 1975; Detwiler *et al.,* 1980). Absence of the transient was a consistent observation in a large number of current-recording experiments (see Baylor *et al.*, 1979a). The different shapes of the current and voltage responses have recently been confirmed in simultaneous recordings of internal voltage and membrane current from nearby rods in the same piece of retina (Nunn *et al.*, 1980); this establishes that the relaxation of potential from the initial peak hyperpolarization is not due to a reopening of light-sensitive channels. The channels delivering the inward current during the voltage relaxation are presumably not localized in the outer segment.

C. Longitudinal Density of Dark Current

When bright diffuse flashes were given while varying the length of outer segment in the recording suction electrode, the size of the maximum light response varied linearly with recorded length. This is consistent with the notion that the sites through which the inward dark current flows are present at relatively uniform longitudinal density in the outer segment envelope.

D. Kinetic Difference between Flash Responses from Base and Tip of Outer Segment

Excitation by narrow transverse slits applied at various longitudinal positions on the outer segment showed the dim flash response to be faster in time to peak for stimuli at the basal end than at the distal tip of the outer segment (see Baylor *et al.*, 1979b). This kinetic difference presumably arises from a difference in internal transmitter kinetics at the two positions; it cannot be explained by the rod's properties as an electrical cable. The constant longitudinal density of the dark current also rules out the possibility that transmitter released at the rod tip must simply diffuse further, to a collection of light-sensitive channels at the base. Transmitter kinetics might depend on longitudinal disk location if there were a longitudinal concentration gradient of a diffusible internal metabolite or ion involved in the intermediary processes of transduction; alternatively, the composition of the disks might differ in some way connected with their aging and renewal.

III. THE SINGLE-PHOTON EFFECT

A. Identification

Figure 3, from Baylor *et al.* (1979b), shows the rod photocurrents evoked by a series of dim flashes of fixed applied intensity. The currents fluctuated between failures and responses of varying size in a manner reminiscent of synaptic transmission at the neuromuscular junction when the quantum content is low. Amplitude histograms of such response series were consistent with the Poisson-

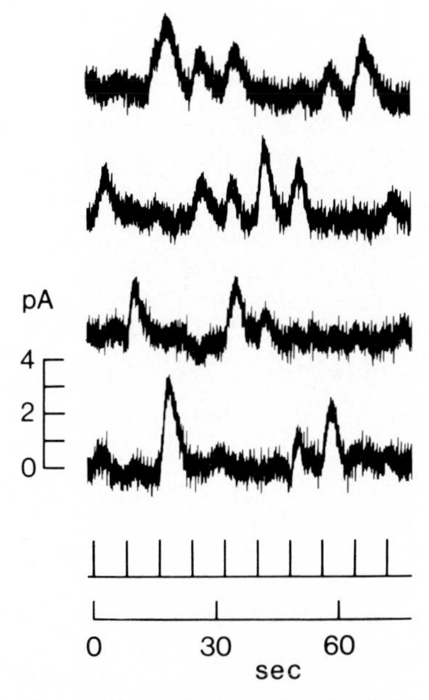

FIG. 3. Responses of a rod to a series of 40 consecutive dim flashes. Ticks below traces indicate timing of 20-msec 500-nm local flashes. Responses are plotted as changes in membrane current from its dark level, with outward current upward. Low-pass-filtered at 30 Hz. Flash photon density was 0.029 photons per square micrometer. Figure reproduced with permission from Baylor *et al.* (1979b).

distributed occurrence of a unitary event having an amplitude of about 1 pA. The probability of obtaining one or more such unitary responses was determined as a function of flash intensity and could be fitted by the cumulative Poisson distribution describing the probability of occurrence of one or more photoisomerizations in the cell. This statistical behavior gives strong evidence that the unitary response, with an amplitude of about 1 pA, is triggered by absorption of a single photon.

B. Kinetics

The kinetics of the average response to a single photon were determined by curve-fitting procedures similar to those used by Baylor *et al.* (1974). Quantitative models based on a series of four low-pass filters (or, in chemical terms, four first-order sequential reactions) gave a satisfactory description of the shape of the response as a function of time (Baylor *et al.*, 1979a,b, 1980); similar findings for rat rods were reported earlier by Penn and Hagins (1972). In many cells the response was fitted by the "Poisson" expression

$$r(t) = ik(\alpha t)^3 e^{-\alpha t} \tag{1}$$

where $r(t)$ is the reduction in membrane current as a function of time, i is the flash photon density, k is a sensitivity constant, and α is a rate constant characteristic of the cell, usually with a value between 1.5 and 3 sec^{-1} near 20°C. This formula gives the output expected if the excitation generated by a brief flash were passed through a series of four linear first-order delay processes with identical time constants $1/\alpha$. The implication is that four slow sequential processes may determine the final excitation of the outer membrane. It should be pointed out that the procedure used to determine the number of delay stages is relatively insensitive to the presence of additional short delay steps. Furthermore, a satisfactory fit to Eq. (1) does not require that the cell delays be identical, only that they be similar. Indeed, in some cells delay in the initial rise of the response was consistent with four processes, but the decline was too slow for a system with four identical (or very similar) time constants. A simple expression useful for fitting the flash response of these cells is the "independence" expression

$$r(t) = ike^{-\alpha t}(1 - e^{-\alpha t})^3 \tag{2}$$

where the symbols have the same meanings as in Eq. (1). This equation describes the response of a system having four time constants of value $1/\alpha$, $1/2\alpha$, $1/3\alpha$, and $1/4\alpha$ in any arbitrary sequence. A general interpretation is that in many cells the four processes have similar time constants of 300–500 msec, whereas in some cells there is more variation in the time constants. It remains for future work to determine the physical nature of the four slow processes that shape the single-photon effect.

C. Estimate of Number of Transmitter Particles Mediating the Single-Photon Effect

Successive single-photon effects had very similar amplitudes and waveforms. Histograms of the peak amplitude were approximately Gaussian and showed a ratio of standard deviation to mean value of about 0.2, and the kinetics of individual responses were similar to the kinetics of the average response. Assuming that the quantity of transmitter mediating the response varies probabilistically in successive trials, the relatively small response variability suggests a relatively large number of particles of transmitter. Using the specific model in which a photon releases a Poisson-distributed number of transmitter particles that block sodium channels, Baylor *et al.* (1979b) estimated that a minimum of 100 particles must mediate the response. For a "negative transmitter," which opens channels in darkness and is reduced by light, similar considerations apply. Thus, if at any instant in darkness there were N Poisson-distributed negative transmitter particles in the cell, the mean value of the dark current (about 25 pA) relative to its root-mean-square continuous noise (about 0.2 pA; see next section) would require that N be of the order of 2×10^4. Because the single-photon response involves a 4% reduction in the dark current, N would have to drop by about 800 particles; again this is almost certainly a lower limit. The general conclusion is that many particles of internal transmitter are affected by one photoisomerization, as required to give reliable, stereotyped signals when single rhodopsin molecules absorb.

IV. DARK NOISE

A. General Description

Intracellular recordings by Simon *et al.* (1975), Lamb and Simon (1976, 1977), and Schwartz (1977) have shown that the membrane potential of cones and rods undergoes spontaneous random fluctuations in darkness. This "dark noise" is blocked by bright light when the cell gives a maintained maximal response. Recordings of membrane current from rods have shown that the dark current also fluctuates randomly about its mean level (Baylor *et al.*, 1980). Sample traces illustrating the current noise are shown in Fig. 4. Traces 1 and 2 were recorded in complete darkness, trace 3 in light bright enough to abolish the dark current completely. The residual noise in light (trace 3) was instrumental, arising from thermal agitation of ions in the leakage resistance of

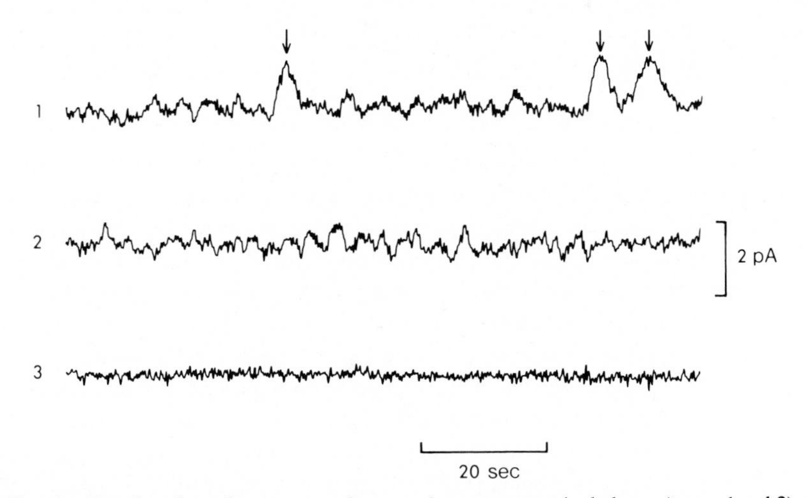

FIG. 4. Samples of membrane current from a rod outer segment in darkness (traces 1 and 2) and during bright light (trace 3) that produced a saturated response. The arrows in trace 1 indicate the times of occurrence of discrete events similar in size and shape to single-photon responses. Low-pass-filtered at 5 Hz. Temperature, 22°C.

the recording electrode. It can be seen that excess noise was present in darkness; this noise was generated by the cell.

The noise in darkness can be seen to consist of two components. The first component is made up of discrete events, three of which are indicated by the arrows in trace 1 in Fig. 4. As will be described shortly, these events are similar in size and shape to the single-photon response and appear to arise from thermal isomerization of rhodopsin. The second component is a smaller-amplitude, faster fluctuation which is continuously present in darkness and which appears throughout traces 1 and 2 in Fig. 4. This noise has an root-mean-square amplitude of about 0.2 pA and seems to consist of the superposition of very small unitary events occurring at a relatively large mean frequency. Evidence to be described indicates that this component of noise is also generated by the transduction mechanism in the outer segment.

B. Discrete Component

The discrete noise events had an amplitude and shape very similar to those of single-photon effects in the same cell. In order to make a more precise comparison, a power spectral analysis was carried out. If in darkness a random time series of independent events occurred, each with the size and form of a single-photon effect, then the power spectrum of the process should have the same form as the spectrum of the single-photon effect. The spectrum of the noise events should have a vertical scaling appropriate for the observed mean event frequency and amplitude. This was tested by comparing the experimental difference spectrum of the discrete component (the spectrum from all current samples in darkness minus the spectrum from dark samples not containing discrete events) with the spectrum calculated from the average single-photon response waveform. In each case there was good agreement between the form of the experimental and calculated spectra, and the scaling of the experimental spectrum was consistent with the observed event frequency and single-photon response amplitude. This analysis indicates that the discrete dark events are indistinguishable from single-photon effects. A simple hypothesis explaining the occurrence of the events is that rhodopsin molecules can undergo spontaneous activation.

Analysis of the intervals between events showed that they occurred randomly and independently (an exponential distribution of intervals was observed) with a mean rate of about 0.02 sec^{-1} at 20°C. Experiments in which the temperature was systematically varied revealed that the mean rate of events was strongly temperature-dependent, and analysis gave the activation energy as 22 kcal $mole^{-1}$. The temperature dependence of the process rules out the possibility that events were triggered by stray light in the opaque experimental box. The apparent activation energy for the process is similar to the 24 kcal $mole^{-1}$ found by Hubbard (1966) for thermal isomerization of 11-*cis*-retinal in solution. Unfortunately the activation energy for thermal isomerization of rhodopsin is not available for comparison. It seems likely that discrete events may be triggered by thermal isomerization of rhodopsin, although more direct tests of the notion would be desirable.

On the assumption that dark events arise from thermal excitation of rhodopsin, the rate constant may be derived from the observed event frequency and the number of rhodopsin molecules in a rod, about 3×10^9 in the toad. The calculated rate constant is 10^{-11} sec^{-1} at 20°C, corresponding to a half-life greater than 1000 yr. It is interesting that the intensity of the absolute "dark light" in psychophysical experiments on human rod vision is very close to that calculated from the rate constant above, its observed temperature dependence, and the pigment content of human rods (see Baylor *et al.,* 1980). This suggests that thermal breakdown of rhodopsin plays an important role in limiting the ability of the human scotopic system to detect very dim lights.

C. Continuous Component

In principle the small, continuously present component of noise could arise within the outer segment, from local conductance fluctuations, or from elsewhere in the cell, if a "distant" noise source generated fluctuations in the intracellular voltage, the driving force on the dark current. Evidence for a local source is provided by the linear dependence of continuous noise variance on the length of outer segment drawn into the recording pipet (Baylor *et al.,* 1980). If the noise shot effects came from a distant source, their amplitude would vary linearly with recorded length and the variance would then show a parabolic dependence on length. For a local source, involving spatially restricted shot effects in a homogeneous outer segment, the apparent event frequency, and thus the variance, would depend linearly on the length of outer segment over which the recording is made. Although this line of evidence depends on untested assumptions about the nature of the noise process and the homogeneity of the outer segment, another property of the noise also points to a local origin. Power spectral analysis of the continuous noise showed that its equivalent spectral bandwidth, a measure of the speed of the underlying shot effects, was correlated with the speed of the light response of the same cell. This correlation could be explained simply if the noise arose in the transduction mechanism of the outer segment.

The form of the power spectrum of the continuous component was fitted by a product of two Lorentzian terms with time constants of 300–500 msec at 20°C. This suggests that the elementary event underlying the noise was generated by a series of two first-order processes; it was clear that a single stage of exponential delay could not fit the spectra. The particular values of the two time constants in a given cell's noise were the same as those of two of the four time constants in the cell's light response; thus, when a rod's light response was fitted by Eq. (1), its noise spectrum could be predicted. Noise and light response kinetics would have this connection if random noise events generated excitation that flowed through the two final stages of the cascade leading to the light response. This is treated further to the final section of this chapter.

To explain the two time constants in the noise spectrum it seems necessary to assume that the shot effect is due to fluctuations in the concentration of the internal transmitter, and that the fluctuations correspond to the sudden appearance or disappearance of multimolecular "puffs" of transmitter particles.

The polarity and amplitude of the noise shot effect are uncertain, but, making certain assumptions, Baylor *et al.* (1980) estimated an amplitude of the order of 2×10^{-15} A and a mean frequency of occurrence in darkness of the order of 6×10^3 sec^{-1}.

V. SENSITIVITY CONTROL IN THE OUTER SEGMENT

Intracellular recordings by Fain (1976) and Bastian and Fain (1979) showed that the voltage signals of toad rods exhibited a phenomenon called adaptation, in which background light reduced the sensitivity to a small superposed test flash. Similar effects have been observed in rod photocurrent (Baylor *et al.*, 1979a, 1980; Yau *et al.*, Chapter 2, this volume), indicating that a significant part of the rod's adaptation mechanism must reside within the outer segment itself.

The observed relation between a rod's incremental flash sensitivity S_F (peak photocurrent amplitude divided by flash intensity) and the steady background intensity I_B can be described by

$$S_F = S_F^D/(1 + I_B/I_0) \tag{3}$$

where the constant S_F^D is the dark value of flash sensitivity and I_0 is the background intensity that halves S_F. In both current and voltage recording experiments (voltage measurements by Bastian and Fain, 1979), I_0 is found to correspond to four to seven photoisomerizations per second. A light of this intensity gives a steady photocurrent of about one-fifth the maximal amplitude.

When a background light reduces the incremental flash sensitivity, the time to peak and duration of the incremental response are also reduced. This means that in steady background light the average effect of a single absorbed photon is briefer than in darkness, as well as smaller. These changes help to prevent saturation (blockage of all the outer segment conductance) and improve time resolution as the background level rises.

The experimental relation between sensitivity and time to peak for incremental flash responses in the presence of background light was of the form

$$S_F \propto (t_{peak})^k \tag{4}$$

where the constant k had a mean value of about 2.6 in different cells. Analyzing the background effect in *Limulus* ommatidia, Fuortes and Hodgkin (1964) have shown that a relation like that in Eq. (4) applies to a multistage low-pass filter in which a background reduces the gain and time constant of each stage by a factor dependent on background intensity. In such a system the exponent k has a value one less than the number of stages in the chain. Thus, for a chain of four filters, corresponding to the Poisson kinetics of Eq. (1), the exponent k is 3 for the Fuortes–Hodgkin model. The similarity of this to the experimental value makes it attractive to think that a mechanism involving their multistage control may be at work in the vertebrate rod. A simpler argument pointing to the same conclusion is that the degree of change observed in t_{peak} (by at least a factor of 3) is too large for a system of four similar delays in which only one delay shortens. This latter system could show no more than about a 30% reduction in t_{peak} if one delay became completely negligible.

Just as individual cells conformed to Eq. (4) in backgrounds of varying intensity, values of S_F and t_{peak} collected from many cells in darkness were correlated by the same relation. This points to the conclusion that the sensitivity control mechanism is already active in darkness, to a degree that varies among different cells.

VI. SUMMARY AND CONCLUSIONS

Electrical experiments of the kind described here do not identify the chemical constituents of the rod transduction system. Nevertheless, no other present method gives such precise, detailed information about the performance of the mechanism under physiological conditions. Information of this kind is useful in defining the properties that need to be explained in chemical terms and in suggesting certain classes of mechanisms likely to be at work.

Many of our results on the kinetics of the transduction mechanism can be brought together with a speculative scheme of the kind outlined in Fig. 5. No more is claimed for this scheme than that it gives a compact way of expressing how the outer segment behaves when processing small incremental stimuli. Isomerization of rhodospin is assumed to set into motion a cascade of events culminating in a reduction in the conductance of the surface membrane. Each stage in the cascade is assumed to involve production of an active substance (e.g., A*) which during its mean lifetime τ catalyzes formation of the next active substance (e.g., B*). The active states end by conversion into substances A′, B′, etc. (which could be the same as A, B, etc.). Four slow steps are depicted, as suggested from the form of the single-photon response; at 20°C, the time constants τ have values of 300–500 msec. Because each reaction produces a catalyst for the next reaction, A* is not consumed in generating B*, etc. This kind of

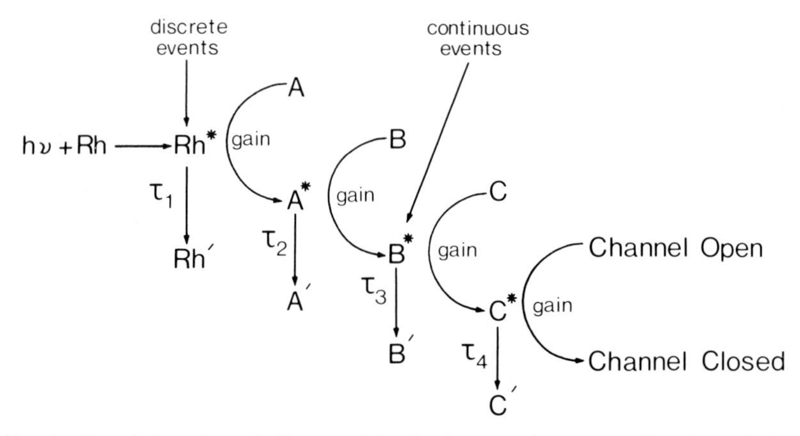

FIG. 5. Speculative schematic diagram of the visual transduction process. Photoisomerization of rhodopsin sets in motion a sequence of four first-order processes culminating in a reduction in inward membrane current. In the diagram, the activated form of a substance is indicated by an asterisk (e.g., A*), and the inactivated form is indicated by a prime (e.g., A′). The activated form of a substance has an exponentially distributed lifetime with an average value ρ. Discrete events and continuous events refer to the two components of dark noise in the outer segment membrane current (see text).

system leads to the simple kinetic equations in Eqs. (1) and (2) and also provides for gain in the steps leading up to a reduction in the membrane conductance (i.e., one Rh* can generate many particles of C*). Gain is indicated in these steps both by the restricted dispersion of the single-photon response amplitude distribution and by the large ratio (about 400) between the amplitudes of the single-photon response and the shot effect underlying the continuous noise. In Fig. 5 the continuous noise corresponds to spontaneous appearances of single B*'s, each of which then produces a puff of C*. The distribution of gain over the cascade is uncertain; not every step need have a gain greater than unity.

The influence of background light on the sensitivity and time scale of the transduction mechanism corresponds to a reduction in the mean lifetimes τ of the active states A*, B*, etc. Because of the multiplicative nature of the cascaded gains, the change in sensitivity in background light will be larger than the change in time scale (see Fuortes and Hodgkin, 1964). For example, when the values of the τ's are halved, the flash sensitivity, which is proportional to $(\tau)^3$, will drop by a factor of 8, whereas the time-to-peak flash response, which is proportional to τ, will drop by only a factor of 2. It would be simple and elegant if a single substance, formed as a product of the chain, controlled the lifetimes of the active intermediates.

ACKNOWLEDGMENT

Supported by grant EY01543 from the National Eye Institute, USPHS.

REFERENCES

Bader, C. R., MacLeish, P. R., and Schwartz, E. A. (1979). A voltage-clamp study of the light response in solitary rods of the tiger salamander. *J. Physiol. (London)* **296,** 1–26.

Bastian, B. L., and Fain, G. L. (1979). Light adaptation in toad rods: Requirement for an internal messenger which is not calcium. *J. Physiol. (London)* **297,** 493–520.

Baylor, D. A., and Fuortes, M. G. F. (1970). Electrical responses of single cones in the retina of the turtle. *J. Physiol. (London)* **207,** 77–92.

Baylor, D. A., Hodgkin, A. L., and Lamb, T. D. (1974). The electrical response of turtle cones to flashes and steps of light. *J. Physiol. (London)* **242,** 685–727.

Baylor, D. A., Lamb, T. D., and Yau, K.-W. (1979a). The membrane current of single rod outer segments. *J. Physiol. (London)* **288,** 589–611.

Baylor, D. A., Lamb, T. D., and Yau, K.-W. (1979b). Responses of retinal rods to single photons. *J. Physiol. (London)* **288,** 613–634.

Baylor, D. A., Matthews, G., and Yau, K.-W. (1980). Two components of electrical dark noise in toad retinal rod outer segments. *J. Physiol. (London)* **309,** 591–621.

Bortoff, A. (1964). Localization of slow potential responses in the *Necturus* retina. *Vision Res.* **4,** 627–636.

Brown, J. E., and Pinto, L. H. (1974). Ionic mechanism for the photoreceptor potential of the retina of *Bufo marinus*. *J. Physiol. (London)* **236,** 575–591.

Cervetto, L., Pasino, E., and Torre, V. (1977). Electrical responses of rods in the retina of *Bufo marinus*. *J. Physiol. (London)* **267,** 17–51.

Cohen, A. I. (1970). Further studies on the question of the patency of saccules in the outer segments of vertebrate photoreceptors. *Vision Res.* **10,** 445–453.

Copenhagen, D. R., and Owen, W. G. (1976a). Coupling between rod photoreceptors in a vertebrate retina. *Nature (London)* **260,** 57–59.

Copenhagen, D. R., and Owen, W. G. (1976b). Functional characteristics of lateral interactions between rods in the retina of the snapping turtle. *J. Physiol. (London)* **259,** 251–282.

Detwiler, P. B., Hodgkin, A. L., and McNaughton, P. A. (1980). Temporal and spatial characteristics of the voltage response of rods in the retina of the snapping turtle. *J. Physiol. (London)* **300,** 213–250.

Fain, G. L. (1976). Sensitivity of toad rods: Dependence on wavelength and background illumination. *J. Physiol. (London)* **261,** 71–101.

Fain, G. L., Gold, G. H., and Dowling, J. E. (1975). Receptor coupling in the toad retina. *Cold Spring Harbor Symp. Quant. Biol.* **40,** 547–561.

Fuortes, M. G. F., and Hodgkin, A. L. (1964). Changes in time scale and sensitivity in the ommatidia of *Limulus*. *J. Physiol. (London)* **172,** 239–263.

Harosi, F. I. (1975). Absorption spectra and linear dichroism of some amphibian photoreceptors. *J. Gen. Physiol.* **66,** 357–382.

Hubbard, R. (1966). The stereoisomerization of 11-*cis*-retinal. *J. Biol. Chem.* **241,** 1814–1818.

Lamb, T. D., and Simon, E. J. (1976). The relation between intercellular coupling and electrical noise in turtle photoreceptors. *J. Physiol. (London)* **263,** 257–286.

Lamb, T. D., and Simon, E. J. (1977). Analysis of electrical noise in turtle cones. *J. Physiol. (London)* **272,** 435–468.

Neher, E., and Sakmann, B. (1976). Single-channel currents recorded from membrane of denervated frog muscle fibres. *Nature (London)* **260,** 799–802.

Nunn, B. J., Matthews, G. G., and Baylor, D. A. (1980). Comparison of voltage and current responses of retinal rod photoreceptors. *Fed. Proc., Fed. Am. Soc. Exp. Biol.* **39,** 2066.

Penn, R. D., and Hagins, W. A. (1972). Kinetics of the photocurrent of retinal rods. *Biophys. J.* **12,** 1073–1094.

Schwartz, E. A. (1973). Responses of single rods in the retina of the turtle. *J. Physiol. (London)* **232,** 503–514.

Schwartz, E. A. (1975). Rod-rod interaction in the retina of the turtle. *J. Physiol. (London)* **246,** 617–638.

Schwartz, E. A. (1976). Electrical properties of the rod syncytium in the retina of the turtle. *J. Physiol. (London)* **257,** 379–406.

Schwartz, E. A. (1977). Voltage noise observed in rods of the turtle retina. *J. Physiol. (London)* **272,** 217–246.

Simon, E. J., Lamb, T. D., and Hodgkin, A. L. (1975). Spontaneous voltage fluctuations in retinal cones and bipolar cells. *Nature (London)* **256,** 661–662.

Tomita, T. (1965). Electrophysiological study of the mechanisms subserving color coding in the fish retina. *Cold Spring Harbor Symp. Quant. Biol.* **30,** 559–566.

Tomita, T. (1970). Electrical activity of vertebrate photoreceptors. *Q. Rev. Biophys.* **3,** 179–222.

Toyoda, J., Nasaki, H., and Tomita, T. (1969). Light-induced resistance changes in single photoreceptors of *Necturus* and *Gekko*. *Vision Res.* **9,** 453–463.

Werblin, F. S., and Dowling, J. E. (1969). Organization of the retina of the mudpuppy, *Necturus maculosus*. II. Intracellular recording. *J. Neurophysiol.* **32,** 339–355.

Yau, K.-W., Lamb, T. D., and Baylor, D. A. (1977). Light-induced fluctuations in membrane current of single toad rod outer segments. *Nature (London)* **269,** 78–80.

Yau, K.-W., Matthews, G., and Baylor, D. A. (1979). Thermal activation of the visual transduction mechanism in retinal rods. *Nature (London)* **279,** 785–786.

Yoshikami, S., Robinson, W. E., and Hagins, W. A. (1974). Topology of the outer segment membranes of retinal rods and cones revealed by a fluorescent probe. *Science* **185,** 1176–1179.

CURRENT TOPICS IN MEMBRANES AND TRANSPORT, VOLUME 15

Chapter 2

Spread of Excitation and Background Adaptation in the Rod Outer Segment

K.-W. YAU,[1] *T. D. LAMB, AND P. A. McNAUGHTON*

Physiological Laboratory
Cambridge University
Cambridge, England

I. INTRODUCTION

There is evidence that both excitation (i.e., light-induced conductance decrease) and background adaptation (or desensitization) in a rod outer segment spread from the site of photoisomerization. Visual transduction is now thought to involve a diffusible substance in the outer segment, which links photon absorption in the disk membranes to the light-induced conductance decrease in the plasma membrane (see Matthews and Baylor, Chapter 1, this volume). In toad rods the peak amplitude of the response to one photon is about 1/30 of the saturating level; this in conjunction with the fact that the light-suppressible dark

[1]Present address: Department of Physiology and Biophysics, The University of Texas Medical Branch, Galveston, Texas.

ISBN 0-12-153315-8

current is fairly uniform along the outer segment (Baylor *et al.*, 1979a,b; see Matthews and Baylor, Chapter 1, this volume) suggests that one photon has to affect at least 1/30 of the outer segment length, or about 2 μm, corresponding to the domain of about 50 disks. Hagins *et al.* (1970) have estimated that in rat rods (outer segment length $\simeq$25 μm) the spread of excitation is about 12 μm; this, however, is likely to be an upper estimate in view of the spatial resolution of their recording method.

The conclusion that the desensitization resulting from one photoisomerization also spreads beyond one disk was drawn by Donner and Hemilä (1978) and Bastian and Fain (1979). Their argument was that adaptation occurred in an outer segment at such low background intensities that at any time instant only a small fraction of the membranous disks could have absorbed photons. The spread of desensitization suggests that it also involves a diffusible substance which may or may not be identical to the internal transmitter postulated to mediate excitation.

We have examined the spatial spread of both excitation and desensitization in more detail in toad rods, and the results are outlined here. The experiments involved recording membrane current from a single rod outer segment with a suction pipet and stimulating it with narrow slits or spots of light. The recording method is briefly described by Matthews and Baylor (Chapter 1, this volume), and details can be found in Baylor *et al.* (1979a). A more complete description of the experiments and an analysis of the spread of excitation and background adaptation are presented elsewhere (Lamb *et al.*, 1981).

II. SPREAD OF EXCITATION

Two kinds of experiments are described here, the first giving an upper estimate of the longitudinal spread of excitation and the second attempting to define the spread in more detail.

A. Experiments with the Outer Segment Partly Drawn into the Recording Pipet

An estimate of the longitudinal spread of excitation can be made by recording from only part of the outer segment and stimulating the unrecorded part with light. If excitation spreads extensively from the site of photoisomerization, the recorded response should not vary much with stimulus position, whereas if excitation is confined to near the photoisomerization site, little or no response should be recorded under these conditions. Figure 1 shows such an experiment in which only the distal part of the outer segment was recorded from by drawing it partly into a suction pipet; a narrow transverse light slit was then displaced along the outer segment, and dim flashes were delivered. The recorded response (open squares in Fig. 1, middle) declined fairly rapidly as the light slit was moved off the pipet tip. In the control experiment with the entire outer segment inside the pipet (solid circles in Fig. 1, middle) the middle of the outer segment nonetheless showed fairly uniform sensitivity. Thus the results suggest restricted spread of excitation from the site of photoisomerization.

In Fig. 1 (bottom) the open circles show the ratio of the recorded responses with the outer segment partly and fully in the pipet plotted against the stimulating slit position; they represent the profile of recorded photocurrent corrected for any nonuniform sensitivity along the outer segment. If the pipet collected all the membrane current from inside its point of narrowest constriction and none from outside, this profile would represent the integral of the membrane current den-

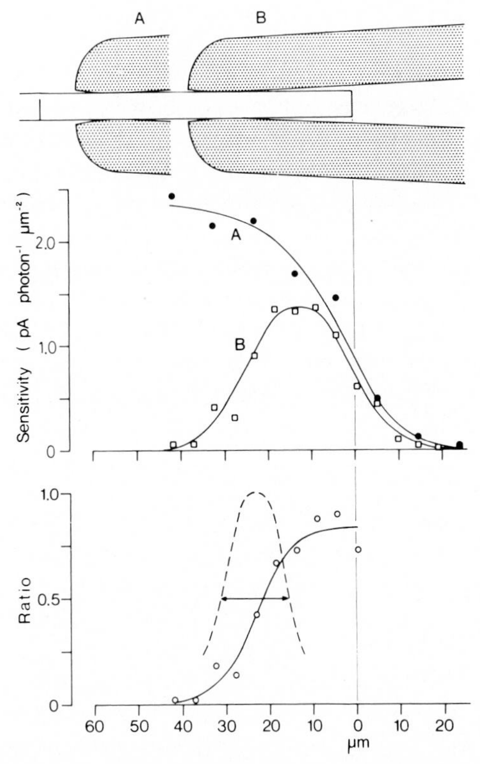

FIG. 1. Spread of excitation in a rod outer segment. (Top) Outer segment almost completely (A) and partly (B) drawn into the pipet. (Middle) Averaged recorded current, divided by flash intensity, as a function of displacement of a transverse slit from the tip of the outer segment, for the two positions of the cell; nominal slit width 6.9 μm. (Bottom) Ratio of the two profiles; ○ are the ratio of the squares to curve A near the solid circles; the broken curve is the normalized spatial derivative of the solid curve. [From McNaughton *et al.*, 1980. Reprinted from *Nature (London)* **283**, 85–87. Copyright © 1980 by Macmillan Journals, Ltd.]

sity. The spatial derivative (broken curve) would then give the current density profile, hence the spread of excitation. In this experiment the estimated spread was about 7.5 μm at half-height on either side. Similar values were obtained in other experiments.

The main limitation of the above analysis is the assumption that the pipet acts as a perfect step collector of current. Typical pipets do not have a sharp constriction point at the tip (see, for example, the scale diagram in Fig. 1, top), and so the estimate of 7.5 μm is likely to be an upper limit on the spread of excitation. In the next section we describe another experiment which suggests that excitation is much more localized.

B. Intensity–Response Relations with Diffuse and Slit Illumination

Figure 2 shows an experiment in which an outer segment, fully drawn into the recording pipet, was stimulated with light of different intensities applied either diffusely or in the form of a narrow transverse slit positioned at the middle of the outer segment. A prominent difference between the two response families is that with slit illumination the response amplitude grew much more slowly as the flash intensity was increased. This observation is consistent with a limited spread of activation, suggesting that it was primarily the ionic channels under the slit that were closed by light and that those further away remained open even at fairly bright slit intensities. Eventually very bright slit flashes produced a saturating response, presumably because the internal transmitter had diffused throughout the outer segment or because light scattered from the slit had generated transmitter throughout the cell.

The relationship between the response amplitude and the flash intensity is expected to depend on the extent of spread of the internal transmitter. In the following we will assume that the concentration of this diffusible messenger is increased by light (the "positive transmitter" hypothesis), but in qualitative terms similar results will be expected for the case where the concentration of the diffusible messenger is reduced by light (the "negative transmitter" hypothesis). At early times after the flash, when the spread of the internal transmitter is limited, the relation is expected to be considerably less steep than the corresponding relation obtained with diffuse illumination. At later times, when the spread becomes more extensive, the relation should steepen, and in the limit of free diffusion throughout the cell it should approach the relation obtained with diffuse illumination.

The relation at various times can be predicted quantitatively if it is assumed that the internal transmitter is released in proportion to the flash intensity at each point along the outer segment (cf. Baylor *et al.*, 1974). With this assumption the

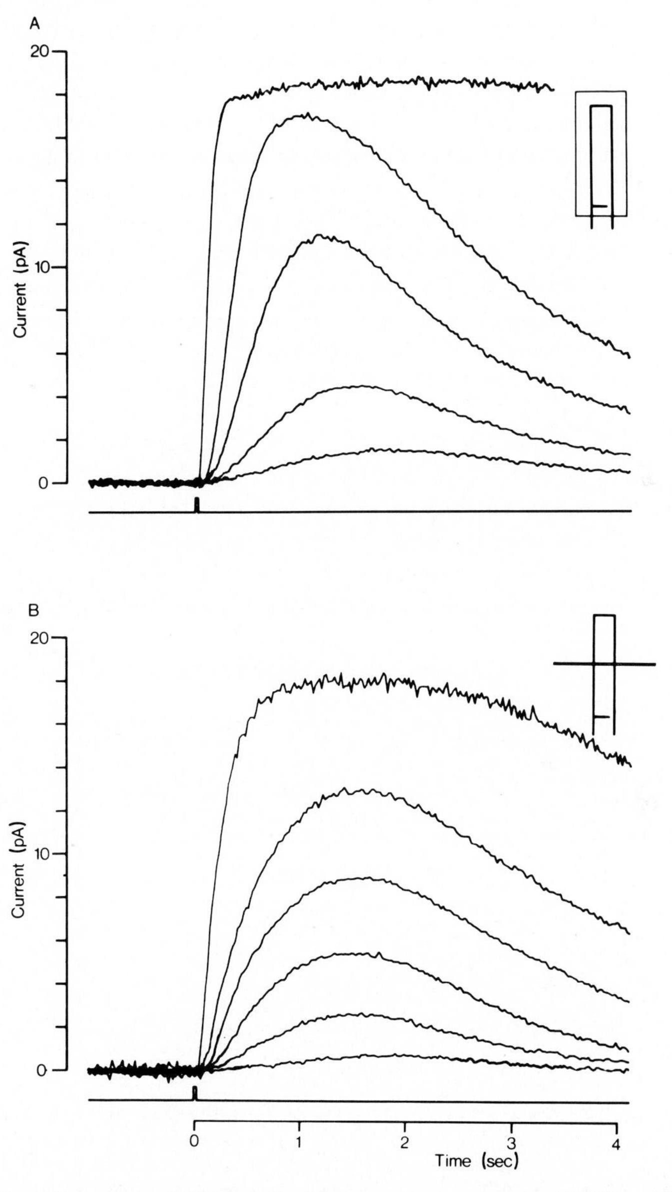

FIG. 2. Intensity–response families obtained from a rod outer segment with diffuse (A) and slit (B) illumination. The outer segment was drawn fully into the recording pipet. The responses to flashes at dim and intermediate intensities have been averaged. Flash intensities were increased by factors of approximately 4, except for the brightest intensity in (A) and in (B), which were, respectively, increased by factors of approximately 8 and 16. (See Lamb *et al.*, 1981.)

profile of internal transmitter concentration as a function of distance along the outer segment can be obtained by convolving the profile of the light stimulus with the profile of transmitter released from an infinitely narrow line source. The response resulting from this transmitter profile can then be calculated by assuming that the response at each point along the outer segment saturates in the same way that the diffuse response saturates with bright light. We have measured the profile of the light stimulus in each experiment by scanning a small aperture connected to a photomultiplier tube across the image of the light slit formed at the position of the microscope eyepiece. The profile of transmitter released by a line source can be calculated by assuming that transmitter diffusion obeys the diffusion equation and that the rate of transmitter release is that required to generate the kinetics of the single-photon response (for further details, see Lamb *et al.*, 1981).

In Fig. 3 the dashed curve shows the relation between the diffuse flash intensity and the response at any given instant in the rising phase of the response. This curve also represents the intensity–response relation with slit illumination in the

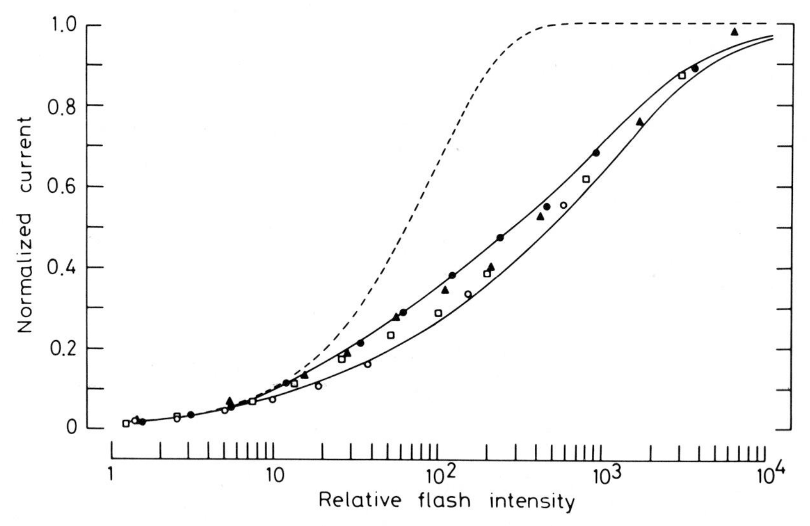

FIG. 3. Intensity–response relations at fixed times, from the experiment in Fig. 2. With diffuse illumination the form of the relation was independent of the time at which it was plotted (up to 1000 msec) and is shown as the dotted line. Relations obtained with slit illumination are plotted at 200 msec (○), 400 msec (□), 600 msec (▲), and 1000 msec (●) and have been shifted along the log intensity axis so that the linear regions observed with dim flashes coincide. The lower solid line is the intensity–response relation computed from the profile of the slit of light and assuming no diffusion of transmitter; the upper solid line is computed assuming that a positive transmitter has a diffusion coefficient of $3 \times 10^{-7} cm^2 sec^{-1}$ and that 1000 msec has elapsed after the flash. (See Lamb *et al.*, 1981.)

limiting case of rapid transmitter diffusion. The symbols show the experimental intensity–response relations with slit illumination at various times after the flash. The slit relations were always found to be considerably less steep than the diffuse relation, although a slight steepening of the slit relations with time was sometimes observed. For instance, in Fig. 3 the solid circles (1 sec after the flash) are slightly closer to the diffuse intensity–response relation than the open circles (200 msec after the flash). The lower solid curve in Fig. 3 is the calculated relation for zero spread of the internal transmitter, and it fits the experimental relation at 200 msec after the flash reasonably well. The upper solid curve corresponds to the spread of transmitter from a narrow line source in exponential fashion, with a space constant of 3.0 μm, and fits the experimental relation at a time of 1 sec after the flash, which is near the time to peak of the response. Results indicating a similarly restricted spread were obtained in other experiments.

For the case of an internal transmitter produced from a line source according to Poisson kinetics (Fuortes and Hodgkin, 1964; Baylor *et al.*, 1974) it can be shown that the concentration profile at any instant decays roughly exponentially with a space constant $\lambda \simeq k(D_e t)^{\frac{1}{2}}$, where D_e is the effective diffusion coefficient, t is time, and k is a constant whose value depends on the number of delay stages in the kinetics (see Lamb *et al.*, 1981). For a Poisson transduction process with four stages of delay (Baylor *et al.*, 1979a) $k \simeq 0.55$. Putting $\lambda \simeq 3.0\ \mu$m and $t = 1$ sec we obtain $D \simeq 3 \times 10^{-7}\ \text{cm}^2\ \text{sec}^{-1}$. This is a very small diffusion coefficient, being approximately 100 times less than the free diffusion coefficient of small ions or molecules in aqueous solution.

The above analysis has been developed in terms of an internal transmitter that is released by light and blocks the light-sensitive channels. There is the possibility, however, that light actually reduces a substance that in darkness opens the channels (see, for example, Hubbell and Bownds, 1979). Although a quantitative analysis has not been made for the case of a negative transmitter, it seems from the large difference between the diffuse and slit intensity–response families that diffusion must again be very restricted.

III. SPREAD OF DESENSITIZATION

A. Choice of Desensitization Parameter

To measure the spread of desensitization a parameter is needed to characterize the degree of desensitization at different positions on the outer segment. In toad rods it has been shown that the reduction in flash sensitivity by diffuse background light is consistent with the Weber–Fechner relation (Fain, 1976; Bastian and Fain, 1979; Baylor *et al.*, 1980; see Matthews and Baylor, Chapter 1, this volume). This can be written as

$$S_F = S_F^D/(1 + I_B/I_0) \quad (1)$$

where S_F is flash sensitivity in the presence of background light of intensity I_B, S_F^D is flash sensitivity in darkness, and I_0 is a constant corresponding to the background that would halve sensitivity. Transposing terms in Eq. (1), we obtain

$$S_F^D/S_F - 1 = I_B/I_0 \quad (2)$$

The expression $S_F^D/S_F - 1$ increases in direct proportion to background intensity I_B and is a convenient variable for measuring desensitization because it allows linear summation.

The expression $S_F^D/S_F - 1$ may have a simple interpretation. On the hypothesis that light generates a diffusible desensitizing substance whose concentration rises linearly with intensity, and that desensitization at any point on the outer segment depends on the local concentration of this substance, then $S_F^D/S_F - 1$ can be interpreted to represent the relative concentration of the substance.

B. Longitudinal Spread of Desensitization

Figure 4 shows an experiment measuring the spread of desensitization from steady illumination. In this experiment a narrow stationary light slit was positioned across the middle of an outer segment, and the resulting desensitization was measured with another slit which delivered dim test flashes at different positions along the outer segment. In Fig. 4, A–E represent consecutive positions of the test slit displaced 7 μm from each other, the stationary adapting slit being at position C. At each position the lighter trace represents the average response to a dim flash in the absence of the adapting slit, and the heavier trace is the average response scaled to the same test flash intensity in the presence of the adapting slit. In darkness, sensitivity was fairly uniform along the outer segment, as indicated by the similar amplitudes of the test responses at different positions. In the presence of the adapting slit, however, there was substantial desensitization at the position of the adapting slit, but this declined on either side. At 14 μm away (positions A and E) there was negligible or no desensitization. Thus, although desensitization spreads longitudinally away from the site of photoisomerization, the spread is restricted to a fraction of the outer segment.

Figure 5 shows the results of a similar experiment with another cell. Here the spread of desensitization was mapped in more detail on one side of the adapting slit. In this figure the parameter $S_F^D/S_F - 1$ is plotted against distance away from the adapting slit. As in Fig. 4, the spatial decline in desensitization was quite steep, dropping to a low value within a few micrometers along the outer segment. The figure also shows the light profile of the adapting slit subsequently measured. The light profile is significantly narrower than the desensitization profile, suggesting that the spread of desensitization is not caused simply by light scattering.

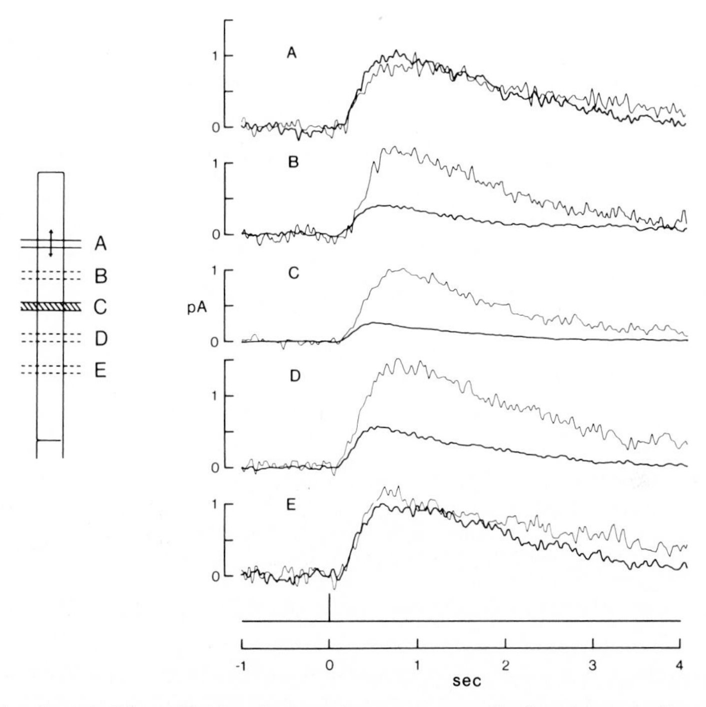

FIG. 4. Spread of desensitization along a rod outer segment. The inset is a scale diagram of the outer segment with a stationary slit at C to deliver steady adapting light and a testing slit at successively A, B, C, D, and E to deliver test flashes. Both slits had a nominal width of 1.7 μm, and the separation between adjacent positions was 7 μm. For each position the lighter trace represents the averaged test response to a dim flash in the absence of the adapting slit, and the heavier trace represents the averaged test response to the same flash in the presence of the adapting slit. (See Lamb *et al.*, 1981.)

It was found that the shape of the desensitization profile obtained in the above manner did not vary greatly among cells, nor did it seem to depend on the intensity of the adapting slit, although we used only dim lights. Thus the separate profiles obtained from individual cells can be normalized and averaged, as shown in Fig. 6. The desensitization profile is quite symmetrical about the adapting slit, and the decline is approximately exponential, with a space constant of about 6 μm on either side (corresponding to approximately $^1/_{10}$ the total outer segment length).

C. Transverse Spread of Desensitization

Figure 7 shows an experiment measuring the transverse spread of desensitization. A small spot of steady light was positioned at the edge of an outer

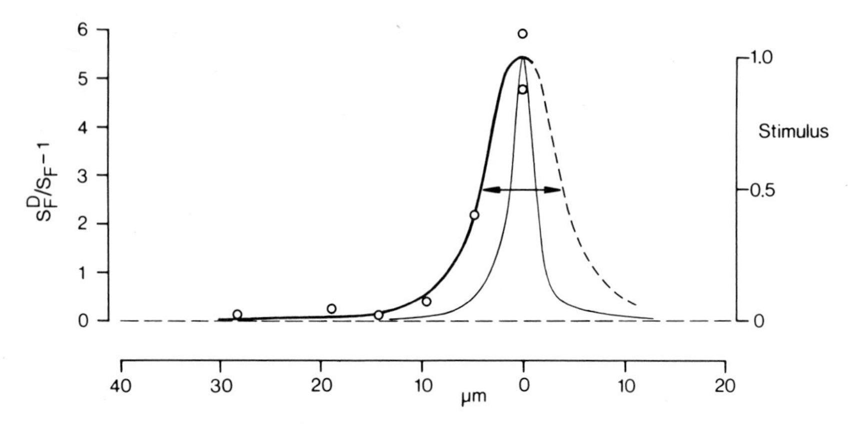

FIG. 5. The same kind of experiments as in Fig. 4. $S_F^D/S_F - 1$ is defined in the text and calculated from the ratio of peak amplitudes of the test responses in the absence and presence of the adapting light. The heavier curve has been fitted to the experimental points (open circles) by eye, and the dashed curve drawn by assuming symmetry about the adapting slit. The lighter curve is the measured light profile of the adapting slit. [From McNaughton *et al.*, 1980. Reprinted with permission from *Nature (London)* **283,** 85–87. Copyright © 1980 by Macmillan Journals, Ltd.]

segment, and the resulting desensitization at the same position (A), across (B), and along the outer segment (C) was measured with a test spot delivering dim test flashes. The last position, D, which was identical to A, acted as a control to make sure that the condition of the cell did not deteriorate during the experiment. The distances AB and AC were both 5 μm. In the absence of the adapting spot the test responses were approximately the same at all three positions. In the presence of the adapting spot, the test responses were about the same at the position of the

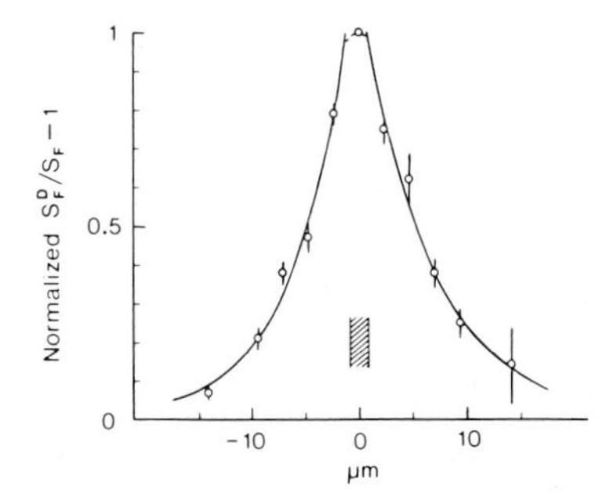

FIG. 6. Steady-state spread of desensitization along the rod outer segment averaged from nine experiments of the kind shown in Figs. 4 and 5. Open circles are the mean, and vertical bars indicate the standard error of the mean; not every experiment gave measurement at each position. Hatched bar indicates position and nominal width (1.7 μm) of the adapting slit. The curves are exponential declines with space constants of 5.6 μm and 6.4 μm on the two sides. (See Lamb *et al.*, 1981.)

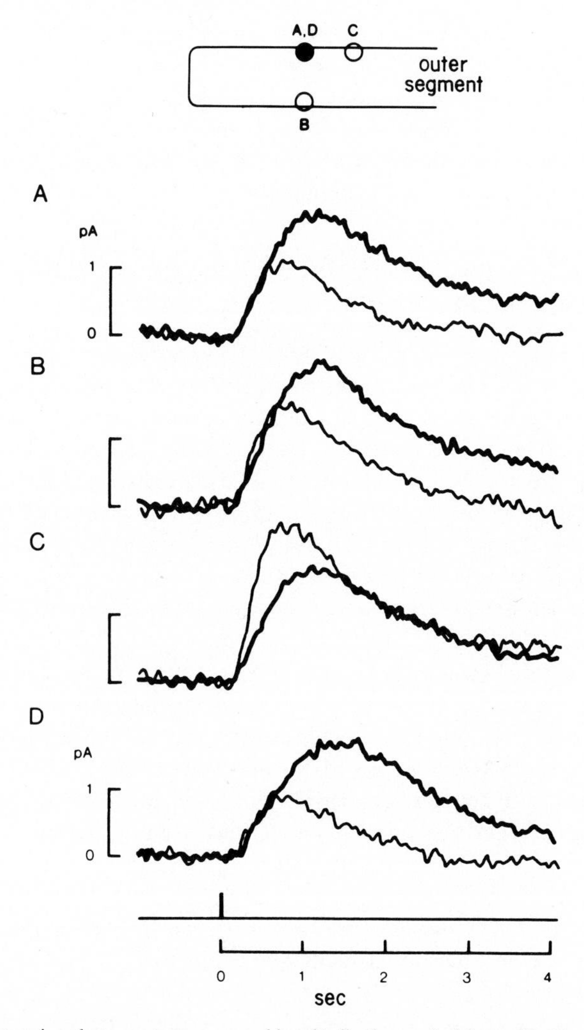

FIG. 7. Comparison between transverse and longitudinal spread of desensitization in the rod outer segment. At the top is a scale diagram of an outer segment with a steady adapting light spot at A and a test spot successively at A, B, C, and D (or A). Both spots had a nominal diameter of 1.6 μm, and the distances AB and AC were 5 μm. At each position the heavier trace is the averaged test response to a dim flash in the absence of the adapting spot, and the lighter trace is the averaged test response to a dim flash 3.87 times brighter and in the presence of the adapting spot. (See Lamb *et al.*, 1981.)

adapting spot and across the outer segment; on the other hand, the test response at the same distance along the outer segment was significantly larger. In other words, desensitization spreads more efficiently across than along the outer segment. The longitudinal decline in the desensitization parameter $S_F^D/S_F - 1$ measured in such a spot experiment was consistent with the space constant of 6 μm obtained from the previous slit experiments.

D. Effective Diffusion Coefficient for Longitudinal Spread of Desensitization

The space constant of 6 μm described above applies only to the steady-state spread of desensitization, i.e., under conditions of steady background illumination at a point. In the event of a single photoisomerization where adaptation is transient and lasts only a few seconds, the longitudinal spread will be less than 6 μm. As in the case of excitation, a useful index for characterizing the spread of desensitization is an effective diffusion coefficient D_e from which both transient and steady-state spread can be constructed. To estimate D_e, we consider the situation where the hypothetical desensitizing substance is being steadily produced under a narrow slit and diffuses longitudinally with an effective diffusion coefficient D_e. We also assume that along its diffusion path it is being uniformly removed by a first-order process of time constant τ. In the steady state, the concentration profile of the substance is an exponential decline, with a space constant $\lambda = (D_e\tau)^{\frac{1}{2}}$. This agrees with the experimental observation that in the steady state $S_F^D/S_F - 1$ declines approximately exponentially, with $\lambda \simeq 6$ μm. To compute D_e it is still necessary to know τ, for which there is no a priori information. An estimate of τ can be obtained from the time course of decline in desensitization at the turning off of a diffuse background light, giving a time constant of about 3 sec. Thus, taking $\tau \simeq 3$ sec, we obtain $D_e \simeq 10^{-7}$ cm^2 sec^{-1}. This is very similar to the effective diffusion coefficient for the longitudinal spread of the internal transmitter mediating excitation.

IV. SUMMARY AND CONCLUSIONS

The general conclusion of the experiments described here is that both excitation and desensitization in the rod outer segment are confined to near the site of photon absorption. At the peak of the response to a single photoisomerization neither effect appears to spread more than about 3 μm along the outer segment; thus the total region affected is about 6 μm. On the other hand, the transverse spread of desensitization (and probably also excitation) seems to be more efficient, so that cross-sectional homogeneity can reasonably be assumed. This anisotropy is perhaps not too surprising if both processes spread by simple aqueous diffusion.

The inside of the outer segment is known to be tightly packed with membranous disks arranged in a stack (see Rodieck, 1973), each disk spanning the entire cross-sectional area of the outer segment. A substance diffusing across the outer segment is relatively unobstructed, but longitudinal diffusion can occur only around the edges of the disks. It can be shown that the longitudinal diffusion coefficient is reduced by a factor of about $r/2g$, where g is the gap distance between the edge of the disk and the plasma membrane and r is the outer segment radius. From electron microscopy $g \sim 100$ Å (see, for example, Nilsson, 1965) and $r \sim 3$ μm for toad rods, so $r/2g \sim 100$. Thus the physical obstruction due to the membranous disks should slow down longitudinal diffusion by about 100-fold. This simple calculation ignores disk incisures and any possible binding of the diffusing substances within the outer segment; these factors, nonetheless, would tend to have counteracting influences on longitudinal diffusion. The free diffusion coefficients for the substances mediating both excitation and desensitization are therefore about 10^{-5} cm^2 sec^{-1}, similar to that for small ions or molecules in aqueous solution. Whether the same substance underlies both processes is still a matter for speculation.

REFERENCES

Bastian, B. L., and Fain, G. L. (1979). Light adaptation in toad rods: Requirement for an internal messenger which is not calcium. *J. Physiol. (London)* **297,** 493–520.

Baylor, D. A., Hodgkin, A. L., and Lamb, T. D. (1974). The electrical response of turtle cones to flashes and steps of light. *J. Physiol. (London)* **242,** 685–727.

Baylor, D. A., Lamb, T. D., and Yau, K.-W. (1979a). The membrane current of single rod outer segments. *J. Physiol. (London)* **288,** 589–611.

Baylor, D. A., Lamb, T. D., and Yau, K.-W. (1979b). Responses of retinal rods to single photons. *J. Physiol. (London)* **288,** 613–634.

Baylor, D. A., Matthews, G., and Yau, K.-W. (1980). Two components of electrical dark noise in retinal rod outer segments. *J. Physiol. (London)* **309,** 591–621.

Donner, K. O., and Hemilä, S. (1978). Excitation and adaptation in the vertebrate rod photoreceptor. *Med. Biol.* **56,** 52–63.

Fain, G. L. (1976). Sensitivity of toad rods: Dependence on wavelength and background illumination. *J. Physiol. (London)* **261,** 71–101.

Fuortes, M. G. F., and Hodgkin, A. L. (1964). Changes in time scale and sensitivity in the ommatidia of *Limulus*. *J. Physiol. (London)* **172,** 239–263.

Hagins, W. A., Penn, R. D., and Yoshikami, S. (1970). Dark current and photocurrent in retinal rods. *Biophys. J.* **10,** 380–412.

Hubbell, W. L., and Bownds, M. D. (1979). Visual transduction in vertebrate photoreceptors. *Annu. Rev. Neurosci.* **2,** 17–34.

Lamb, T. D., McNaughton, P. A., and Yau, K.-W. (1981). Spread of activation and desensitization in toad rod outer segments. *J. Physiol. (London)* (in press).

McNaughton, P. A., Yau, K.-W., and Lamb, T. D. (1980). Spread of activation and desensitization in rod outer segments. *Nature (London)* **283,** 85–87.

Nilsson, S. E. G. (1965). The ultrastructure of the receptor outer segments in the retina of the leopard frog (*Rana pipiens*). *J. Ultrastruct. Res.* **12,** 207.

Rodieck, R. W. (1973). "The Vertebrate Retina." Freeman, San Francisco, California.

Chapter 3

Ionic Studies of Vertebrate Rods

W. GEOFFREY OWEN[1] *AND VINCENT TORRE*

Physiological Laboratory
Cambridge University
Cambridge, England

I. INTRODUCTION

The ionic mechanism by which the vertebrate rod generates an electrical response to light has been the focus of considerable interest during the last decade. That so much about it remains to be understood is testament both to the formidable technical difficulties confronting the investigator and to the complex-

[1]Present address: Department of Biophysics and Medical Physics, University of California, Berkeley, California.

ISBN 0-12-153315-8

ity, which we are only now beginning to appreciate, of the mechanism itself.

Any short review of this subject must suffer from one or more deficiencies. Since it cannot hope to be comprehensive, it must be selective in what it covers, and in being selective, almost inevitably it will be biased. In this chapter we make no attempt to be comprehensive. Discussion is restricted to studies of vertebrate rods. There have been few ionic studies of cones, too few to have established with certainty any differences between cones and rods, though differences may well exist. Our aim is to provide a general picture of the distribution of ions across the plasma membrane of the rod, of the mechanisms responsible for maintaining this distribution, and of the conductances that contribute directly or indirectly to the light response. The focus is on the plasma membrane, therefore. Light-induced events occurring prior to the modulation of membrane conductance will not be discussed.

This chapter is biased, as will quickly become apparent, toward electrophysiological results in general and intracellular studies in particular. We try to emphasize what, on the basis of good experimental evidence, "is" and to avoid speculation concerning what "might be."

Before proceeding, however, it will be helpful to discuss briefly the limitations of the various technical approaches that have been adopted in published studies.

II. THE EXPERIMENTAL APPROACH

Vertebrate photoreceptors are sandwiched between the remainder of the neural retina and the retinal pigmented epithelium (RPE). The retina constitutes a significant barrier to diffusion with the result that, although the *vitreal* surface of the retina in an eyecup preparation can be perfused, the effects of rapid changes in ionic concentration take several minutes before they are fully evident at the level of the receptors (see, for example, Cervetto, 1973). Any *quantitative* study of ionic photoreceptor mechanisms by this technique is therefore unlikely to be reliable, since internal concentrations will have time to change too.

Most recently, the standard approach has been to remove the retina and epithelium from the eye and then to peel off the RPE gently, exposing the receptor outer segments which can thus be perfused directly. While such an approach allows us to study the membrane in a reliable way, it must be recognized that under normal conditions in the intact eye the functional environment may be quite different from the experimental one, with consequent differences in behavior. By removing the RPE, a much larger than normal free volume is created around the receptors. This is experimentally desirable because we wish to minimize the effects of "unstirred layers." Local changes in ionic concentration that normally occur in the external medium will be greatly diminished, and yet such changes may be of functional importance. Certainly, such changes occur and can be significant (Oakley and Green, 1976; Oakley, 1977; Steinberg *et al.*, 1980). Furthermore, the RPE is not simply a passive barrier to the diffusion of ions and molecules out of the retina. It actively transports substances to and from the retina and serves to regulate ionic concentrations around the receptor outer segments (Steinberg and Miller, 1973; Miller and Steinberg, 1976, 1977a,b, 1979). In the isolated retina therefore we study cells in a quite unnatural environment, and this must be taken into account when assessing the *functional* importance of ionic mechanisms in the photoreceptor membrane.

In studying excitable cells, the ideal physiological approach involves voltage-clamping and current-clamping the membrane while varying the ionic concentrations in the external medium. The power of such an approach is exemplified by the advancement in understanding of the nerve action potential that resulted from the work of Hodgkin and Huxley in the early 1950s. The squid axon was an almost ideal preparation for such an approach. Vertebrate photoreceptors are not. In 1971 it was shown that cones in the retina of the turtle were coupled together electrotonically in a network (Baylor *et al.*, 1971). Rods, too, were later found to be similarly coupled (Schwartz, 1975, 1976; Copenhagen and Owen, 1976a,b; Fain *et al.*, 1976). As a result of this, current, whether resulting from photostimulation or extrinsic injection, can spread from one receptor to another limited only by the relative impedances of the receptor membrane and junctional complex. Attempts to control voltage across the receptor membrane by point current injection, so that membrane conductances can be studied, are doomed to failure because the membrane voltage of neighboring coupled receptors cannot also be controlled. There is an inevitable failure of the space clamp. More elaborate techniques of voltage control, such as that described by Trifanov and Chailakhyan (1975), must be used. Using a standard isolated, perfused retina preparation, we are therefore restricted to measuring the effects of ionic concentration changes on membrane potentials and on network properties such as the input impedance and the space constant.

A recently developed technical modification circumvents this problem in an elegant way and raises exciting possibilities for the immediate future. Photoreceptors are isolated from the remainder of the retina, either by gentle shaking (Werblin, 1978) or by digesting away the connective tissue with proteolytic enzymes (Bader *et al.*, 1978). It is not yet clear to what extent the receptors may be damaged by these treatments, but what is important is that when isolated they can be current-clamped and voltage-clamped at will. There seems little doubt that, once refined, this approach is likely to provide the best chance of elucidating the ionic mechanisms of the photoreceptor membrane.

The data described in the remainder of this chapter were obtained from receptors in intact retina except where it is specifically stated to the contrary.

III. EARLY IONIC STUDIES

The work of Sillman *et al.* (1969) on the aspartate-isolated late receptor potential of the frog established that the rod's response to light is highly sensitive to the concentrations of Na^+ and K^+ in the extracellular medium. It was well known from earlier intracellular recordings that photoreceptors hyperpolarized in response to light (Bortoff, 1964; Tomita, 1965), and this had been shown to be accompanied by a decrease in input conductance (Toyoda *et al.*, 1969). In view of this, Sillman *et al.* proposed that the rod membrane is permeable to both Na^+ and K^+ and that the effect of light is to reduce the conductance of the membrane to Na^+.

By analyzing the radial distribution of current flowing in the extracellular space between rat rods, Hagins *et al.* (1970) demonstrated the existence of a standing positive current which, in darkness, entered the rod across the outer segment plasma membrane and was actively extruded by a metabolically driven pump. The effect of light was to reduce this "dark current" by altering the permeability of the outer segment membrane. Their technique allowed only the measurement of *net* current, so that they could not distinguish among a light-induced reduction in inward current, a light-generated outward current, or indeed a current carried by anions. In view of the known effects of light on membrane potential and the strong dependence they found of the dark current on the external Na^+ concentration, they concluded that the dark current was an inward Na^+ current modulated by light.

This view received support from the ingenious experiments of Korenbrot and Cone (1972) who osmotically shocked isolated rod outer segments in suspension. They found the rod outer segment membrane to be highly permeable to inward Na^+ movement. K^+ was only about 10^{-2} as permeable in the inward direction. Isosmotic exposure to low $[Na^+]_{out}$ revealed no detectable outward flow of Na^+, leading them to suggest that the rod outer segment Na^+ conductance was highly rectified. Indeed, the rectification appeared to be on the order of 10^3 in favor of inward Na^+ movement (see, however, Section V,A). The ability of K^+ to flow out of the rod outer segment was not checked, but Cl^- was found to move freely in either direction. The effect of light was to reduce the permeability of the membrane to Na^+, other ion permeabilities being unaffected.

In summary, these early studies all supported the view that the response of the rod to light results from a decrease in the Na^+ conductance of the outer segment membrane. Since the rod membrane is permeable to K^+ also, this drives the transmembrane potential toward the K^+ equilibrium potential (E_K), yielding the familiar hyperpolarization.

While the osmotic experiments of Korenbrot and Cone (1972) contributed greatly to the development of this view, it should be mentioned that they have excited considerable controversy since their publication. Cobbs and Hagins

(1974) were unable to duplicate the results, and Bownds and Brodie (1975) found only a very slow osmotic response requiring many minutes and even hours, which was nonetheless light-modulated. Despite this, the experiments were successfully duplicated by Wormington and Cone (1978), and they developed interesting new results using the same technique. They pointed out, however, that even small departures from their recommended experimental conditions could lead to a drastic reduction in the osmotic behavior of the rod outer segment suspension. It is this critical dependence of the phenomenon on experimental technique that has led to controversy and which must be taken into account when assessing the reliability of findings obtained using this approach.

Since these early studies were carried out, a number of questions have attracted considerable attention, many of which have yet to be satisfactorily answered. The early work had not established beyond doubt that Na^+ was the only ion whose passage across the membrane was light-modulated. If Na^+ crossed the rod outer segment membrane through a channel like that in the squid axon, some K^+ movement would be expected too. The selectivity of the light-modulated conductance remains undefined. Nor is it known whether Na^+ crosses through a channel or via a carrier molecule. And what role do voltage-dependent mechanisms play in the generation of the light response?

In the sections that follow we shall discuss what is known about ionic mechanisms in the rod membrane and attempt to point out, where necessary, the weaknesses in some of the current hypotheses.

IV. DEPENDENCE OF THE DARK POTENTIAL ON NA$^+$, K$^+$, AND Cl$^-$

Brown and Pinto (1974), recording intracellularly from rods in the isolated, perfused retina of the toad, showed that replacement of all external Na^+ with lithium, choline, or sucrose caused the rod to hyperpolarize in darkness by ~30 mV. Light-evoked potentials were abolished. These effects were rapidly reversed when the normal external Na^+ concentration was restored. Lowering the external K^+ had a relatively small effect on the dark potential but made the potential at the peak of the light response much more negative, thereby increasing response amplitude. Raising the external potassium also had a relatively small effect on the dark potential and reduced the potential at the peak of the light response. Changes in the external Cl^- and Mg^{2+} had negligible effects.

It was clear from these experiments that the membrane potential of the rod depended primarily on Na^+ and K^+, the K^+ dependence at the peak of the light response being greater than in darkness. This is consistent with the idea that light reduces the Na^+ conductance of the rod membrane. Furthermore, the weak effects of changes in $[Cl^-]_{out}$ suggested either that chloride was quickly and passively distributed across the membrane or that chloride conductance was small.

Somewhat surprisingly, perhaps, these basic properties were not reexamined more quantitatively until the recent work of Capovilla *et al.* (1980b), again on toad rods. They, too, found the dependence of the dark potential on external $[Cl^-]$ to be small. Reducing $[Cl^-]_{out}$ from 120.6 to 10.6 m*M*, by replacing it with methane sulfate, isethionate, or propionate, depolarized the membrane by ~5 mV after a correction for liquid junction potentials had been made. The equilibrium potential of chloride should have been changed by ~60 mV. The most likely explanation of this result is that, in darkness, the chloride conductance represents less than 1/10 of the *total* membrane conductance. A similar conclusion was reached by Pinto and Ostroy (1978).

Capovilla *et al.* (1980a) reexamined the dependence of the membrane potential on external Na^+ concentrations over the range 0–13 m*M* using choline as the principal substitute. These results are plotted in Fig. 1. The relation between the dark potential and the logarithm of the external Na^+ concentration is nonlinear, approaching asymptotically a straight line of slope 27 mV log unit^{-1}, which fits the data in the range 5–132 m*M*. At lower concentrations the relation is less steep, possibly because of imperfect selectivity of the Na^+ conductance or because of the activation of a voltage-dependent conductance as the rods hyperpolarized. If it is assumed that at higher concentrations the effects of voltage-

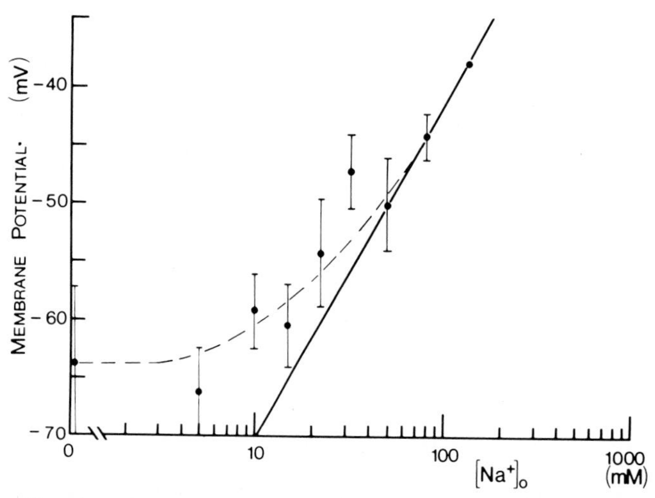

FIG. 1. The relation between membrane potential in darkness (dark potential) and the external concentration of Na^+ (log scale) in rods of the toad *Bufo marinus*. In each experiment the displacement of the dark potential from its value in control Ringer's solution was measured. All values were then normalized against the *mean* dark potential in control Ringer's solution (−38 mV) and averaged. The bars represent ±1 SD. Data from Capovilla *et al.* (1980a).

dependent conductances are small and that conductances are independent of external $[Na^+]$, the slope of the relation suggests that, under normal physiological conditions in darkness, the Na^+ conductance accounts for approximately half of the *total* membrane conductance.

If the entire Na^+ conductance of the membrane is accounted for by the light-modulated conductance, then, given that the chloride conductance is relatively small, the rod membrane should behave rather like a K^+ electrode at the peak of a saturating light response. Capovilla *et al.* (1980b) found that this was only approximately true, because under normal circumstances the peak of the response was limited by the action of a time-varying, voltage-dependent conductance (Fain *et al.*, 1978).

We repeated their experiments adding 2 m*M* Cs^+ to the Ringer's solution, since this is known to block the relaxation from peak to plateau of such responses (Fain *et al.*, 1978). In Fig. 2, the potential at the peak of the light response has been plotted against the external K^+ concentration (log scale). With concentrations between 2.6 and 10 m*M*, there was evidence that the slope of the line that best fitted the data might be *greater* than 58 mV log unit^{-1}. For concentrations above 10 m*M*, however, the data were well fitted by a line of this slope,

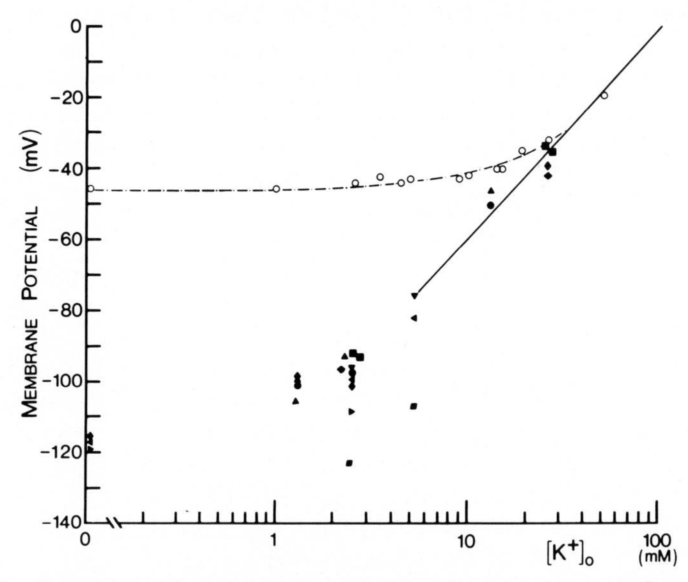

FIG. 2. Dependence of the rod membrane potential on the external concentration of K^+ (log scale) in rods of *B. marinus* in Ringer's solution containing 2 m*M* Cs^+. The solid symbols plot the potential at the peak of the rod's response to a saturating light flash. Each symbol represents an unnormalized value from one of eight rods. The solid line has a slope of 58 mV log unit^{-1} concentration. For comparison, the open circles plot average values of the dark potential in the same cells. The broken line was drawn by eye through these points.

indicating that the membrane behaved almost perfectly as a K^+ electrode in that range. From these data we estimate the internal concentration of K^+ to be $\simeq$94 m*M*, which corresponds to an equilibrium potential E_K of -90 mV under normal physiological conditions.

In our experiments, the mean value of the dark potential of the rod in normal Ringer's solution was -44 mV $\pm$ 3.4 (mean $\pm$ SE from 47 rods). If, as Capovilla *et al.* (1980a) suggested, the Na^+ conductance in darkness represents half of the total membrane conductance, the equilibrium potential for Na^+ should lie somewhere between 0 and 10 mV.

We do not have a reliable estimate of E_{Na}. Of course, if the light-modulated current is carried exclusively by Na^+, an estimate can be obtained from a measure of the reversal potential of the light response. Unfortunately, attempts to determine the reversal potential by point injection of polarizing currents are incapable of yielding a reliable value because of the effects of rod–rod coupling. Recent experiments by Werblin (1978) and by Bader *et al.* (1979), in which isolated rods were voltage-clamped, avoided this problem, and the reversal potential for the photocurrent in rods of tiger salamander was found to lie between 0 and 10 mV. If the photocurrent is exclusively carried by Na^+, the internal Na^+ concentration should lie in the range 75 m$M \leqslant [Na^+]_{in} \leqslant$ 110 m*M*. At first sight, such a high value might be hard to accept. To satisfy osmotic equilibrium and electroneutrality, the total concentrations of cations inside and outside the cell should be nearly equal. The external cations totaled 120 m*M* in the experiments of Bader *et al.* (1979). If the internal K^+ concentration were as high as it is in toad rods, it would be unlikely that $[Na^+]_{in}$ could be much greater than 30–40 m*M*.

The most likely explanation of this apparent discrepancy is that the internal K^+ in tiger salamander rods is lower than in the toad. The external K^+ concentration in the experiments of Bader *et al.* was 1.5 m*M*. With the lower limit $[Na^+]_{in}$ = 75 m*M*, which allows $[K^+]_{in}$ to be 45 m*M*, the value of E_K is -86 mV, in reasonable agreement with that in the toad.

Another factor that must be considered is that the light-modulated conductance may be permeable to ions other than Na^+. Studies of the selectivity of the light-modulated conductance have recently begun, but as yet little information is available. If the selectivity were similar to that of Na^+ channels in the squid axon, some K^+ would permeate through and the reversal potential of the photocurrent would be 7 or 8 mV less positive than E_{Na}. The internal Na^+ concentration would then be somewhat lower than we have estimated.

In summary:

1. The value of E_K lies close to -90 mV in toad rods.
2. The internal concentration of K^+ is high, ~94 m*M* in toad rods, but it may vary from species to species.

3. The membrane is highly permeable to K^+, the K^+ conductance accounting for nearly half of the *total* membrane conductance in darkness.

4. The value of E_{Na} probably lies in the range 0–10 mV, though it may be higher. We cannot be certain, however, until the selectivity of Na^+ conductance has been established and we know whether or not the light-modulated conductance is permeable to ions other than Na^+.

5. The Na^+ conductance in darkness accounts for about half of the total conductance of the membrane.

6. The Cl^- conductance is probably low.

7. The Mg^{2+} conductance is probably low.

In the following section we shall discuss mechanisms by which the distribution of the major ions across the plasma membrane is maintained.

V. ION-EXCHANGE PUMPS

A. Na^+–K^+ Exchange

Electrophysiological evidence that the internal concentrations of Na^+ and K^+ are maintained by a Na^+–K^+ exchange pump in the rod's plasma membrane came first from an experiment by Sillman *et al.* (1969) with the frog retina. They found that the aspartate-isolated late receptor potential (LRP) rapidly disappeared after exposure of the retina to the cardioactive steroid ouabain. Responses could be elicited in the presence of ouabain so long as a Na^+ concentration gradient was artifically maintained, but these, too, disappeared when internal and external concentrations were allowed to equilibrate. Similar results were obtained in rat rods by Yoshikami and Hagins (1970). These results clearly indicate that the Na^+ concentration gradient is normally maintained by a Na^+–K^+ exchange pump which takes no direct part in generating the rod's response to light.

By analyzing the distribution of current flowing around the rat rod, Hagins *et al.* (1970) showed that the source of the dark current was located near the distal end of the rod's inner segment, a region densely packed with mitochondria. This is consistent with the finding of Bownds *et al.* (1974) that the rod outer segment contained no Na^+–K^+-ATPase.

Na^+–K^+ exchange pumps are, in general, electrogenic to some degree (Glynn and Karlish, 1975). Using a technique similar to that of Hagins and co-workers, Zuckerman (1973) found that ouabain altered the distribution of current sources and sinks along the frog rod. He argued that the Na^+–K^+ exchange must be significantly electrogenic so that the pump contributed directly to the dark potential. While this idea is not unreasonable, it was not supported by the later experiments of Hagins and Yoshikami (1975) who were unable to reproduce

Zuckerman's results. Thus the degree of electrogenicity of the Na^+–K^+ exchange pump is at present uncertain.

B. Na^+–Ca^{2+} Exchange

The internal free calcium concentration of nerve and muscle cells is generally very low, near 0.1 μM. Photoreceptors, too, are thought to have a low cytoplasmic concentration of Ca^{2+}, at least in their outer segments (Hagins and Yoshikami, 1977). The extracellular Ca^{2+} concentration is about 2 m*M*, so that the concentration gradient across the plasma membrane is probably very large. Maintaining it requires either an enormous buffering capacity in the rod cytoplasm or a highly efficient means of pumping Ca^{2+} out of the rod. The mechanisms believed to effect this in other nerve and muscle cells have been reviewed recently by McNaughton (1978). They include an ATP-dependent, uncoupled Ca^{2+} extrusion and an ATP-dependent Na^+–Ca^{2+} exchange mechanism which can reverse when external Na^+ is greatly reduced.

There has been recent speculation that both these mechanisms exist in the plasma membrane of the rod (see, for example, Fain and Lisman, 1981). A Na^+–Ca^{2+} exchange is a particularly attractive possibility because, if present, it could explain some of the puzzling contradictions in the literature. (An example is given in the following section.) Unfortunately, there is little firm evidence that can be cited either for or against it.

Daeman *et al.* (1977) showed that the accumulation of ^{45}Ca by a suspension of rod outer segments that had been depleted of Ca^{2+} with EGTA was ATP-dependent and was inhibited (slightly) by external Na^+. The efflux of ^{45}Ca from preloaded outer segments, on the other hand, was stimulated by external Na^+. These observations are consistent with the hypothesis that the rod outer segment contains a Na^+–Ca^{2+} exchange mechanism. The difficulty is that most techniques used for making bulk suspensions of rod outer segments yield significant fractions of shattered outer segments and outer segments to which are attached fragments of inner segments. Since these fragments are likely to be rich in mitochondria, there must remain some doubt when movements of small amounts of Ca^{2+} are interpreted entirely in terms of outer segment mechanisms. Nonetheless, the possibility of Na^+–Ca^{2+} exchange must be borne in mind when interpreting the effects of changes in the external concentrations of Na^+ and Ca^{2+}.

VI. THE LIGHT-MODULATED CONDUCTANCE

A. Channel or Carrier?

Studies on ionic selectivity of the light-modulated conductance are only in their early stages. Without accurate and detailed information on this phenomenon, the likely nature of the conductance mechanism remains a matter for speculation.

Three experimental results are frequently invoked as evidence that Na^+ does not pass through Na^+ channels like those in other systems such as the squid axon. First, agents that have been found to block Na^+ channels, including tetrodotoxin (squid axon) and amiloride (epithelial cells), have little or no effect on the rod's light response (Fain *et al.*, 1980; W. G. Owen and V. Torre, unpublished observations). This may simply mean that the site(s) at which these agents act is inaccessible from the external medium, however.

Second, the fact that Li^+ does not support either the dark potential or the light response when exchanged for Na^+ suggests an unusually high selectivity for Na^+. The Na^+ channel of the squid axon does not discriminate between Li^+ and Na^+. Third, the osmotic experiments of Korenbrot and Cone (1972) appeared to show an extremely high degree of rectification. They compared the rate at which Na^+ flowed into the rod outer segment during a hyperosmotic shock with the rate at which it flowed out during an equivalent isosmotic exposure to low external Na^+ and found the inward flow to be greater than the outward flow by a factor of 10^3. Such a high degree of inward rectification is not easily explained in terms of a channel mechanism.[2]

Both these latter observations, however, have an alternative explanation (see Fain and Lisman, 1981). There is evidence that the light-modulated conductance can be blocked by internal Ca^{2+} (see the following section). Many nerve and muscle cells contain a Ca^{2+}–Na^+ exchange mechanism that is ATP-dependent (see McNaughton, 1978). In such a mechanism, as in some other metabolically driven pumping mechanisms, Li^+ is a very poor substitute for Na^+. Under normal circumstances Ca^{2+}–Na^+ exchange acts to expel Ca^{2+} from the cell. Replacing external Na^+ with Li^+ reverses the exchange so that Ca^{2+} can then move inward across the membrane (Baker *et al.*, 1969).

If the rod's plasma membrane contained such a mechanism, the influx of Ca^{2+} resulting from the replacement of external Na^+ would result in blockage of the light-modulated conductance (see the following section). This would account for the effects of substituting Li^+ for Na^+ on both the dark potential and the light response. It would also account for the failure of Na^+ to flow out of the rod outer segment during isosmotic exposure to low external Na^+ in the experiments of Korenbrot & Cone. Of course, the evidence for the existence of Na^+–Ca^{2+} exchange across the rod plasma membrane is weak, and so this explanation, attractive though it is, remains speculative.

In summary, the evidence at present available cannot be said to favor either mechanism. Resolution of this question must await more precise, quantitative

[2]Recent voltage-clamp measurements made in isolated rods by Bader *et al*. (1979) indicate that the light-modulated conductance exhibits a significant degree of *outward* rectification, however (see Section VI, C).

techniques for studying ion movements across the rod outer segment plasma membrane.

B. The Action of Ca^{2+} on the Light-Modulated Conductance

Yoshikami and Hagins (1972) found that adding Ca^{2+} to the Ringer's solution bathing the retina caused a marked reduction in the dark current. They concluded that Ca^{2+} was capable of blocking the light-modulated conductance, a view that received quick support from the osmotic experiments of Korenbrot and Cone (1972). They noted that the influx of Na^+ during a hyperosmotic Na^+ shock was blocked when Ca^{2+} was added to the external medium. Thus Ca^{2+} mimicked the action of light.

When Ca^{2+} is removed from the external medium ($[Ca^{2+}]_{out} \rightarrow 10^{-5}$ *M*), the rod *depolarizes* by about 30 mV in darkness. The potential attained at the peak of the response to a saturating light flash remains unchanged, however, so that the response amplitude is increased by this amount (Brown and Pinto, 1974; Owen and Torre, in preparation). Moreover, reducing external Ca^{2+} in this way results in a very large increase in membrane conductance (Owen and Torre, 1981). These observations are consistent with the action of Ca^{2+} being to reduce the light-modulated conductance. If we assume this to be the only action of Ca^{2+}, the magnitude of the depolarization induced by (nominally) zero $[Ca^{2+}]_{out}$ and the estimated increase in membrane conductance will imply that, under normal conditions in darkness, more than 90% of the total light-modulated conductance is blocked. A similar suggestion was made for cones on the basis of similar evidence (Bertrand *et al.*, 1979). The possible functional advantages of this are obscure.

That the blocking action of Ca^{2+} is a direct one involving few, if any, intermediate steps, received support from a recent report by Yoshikami and Hagins (1980). Using a technique by which the external Ca^{2+} concentration could be raised within 90 msec, they showed that the dark current was reduced with no detectable delay.

The osmotic experiments of Wormington and Cone (1978) showed that the light-modulated conductance could be reduced by Ca^{2+} much more effectively from inside the membrane than from outside. Using the Ca^{2+} ionophores X537A and A23187, they increased Ca^{2+} permeability so that the internal and external concentrations could be regarded as equal. They found Na^+ influx to be completely blocked when the external (hence internal) concentration of Ca^{2+} was greater than $\simeq 10\ \mu M$. Without the ionophores, 10 m*M* Ca^{2+} was required in the external medium to block Na^+ influx, suggesting that cytoplasmic Ca^{2+} was about 1000 times as effective as external Ca^{2+}.

Using conventional intracellular recording techniques, we have found that raising $[Ca^{2+}]_{out}$ to 10 m*M* reduces the rod's response to a bright flash by about 8 mV but does not abolish it (Owen and Torre, in preparation), suggesting that under physiological conditions the light-modulated conductance may be rather less sensitive to $[Ca^{2+}]_{out}$ than Wormington and Cone's experiments indicated.

In summary, the evidence at present available indicates that the light-modulated conductance can be reduced by Ca^{2+}, and that this action is more efficient when Ca^{2+} acts from inside of the membrane than when it acts from outside.

C. Effects of pH

Early studies on the effects of pH on the rod response, using the aspartate-isolated LRP as a monitor of rod activity, were in agreement that raising the pH in the range 6–8 caused a large increase in response amplitude (Sillman *et al.*, 1972; Ward and Ostroy, 1972; Gedney and Ostroy, 1974). The exact relation between LRP amplitude and pH appeared to be buffer-dependent (Gedney and Ostroy, 1974). Such results are difficult to interpret, because the site of action of the proton is unknown. It was argued in terms of a pH effect on photopigment bleaching kinetics. Two more recent studies, however, have shown that protons may bind to anionic sites on the plasma membrane, directly affecting membrane conductance.

Wormington and Cone (1978) found evidence that protons could bind to such a site, having a pK_a of 5.8, and thereby block a Na^+ channel with which it was associated. Since this occurred in the rod outer segment, it raised the possibility that the light-induced decrease in Na^+ conductance might be mediated by an internal release of protons.

Pinto and Ostroy (1978), recording intracellularly from rods in the isolated retina, concluded that the principal effect of protons was to block K^+ conductance, presumably located in the inner segment of the rod.

While there is therefore agreement that protons can affect membrane conductance, the identity and location of this conductance within the cell is uncertain.

VII. VOLTAGE-DEPENDENT CONDUCTANCES

If the rod membrane contained only ohmic conductances, the voltage response would have a time course that mirrored exactly the time course of the photocurrent. This is clearly not the case. In Fig. 3 a set of intracellularly recorded voltage responses, elicited with flashes of increasing intensity, is compared with photocurrents flowing across the rod outer segment membrane measured by the technique of Baylor *et al.* (1979). (See Chapter 1, this volume.) The retina in each case was that of the toad *Bufo marinus* bathed in identical Ringer's solutions buffered with bicarbonate and CO_2 at pH 7.8.

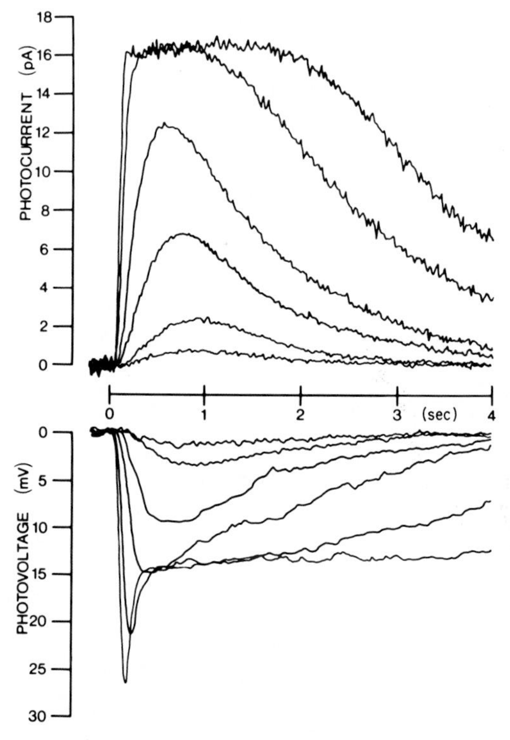

FIG. 3. Comparison between the time courses of photocurrents and photovoltages elicited by diffuse illumination in rods of the toad *B. marinus*. The photocurrents were recorded using a suction electrode according to the technique of Baylor *et al.* (1979). Photovoltages were recorded in a second retina using conventional intracellular techniques. The rods were bathed in identical Ringer's solutions (bicarbonate/CO_2 buffer) and stimulated with flashes of six intensities 0.6 log units apart. The intensities used in the two recording situations were closely similar but not identical. (We thank Drs. T. D. Lamb, P. A. McNaughton, and K. -W. Yau for permission to reproduce their recordings of the photocurrents.)

At low intensities both the photovoltage and the photocurrent have a roughly similar time course. The photocurrent reaches its peak ~*960* msec after the light flash, whereas the photovoltage peaks after ~*850* msec.

At high intensities the discrepancy between voltage and current is much more obvious. The photocurrent reaches a peak and then slowly decays with a roughly exponential time course. The photovoltage, on the other hand, decays rapidly after the peak to a plateau level which is maintained for several seconds before gradually returning to the dark potential. These differences in the time course of the photocurrent and the photovoltage clearly indicate that the membrane contains time-varying, voltage-dependent conductances, active in the physiological range, that play a major role in shaping the voltage response to light.

In addition, studies on the input current–voltage relation of the rod network (Lasansky and Marchiafava, 1974; Werblin, 1975; Copenhagen and Owen, 1980; Fain and Quandt, 1980) and of isolated rods (Werblin, 1979; Bader *et al.*, 1979; Attwell and Wilson, 1980) have revealed that the rod membrane exhibits a high degree of outward rectification.

In this section we shall discuss the evidence for the existence of four different voltage-dependent conductances and review what is known about the ionic species associated with each one.

A. Peak–Plateau Relaxation: The Cs^+-Sensitive Conductance

The relaxation from peak to plateau of the rod's voltage response to an intense flash can be blocked by the addition of a small concentration of Cs^+ to the Ringer's solution (Fain *et al.*, 1978). This is illustrated in Fig. 4. In the presence of Cs^+, the voltage response increases to a peak, reached after ~2 sec, and slowly falls back toward the dark potential. The time to peak of the photovoltage in Cs^+ Ringer's solution is thus similar to that of the photocurrent elicited by stimuli of this intensity. Fain *et al.* (1978) also found that, when peak-plateau relaxation was blocked in this way, a large light-induced increase in input impedance was revealed. Since in normal Ringer's solution this was not seen, they concluded that peak–plateau relaxation resulted from a substantial increase in membrane conductance, which caused membrane potential to decrease. Since the effects of adding Cs^+ and of removing all external Na^+ during the plateau phase of the response to a bright, steady light were similar, they concluded that the relaxation was brought about by a Na^+ conductance that increased with hyperpolarization.

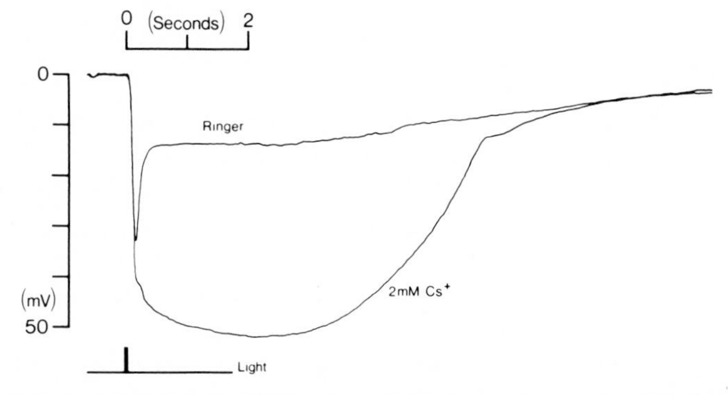

FIG. 4. The effect of 2 m*M* Cs^+ on the rod's response to a flash of near-saturating intensity (6089 Rh* per rod).

Since 2 mM Cs^+ has no effect on the dark potential, it is easy to compare responses elicited in normal Ringer's solution with similar responses in the presence of Cs^+. We have done this and find that the response time courses in the two cases are identical provided they do not exceed about 8 mV peak amplitude. Above this, substantial differences quickly develop (W. G. Owen and V. Torre, unpublished). This is also evident in the data published by Fain *et al.* (1978). Since the dark potential in our experiments was close to -44 mV, this suggests that activation of the Cs^+-sensitive mechanism occurs when membrane potential becomes more negative than about -52 mV. There is some uncertainty in this conclusion, however, because the blocking action of Cs^+ may itself be voltage-dependent (Coronado and Miller, 1979).

The conclusion that a Na^+ conductance increase is responsible for the relaxation in the response deserves attention. The magnitude of the conductance change inferred from the measurements of Fain *et al.* (1978) is so great that, if the conductance were permeable only to Na^+, the membrane potential would be expected to approach E_{Na}. Rather, it suggests that the reversal potential of this mechanism must be much more negative than E_{Na}. Furthermore, Capovilla *et al.* (1980a) found the plateau potential to be relatively insensitive to external $[Na^+]$ for concentrations above ~25 mM. Reducing $[Na^+]_{out}$ below this value caused the plateau to hyperpolarize, though the total response amplitude had decreased to less than 4 mV by this point.

It seems unlikely, in view of this, that the Cs^+-sensitive conductance is simply permeable to Na^+. It may be that it is permeable to more than one ion, perhaps Na^+ and K^+. This would permit a reversal potential negative with respect to E_{Na}. However, one must also assume that the net current flowing through the conductance is independent of the driving force on Na^+ until it drops below a critical value, consistent with $[Na^+]_{out}$ = 25 mM, if the findings of Capovilla *et al.* are to be explained.

B. High-Pass Filtering of Small Signals: The "Small-Signal Inductance"

When a long, narrow bar of low-intensity light is flashed on the retina at increasing distances from an impaled rod, the time to peak of the recorded response becomes progressively shorter, as illustrated in Fig. 5a. This was first noticed by Detwiler *et al.* (1978) who pointed out its implication that the rod network acts as a high-pass filter to laterally propagating small signals. They showed that this behavior could be modeled by an electrical circuit in which each of the transverse elements contained an inductance. For small perturbations, the arm of the circuit containing the inductance is the electrical analogue of a conductance that varies with time and voltage in a manner described by the Hodgkin–Huxley equations. The fact that the circuit contains only linear elements should not be taken to imply that the mechanism it represents is an inherently linear one.

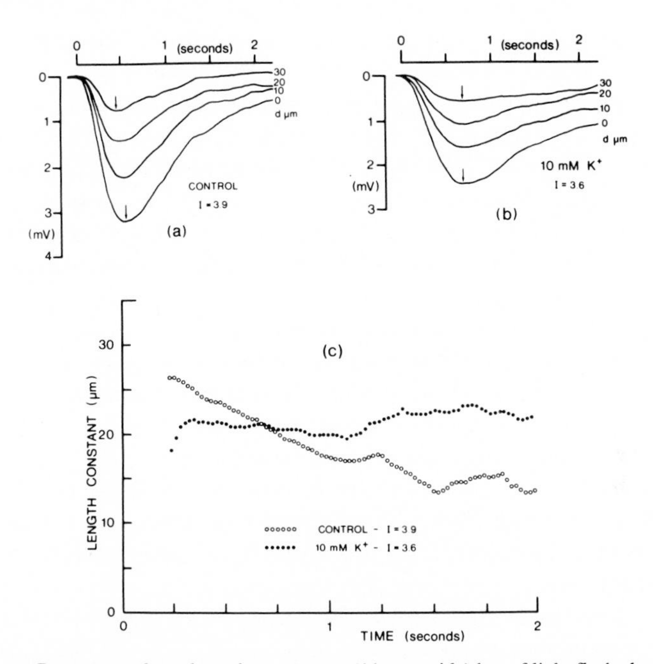

FIG. 5. (a) Responses of a rod to a long, narrow (11-μm-wide) bar of light flashed at successive 10-μm displacements from the centered position. Recordings made in control Ringer's solution which contained 2.6 m*M* K^+. Stimuli bleached 28.7 rhodopsin molecules per rod. (b) The Ringer's solution was changed for one containing 10 m*M* K^+, and the experiment repeated. A slightly more intense stimulus (65.7 rhodopsin molecules per rod) was necessary to elicit responses of comparable magnitude. (c) The length constant of the rod network calculated on the basis of the model of Detwiler, Hodgkin, and McNaughton and plotted as a function of time during the response.

Any process behaves in a linear fashion provided pertubations are sufficiently small. Such a circuit therefore is useful *only* when small signals are elicited.

This voltage-dependent mechanism, though having perhaps a less dramatic effect on the time course of the rod's response than that described in the preceding section, could well be functionally more important. Rods subserve detection of the dimmest stimuli. Such stimuli elicit small responses, and these are shaped by this mechanism.

We studied the ionic basis of this behavior in the isolated perfused retina of the toad, and the results are summarized in Fig. 5b. In normal Ringer's solution the high-pass filtering by the rod network was characterized by a fall in the length constant during the response from a high-frequency value of ~30 μm to a low-frequency value of ~15 μm (open circles in Fig. 5c). This was not affected by the addition of up to 10 m*M* Cs^+ to the perfusion medium, even though a concentration of only 2 m*M* was sufficient to block the peak–plateau relaxation described in the previous section. The blocking action of Cs^+ is known to be voltage-dependent in some systems (Coronado and Miller, 1979), so this result by itself does not conclusively rule out the possibility that the two phenomena depend on the same mechanism.

We were able to abolish the high-pass filtering behavior by raising the external K^+ concentration. In 10 m*M* K^+ the time to peak of linear range responses was increased by about 30% (Fig. 5b). The shortening of the time to peak with progressive displacement of bar-shaped stimuli was no longer seen. The length constant of the network throughout the response remained at a fixed value slightly below the high-frequency value in normal Ringer's solution (solid circles in Fig. 5c).

Analysis of the network (Owen and Torre, in preparation) showed that the time course of responses in 10 m*M* K^+ was closely similar to that of the photocurrent computed from responses obtained in normal Ringer's solution.

Since raising the external K^+ concentration to 10 m*M* only depolarized the membrane in darkness by 2 or 3 mV, the driving force on K^+ was lowered by about 60%. This reduced the effects of time-dependent changes in K^+ conductance. It was concluded therefore that the high-pass filtering of small signals depended on a K^+ conductance that decreased with a delay when the membrane was hyperpolarized.

Raising the external $[K^+]$ to 10 m*M* also affected the responses to near-saturating stimuli. The voltage at the peak of the response was substantially reduced, whereas the plateau depolarized by only 2 or 3 mV. This is consistent with the results of Capovilla *et al.* (1980b), which showed that the K^+ permeability represented a much smaller fraction of the total membrane permeability at the plateau than at the peak of the response. Nonetheless, in 10 m*M* K^+ a relaxation from peak to plateau was still present, and this could be blocked by adding 2 m*M* Cs^+ to the perfusion medium, strongly suggesting that the mechanisms underlying the two phenomena were indeed separate and independent.

C. Outward Rectification of the Rod Membrane

When steps of outward (depolarizing) current are injected into a rod, an outward rectification of the membrane is revealed (Lasansky and Marchiafava, 1974; Werblin, 1974, 1978, 1979; Owen and Copenhagen, 1977; Copenhagen and Owen, 1980; Fain *et al.*, 1977; Fain and Quandt, 1980; Attwell & Wilson, 1980). During step depolarizations to potentials less positive than about -20 mV, one generally sees a slow relaxation in the induced potential from an initial value to one that is rather more negative. The time constant of this relaxation may be several hundred milliseconds (Copenhagen and Owen, 1980). This behavior probably reflects the action of time-varying, voltage-dependent conductances similar to that described in the previous section.

Depolarizing the rod beyond about -20 mV reveals a very strong outward rectification which appears to be instantaneous because its time constant is generally shorter than that of the recording apparatus ($<\sim 10$ msec). Fain and

Quandt (1980) found this instantaneous rectification to be greatly reduced and in some cases abolished when 12 m*M* tetraethylammonium chloride (TEA) was added to the perfusing medium. Recording from isolated rods under both current-clamped and voltage-clamped conditions, however, Werblin (1979) found that, whereas TEA reduced the outward rectification of the membrane, the effect was considerably smaller than that reported by Fain and Quandt (1980).

Certain voltage-dependent potassium conductances in other systems are blocked by TEA (Hagiwara *et al.*, 1974; Narahishi, 1974; Armstrong, 1975; Meves and Pichon, 1977; Atwater *et al.*, 1979) and, largely because of this, Fain & Quandt have suggested that the ''instantaneous'' outward rectification reflects the action of a voltage-dependent potassium conductance activated at potentials more positive than about −20 mV. It must be emphasized, however, that there is no *direct* experimental evidence that TEA affects a K^+ conductance in the rod membrane, and therefore the assertion that a K^+ conductance is responsible, even in part, for the outward rectification remains unproved.

There is evidence that the outward rectification involves another component, however. In their recent voltage-clamp study on isolated rods, Bader *et al.* (1979) found that the light-modulated conductance itself exhibited a pronounced outward rectification. Indeed, this rectification was sufficient to account for all the outward rectification they measured in darkness. This is surprising because, if generally true, one would have to suppose that the effects of TEA observed by Fain and Quandt and by Werblin resulted from its action on the light-modulated conductance. Since Fain and Quandt (1980) reported that TEA *decreased* the membrane conductance, we would expect the rod to hyperpolarize in this case. In fact, it depolarizes, which is consistent with its blocking of a K^+ conductance but inconsistent with blocking of the light-modulated conductance.

The degree of rectification seen by Bader *et al.* is probably too small to account for that measured in the intact retina (Copenhagen and Owen, 1980), and one must wonder whether the proteolytic enzymes used to isolate their rods were entirely benign in their effect on other voltage-dependent conductances in the membrane.

In summary, the instantaneous outward rectification of the rod membrane is not yet understood. The suggestion that a voltage-dependent potassium conductance may be involved, while plausible, is not yet supported by experimental evidence. The light-modulated conductance itself appears to be outwardly rectifying and must contribute to the overall rectification we see.

D. Regenerative Behavior of the Rod Membrane

THE Ca^{2+} CONDUCTANCE

As first reported by Fain *et al.* (1977) and more recently by Fain *et al.* (1980), when rods are perfused with Ringer's solution to which a small concentration

(>5 m*M*) of TEA ions has been added, the rod membrane depolarizes by 8–10 mV and large oscillatory potentials appear during the recovery from a bright flash of light. These potentials are slow, much slower than nerve action potentials, and die away soon after the membrane has repolarized to the dark potential. With prolonged exposure, however, the oscillatory potentials often become spontaneous, and much faster spikelike potentials may be seen. The action of light is then to suppress these spontaneous potentials.

A similar sequence of events occurs when 1–2 m*M* Ba^{2+} or 5–10 m*M* 4-aminopyridine is added. In all cases this oscillatory behavior is blocked by agents that in other systems are known to block Ca^{2+} channels: cadmium (25 μM), cobalt (400 μM), magnesium (5 m*M*), D-600 (100 μM) (Fain and Quandt, 1980).

If strontium is added to Ringer's solution containing TEA, large spikelike potentials, spontaneous in darkness and suppressed by light, quickly develop. Indeed, we have found (Owen and Torre, in preparation) that adding Sr^{2+} alone in concentrations of 10 m*M* or greater results in spontaneous spiking, the spike amplitude increasing with increasing Sr^{2+} concentration. These spikes are blocked by the same agents at the same concentrations that block the oscillatory potentials (Fain *et al.*, 1980). Oscillatory potentials and spiking behavior induced by techniques similar to these have been studied in a variety of nerve and muscle cells (see Reuter, 1973, for a review of this topic). There is considerable evidence, in these cases, suggesting that these phenomena reflect the activation of a voltage-gated Ca^{2+} channel whose conductance is negative over a certain range of potentials. Because of this, depolarization of the membrane causes an inward Ca^{2+} current to develop, which may or may not become regenerative depending on how large an outward current opposes it. Regenerative behavior may be induced either (1) by sufficiently decreasing the magnitude of the outward component of transmembrane current or (2) by augmenting the inward component of transmembrane current.

Both TEA and 4-aminopyridine have been found to block outward K^+ currents in many preparations, and it is reasonable to suppose that they induce oscillatory behavior by mechanism 1. On the other hand, Sr^{2+} can be substituted for Ca^+ in many preparations and, in Purkinje fibers and certain ventricular myocardial preparations, it has been shown to be considerably more permeable (Reuter, 1973). It seems likely therefore that Sr^{2+} induces spiking by augmenting the inward current so that it becomes fully regenerative. Interestingly, Ba^{2+} is both a carrier of inward current and a blocker of K^+ channels and probably exerts a dual action which destabilizes the membrane.

While it is not unreasonable to suppose that the regenerative behavior described by Fain *et al.* in the toad rod is explained in this way, neither is it certain. Capovilla *et al.* (1980b) noted that oscillations could be induced by reducing the external Cl^- concentration and emphasized that it was only necessary to suppose that reducing $[Cl^-]_{out}$ caused a reduction in the membrane's shunting conductance to explain this finding. Thus the K^+ conductance, specifically, need not be the one affected by the blocking agent.

It should also be mentioned that spiking behavior can be elicited in cones when they are bathed in Sr^{2+} or Ba^{2+} (Piccolino and Gerschenfeld, 1980). Although some spontaneous spiking was observed, spikes were most effectively elicited when the periphery of the cone's receptive field was illuminated with a bright flash of light. Piccolino and Gerschenfeld suggested that such spikes resulted from a regenerative increase in a Ca^{2+} conductance localized close to the synaptic terminal and triggered by synaptic feedback from horizontal cells. That spikes cannot be elicited in rods by this type of illumination is consistent with the view that rods do not receive any feedback from horizontal cells (Owen and Copenhagen, 1977).

VIII. SUMMARY

The plasma membrane of the vertebrate rod is clearly no less complicated a barrier to ion movement than the membranes of other excitable cells. Indeed, there is the added complication that the cell is highly differentiated into an outer and an inner segment and ionic permeabilities may be different in each segment.

In its resting state, in darkness, the rod is about equally permeable to Na^+ and K^+, the permeability to Cl^-, Mg^{2+}, and Ca^{2+} being much lower. The internal concentration of K^+ is high, near 90 m*M* in the toad rod, whereas that of Na^+ is lower. The reverse is true in the extracellular medium. This distribution of Na^+ and K^+ is maintained by a Na^+-K^+, ATP-dependent pump located toward the distal end of the rod's inner segment.

The direct effect of light is to reduce the Na^+ permeability of the outer segment membrane, causing the rod to hyperpolarize. A small hyperpolarization triggers a delayed reduction in K^+ conductance, causing the photovoltage to peak rather sooner than the photocurrent (the small-signal inductance). Hyperpolarization by more than $\sim$8 mV triggers, in addition, a delayed conductance increase, limiting the peak potential and causing relaxation to a plateau. This relaxation can be blocked by Cs^+. It is likely that the Cs^+-sensitive conductance is permeable not to Na^+ alone but perhaps to both Na^+ and K^+.

What are the functional consequences of these voltage-dependent, time-varying conductance changes? The initial phototransduction process involves an energy gain of about 10^4. To achieve such a gain requires that it consist of several sequential stages. Indeed, at least four such stages were found necessary to account for the delay in the onset of the light response (Cervetto *et al.*, 1977). In consequence, the modulation of the Na^+ conductance is very slow compared with the time course of the initial photochemical events. Under scotopic conditions an animal may rely for its survival on the visual information provided by its rods. In these terms, the very slowness of the transduction process, if not compensated for, could be a disadvantage. The small-signal inductance in effect differentiates the photocurrent. Since the modulation of synaptic transmitter release is mediated by voltage, not current, the effect of this differentiation is communicated to the bipolar cell. Thus partial compensation for the slowness of the phototransduction process may be achieved by this mechanism.

There is little that one can say concerning the functional consequences of the voltage-dependent Ca^{2+} conductance. One can speculate that this mechanism might be localized to the synaptic terminal and play a role in modulating transmitter release. No doubt, as new techniques for examining these mechanisms are perfected and they can be localized to the outer segment, inner segment, or synaptic terminal, their roles in the complex machinery of the rod will be properly understood.

ACKNOWLEDGMENTS

We wish to thank Professor Sir Alan Hodgkin, F. R. S., and Drs. T. D. Lamb, P. A. McNaughton, and K.-W. Yau for many helpful discussions and comments during the writing of this chapter. We are especially grateful to Drs. T. D. Lamb, P. A. McNaughton, and K. -W. Yau for generously allowing us to reproduce some of their data in Fig. 3. This chapter was written while W. Geoffrey Owen was in receipt of a research career development award (EY 00113) and research grant (EY 02493) from the National Institutes of Health (U.S.P.H.S.) and Vincent Torre was in receipt of an EMBO fellowship.

REFERENCES

Armstrong, C. M. (1975). Channels and voltage-dependent gates in nerve. *In* "Membranes—A Series of Advances: Artificial and Biological Membranes" (G. Eisenman, ed.), Vol. 3, pp. 325–358. Dekker, New York.

Attwell, D., and Wilson, M. (1980). Behaviour of the rod network in the tiger salamander retina mediated by membrane properties of individual rods. *J. Physiol. (London)* **309,** 287–310.

Atwater, I., Ribalet, B., and Rojas, E. (1979). Mouse pancreatic β-cells: Tetraethylammonium blockage of the potassium increase induced by depolarization. *J. Physiol. (London)* **288,** 561–574.

Bader, C. R., MacLeish, P. R., and Schwartz, E. A. (1978). Responses to light of solitary rod photoreceptors isolated from tiger salamander retina. *Proc. Natl. Acad. Sci. U.S.A.* **75,** 3507–3511.

Bader, C. R., MacLeish, P. R., and Schwartz, E. A. (1979). A voltage-clamp study of the light response in solitary rods of the tiger salamander. *J. Physiol. (London)* **296,** 1–26.

Baker, P. F., Blaustein, M. P., Hodgkin, A. L., and Steinhardt, R. A. (1969). The influence of calcium on sodium efflux in squid axons. *J. Physiol. (London)* **200,** 431–458.

Baylor, D. A., Fuortes, M. G. F., and O'Bryan, P. M. (1971). Receptive fields of single cones in the retina of the turtle. *J. Physiol. (London)* **214,** 265–294.

Baylor, D. A., Lamb, T. D., and Yau, K.-W. (1979). The membrane current of single rod outer segments. *J. Physiol. (London)* **288,** 589–611.

Bertrand, D., Fuortes, M. G. F., and Pochabradsky, J. (1978). Actions of EGTA and high calcium on the cones of the turtle retina. *J. Physiol. (London)* **275,** 419–437.

Bortoff, A. (1964). Localization of slow potential responses in the *Necturus* retina. *Vision Res.* **4,** 627–636.

Bownds, D., and Brodie, A. E. (1975). Light-sensitive swellings of isolated frog rod outer segments as in *in vitro* assay for visual transduction and dark adaptation. *J. Gen Physiol.* **66,** 407–425.

Bownds, D., Brodie, A., Robinson, W. E., Palmer, D., Miller, J., and Shedlovsky, A. (1974). Physiology and enzymology of frog photoreceptor membranes. *Exp. Eye. Res.* **18,** 253–266.

Brown, J. E., and Pinto, L. H. (1974). Ionic mechanism for the photoreceptor potential of the retina of *Bufo marinus. J. Physiol. (London)* **236,** 575–591.

Capovilla, M., Cervetto, L., Pasino, M., and Torre, V. (1980a). The sodium current underlying responses to light of rods. *J. Physiol. (London)* (in press).

Capovilla, M., Cervetto, L., and Torre, V. (1980b). Effects of changing the external [K^+] and [Cl^-] on the photoresponses of *Bufo bufo. J. Physiol. (London)* **307,** 529–551.

Cervetto, L. (1973). Influence of sodium, potassium and chloride ions on the intracellular responses of turtle photoreceptors. *Nature (London)* **241,** 401–403.

Cervetto, L., Pasino, M., and Torre, V. (1977). Electrical responses of rods in the retina of *Bufo marinus. J. Physiol. (London)* **267,** 17–51.

Cobbs, W. H., and Hagins, W. A. (1974). Are isolated frog rod outer segments light-sensitive osmometers? *Fed. Proc.* **33,** 1576.

Copenhagen, D. R., and Owen, W. G. (1976a). Functional characteristics of lateral interactions between rods in the retina of the snapping turtle. *J. Physiol. (London)* **259,** 251–282.

Copenhagen, D. R., and Owen, W. G. (1976b). Coupling between rod photoreceptors in a vertebrate retina. *Nature (London)* **260,** 57–59.

Copenhagen, D. R., and Owen, W. G. (1980). Current-voltage relations in the rod photoreceptor network of the turtle retina. *J. Physiol. (London)* **308,** 159–184.

Coronado, R., and Miller, C. (1979). Voltage-dependent blockage of a cation channel from fragmented sacroplasmic reticulum. *Nature (London)* **280,** 807–810.

Daeman, F. J. M., Schnetkamp, P. P. M., Hendricks, T., and Bonting, S. L. (1977). Calcium and rod outer segments. *In* "Vertebrate Photoreception" (H. B. Barlow and P. Fatt, eds.), pp. 29–40. Academic Press, New York.

Detwiler, P. B., Hodgkin, A. L., and McNaughton, P. A. (1980). Temporal and spatial characteristics of the voltage response of rods in the retina of the snapping turtle. *J. Physiol. (London)* **300,** 213–250.

Fain, G. L., and Lisman, J. E. (1981). Membrane conductances of photoreceptors. *Prog. Biophys. Mol. Biol.* **37,** 91–147.

Fain, G. L., and Quandt, F. N. (1980). The effects of TEA and cobalt ions on responses to extrinsic current in toad rods. *J. Physiol. (London)* **303,** 515–533.

Fain, G. L., Gold, G. H., and Dowling, J. E. (1976). Receptor coupling in the toad retina. *Cold Spring Harbor Symp. Quant. Biol.* **40,** 547–561.

Fain, G. L., Quandt, F. N., and Gerschenfeld, H. M. (1977). Calcium-dependent regenerative responses in rods. *Nature (London)* **269,** 707–710.

Fain, G. L., Quandt, F. N., Bastian, B. L., and Gerschenfeld, H. M. (1978). Contribution of a caesium-sensitive conductance increase to the rod photoresponse. *Nature (London)* **272,** 467–469.

Fain, G. L., Gerschenfeld, H. M., and Quandt, F. N. (1980). Ca^{++} spikes in rods. *J. Physiol. (London)* **303,** 495–513.

Gedney, C., and Ostroy, S. E. (1974). Hydrogen ion changes in the visual system and the membrane permeability of the vertebrate photoreceptor. *Fed. Proc., Fed. Am. Soc. Exp. Biol.* **33,** 1472.

Glynn, I. M., and Karlish, S. J. D. (1975). The sodium pump. *Annu. Rev. Physiol.* **37,** 13–53.

Hagins, W. A., and Yoshikami, S. (1977). Intracellular transmission of visual excitation in photoreceptors: Electrical effects of chelating agents introduced into rods by vesicle fusion. *In*

"Vertebrate Photoreception" (H. B. Barlow and P. Fatt, eds.), pp. 97–139. Academic Press, New York.

Hagins, W. A., Penn, R. D., and Yoshikami, S. (1970). Dark current and photocurrent in retinal rods. *Biophys. J.* **10,** 380–412.

Hagiwara, S., Fukuda, J., and Eaton, D. C. (1974). Membrane currents carried by Ca, Sr, and Ba in barnacle muscle fibre during voltage clamp. *J. Gen. Physiol.* **63,** 564–578.

Korenbrot, J., and Cone, R. A. (1972). Dark ionic flux and the effects of light in isolated rod outer segments. *J. Gen. Physiol.* **60,** 20–45.

Lasansky, A., and Marchiafava, P. O. (1974). Light induced resistance changes in retinal rods and cones of the tiger salamander. *J. Physiol. (London)* **236,** 171–191.

McNaughton, P. A. (1978). Calcium transport in excitable membranes. *In* "Biophysical Aspects of Cardiac Muscle" (M. Morad and M. Tabatabai, eds.), pp. 107–128. Academic Press, New York.

Meves, H., and Pichon, Y. (1977). The effect of internal and external 4-aminopyridine on the potassium currents in extracellulary perfused squid giant axons. *J. Physiol. (London)* **268,** 511–532.

Miller, S. S., and Steinberg, R. H. (1976). Transport of taurine, L-methionine and 3-*O*-methyl-D-glucose across frog retinal pigment epithelium. *Exp. Eye. Res.* **23,** 177–189.

Miller, S. S., and Steinberg, R. H. (1977a). Passive ionic properties of frog retinal pigment epithelium. *J. Membr. Biol.* **36,** 337–372.

Miller, S. S., and Steinberg, R. H. (1977b). Active transport of ions across frog retinal pigment epithelium. *Exp. Eye. Res.* **25,** 235–248.

Miller, S. S., and Steinberg, R. H. (1979). Potassium modulation of taurine transport across the frog retinal pigment epithelium. *J. Gen. Physiol.* **74,** 237–259.

Narahishi, T. (1974). Chemicals as tools in the study of excitable membranes. *Physiol. Rev.* **54,** 813–889.

Oakley, B., II (1977). Potassium and the photoreceptor dependent pigment epithelial hyperpolarization. *J. Gen. Physiol.* **70,** 405–425.

Oakley, B., II, and Green, D. G. (1976). Correlation of light-induced changes in retinal extracellular potassium concentration with C-wave of the electroretinogram. *J. Neurophysiol.* **39,** 1117–1133.

Owen, W. G., and Copenhagen, D. R. (1977). Characteristics of the electrical coupling between rods in the turtle retina. *In* "Vertebrate Photoreception" (H. B. Barlow and P. Fatt, eds.), pp. 169–192. Academic Press, New York.

Piccolino, M., and Gerschenfeld, H. M. (1980). Characteristics and ionic processes involved in feedback spikes of turtle cones. *Proc. R. Soc. London* **206,** 439–463.

Pinto, L. H., and Ostroy, S. E. (1978). Ionizable groups and conductances of the rod photoreceptor membrane. *J. Gen. Physiol.* **71,** 329–345.

Reuter, H. (1973). Divalent cations as charge carriers in excitable membranes. *Prog. Biophys. Mol. Biol.* **26,** 1–43.

Schwartz, E. A. (1975). Rod-rod interaction in the retina of the turtle. *J. Physiol. (London)* **246,** 617–638.

Schwartz, E. A. (1976). Electrical properties of the rod syncitium in the retina of the turtle. *J. Physiol. (London)* **257,** 379–406.

Sillman, A. J., Ito, H., and Tomita, T. (1969). Studies on the mass receptor potential of the isolated frog retina. II. On the basis of the ionic mechanism. *Vision Res.* **9,** 1443–1451.

Sillman, A. J., Owen, W. G., and Fernandez, H. R. (1972). The generation of the late receptor potential: An excitation-inhibition phenomenon. *Vision Res.* **12,** 1519–1531.

Steinberg, R. H., and Miller, S. S. (1973). Aspects of electrolyte transport in frog pigment epithelium. *Exp. Eye. Res.* **16,** 365–372.

Steinberg, R. H., Oakley, B., II, and Niemeyer, G. (1980). Light evoked changes in $[K]_o$ in the retina of the intact cat eye. *J. Neurophysiol.* **44,** 897-921.

Tomita, T. (1965). Electrophysiological study of the mechanisms subserving colour coding in the fish retina. *Cold Spring Harbor Symp. Quant. Biol.* **30,** 559-566.

Toyoda, J., Nosaki, H., and Tomita, T. (1969). Light-induced resistance changes in single photoreceptors of *Necturus* and *Gekko. Vision Res.* **9,** 453-463.

Trifanov, Y. A., and Chailakhyan, L. M. (1975). Uniform polarization of fibres and syncitial structures by extracellular electrodes. *Biofizika* **20,** 107-112.

Ward, J. A., and Ostroy, S. E. (1972). Hydrogen ion effects and the vertebrate late receptor potential. *Biochim. Biophys. Acta* **468,** 194-208.

Werblin, F. S. (1975). Regenerative hyperpolarization in rods. *J. Physiol. (London)* **244,** 53-81.

Werblin, F. S. (1978). Transmission along and between rods in the tiger salamander retina. *J. Physiol. (London)* **280,** 449-470.

Werblin, F. S. (1979). Time- and voltage-dependent ionic components of the rod response. *J. Physiol. (London)* **294,** 613-626.

Wormington, C. M., and Cone, R. A. (1978). Ionic blockage of the light-regulated sodium channels in isolated rod outer segments. *J. Gen. Physiol.* **71,** 657-681.

Yoshikami, S., and Hagins, W. A. (1970). Ionic basis of dark current and photocurrent of retinal rods. *Biophys. Soc. Annu. Meet. A* Vol. 10, p. 60a.

Yoshikami, S., and Hagins, W. A. (1972). Control of dark current in vertebrate rods and cones. *In* "Biochemistry and Physiology of Visual Pigments" (H. Langer, ed.), pp. 245-255. Springer-Verlag, Berlin and New York.

Yoshikami, S., and Hagins, W. A. (1980). Kinetics of control of the dark current of retinal rods by Ca^{++} and by light. *Fed. Proc., Fed. Am. Soc. Exp. Biol.* **39,** No. 6, 1092.

Zuckerman, R. (1973). Ionic analysis of photoreceptor membrane currents. *J. Physiol. (London)* **235,** 333-354.

Chapter 4

Photoreceptor Coupling: Its Mechanism and Consequences

GEOFFREY H. GOLD[1]

Department of Physiology
University of California School of Medicine
San Francisco, California

I. INTRODUCTION

Rod and cone photoreceptors occur in very high density in the vertebrate retina and therefore can sample the retinal image with very high resolution. However, this fine-grained neural representation is rarely utilized by the brain, because in most species each retinal ganglion cell conveys the responses of many photoreceptors.[2] This convergence results largely from spatial summation provided by the dendritic processes of retinal bipolar and ganglion cells. However, it is now well established in a number of vertebrates that convergence can also occur at the level of the receptors.

[1]Present address: Department of Physiology, Yale University School of Medicine, New Haven, Connecticut 06510.

[2]This statement does not apply to the foveal regions of the primate and some avian retinas. In these regions each photoreceptor has a ''private line'' to the brain via the midget bipolar and ganglion cells (Polyak, 1941), and acuity appears to be limited only by the distance between adjacent cells (Green, 1970).

ISBN 0-12-153315-8

Direct interactions between photoreceptors were discovered in the turtle retina, where intracellular recordings revealed that a cone's electrical response was enhanced by the illumination of other cones up to 70 μm away (Baylor *et al.*, 1971).[3] This phenomenon, termed photoreceptor coupling, has subsequently been observed between cones in the cat (Nelson, 1977) and perch (Burkhardt, 1977) and between rods in the retina of the toad (Fain, 1975a), tiger salamander (Werblin, 1978), and turtle (Schwartz, 1973). Indeed, all vertebrate photoreceptors which have thus far been studied electrophysiologically exhibit photoreceptor coupling. Coupling is strongest between receptors of the same type, i.e., between rods or between cones with the same photopigment. Therefore information about the color of incident light is generally preserved in the presence of receptor coupling. However, in most species studied, a quantitatively weaker interaction has been detected between rods and cones (Schwartz, 1975b; Copenhagen and Owen, 1976; Fain, 1976; Nelson, 1977; Attwell and Wilson, 1980). Thus rod–cone coupling, in contrast to coupling between rods and between cones, may cause some loss of spectral information.

The existence of photoreceptor coupling has modified the classical notion that vertebrate photoreceptors are independent light transducers. In addition, the existence of this phenomenon has raised several questions: What is the mechanism of interaction between photoreceptors? What is the functional significance of coupling for vision? Does photoreceptor coupling occur in all vertebrate retinas, particularly in the human retina? In regard to the first question, the vast majority of evidence indicates that this interaction is mediated by electrical coupling. Much of this evidence comes from studies on coupling between rods in the toad *Bufo marinus*. This work will be reviewed here and compared with studies on receptor coupling in other species. The functional significance of coupling remains to be established. However, a comparison of data from several species suggests that the presence of coupling may be correlated with the membrane properties of individual receptors. Finally, it appears that photoreceptor coupling may not occur in all vertebrate retinas. Its presence in the human retina is not yet certain, but the anatomical criteria reviewed here may help to establish this in future studies.

II. RECEPTOR COUPLING IN THE TOAD RETINA

The discovery of photoreceptor coupling in the toad retina arose from an attempt to observe, intracellularly, the response of a rod to single photons of light (Fain, 1975a). Because of the quantal nature of light, it was anticipated that rod

[3]An inhibitory surround was also found that was due to feedback from horizontal cells, but it will not be discussed here (see also Fuortes *et al.*, 1973).

responses to dim flashes of light would exhibit quantal fluctuations in peak amplitude, and that the smallest observable responses would reflect the absorption of a single photon of light. Although quantal fluctuations should have been readily detectable in the toad (based on the high sensitivity of toad rods), the observed fluctuations were about 10 times smaller than predicted. Fain concluded that this reduction in response variability resulted from the pooling of the responses of many rods; this conclusion has been confirmed by subsequent studies (Leeper *et al.*, 1978; Gold, 1979; Griff and Pinto, 1981).

The coupling observed in the toad retina occurs predominantly between red rods, since their action spectrum fits the red rod photopigment absorbance spectrum under most conditions of light and dark adaptation (Fain, 1976). However, a small cone input into red rods has been observed (Schwartz, 1975b; Fain, 1976), suggesting that, to a lesser extent, coupling can also occur between rods and cones. Based on the specificity of these interactions, an anatomical study was made to identify the structures that mediate receptor coupling in the toad retina. The site of coupling between red rods was identified and found to be located at the level of the rod inner segments (Gold and Dowling, 1979). Here red rods were found to be interconnected by a network of gap (electrical) junctions. In agreement with the physiological data, these gap junctions were found to interconnect only red rods and were never seen between red rods and other receptor types. However, focal gap functions, which are smaller and less numerous than the gap junctions that interconnect the red rods, were observed between red rods and cones and may partially mediate rod–cone coupling.

A. Anatomy

The site of the interreceptor junctions is illustrated in the scanning electron micrograph of the photoreceptor layer of the toad retina in Fig. 1a. Three basic types of photoreceptors are found in the toad: red rods, green rods, and cones (single and double cones are not distinguished here). The naming of red and green rods derives from their red and green appearance when seen in the light microscope, and this reflects the different photopigments in these two cell types [λ_{max} = 502 and 433 nm, respectively (Harosi, 1975)]. The different photoreceptor types can also be distinguished by their overall shapes and sizes (Fig. 1a): Cones have a short, conical outer segment, whereas rods have a larger, cylindrical outer segment. Green rods are distinguished from red rods by their more distally positioned outer segments which project about 20 μm beyond the tips of red rods. In addition, green rod inner segments are narrower (2–3 μm in diameter) than inner segments of red rods (6–7 μm in diameter; see also Fig. 4). The junctions between red rods are found at the level of the inner segments, just distal to (above in Fig. 1a) the external limiting membrane. The actual site of the junctions is indicated by the arrow in Fig. 1b, an enlargement of a red rod inner

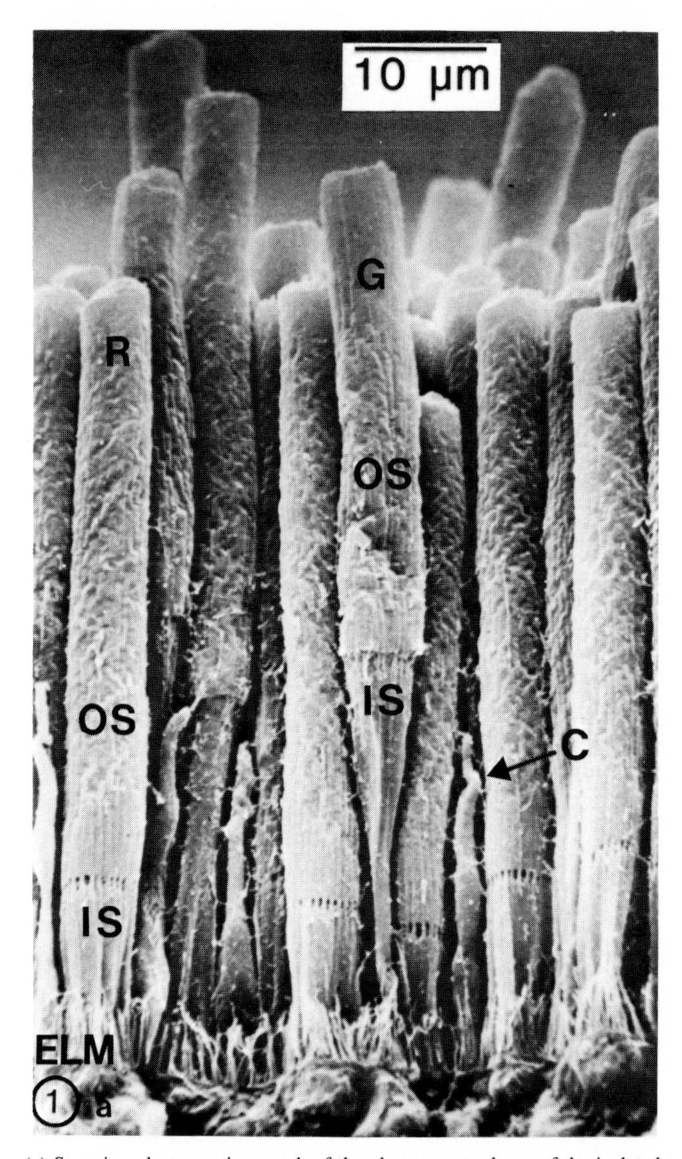

FIG. 1. (a) Scanning electron micrograph of the photoreceptor layer of the isolated toad retina. The morphology of the red rods (R), green rods (G), and cones (C) is described in the text. The inner and outer segment regions of the rods are indicated by IS and OS, respectively. The external limiting membrane (ELM) is a horizontal layer of junctions between glial cells and between glia and receptors. Reproduced with permission from Brown and Flaming, 1977.

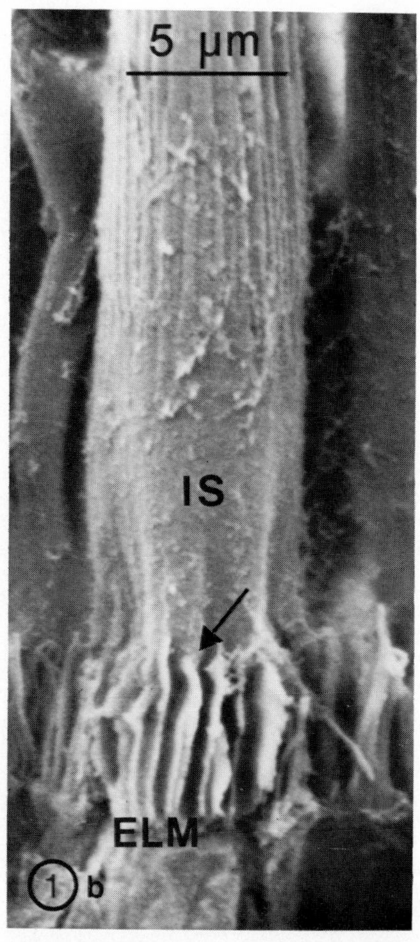

FIG. 1. (b) Scanning electron micrograph of a red rod inner segment from the frog retina, illustrating the rod fins (arrow). The fins are located just above the ELM. Fins in the frog retina appear identical to fins in the toad retina, although gap junctions have not been identified in the frog. Micrograph kindly provided by R. H. Steinberg.

segment. At this level, the surface of the red rod is not smooth, but rather exhibits a number of outward projections called fins, and it is between the fins of adjacent red rods that gap junctions are found. The fins give the rod inner segment a gearlike appearance, which is also evident when inner segments are seen in cross section in the transmission electron microscope, as in Fig. 2. This micrograph shows that the fins extend outward only 1–2 μm before encountering a neighboring receptor. In the space between the photoreceptors, the fins are in close proximity to each other and to glial processes (called the fiber basket). Gap junctions between the fins are readily identified (solid arrows in Fig. 2).

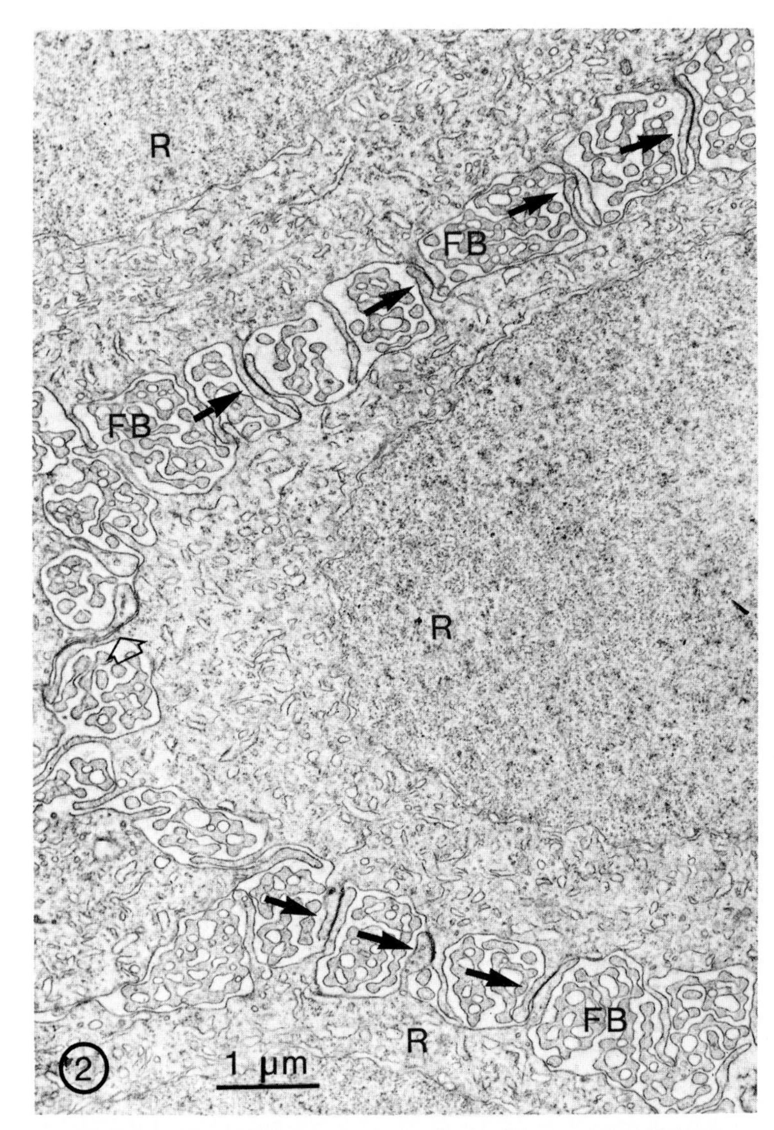

FIG. 2. Transmission electron micrograph of several red rod inner segments (R) in cross section. In this orientation the fins are seen to be thin outward projections which extend only as far as the neighboring cell. The fins of adjacent cells are in close proximity to each other, and several gap junctions between red rod fins are noted (solid arrows). At the left, a fin from a neighboring cone makes a focal gap junction onto a red rod fin (open arrow). Note that most of the extracellular space is taken up by the fiber basket (FB) which consists of fibers from retinal glial cells. Reproduced with permission from Gold and Dowling, 1979.

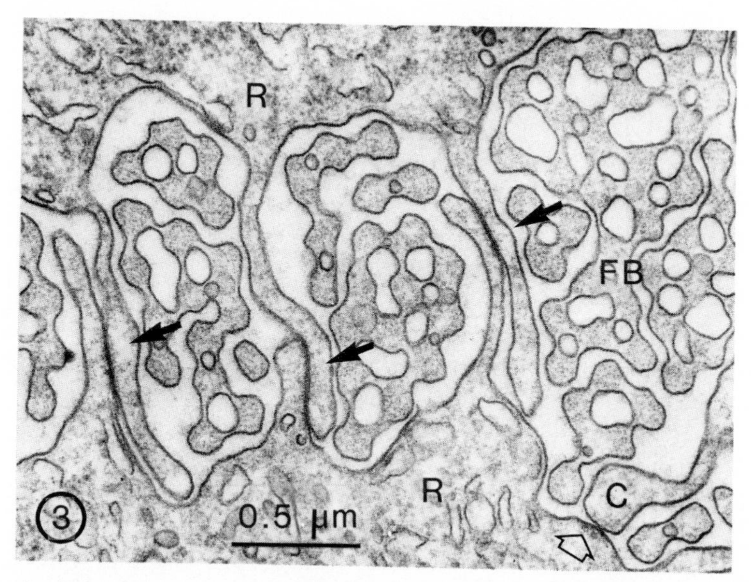

FIG. 3. Transmission electron micrograph of rod fins, same orientation as Fig. 2. Gap junctions between red rod fins are indicated by solid arrows. A focal gap junction between a fin from an adjacent cone (C) and a rod inner segment (R) is indicated by an open arrow. FB, Fiber basket. Reproduced with permission from Gold and Dowling, 1979.

Gap junctions are scattered over the total length (about 5 μm) of the fins and therefore are only occasionally seen in random thin sections. However, fortuitous sections, like the one in Fig. 3, demonstrate that every fin can make a junction. At this magnification, the junctions appear similar to gap junctions described elsewhere; their ultrastructure will be described in further detail below. The total junctional area connecting each red rod to its neighbors was quantitated by serially reconstructing this level of the rods. It was found that 82% of the fins made junctions and that a rod was connected to each of 4.6 (average) neighboring rods by 4.7 (average) fins with a total junctional area of 0.7 μm^2. Thus red rods are interconnected by a dense, regular network of gap junctions. Red rods were also examined at other potential sites of interaction, and no other junctions were found (Gold and Dowling, 1979). Therefore the junctions between red rod inner segments are the only candidates for mediating the coupling observed between red rods.

These gap junctions interconnect only red rods and have not been observed between red rods and adjacent green rods or cones. The high specificity of gap junctions is also illustrated in Figs. 4 and 5. Figure 4 shows that two red rods, separated by a green rod and cone, still contact each other. Despite the large area of apposition between the red rod fins and the green rod inner segment, no gap

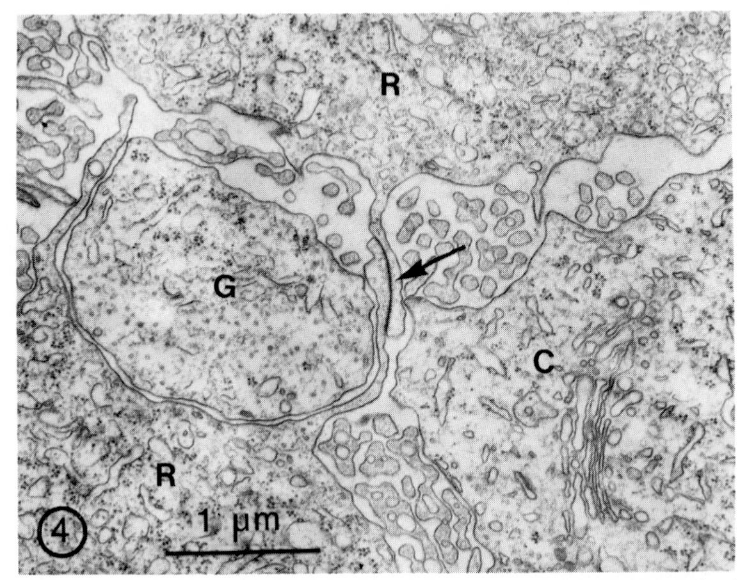

FIG. 4. Transmission electron micrograph of a gap junction (arrow) connecting two red rods (R) which are separated by a green rod (G) and a cone (C). Reproduced with permission from Gold and Dowling, 1979.

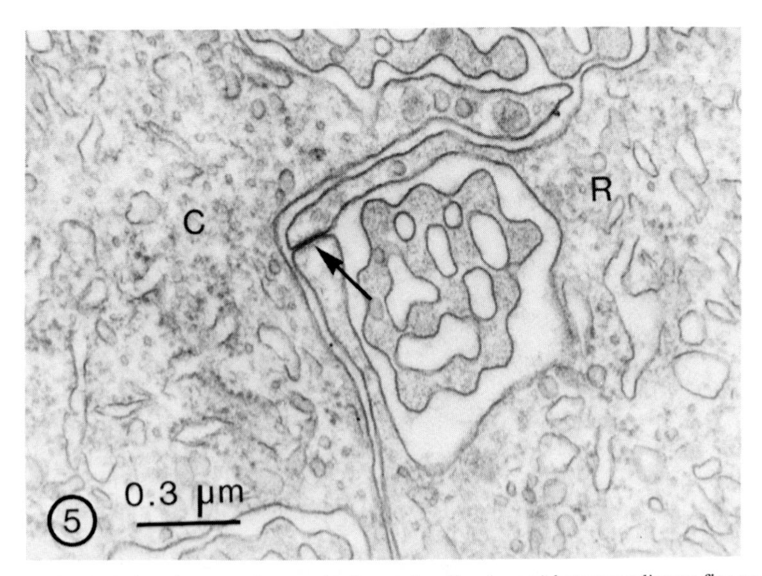

FIG. 5. Transmission electron micrograph of a gap junction (arrow) between adjacent fins on the same red rod (R). Note that the fins are completely enclosed by fins from an adjacent cone (C). Reproduced with permission from Gold and Dowling, 1979.

junctions were found between them, suggesting that junction formation is essentially forbidden between red rods and other receptor types. This point is also made concerning the cones in Fig. 5. Here two red rod fins are completely enclosed by adjacent cone fins, yet the red rod fins contact each other in preference to the cone.

In addition to the gap junctions observed between red rods, morphologically different focal gap junctions (Raviola and Gilula, 1973) are found to interconnect red rods and cones of the toad retina. Figure 3 shows such a junction between a red rod inner segment and a fin from an adjacent cone (open arrow). This contact looks like a very small gap junction or a larger gap junction that is just barely in the section. But rod–cone junctions always have this appearance and therefore represent a morphologically distinct type of interreceptor connection.

To further investigate the nature of coupling, the ultrastructure of these junctions was examined with both thin sections and the freeze-fracture technique. Figures 6 and 7 show the appearance of rod–rod and rod–cone junctions as revealed by these methods. In thin sections (Fig. 6a), rod–rod junctions consisted of a very close apposition of the adjacent rod membranes. The membranes were frequently so close that no distinct gap was apparent between them; the existence of a gap was confirmed by observing the penetration of colloidal lanthanum into the junctions (Gold and Dowling, 1979). The importance of the gap is that at tight junctions, where the membranes of adjacent cells fuse and there is no detectable gap, electrical coupling is not observed (Brightman and Reese, 1969). Rod–rod junctions were also identified in freeze-fracture replicas (Fig. 6b) and consisted of a dense patch of intramembrane particles along the inner (cytoplasmic) leaflet of the membrane. As seen with both techniques, these junctions were indistinguishable from gap junctions that mediate electrical coupling elsewhere in the nervous system (Bennett, 1973). By comparison, rod–cone junctions appeared almost as point contacts between adjacent rods and cones (Fig. 7a). At these focal junctions, the adjacent membranes came into very close proximity, and there appeared to be increased density along both membranes. Because of the small size of the focal junctions it was difficult to say whether a gap existed or not. Rod–cone junctions always appeared as point contacts when the inner segments were viewed in cross section, suggesting that they may represent linear junctions. Linear arrays of membrane particles were observed at this level of the retina (Fig. 7b), but it was not confirmed that these particles corresponded to rod–cone junctions.

Rod–rod and rod–cone junctions differ not only in arrangement but also in the total number of junctional particles; a rod–cone junction (average length, 0.7 μm) consists of about 50 particles, compared with a rod–rod junction (average area, 0.15 μm^2) which contains about 750 particles. Serial reconstructions have revealed that each cone contacts 4 neighboring rods with a total of 12 junctions. Therefore, because red rods and cones (including both single and double cones)

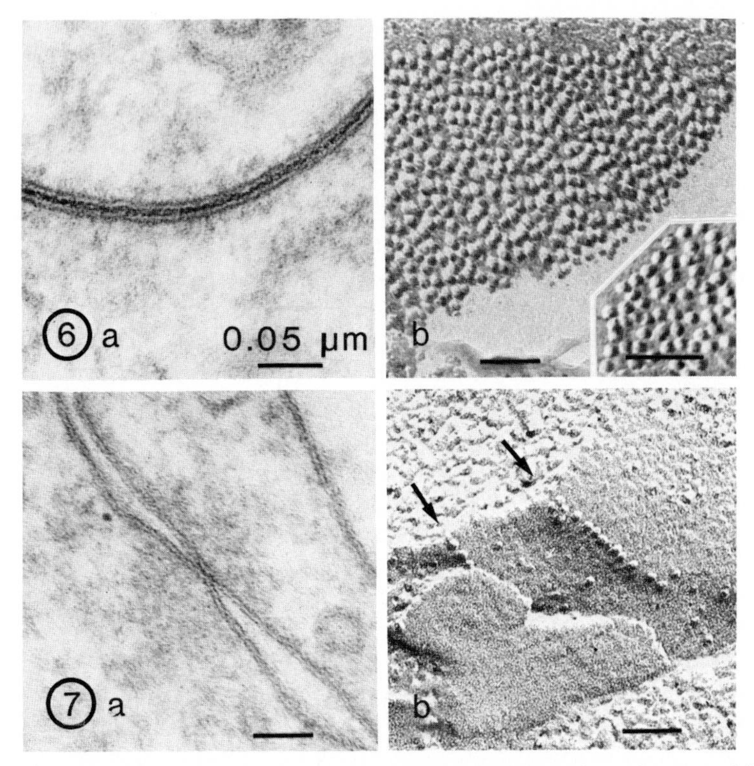

FIG. 6. The ultrastructure of rod–rod gap junctions as revealed in thin sections (a) and freeze-fracture replicas (b). The calibration bars are 0.05 μm. Reproduced with permission from Gold and Dowling, 1979.

FIG. 7. The ultrastructure of rod–cone focal gap junctions as revealed in thin sections (a) and freeze-fracture replicas (b). The calibration bars are 0.05 μm. Reproduced with permission from Gold and Dowling, 1979.

occur in a ratio of about 2:1 (Fain, 1976), each rod is coupled to two adjacent cones by approximately 300 particles, and to 4.6 adjacent rods by a total of 3500 particles. If the conductances provided by the particles are all equal, then the rod–rod coupling conductance will be approximately 12 times greater than the rod–cone coupling conductance.

B. Mechanism of Rod Coupling

These observations provide an anatomical basis for the pooling of red rod signals observed by Fain. In addition, the strong correlation between gap junctions and electrical coupling elsewhere in the nervous system (Peters *et al.*, 1970) provides evidence that the pooling is mediated by electrical coupling. The electrical nature of this coupling has been further demonstrated by the fact that cobalt ions, which block chemical but not electrical synapses, have no effect on the coupling between rods (Griff and Pinto, 1981). Therefore, the toad's red rods constitute a network of electrically coupled cells. In the next section, the theoretical properties of such a network will be described and compared with experimental observations. This comparison will serve to confirm the electrical nature of receptor coupling and also to specify the resistance of the gap junctional membrane.

C. Electrophysiology of Rod Coupling

A necessary consequence of this coupling is that a single rod should respond to illumination of nearby rods or that illumination of one rod will cause a signal to spread to its neighbors. How far these signals spread through the network depends on the network length constant λ, a parameter that can be measured electrophysiologically. In addition, if the geometry of the interconnections is known, then the length constant can be related to the resistances of the gap junctions and the rod plasma membrane. Therefore, in light of the anatomical results described above, the toad retina has provided a unique opportunity to compare quantitatively electrophysiological measurements of coupling with the properties of an equivalent electrical model.

The form of the network used to describe toad red rods is shown in Fig. 8. In the network, each rod is represented by a single resistor r_m [the inner and outer segments are assumed to be isopotential, which has been demonstrated by Werblin (1978) and Attwell and Wilson (1980)]. The interior of each rod is connected to its four neighbors by a single resistor r_s and therefore closely approximates the real network in which the gap junctions couple each rod to an average of 4.6 nearest neighbors. Capacitance in both the cell and gap junctional membranes is ignored because of the slow kinetics of the rod response to dim lights. The network length constant λ is given by

$$\lambda = D\,(r_m/r_s)^{\frac{1}{2}} \tag{1}$$

where D is the center-to-center spacing between the cells when placed in a square array (i.e., D = (cell density)$^{-\frac{1}{2}}$).[4]

This network is linear because all of its elements exhibit linear current–voltage relationships. Therefore the network obeys the principle of linear superposition; i.e., the voltage distribution that results when a number of rods are stimulated is determined simply by adding the voltages produced when each rod is stimulated separately. This feature considerably simplifies the analysis of the network's

[4]This expression is valid for $\lambda/D > 1$, but for $\lambda/D < 1$ a more complex expression is required (Eq. 14a; Lamb and Simon, 1976).

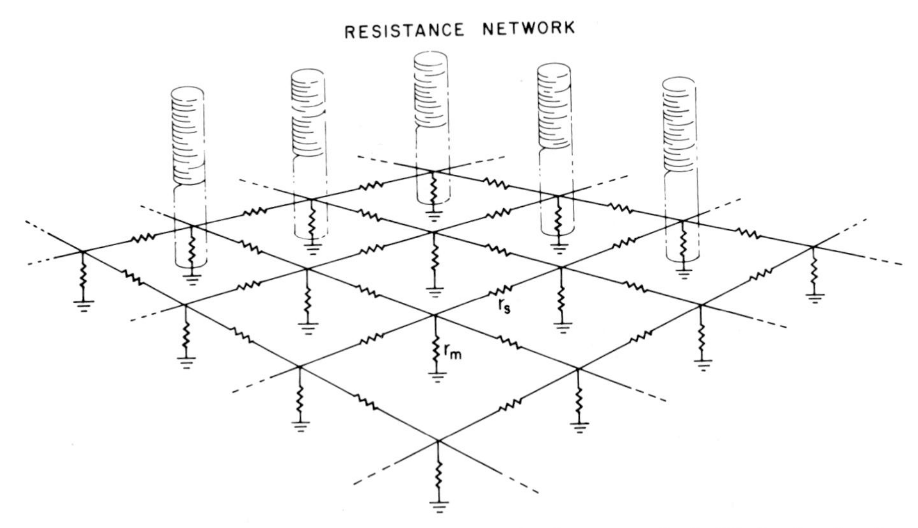

FIG. 8. The equivalent circuit used to describe coupling between toad rods. The interior of each rod is connected to four adjacent rods by a resistance r_s. The total membrane resistance of single rod is represented by r_m. The extracellular space is assumed to be isopotential and is represented by a ground symbol. Reproduced with permission from Gold, 1979.

properties and is the main motivation for studying retinal rods electrophysiologically under conditions where they behave linearly. Analysis of this network's properties has shown how λ is related to the network input resistance and also to the voltage distributions produced when the retina is illuminated with various geometrical patterns (Lamb, 1976; Lamb and Simon, 1976; Schwartz, 1976; Gold, 1979). A one-dimensional stimulus, such as an edge or bar of light, is particularly convenient to use because centering of the stimulus is unnecessary and also because responses then vary only in the direction perpendicular to the stimulus. In this situation, voltage spread is described by a simple exponential function with a length constant equal to λ. Therefore electrophysiological measurements can be used to estimate the network length constant, and consequently the values of r_s and r_m.

One measurement of λ was derived from the voltage distribution resulting from stimulation with an edge of light. This voltage distribution, or edge response profile, was measured in the isolated retina by illuminating the retina with an edge of light at various positions in the receptive field of an impaled rod. A dim flash of light was presented at each position. Dim intensities were used to ensure that rod response amplitudes were proportional to light intensity. Under these conditions, the rod network behaved linearly (Schwartz, 1976; Leeper *et al.*, 1978), so that electrophysiological measurements could be analyzed in terms of the linear network model described above. The peak response amplitudes were recorded at each position and are plotted as open circles in Fig. 9. The family of curves (labeled 0, 20, 40 μm, etc.) represents the response profiles predicted when the network length constant is equal to 0, 20, 40 μm, etc. These curves have been calculated from the exponential voltage spread function of the network, assuming that the photocurrents from the various stimulated rods sum linearly. The longer the network length constant, the broader the edge response profile. In the absence of coupling ($\lambda = 0$ μm), the edge response profile corresponds to the intensity profile of the edge. This was measured by displacing the edge past a photodiode covered with a 12-μm aperture. No correction was made for light scattering, since the stimulus was incident directly on the rod outer segments. However, when the retina is illuminated from the vitreal surface, as is necessary with an eyecup preparation, scattering by inhomogeneities within the retina must be corrected for; these corrections have been a difficult experimental problem (e.g., Copenhagen and Owen, 1976; Detwiler and Hodgkin, 1979).

The edge data shown in Fig. 9 exhibited a length constant of about 20 μm, and the average value for 11 different cells was 19 ± 7 μm (mean ± SD; Gold, 1979). This result agrees with estimates based on bar profile measurements (λ = 24 ± 5 μm from Leeper *et al.*, 1978 and 22 ± 5 μm from Griff and Pinto, 1981) and receptive field size measurements using circular spots of light (λ = 21 ± 6 μm, Leeper *et al.*, 1978). The agreement between these different methods for

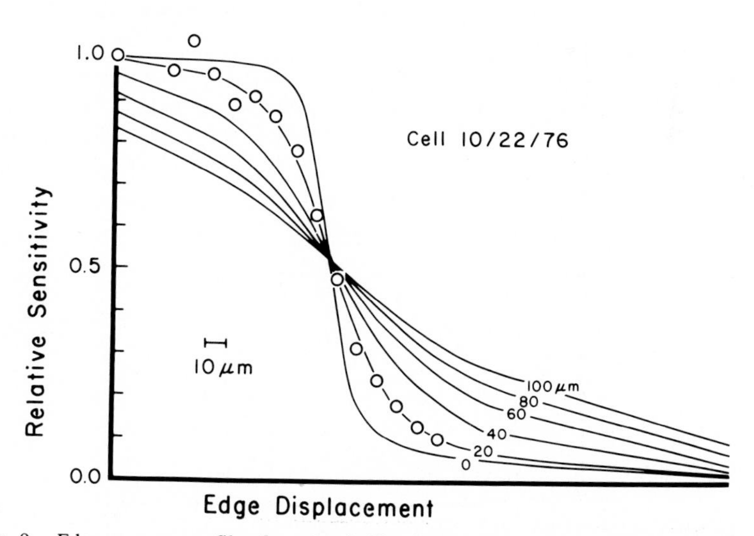

FIG. 9. Edge response profile of a red rod. The circles are the peak response amplitudes (normalized to the maximum) observed as a function of edge displacement. The solid line labeled 0 μm is the intensity profile of the edge. The solid lines labeled 20, 40, 60 μm etc., are the predicted response profiles if the network had a length constant of 20, 40, 60 μm, etc. Reproduced with permission from Gold, 1979.

determining λ supports the validity of the network model and the assumption that small signals sum linearly within the network.

Analysis of the resistance network model has also shown that the input resistance of a rod in the network is a function of the length constant (Lamb and Simon, 1976; Schwartz, 1976; Werblin, 1978; Gold, 1979). The input resistance r_{in} of an isolated cell (defined as the voltage across the cell membrane divided by the current injected into the cell) is equal to the membrane resistance r_m. However, in the presence of the coupling resistors r_s, part of the injected current flows out of the cell, so that less current will generate a voltage across r_m. Consequently, r_{in} will be smaller than r_m in the presence of electrical coupling, and the quantitative dependence of r_{in} on λ is plotted as r_{in}/r_m versus λ/D in Fig. 10. The input resistance of a rod within the *Bufo* network was measured by passing current through a single electrode in a Wheatstone bridge circuit. The average value of 21 measurements was 102 ± 36 MΩ (Gold, 1979), which agrees with more recent measurements using double-barreled electrodes (114 ± 51 MΩ, Griff 1979). When the spacing of the rods is taken as 7.6 μm (Fain, 1976), λ/D has a value of 2.5, which from the curve in Fig. 10 predicts that $r_{in}/r_m = 0.07$. Therefore the input resistance measurement predicts that r_m = 1300 MΩ. The measurement of r_m for *Bufo* rods has not yet been reported,

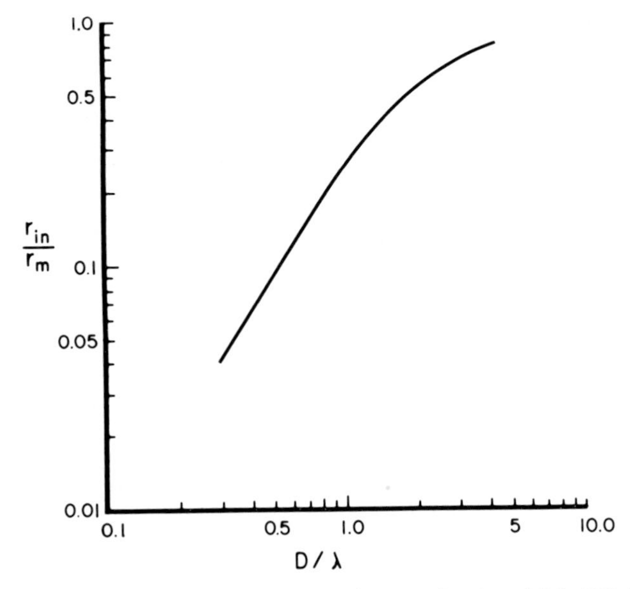

FIG. 10. Network input resistance, defined by r_{in}/r_m, as a function of D/λ. This function was derived from computer simulations of the discrete network in Fig. 8. Reproduced with permission from Gold, 1979.

although a value of about 700 MΩ is predicted from the measured value of r_m in solitary tiger salamander rods (Bader *et al.*, 1978; Attwell and Wilson, 1980). The reason for this discrepancy is unknown, but it may be that the experimental measurement of r_m underestimates the predicted value because of electrode-induced leakage in the cell membrane (e.g., Attwell and Wilson, 1980). With the values of 700 and 1300 MΩ as lower and upper limits for r_m, Eq. (1) predicts that the value of r_s will be 100–200 MΩ. This is the resistance provided by 0.7 μm^2 of gap junction, so that the resistivity of the junctional membrane is 0.75–1.5 $\Omega\ cm^2$. In addition the density of membrane particles (presumably the trans-junctional channels), which is $5 \times 10^3\ \mu m^{-2}$ (Gold and Dowling, 1979), predicts the conductance of a single channel to be 2–4 pS. These results compare well with estimates of gap junction resistivity from heart muscle (1.4–3 $\Omega\ cm^2$, Weidmann, 1966; Spira, 1971), but not with estimates from coupled embryonic cells (0.01 $\Omega\ cm^2$, Ito and Loewenstein, 1969). However, this difference correlates with the higher permeability of the dye Procion yellow through the embryonic junctions (Gold, 1979; Loewenstein, 1975), suggesting that the size of gap junctional channels may vary depending on the function they subserve.

D. Effects of Voltage-Dependent Conductances

The agreement between the network properties and the electrophysiological data demonstrated that the spread of rod signals near threshold could be adequately described by a linear resistance network. However, the validity of a purely resistive network was recently questioned by the observation that signal spread in the turtle rod network (Detwiler *et al.*, 1978), and also in the toad (Torre and Owen, 1981), was modified by the voltage-dependent properties of the rod membrane. Detwiler *et al.* found that, even within the linear response range, i.e., the range of intensities over which response amplitude is proportional to stimulus intensity, the kinetics of a rod's response to a flashed bar of light varied with the separation between the bar and the impaled cell. In particular, the time to peak of the response decreased with increasing separation between the bar and the rod. Such an effect is not predicted by the purely resistive model described above. Phenomenologically, Detwiler *et al.* described this effect by placing an inductance in parallel with r_m, and they proposed that a voltage-dependent conductance in the rod membrane might underlie the inductive behavior. This hypothesis has since been proven to be correct by Attwell and Wilson (1980), who have examined this phenomenon in tiger salamander rods. Their elegant analysis has demonstrated that voltage-dependent conductances affect the spread of rod signals over the full range of rod response amplitude.

By voltage-clamping isolated tiger salamander rods, Attwell and Wilson fully characterized the voltage-dependent properties of the rod membrane. They found that it exhibited a voltage- and time-dependent current I_A which was voltage-dependent over a range including the dark resting membrane potential V_{rest}. Therefore the presence of I_A can affect the rod network even under conditions where the network has been assumed to behave in a passive and linear fashion. However, the presence of I_A does not substantially alter the linear nature of the rod network near threshold, because I_A itself varies almost linearly with the membrane potential within a few millivolts of V_{rest}.[5] This explains why linear network analysis has successfully described the spread of signals through the rod network, even when ignoring the existence of I_A (Schwartz, 1976; Leeper *et al.*, 1978; Gold, 1979; Copenhagen and Owen, 1980; Griff and Pinto, 1981). The presence of I_A is most clearly revealed by the fact that its time course lags slightly behind the time course of the photocurrent[6]; it is this lag that causes the shortening of the time to peak that was discovered by Detwiler *et al.* The existence of a lag was demonstrated by Attwell and Wilson's voltage-clamp data; following small voltage steps away from the resting potential, they found that I_A reached a new steady value with a time constant of about 200 msec.

Attwell and Wilson also used their voltage-clamp data to demonstrate the precise effect I_A had on the spread of rod signals near threshold. This was done by modeling the receptor network with a square grid of resistively coupled cells, as in Fig. 8, in which I_A was included in the membrane properties of the rods. To simulate the experiment with a flashed bar of light, they applied to one row of cells (row 0) a voltage waveform that mimicked the time course of the rod response near threshold. The voltages that would result in other rows of the network were calculated and have been plotted as a function of time in Fig. 11a (curve 0 indicates the voltage in the stimulated row; solid curves 1 and 2 indicate the voltages in the adjacent and subadjacent rows, respectively). With increasing separation from the stimulated row, the responses become smaller and also exhibit the decreased time to peak reported by Detwiler *et al.* The effect of I_A in this situation can be illustrated by calculating the voltages that would have occurred if I_A had been omitted from the network. In this case, the network would have consisted of only r_s and r_m, which had values of 300 and 464 MΩ, respectively, in Attwell and Wilson's simulation (464 MΩ is the instantaneous slope resistance at V_{rest}). These values of r_s and r_m predict a network length constant, using Eq. (1), of 1.24*D*. In the absence of I_A, voltages in all rows follow the same time course (neglecting capacitative effects) and decay exponentially in amplitude with increasing separation from the stimulated row (dashed lines in Fig. 11a).[7] The difference between the dashed and solid lines for a

[5]The steady-state value of I_A is proportional (within 10%) to voltage excursions of up to 3 mV from V_{rest} (see Eq. 7, Attwell and Wilson, 1980).

[6]Here, the term "photocurrent" denotes the light-dependent currents generated in the outer segment, in contrast to the voltage-dependent currents which probably are generated in the inner segment (Baylor *et al.*, 1979).

[7]Attenuation was determined from Fig. 2 of Gold (1979). Note that the parameter in this figure is D/λ, not λ/D, as was incorrectly stated in the original figure legend. The curves in Fig. 2 are

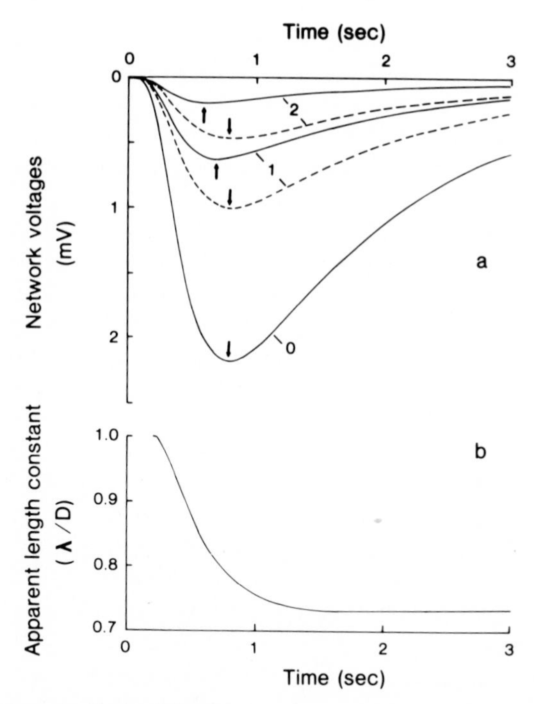

FIG. 11. (a) Network voltage responses to a voltage transient simulating a flashed bar of light. The voltage responses were calculated in the presence (solid lines) and absence (dashed lines) of the voltage-dependent current I_A. A voltage transient was applied to row 0, and the resulting voltages were calculated in the adjacent and subadjacent rows (labeled 1 and 2, respectively). The calculated responses in the presence of I_A were kindly provided by D. Attwell and M. Wilson [see Attwell and Wilson (1980) for a complete description of these calculations]. The responses in the absence of I_A were calculated simply by multiplying the voltage in row 0 by the attenution calculated for a purely resistive network with the same values of r_s and r_m as were used by Attwell and Wilson. The peaks of the responses are denoted by arrows; the time to peak decreases with increasing displacement in the presence of I_A but not in its absence. (b) Apparent length constant as a function of time, calculated from the voltage distributions in Fig. 10a. Values of the apparent length constant are plotted only between 0.2 and 3.0 sec; values between 0 and 200 msec were not plotted because the computed results did not contain sufficient significant digits to do so reliably. Voltage was empirically found to decay exponentially with distance, so λ was defined by $V(1) = V(0)\ e^{-D/\lambda}$.

particular row indicates the effect of I_A on the response: It decreases the amplitude and the time to peak of the response, having a relatively greater effect on cells more distant from the stimulus. The voltages are smaller in the presence of I_A because the sign of the voltage-induced current is opposite that of the

---

equivalent to Eq. 14a of Lamb and Simon (1976). The different slopes of the curves labeled 1, 2, and 3 reflect the deviations of Eq. (1), this chapter, from Eq. 14a of Lamb and Simon (1976). Both methods are equivalent and yield the same numerical results.

light-induced photocurrent (I_A is an inward current which is activated by hyperpolarization). The responses in more distant cells peak earlier because I_A lags slightly behind the photocurrent; this shifts the peak of the resulting current, and therefore the photovoltage, to earlier times. These effects due to I_A become more pronounced with increasing separation from the stimulus, because the ratio of I_A to the photocurrent is greater in more distant cells. This occurs because I_A is generated actively by all cells in which a voltage change occurs, whereas the photocurrent is generated only in the stimulated row. Therefore, as I_A is activated, it attenuates the voltage responses throughout the network but has a relatively larger effect on cells more distant from the stimulus. The result of this is a reduction in the lateral spread of the photovoltage, and hence the magnitude of the apparent length constant (the term "apparent length constant" is used because it reflects the presence of I_A and not changes in r_m/r_s). This effect of I_A on the voltage distribution was also observed by Detwiler *et al.* and is illustrated by calculating the apparent length constant for the responses in Fig. 11a as a function of time (Fig. 11b). The values of the apparent length constant are shown between 0.2 and 3.0 sec. Over this time interval, the apparent length constant decays from about $1.0D$ to $0.74D$, with a value of about $0.8D$ at the peak of the response. This is in contrast to the true network length constant, as defined by Eq. (1), which has a value of $1.24D$. Therefore a measurement of λ in the tiger salamander, based on the peak amplitudes of linear range responses, would underestimate the true network length constant by about 30%.

These results in the tiger salamander, if quantitatively applicable to *Bufo*, suggest that previous measurements of λ may considerably underestimate the true network length constant. Therefore, for *Bufo* rods, λ may be closer to 30 μm rather than 20 μm. This suggestion is supported by the measurements of Torre and Owen (1981; see also Chapter 3, this volume), who found that λ was initially about 30 μm and decayed to about 15 μm during the course of the photoresponse. However, the quantitative effects of I_A may be different in the toad and tiger salamander, because of differences in the extent of coupling and the magnitude of I_A. Consequently the ratio of the true length constant to the apparent length constant at peak may be different in the two species. This possibility should be considered, because a length constant of 30 μm in *Bufo* predicts a rather large value for r_m. If $\lambda = 30$ μm, $\lambda/D = 4$, so that Fig. 10 predicts that $r_{in}/r_m = 0.03$. Therefore the reported value of r_{in} (about 100 MΩ) predicts that $r_m = 3300$ MΩ, which for a cylindrical cell 7 μm in diameter and 100 μm long predicts a membrane resistivity of 7.3×10^4 Ω cm^2. This is considerably greater than the measured membrane resistivity of isolated tiger salamander rods (1.5×10^4 Ω cm^2; Bader *et al.*, 1978). This discrepancy will need to be resolved before the true network length constant and membrane resistivity are established in *Bufo*. Such a discrepancy does not arise in the tiger salamander, where the true network length constant predicts that $r_{in}/r_m = 0.15$. This is consistent with the observed values of r_{in} (90 MΩ, Attwell and Wilson, 1980) and r_m (600 MΩ, Bader *et al.*, 1978; Attwell and Wilson, 1980).

Another potential conflict between the resistive and inductive networks arises because the shifts of time to peak in turtle rods are not seen in all cells [about 50% of the rods examined by Copenhagen and Owen (1980) exhibited shifts]. Copenhagen and Owen found no significant difference between the length constants (measured from the peak amplitudes) seen in rods that did or did not exhibit shifts. This suggests that voltage-dependent currents do not significantly affect the peak amplitude of the photovoltage, and therefore the measurements of length constant, based on the amplitudes at peak. However, the simulation in Fig. 11a shows that the effect of I_A on voltage amplitude is much larger than might be anticipated from the small shifts in time to peak. Since these shifts result from small differences between the kinetics of I_A and the photocurrent, making I_A slightly faster or the photocurrent slightly slower would drastically reduce their visibility. Therefore, it is suggested that length constant measurements (as a function of time) may be a more reliable indicator of the presence of I_A during dim illumination than are the shifts in time to peak.

Although these changes in λ substantially alter the predicted value of r_m, they do not significantly affect the estimate of r_s. This is because r_s is a slowly varying function of λ, which can be shown in the following way. Since $r_{in}/r_s = (r_{in}/r_m)(r_m/r_s)$, r_{in}/r_s as a function of λ can be derived from Fig. 10 and Eq. (1). For the reported value of $\lambda/D = 2.5$, $r_{in}/r_m = 0.066$ and $r_m/r_s = 6.25$, so that $r_s = 2.4r_{in}$. However, if λ/D increases to 4, as suggested above, $r_{in}/r_m = 0.031$ and $r_m/r_s = 16$, so that $r_s = 2.0r_{in}$, which results in only a 20% decrease in r_s. Therefore the presence of the inductive behavior does not significantly alter the estimate of r_s, and therefore the conductance of the junctional channels.

The preceding results demonstrate that the properties of the rod network are complicated by the presence of voltage-dependent conductances in the rod membrane. In particular, it has been shown how these voltage-dependent conductances can affect measurement of the network length constant of tiger salamander rods. It appears that a passive resistance network may be an inadequate model for the rod network in the tiger salamander. However, it is not yet established how large the voltage-dependent effects are in other species. The agreement among the various measurements of coupling in *Bufo* argues that a resistance network is a reasonably good approximation to the real network. However, this point will need to be addressed in future investigations.

D. Rod–Cone Coupling

In contrast to the mechanism of rod–rod coupling, the mechanism of rod–cone coupling remains much less certain in *Bufo*. One reason is that the physiological properties of the rod–cone interaction have not been characterized as yet.

Schwartz (1975b) has described an apparently additive input from cones into rods in the turtle retina at long wavelengths (620–700 nm). He states that a similar phenomenon exists in the toad but has not reported its magnitude. However, this cone input is different in kinetics and sign from the cone signals Fain (1976) recorded in saturated rods, suggesting two types of rod–cone interaction. Part of this interaction may be mediated by the focal junctions reported by Gold and Dowling. However, it may be that cone processes synapse onto rod receptor terminals, a possibility not investigated by Gold and Dowling. Therefore additional information on the physiology and anatomy of rod–cone coupling is needed before the mechanism and significance of this phenomenon can be established in *Bufo*.

III. GENERAL FEATURES OF RECEPTOR COUPLING

The evidence presented above establishes that the coupling between toad rods is mediated by passive electrical coupling but that the spread of the rod signal is modulated by voltage-dependent conductances in the rod membrane. This coupling occurs predominantly between red rods, although there is also evidence for a weaker interaction between red rods and cones. There is also electrophysiological evidence for receptor coupling in the turtle, tiger salamander, perch, and cat. Are the features of coupling in *Bufo* common throughout these species, and do these features suggest functional roles for receptor coupling? The physiological data will be reviewed as they pertain to these questions. The accumulating evidence concerning coupling also raises the question of whether or not receptor coupling is a feature of all vertebrate retinas and, in particular, whether or not receptor coupling occurs in the human retina. There are no electrophysiological data regarding coupling in primates or in any other warm-blooded vertebrates except the cat. However, anatomical reports of interreceptor junctions have suggested the existence of receptor coupling in the human and other retinas. These reports will be reviewed and compared with the anatomical substrate of coupling where it is known to exist.

Many of the present data on photoreceptor coupling have been obtained from the turtle retina where coupling occurs between rods and also between red and green sensitive cones (Baylor and Hodgkin, 1973; Schwartz, 1975a,b; Copenhagen and Owen, 1976; Detwiler and Hodgkin, 1979). Coupling between turtle rods is quite similar to the rod coupling observed in the toad. For example, rods are coupled primarily to other rods since the dark-adapted rod action spectrum fits the rod photopigment absorbance spectrum over the entire range of wavelengths tested (450–680 nm; Copenhagen and Owen, 1976). A small cone input was observed at long wavelengths (Schwartz, 1975b; Copenhagen and Owen, 1976); however, it did not affect the rod's peak amplitude (i.e., sensitivity) in fully dark-adapted retinas (Copenhagen and Owen, 1976). As in the toad, the spread of the peak amplitude of rod responses near threshold is adequately described by a purely resistive network (Schwartz, 1976; Copenhagen and Owen, 1980). However, as described above, this spread is affected by the voltage-dependent properties of the rod membrane. Coupling is unaffected by the presence of cobalt ions, which block chemical synaptic transmission between receptors and second-order cells (Schwartz, 1976; Owen and Copenhagen, 1977). Therefore the existing data indicate that rod–rod coupling in the turtle is electrical. Coupling between turtle cones also appears to be electrical, since the spread of cone signals is well described by a resistive network (Lamb and Simon, 1976; Detwiler and Hodgkin, 1979). In fact, cones do not exhibit inductive behavior, so that a purely passive network describes all the features of the cone signal spread (Detwiler and Hodgkin, 1979).

The anatomical substrate that mediates coupling in the turtle retina remains to be identified. Gold and Dowling (1979) have found that there are no interreceptor junctions at the level of the inner segments, as there are in *Bufo*. This is not surprising, since like receptors are not frequently adjacent to each other in the turtle and therefore contacts between adjacent cells would not be an effective means of coupling. Lasansky (1971) has reported junctions between adjacent cone terminals, but for the above reason and also because these junctions do not resemble gap junctions, their functional significance remains obscure. Therefore coupling between turtle receptors must be mediated by processes such as the receptor teleodendria that extend laterally in the outer plexiform layer (Lasansky, 1971; Copenhagen and Owen, 1976). The strong electrophysiological evidence in favor of electrical coupling argues that gap junctions will eventually be found on these processes.

Evidence in other species is less extensive but suggests that photoreceptor coupling is mediated electrically. In the tiger salamander, Werblin (1978) and Attwell and Wilson (1980) have found that the spread of rod signals can be modeled by networks in which the rods are electrically coupled. In addition, Werblin has found that rod coupling is not affected by cobalt. The anatomical substrate for rod–rod coupling in the tiger salamander has not been identified, but in a closely related species, the axolotl, rods are coupled by gap junctions between the rod fins, as in *Bufo* (Custer, 1973).

In the cat retina, Nelson (1977) has reported that cones are coupled to each other and to rods, and that cone coupling can be described by a resistance network. A likely anatomical substrate for this coupling has been found by Kolb (1977), who has demonstrated that cat cones are interconnected by gap junctions at the level of the receptor terminals. Based on the retinal eccentricities of Nelson's recordings, it can be estimated that his values of λ/D would vary between 5 and 15 (D is estimated from the data of Steinberg *et al.*, 1973). However, such large values are quite surprising, since λ/D varies only between 1 and 3 in all other species (Table I). Based on the conductance of junctional membrane estimated in *Bufo,* this would require that as much as 1.5% of the cat cone membrane be taken up by gap junctions, as compared with 0.01% in *Bufo*. Such a large extent of coupling is surprising and may require further quantitation.

Our understanding of rod–cone coupling is much less complete, largely because of the scarcity of data concerning this phenomenon. Cone inputs into rods have been observed in the toad, turtle, and tiger salamander (Schwartz, 1975b; Copenhagen and Owen, 1976; Fain, 1976), and rod inputs into cones have been seen in the tiger salamander (Attwell and Wilson, 1980) and cat (Nelson, 1977). There are no electrophysiological data concerning the mechanism of this interaction. Focal junctions have been observed between rods and cones in the toad (Gold and Dowling, 1979) and cat (Kolb, 1977), suggesting that rod–cone coupling is electrical. However, chemical synapses have been reported between rods and cones in the axolotl (Custer, 1973) and tiger salamander (Lasansky, 1973), suggesting alternative sites of interaction. There is little evidence concerning the reciprocity of rod–cone coupling. In the cat retina, Nelson *et al.* (1975) have found that responses recorded from horizontal cell axon terminals, which contact only rods, exhibit a small cone input. This suggests that rods receive an input from cones that is smaller than the input cones receive from rods. The apparent asymmetry could result from rectification at rod–cone junctions but could also reflect an impedance mismatch between rods and cones.

The strength of the rod–cone interaction has not been quantitated as yet. Schwartz (1975b) found that at 690 nm the cone input in the turtle increased the rod's sensitivity by 1.7 log units, and Nelson (1977) found that at 441 nm the rod input increased cone sensitivity by over 2 log units. However, these figures do not reflect the extent of coupling alone, because they also depend on the absolute sensitivities of the cells and the relative absorbance of the two pigments. Atwell and Wilson have found rod–cone coupling in the tiger salamander to be one-fourth to one-half as strong as rod–rod coupling, but this measurement is uncertain because of possible disruption of interreceptor junctions in their retinal slice preparation. Therefore the extent of rod–cone coupling remains to be measured in an intact retinal network. Nevertheless, there can be little doubt that rod–cone coupling is weaker than rod–rod or cone–cone coupling, since the spectral properties of most visual pigments are preserved in the responses of individual photoreceptors.

Direct evidence for receptor coupling is not practically obtainable in many higher vertebrates because of the difficulty of intracellular recording in these species. However, there are a number of reports of interreceptor junctions in higher vertebrates, and these findings will now be reviewed as they relate to receptor coupling. Possible sites of interaction include photoreceptor fins (as in *Bufo*), receptor terminals (as in the cat), and teleodendria (as is likely in the turtle). The first report of gap junctions between receptor fins was made by

Custer (1973), who found interconnections between rod fins in the axolotl retina. Although rod coupling has not been examined electrophysiologically in the axolotl, the similarity of Custer's observations to those in *Bufo* argues strongly in favor of coupling in the axolotl. In addition, Custer has observed chemical synapses from rods onto cones, the significance of which is not known. The only other report of interreceptor junctions between fins comes from the human retina, where Uga *et al.* (1970) found junctions between rods and between rods and cones. However, these junctions are quite different in appearance from the junctions found in the toad and cat and closely resemble desmosomal junctions which are not believed to mediate electrical coupling (Farquhar and Palade, 1963; Kuffler and Potter, 1964). Junctions between fins have not been reported in any other species and are notably absent on fins from the turtle, skate, rat, mud puppy (Gold and Dowling, 1979), pigeon (Cohen, 1963), lizard, and house finch (Dunn, 1972). Therefore the fins are not a general site for interreceptor coupling. This may reflect the fact that coupling at this level involves only neighboring cells and therefore would not be effective for interconnecting minority receptor types. Therefore the functional role of the fins must be related to something other than interreceptor coupling. The large area of apposition between the fins and glial processes has suggested that interactions between these two structures may be important (Dunn, 1972).

A variety of interreceptor junctions has been reported between receptor terminals and also between teleodendria or basal processes. However, the only junctions that closely resemble the gap junctions in toad and cat are those found between rods and also between cones in carp and catfish (Witkovsky *et al.*, 1974). Most other junctions resemble either the focal junctions found in *Bufo* (Raviola and Gilula, 1973; Kolb, 1977) or desmosomes (Cohen, 1964, 1965, 1969; Dowling and Boycott, 1966; Sjöstrand, 1969; Lasansky, 1971; West and Dowling, 1975), the latter not being associated with electrical coupling.[8] The functional significance of these morphological variations in junctional structure is not fully understood. Nevertheless, the strong similarity among the junctions known to mediate coupling (i.e., those in toad and the cat) argues against a synaptic function for many of the reported junctions (e.g., Gold and Dowling, 1979; Detwiler and Hodgkin, 1979).

Another criterion for evaluating the functional significance of interreceptor junctions may be their specificity. Photoreceptor coupling is an exceedingly specific interaction; this is apparent both anatomically and electrophysiologically. Anatomically, this point is made by micrographs like Figs. 4 and 5; not only were aberrant connections never observed, but it is evident that specificity is actively determined by features of the different membranes. Since gap junctions

[8]Raviola and Gilula (1973) also reported clusters of membrane particles, seen in freeze-fracture, coupling primate cones, but corresponding gap junctions were not demonstrated in thin sections.

probably arise from interactions between subunits in the two parent membranes (Loewenstein *et al.*, 1978), the presence of these subunits in only red rod membranes could itself be the feature that determines specificity. Electrophysiologically, there is no evidence for random cross-talk between different receptor types (Baylor and Hodgkin, 1973; Detwiler and Hodgkin, 1979). Note that the existence of rod–cone coupling does not contradict this statement. Rod–cone coupling does not reflect a *lack* of specificity of rod–rod coupling, as would be the case if it were mediated by occasional gap junctions between rods and cones. Rather, rod–cone coupling is itself a specific interaction mediated by a morphologically different junction. Therefore there is no evidence that receptor coupling is ever an ''indiscriminate'' interaction, as suggested by Raviola and Gilula (1973) for the mammalian retina. The above considerations of junctional morphology and specificity provide useful criteria for evaluating anatomical studies. In view of these criteria the anatomical evidence for coupling in many species (including humans) remains equivocal. As Cohen (1964) points out, ''a 'synapse' is a physiological concept and can only be suggested by morphology.''

The accumulating evidence regarding receptor coupling raises the issue of whether or not this interaction plays a role in all vertebrate retinas, and also whether or not coupling is necessary to the functioning of all types of photoreceptors. So far, all receptors studied electrophysiologically have displayed coupling. However, the anatomical picture is not so clear. The failure to find interreceptor junctions cannot be taken as conclusive evidence. Nor, as pointed out above, is the existence of interreceptor junctions conclusive evidence that significant coupling exists. Nevertheless, the failure to find any indication of junctions between rods in the rat retina (Gold and Dowling, 1979) suggests that coupling may not subserve a function that is necessary in all vertebrate photoreceptors.

IV. CONSEQUENCES OF COUPLING

So far, it seems that photoreceptor coupling has had greater consequences for the electrophysiologist than for the toad or turtle. For example, the presence of electrical coupling has obscured several interesting properties of the vertebrate photoreceptor, necessitating the development of techniques for measuring local membrane currents (McBurney and Normann, 1977; Baylor *et al.*, 1979; Jagger, 1979) and for uncoupling photoreceptors (Bader *et al.*, 1978; Werblin, 1979; Attwell and Wilson, 1980). In addition, electrical coupling complicates the quantitative analysis of experiments in which pharmacological agents are injected into single cells in the network. This is because the strong coupling in *Bufo,* for example, should attenuate the effects of injected agents by more than a factor of 10. However, large effects are observed after injecting cyclic guanosine 3′, 5′-monophosphate (Nicol and Miller, 1978; see also Miller and Nicol, Chapter

24, this volume) as well as calcium and EGTA (Brown *et al.*, 1977) into *Bufo* rods. The large size of these effects remains puzzling.

The functional role that receptor coupling has in vision must be reflected by the effect that coupling has on the photoreceptors and on retinal signal transmission. However, photoreceptor coupling has several effects, so that it is difficult to pinpoint the particular role that coupling subserves, if a single role even exists. Nevertheless, the relative importance of some effects can still be considered and will be reviewed below.

Perhaps the most significant feature of receptor coupling (rod–rod and cone–cone) is that it is mediated electrically. This means that the spread of photocurrents is a dissipative process and cannot serve to amplify receptor signals. Therefore one consequence of coupling is averaging of the responses in many receptors. This reduces spatial resolution at the receptor level, but it is not known how much it affects the overall visual acuity in the animals studied. The averaging of receptor signals also reduces the number of synaptic connections required between receptors and second-order neurons (Lamb and Simon, 1976; Gold, 1979). Alternatively, the spread of photocurrents is known to be modified by voltage-dependent conductances in the photoreceptor membrane, so that the role of coupling may be related to this effect. For example, Detwiler *et al.* (1978) have shown that the voltage-dependent conductances in turtle rods cause the rod network to behave like a high-pass temporal filter. Such a network would allow transient signals to be integrated spatially over a large area, whereas prolonged signals could be integrated temporally with higher resolution. However, this hypothesis does not explain why both rods and cones are coupled but only rods exhibit the inductive behavior. Another proposal is that coupling serves to reduce the intrinsic noise level in individual rods and cones by averaging the voltages in adjacent cells (Lamb and Simon, 1976). This suggestion is consistent with the facts that both rods and cones are coupled and both exhibit intrinsic noise.

A discussion of rod–cone coupling must be based on entirely different considerations because, unlike rod–rod and cone–cone coupling, rod–cone coupling allows interactions between two different anatomical and physiological pathways through the retina. Rod–cone coupling could thus enlarge the signal processing capabilities of the retina. However, the discussion of rod–cone coupling is complicated because the mechanism underlying this phenomenon is unknown. There is ample electrophysiological evidence for rod–cone interactions within the retina. In the cat, Nelson (1977) has shown that these interactions are mediated at the receptor level. However, Fain (1975b) has found that rod–cone interactions in the mud puppy retina are *not* mediated at the receptor level. Therefore no general conclusions can be drawn concerning the retinal location of these interactions. There is psychophysical evidence in humans for rod–cone interactions (see MacLeod, 1978, for review), although the site of these interactions also remains unknown. Because rod–cone interactions can be studied both psychophysically

TABLE I
LENGTH CONSTANT DATA[a]

	λ (μm)	D (μm)	λ/D	Plateau/peak ratio
Bufo marinus (toad)				
Rod	21	7.6[b]	2.8	0.56
Chelydra serpentina (snapping turtle)				
Rod	60	27.9[c]	2.2	0.5
Red cone	13	20.1[d]	0.65	0.9[g]
Pseudemys scripta elegans (red-eared turtle)				
Red cone	22	17.4[e]	1.3	0.8
Green cone	26	19.8[e]	1.3	0.8
Ambystoma tigrinum (tiger salamander)				
Rod	17	15.9[f]	1.1	0.62

[a] All entries are averages of the measurements cited in this chapter with the exceptions noted in footnotes *b–g*.
[b] Fain (1976).
[c] Copenhagen and Owen (1980).
[d] G. H. Gold, unpublished observation.
[e] Calculated using the inner segment diameter and relative frequency data of Baylor and Fettiplace (1975). The more comprehensive study of Granda and Haden (1970) gave larger values but was not used because the retinal regions used by Lamb and Simon (1976) and by Detwiler and Hodgkin (1979) were not accurately specified.
[f] Attwell and Wilson (1980).
[g] J. F. Ashmore, D. R. Copenhagen, and J. L. Schnapf, personal communication.

in humans and electrophysiologically in lower vertebrates, combining these two approaches may lead to useful extrapolations about the mechanism and functional significance of this phenomenon.

Many of the ideas discussed above remain speculative. However, some of the hypotheses concerning the functional significance of rod–rod and cone–cone coupling may be more critically evaluated by comparing existing electrophysiological data (Table I). These data consist mainly of the network length constants observed in cold-blooded vertebrates.[9] The length constants are also expressed as λ/D, since this parameter reflects the electrical (and not the spatial) extent of coupling. The difference between these two parameters is apparent in comparing rod coupling in the toad and snapping turtle: The electrical extent of coupling (which reflects r_s/r_m, r_{in}/r_m, and the reduction in intrinsic noise amplitude) is comparable in these two species, yet the length constant is three

[9]The values of λ quoted were all derived from measurements of peak response amplitude. As described above, these values may be modified by the presence of voltage-dependent conductances in the photoreceptor membrane. The size of this modification is known only in the tiger salamander, so no attempt was made to correct data from other species.

times longer in the turtle. This difference simply reflects the different densities of rods in the toad and turtle (and the fact that toad rods are connected by short processes, whereas turtle rods must be connected by long processes).

One approach in elucidating a theory of receptor coupling is to search for correlations between the extent of coupling and related anatomical and physiological features in different species. One feature for which considerable data exist is the presence of voltage-dependent conductances in the photoreceptor membrane. The magnitudes of these conductances are probably reflected by the size of the decay from the peak to the plateau phase of the photoresponse (Fain *et al.*, 1978). Therefore the ratio of the plateau to the peak amplitude will be taken as an approximate measure of the size of the voltage-dependent conductances in different receptors. A plateau/peak ratio of 1, which reflects the absence of a plateau phase, is probably due to the absence of voltage-dependent conductances. Plateau/peak ratios are included in Table I. To see if a correlation exists between the presence of voltage-dependent conductances and the extent of coupling, the plateau/peak ratio is plotted versus λ/D in Fig. 12. Two possible trends in the data are indicated by the solid and dashed lines. If the dashed lines are meaningful, the electrical extent of coupling is not a function of the magnitude of voltage-dependent conductances. Based on this view, the clustering of rods (solid circles) and cones (open circles) into two separate groups suggests that voltage-dependent conductances are more prominent in rods than in cones and

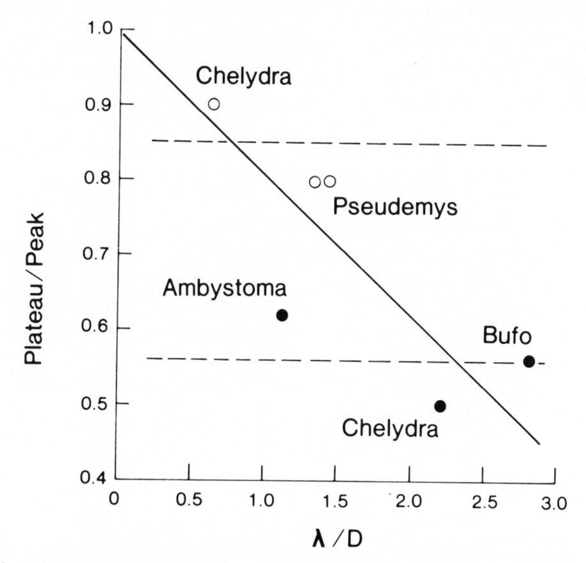

FIG. 12. Plot of the plateau/peak ratio versus λ/D. Data are from Table I. Open circles indicate cones, and filled circles indicate rods. Two possible correlations are indicated by the dashed and solid lines.

that coupling is unrelated to the size of the voltage-dependent conductances. In this case, other roles for coupling must be considered. Another correlation suggested by the data in Fig. 12 is indicated by the solid line. If meaningful, this correlation indicates that, independently of receptor type, the extent of coupling may reflect the size of the voltage-dependent conductances. Such a correlation favors the proposals that the significance of coupling is derived from its effects on the voltage-dependent properties of the photoreceptor (Schwartz, 1976; Detwiler *et al.*, 1978). In addition, the voltage-dependent conductances are sources of noise (Schwartz, 1977), so that this correlation also favors the suggestion that coupling serves to reduce the noise level in individual photoreceptors (Lamb and Simon, 1976).

It is not clear how to distinguish these alternatives at present, but additional measurements of λ should help to confirm or refute the correlations suggested in Fig. 12. For example, according to the solid line theory, any receptor with a plateau/peak ratio of 1 (i.e., a receptor without voltage-dependent conductances) should not exhibit coupling. Since tiger salamander cones appear to meet this criterion (Lasansky and Marchiafava, 1974; Schnapf and McBurney, 1980), a measurement of coupling between these receptors would be a useful test of the hypothesis that coupling is related to the presence of voltage-dependent conductances.

The role of photoreceptor coupling might best be investigated if retinal physiology could be studied both in the presence and the absence of this interaction. Unfortunately, attempts to uncouple toad rods in a retinal preparation, using established techniques for uncoupling electrical junctions, have not been successful (Griff, 1979). Therefore new approaches may be needed before the significance of this curious and pervasive phenomenon is revealed.

ACKNOWLEDGMENTS

I am indebted to D. Attwell, D. R. Copenhagen, J. E. Dowling, J. I. Korenbrot, and M. Wilson for reviewing the manuscript. The completeness of this chapter also benefitted from, at the time, unpublished manuscripts generously provided by D. Attwell, D. R. Copenhagen, E. R. Griff, W. G. Owen, J. L. Schnapf, and M. Wilson. D. Attwell and M. Wilson were particularly helpful in providing their original computations (for Fig. 11), in discussing their work with me, and in criticizing my interpretation of it. I am very grateful to L. M. Masukawa, both for reviewing the manuscript and for many helpful discussions over the course of numerous revisions. Preparation of this manuscript was supported by NIH Program Project Grant GM27057 and National Research Service Award EY05286.

REFERENCES

Attwell, D., and Wilson, M. (1980). Behaviour of the rod network in the tiger salamander retina mediated by membrane properties of individual rods. *J. Physiol. (London)* **309,** 287–316.

Bader, C. R., MacLeish, P. R., and Schwartz, E. A. (1978). Responses to light of solitary rod photoreceptors isolated from tiger salamander retina. *Proc. Natl. Acad. Sci. U.S.A.* **75,** 3507–3511.

Baylor, D. A., and Fettiplace, R. (1975). Light path and photon capture in turtle photoreceptors. *J. Physiol. (London)* **248,** 433–464.

Baylor, D. A., and Hodgkin, A. L. (1973). Detection and resolution of visual stimuli by turtle photoreceptors. *J. Physiol. (London)* **234,** 163–198.

Baylor, D. A., Fuortes, M. G. F., and O'Bryan, P. M. (1971). Receptive fields of single cones in the retina of the turtle. *J. Physiol. (London)* **214,** 265–294.

Baylor, D. A., Lamb, T. D., and Yau, K.-W. (1979). The membrane current of single rod outer segments. *J. Physiol. (London)* **248,** 433–464.

Bennett, M. V. L. (1973). Permeability and structure of electronic junctions and intercellular movement of tracers. *In* "Intracellular Staining in Neurobiology" (S. Kater and C. Nicholson, eds.), pp. 114–134. Springer-Verlag, Berlin and New York.

Brightman, M. W., and Reese, T. S. (1969). Junctions between intimately apposed cell membranes in the vertebrate brain. *J. Cell Biol.* **40,** 648–677.

Brown, J. E., Coles, J. A., and Pinto, L. H. (1977). Effects of injections of calcium and EGTA into the outer segments of retinal rods of *Bufo marinus*. *J. Physiol. (London)* **269,** 707–722.

Brown, K. T., and Flaming, D. G. (1977). New microelectrode techniques for intracellular work in small cells. *Neuroscience* **2,** 813–827.

Burkhardt, D. A. (1977). Responses and receptive-field organization of cones in perch retinas. *J. Neurophysiol.* **40,** 53–62.

Cohen, A. I. (1963). The fine structures of the visual receptors of the pigeon. *Exp. Eye Res.* **2,** 88–97.

Cohen, A. I. (1964). Some observations on the fine structure of the retinal receptors of the American gray squirrel. *Invest. Ophthalmol.* **3,** 198–216.

Cohen, A. I. (1965). Some electron microscopic observations on interreceptor contacts in the human and macaque retinae. *J. Anat.* **99,** 595–610.

Cohen, A. I. (1969). Rods and cones and the problem of visual excitation. *In* "The Retina: Morphology, Function, and Clinical Characteristics" (B. R. Straatsma, M. O. Hall, R. A. Allen, and F. Crescitelli, eds.), pp. 31–62. Univ. of California Press, Berkeley.

Copenhagen, D. R., and Owen, W. G. (1976). Functional characteristics of lateral interactions between rods in the retina of the snapping turtle. *J. Physiol. (London)* **259,** 251–282.

Copenhagen, D. R., and Owen, W. G. (1980). Current-voltage relations in the rod and photoreceptor network of the turtle retina. *J. Physiol. (London)* **308,** 159–184.

Custer, N. V. (1973). Structurally specialized contacts between the photoreceptors of the retina of the axolotl. *J. Comp. Neurol.* **151,** 35–56.

Detwiler, P. B., and Hodgkin, A. L. (1979). Electrical coupling between cones in turtle retina. *J. Physiol. (London)* **291,** 75–100.

Detwiler, P. B., Hodgkin, A. L., and McNaughton, P. A. (1978). A surprising property of electrical spread in the network of rods in the turtle's retina. *Nature (London)* **274,** 562–565.

Dowling, J. E., and Boycott, B. B. (1966). Organization of the primate retina: Electron microscopy. *Proc. R. Soc. London, Ser. B* **166,** 80–111.

Dunn, R. R. (1972). The ultrastructure of the vertebrate retina. *In* "The Ultrastructure of Sensory Organs" (I. Friedmann, ed.), pp. 155–222. Am. Elsevier, New York.

Fain, G. L. (1975a). Quantum sensitivity of rods in the toad retina. *Science* **187,** 838–841.

Fain, G. L. (1975b). Interactions of rod and cone signals in the mudpuppy retina. *J. Physiol. (London)* **252,** 735–769.

Fain, G. L. (1976). Sensitivity of toad rods: Dependence on wavelength and background illumination. *J. Physiol. (London)* **261,** 71–101.

Fain, G. L., Quandt, F. N., Bastian, B. L., and Gerschenfeld, H. M. (1978). Contribution of a caesium-sensitive conductance increase to the rod photoresponse. *Nature (London)* **272,** 467–369.

Farquhar, M. G., and Palade, G. E. (1963). Junctional complexes in various epithelia. *J. Cell Biol.* **17,** 375–412.

Fuortes, M. G. F., Schwartz, E. A., and Simon, E. J. (1973). Colour dependence of cone responses in the turtle retina. *J. Physiol. (London)* **234,** 199–216.

Gold, G. H. (1979). Photoreceptor coupling in retina of the toad, *Bufo marinus*. II. Physiology. *J. Neurophysiol.* **42,** 292–310.

Gold, G. H., and Dowling, J. E. (1979). Photoreceptor coupling in retina of the toad, *Bufo marinus*. I. Anatomy. *J. Neurophysiol.* **42,** 292–310.

Granda, A. M., and Haden, K. W. (1970). Retinal oil globule counts and distributions in two species of turtles: *Pseudemys scripta elegans* (Wied) and *Chelonia mydas mydas* (Linnaeus). *Vision Res.* **10,** 79–84.

Green, D. G. (1970). Regional variations in the visual acuity for interference fringes on the retina. *J. Physiol. (London)* **207,** 351–356.

Griff, E. R. (1979). A study of interactions among rods in the isolated retina of *Bufo marinus*. Ph.D. Thesis, Purdue University, West Lafayette, Indiana.

Griff, E. R., and Pinto, L. H. (1981). Interactions among rods in the isolated retina of *Bufo marinus*. *J. Physiol. (London)* **314,** 237–254.

Harosi, F. I. (1975). Absorption spectra and linear dichroism of some amphibian photoreceptors. *J. Gen. Physiol.* **66,** 357–382.

Ito, S., and Loewenstein, W. R. (1969). Ionic communication between early embryonic cells. *Dev. Biol.* **19,** 228–243.

Jagger, W. S. (1979). Photoresponses of isolated frog rod outer segments. *Vision Res.* **19,** 159–167.

Kolb, H. (1977). The organization of the outer plexiform layer in the cat: Electron microscopic observations. *J. Neurocytol.* **6,** 131–153.

Kuffler, S. W., and Potter, D. D. (1964). Glia in the leech central nervous system: Physiological properties and neuron-glia relationship. *J. Neurophysiol.* **27,** 290–320.

Lamb, T. D. (1976). Spatial properties of horizontal cell responses in the turtle retina. *J. Physiol. (London)* **263,** 239–255.

Lamb, T. D., and Simon, E. J. (1976). The relation between intercellular coupling and electrical noise in turtle photoreceptors. *J. Physiol. (London)* **263,** 257–286.

Lasansky, A. (1971). Synaptic organization of cone cells in the turtle retina. *Philos. Trans. Soc. London, Ser. B* **262,** 365–381.

Lasansky, A. (1973). Organization of the outer synaptic layer in the retina of larval tiger salamander. *Philos. Trans. R. Soc. London, Ser. B* **265,** 471–489.

Lasansky, A., and Marchiafava, P. L. (1974). Light-induced resistance changes in retinal rods and cones of the tiger salamander. *J. Physiol. (London)* **236,** 171–191.

Leeper, H. F., Normann, R. A., and Copenhagen, D. R. (1978). Red rods of the toad retina: Evidence for passive electrotonic interactions. *Nature (London)* **275,** 234–236.

Lowenstein, W. R. (1975). Permeable junctions. *Cold Spring Harbor Symp. Quant. Biol.* **40,** 49–63.

Lowenstein, W. R., Kanno, Y., and Socolar, S. J. (1978). The cell-to-cell channel. *Fed. Proc., Fed. Am. Soc. Exp. Biol.* **37,** 2645–2650.

McBurney, R. N., and Normann, R. A. (1977). Current and voltage responses from single rods in toad retina. *J. Gen. Physiol.* **70,** 12a.

MacLeod, D. J. A. (1978). Visual sensitivity. *Annu. Rev. Psychol.* **29,** 613–645.

Nelson, R. (1977). Cat cones have rod input: A comparison of the response properties of cones and horizontal cell bodies in the retina of the cat. *J. Comp. Neurol.* **172,** 109–136.

Nelson, R., Lytzow, A. v., Kolb., H., and Gouras, P. (1975) Horizontal cells in cat retina with independent dendritic systems. *Science* **189,** 137–139.

Nicol, G. D., and Miller, W. H. (1978). Cyclic GMP injected into retinal rod outer segments increases latency and amplitude of response to illumination. *Proc. Natl. Acad. Sci. U.S.A.* **10,** 5217–5220.

Owen, W. G., and Copenhagen, D. R. (1977). Characteristics of the electrical coupling between rods on the turtle retina. *In* "Vertebrate Photoreception" (H. B. Barlow and P. Fatt, eds.), pp. 169–192. Academic Press, New York.

Peters, A., Palay, S. L., and Webster, H. DeF. (1970). "The Fine Structure of the Nervous System." Harper & Row, New York.

Polyak, S. L. (1941). "The Retina." Univ. of Chicago Press, Chicago, Illinois.

Raviola, E., and Gilula, N. B. (1973). Gap junctions between photoreceptor cells in the vertebrate retina. *Proc. Natl. Acad. Sci. U.S.A.* **70,** 1677–1681.

Schnapf, J. L., and McBurney, R. N. (1980). Light-induced changes in membrane current in cone outer segments of tiger salamander and turtle. *Nature (London)* **287,** 239–241.

Schwartz, E. A. (1973). Responses of single rods in the retina of the turtle. *J. Physiol. (London)* **232,** 503–514.

Schwartz, E. A. (1975a). Rod-rod interaction in the retina of the turtle. *J. Physiol. (London)* **246,** 617–638.

Schwartz, E. A. (1975b). Cones excite rods in the retina of the turtle. *J. Physiol. (London)* **246,** 639–651.

Schwartz, E. A. (1976). Electrical properties of the rod syncytium in the retina of the turtle. *J. Physiol. (London)* **257,** 379–406.

Schwartz, E. A. (1977). Voltage noise observed in rods of the turtle retina. *J. Physiol. (London)* **272,** 217–246.

Sjöstrand, F. S. (1969). The outer plexiform layer and the neural organization of the retina. *In* "The Retina: Morphology, Function and Clinical Characteristics" (B. R. Straatsma, M. O. Hall, R. A. Allen, and F. Crescitelli, eds.), pp. 63–100. Univ. of California Press, Berkeley.

Spira, A. W. (1971). The nexus in the intercalated disc of the canine heart: Quantitative data for an estimation of its resistance. *J. Ultrastruct. Res.* **34,** 409–425.

Steinberg, R. H., Reid, M., and Lacy, P. L. (1973). The distribution of rods and cones in the retina of the cat (*Felis domesticus*). *J. Comp. Neurol.* **148,** 229–248.

Torre, V., and Owen, W. G. (1981). Ionic basis of high pass filtering of small signals by the network of retinal rods in the toad. *Proc. R. Soc. (London), Ser. B* (in press).

Uga, S., Nakao, F., Mimura, M., and Ikui, H. (1970). Some new findings on the fine structure of the human photoreceptor cells. *J. Electronmicrosc. (Tokyo)* **19,** 71–84.

Weidmann, S. (1966). The diffusion of radiopotassium across intercalated discs of mammalian cardiac muscle. *J. Physiol. (London)* **187,** 323–342.

Werblin, F. S. (1978). Transmission along and between rods in the tiger salamander retina. *J. Physiol. (London)* **280,** 449–470.

Werblin, F. S. (1979). Time- and voltage-dependent ionic components of the rod response. *J. Physiol. (London)* **294,** 613–626.

West, R. W., and Dowling, J. E. (1975). Anatomical evidence for cone- and rod-like receptors in the gray squirrel, ground squirrel, and prairie dog retinas. *J. Comp. Neurol.* **159,** 434–460.

Witkovsky, P., Shakib, M., and Ripps, H. (1974). Inter-receptoral junctions in the teleost retina. *Invest. Ophthalmol.* **13,** 996–1009.

Part II
The Cyclic Nucleotide Enzymatic Cascade and Calcium Ion

The cyclic nucleotide system and calcium ion are the major known active components for transduction in the rod outer segment aside from rhodopsin

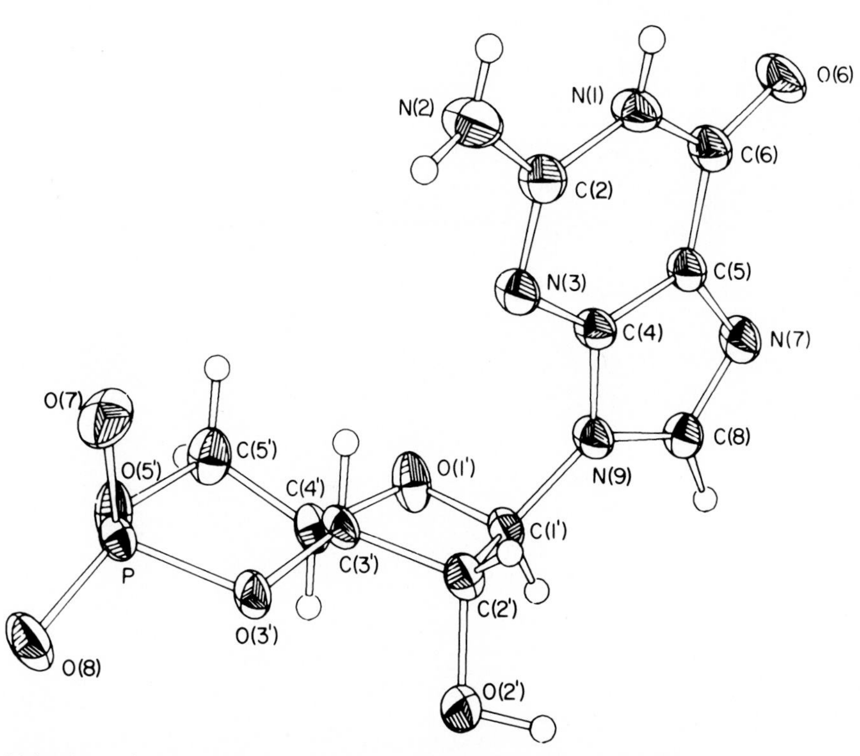

Molecular conformation and atom numbering system for cyclic GMP, depicted here in the syn conformation. The effect of the conformation on biological activity is unknown. Thermal ellipsoids draw at 50% probability level. [From A. K. Chwang and M. Sundaralingam, The crystal and molecular structure of guanosine 3′,5′-cyclic monophosphate (cyclic GMP) sodium tetrahydrate; *Acta Cryst.* (1974) **B30,** 1233–1240; with permission.]

Chapter 5

First Stage of Amplification in the Cyclic-Nucleotide Cascade of Vision

LUBERT STRYER, JAMES B. HURLEY, AND BERNARD K.-K. FUNG[1]

Department of Structural Biology
Sherman Fairchild Center
Stanford University School of Medicine
Stanford, California

I. ACTIVATION OF PHOSPHODIESTERASE BY PHOTOLYZED RHODOPSIN

Light activates two enzymes in retinal rod outer segments (ROS): a phosphodiesterase specific for cyclic guanosine 3′,5′-monophosphate (cyclic GMP) (Chader *et al.*, 1973; Miki *et al.*, 1973; Pannbacker *et al.*, 1972) and a guanosinetriphosphatase (GTPase) (Wheeler and Bitensky, 1977; Robinson and Hagins, 1977). Our interest in the photoenzymology of vision was further stimulated by electrophysiological studies showing that cyclic GMP depolarized the

[1]Present address: Department of Radiation Biology and Biophysics, University of Rochester Medical Center, Rochester, New York.

ISBN 0-12-153315-8

ROS plasma membrane within milliseconds after being injected intracellularly (Nicol and Miller, 1978). Another striking effect of injected cyclic GMP is that it increases the latency of the light-induced hyperpolarization (Nicol and Miller, 1978; Miller and Nicol, 1979). A further impetus came from biochemical studies demonstrating that one photolyzed rhodopsin molecule can lead to the hydrolysis of 4×10^5 cyclic-GMP molecules per second in disk membrane suspensions (Yee and Liebman, 1978). From the known turnover number of phosphodiesterase, it was inferred that the absorption of a single photon could activate hundreds of enzyme molecules. A high gain in the light-triggered hydrolysis of cyclic GMP was also found in freshly detached ROS (Woodruff and Bownds, 1979). These experimental findings strengthened the notion that cyclic nucleotides were important in vision (Miller *et al.*, 1971; Lipton *et al.*, 1977; Pober and Bitensky, 1979) and suggested that the photoactivation of phosphodiesterase was a critical early event in visual excitation (Liebman and Pugh, 1979).

The intriguing questions posed by these studies were: How does a single photolyzed rhodopsin molecule activate a large number of phosphodiesterase molecules? In particular, which proteins participate in this photoactivation? How does information flow from photolyzed rhodopsin to the phosphodiesterase? Most important, what is the molecular mechanism of amplification? In this chapter, we review our experiments showing that photoactivation of phosphodiesterase is mediated by transducin, a protein consisting of three kinds of subunits. Photolyzed rhodopsin catalyzes the exchange of guanosine 5′-triphosphate (GTP) for guanosine 5′-diphosphate (GDP) bound to transducin, which converts it into an activator of phosphodiesterase. The GTP–transducin complex is likely to be the first amplified intermediate in the cyclic-nucleotide cascade of vision (Fung and Stryer, 1980; Fung *et al.*, 1981).

II. CATALYSIS OF GTP–GDP EXCHANGE BY PHOTOLYZED RHODOPSIN IN ROS MEMBRANES

At the start of our experiments, it was known that GTP or a nonhydrolyzable analogue such as guanosine 5′-(β, γ-imido) triphosphate (GNP) was needed in addition to light to activate phosphodiesterase (Bitensky *et al.*, 1978; Yee and Liebman, 1978). This finding pointed to a guanyl nucleotide-binding protein as a likely link in the activation sequence. Our initial experiments were therefore designed to detect the light-sensitive binding of guanyl nucleotides to ROS membranes. We incubated ROS membranes with [α-^{32}P]GTP in the dark, washed the membranes to remove unbound nucleotide, and then analyzed them for radioactive nucleotide. About 1.3 mmoles nucleotide mole^{-1} rhodopsin became bound after a 90-min incubation in the dark. Thin-layer chromatography showed that the incorporated nucleotide was GDP. Brief sonication did not release the radioactive GDP, showing that it was bound to a membrane component instead of being sequestered within the internal aqueous space of the ROS membrane vesicles. The light-dependent binding of guanyl nucleotides to ROS membranes was also observed by Godchaux and Zimmerman (1979). The effect of light on the binding of GDP was then measured. We found that illumination led to the release of GDP, which was half-maximal when 1% of the rhodopsin was photolyzed in the absence of added guanyl nucleotide. ROS are rich in GTP (Robinson and Hagins, 1979), and so it was of interest to determine whether GTP influenced the release of GDP. A striking effect was observed: GTP markedly enhanced the action of light in releasing bound GDP. As shown in Fig. 1, almost no GDP was released on photolysis of 0.06% of the rhodopsin in the absence of GTP or on addition of 1.5 μM GTP in the dark. In contrast, most of the bound GDP was released on photolysis of 0.06% of the rhodopsin in the presence of GTP. The extent of release of GDP depended on the concentration of GTP. Release was half-maximal when the degree of photolysis was 0.023% in the presence of 0.1 μM GTP, and 0.0067% with 1 μM GTP.

We then turned to the effect of light on the uptake of GTP. GNP, a nonhyd-

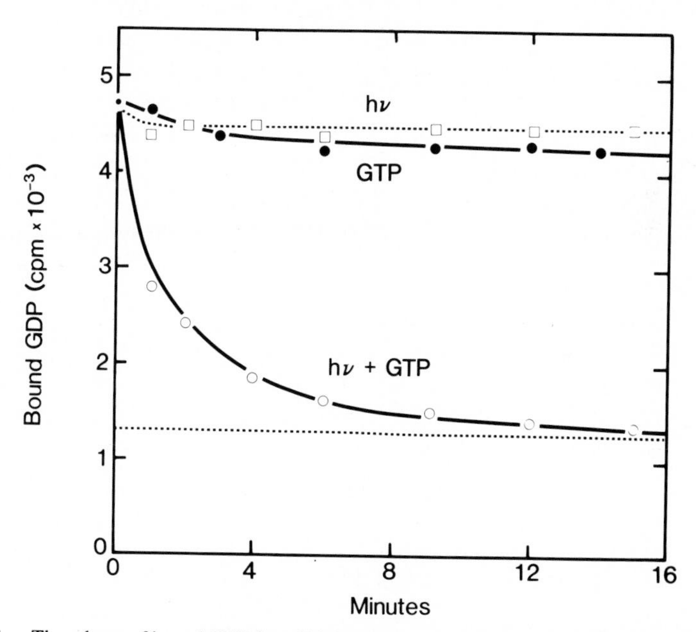

FIG. 1. The release of bound GDP from ROS membranes on illumination in the presence of GTP. The degree of photolysis was 0.06%, and the concentration of GTP was 1.5 μM. Open circles, light plus GTP; open squares, light without GTP; solid circles, GTP without light. The amount of radioactive GDP bound 15 min after 100% photolysis in the presence of GTP is denoted by the lower dashed line.

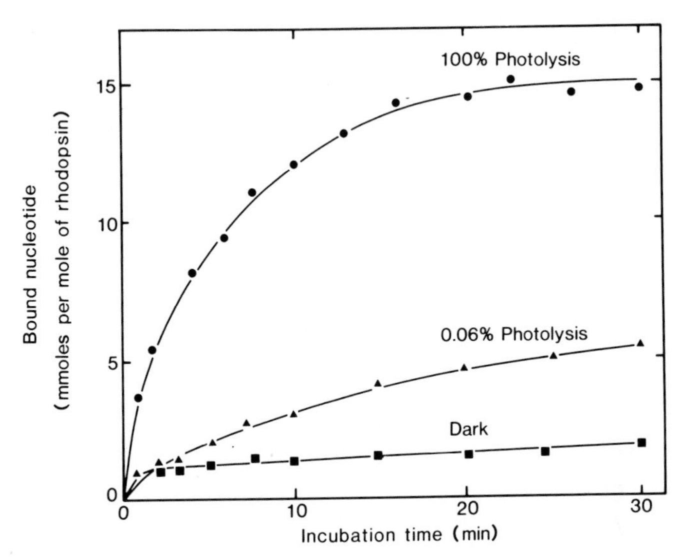

FIG. 2. Light-activated binding of GNP by ROS membranes. Squares, dark; triangles, 0.06% photolysis; circles, 100% photolysis. The concentration of GNP was 10 μM.

rolyzable analogue, was used to simplify the analysis. As shown in Fig. 2, light markedly accelerates the uptake of GNP by ROS membranes. The binding of GNP, like the release of GDP, requires the photolysis of only a small proportion of the rhodopsin molecules.

At this point, we knew that light stimulated the uptake of GNP and the release of GDP, and so we surmised that these processes were linked. This idea was

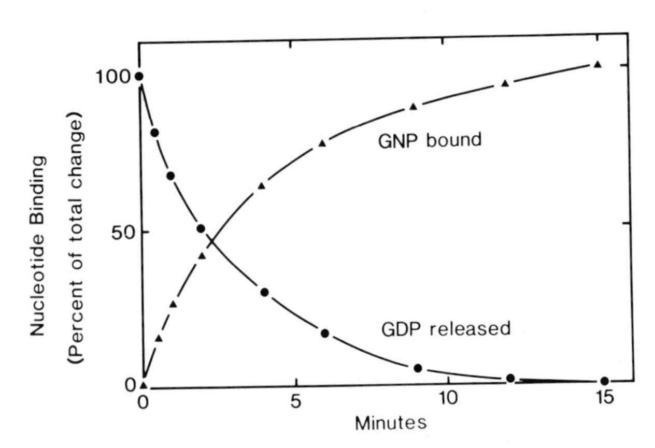

FIG. 3. Exchange of GNP for GDP bound to photolyzed ROS membranes. The uptake of [^{3}H]-GNP (triangles) parallels the release of bound [α-^{32}P]GDP (circles) by ROS membranes containing 0.1% photolyzed rhodopsin.

tested by comparing the kinetics of uptake of [^{3}H]GNP with the kinetics of release of bound [α-^{32}P]GDP. In fact, the percentage of GDP released paralleled the percentage of GNP that was bound to the membranes (Fig. 3). These data demonstrated that GNP was exchanged for bound GDP in the presence of photolyzed rhodopsin.

III. THE LIGHT-TRIGGERED AMPLIFIED EXCHANGE OF GTP FOR BOUND GDP

The effectiveness of very low degrees of photolysis in eliciting the release of bound GDP and the concomitant uptake of GNP suggests that photolyzed rhodopsin acts catalytically. To test this idea, we measured the extent of uptake of GNP as a function of the degree of photolysis and of the concentration of GNP in the incubation mixture. A 30-min incubation with 10 μM GNP resulted in the uptake of 1 GNP molecule per 32 rhodopsin molecules. Unilluminated membranes incorporated 2.7, 6.8, and 17.2 mmoles GNP mole^{-1} rhodopsin after a 30-min incubation with 0.1, 1, and 10 μM GNP, presumably because they contained very small amounts of photolyzed rhodopsin. Binding above these dark background levels was half-maximal when the degree of photolysis was 0.14, 0.014, and 0.0011% for 0.1, 1, and 10 μM GNP, respectively (Fig. 4). These data enabled us to calculate the number of GNP molecules taken up by the

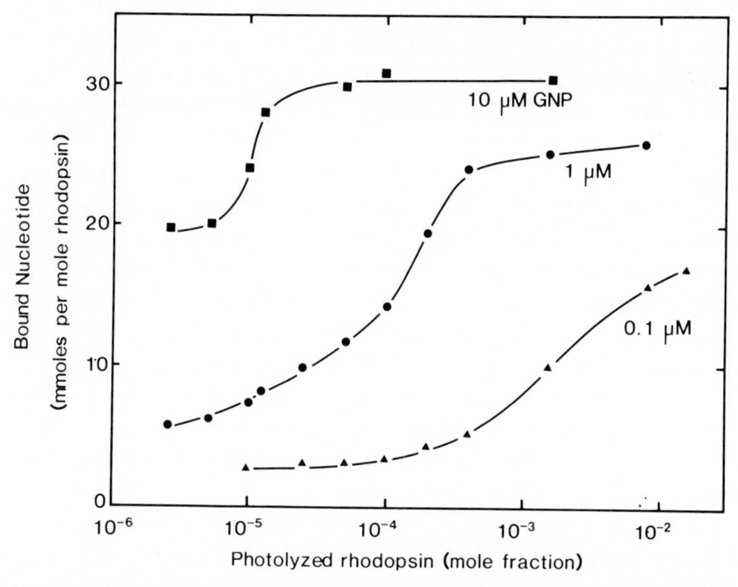

FIG. 4. Dependence of the uptake of GNP by ROS membranes on the mole fraction of photolyzed rhodopsin. The incubation time was 30 min. Triangles, 0.1 μM GNP; circles, 1 μM GNP; squares, 10 μM GNP.

membranes per photolyzed rhodopsin molecule. In particular, incubation with 10 μM GNP led to the binding above the dark background level of 5.5 mmoles GNP $mole^{-1}$ rhodopsin when the concentration of photolyzed rhodopsin was 0.011 mmole $mole^{-1}$ rhodopsin. *Hence one photolyzed rhodopsin molecule catalyzed the binding of 500 GNP molecules.* This stoichiometry indicates that a single photolyzed rhodopsin molecule can catalyze the exchange of about half of the guanyl nucleotides bound to a disk. It is also noteworthy that this high degree of amplification was achieved without hydrolysis of the GTP analogue. As will be discussed later, the role of GTP hydrolysis is to reset the system to the dark state.

IV. TRANSDUCIN, A MULTISUBUNIT PROTEIN THAT BINDS GUANYL NUCLEOTIDES AND HAS GTPASE ACTIVITY

The observation of a high degree of amplification of the GTP–GDP exchange stimulated us to isolate the guanyl nucleotide-binding protein. It seemed likely that the role of the GTP complex of this protein was to activate phosphodiesterase. Consequently, we named this protein transducin. Our working hypothesis (Fung and Stryer, 1980) was that the flow of information in the cyclic-nucleotide cascade is:

$$\underset{1}{R^*} \longrightarrow \underset{\sim 500}{\text{T-GTP}} \longrightarrow \underset{\sim 500}{\text{PDE}^*}$$

where R* denotes photolyzed rhodopsin, T-GTP denotes the GTP–transducin complex, and PDE* denotes the active form of phosphodiesterase. This scheme predicts that (1) T-GTP can be formed in the absence of phosphodiesterase, and (2) phosphodiesterase can be activated by T-GTP in the absence of photolyzed rhodopsin.

Our procedure for the isolation of transducin, a peripheral membrane protein, is shown in Fig. 5. A fraction rich in GTPase activity was prepared by a recently published procedure (Kühn, 1980) which proved to be highly effective. We further purified this GTPase fraction by chromatography on hexylagarose, which also removed unbound nucleotides and concentrated the protein. As shown in Fig. 6, a small amount of protein impurity was eluted by 75 mM NaCl. An increase in the salt concentration to 300 mM led to the elution of a protein fraction with GTPase activity and GNP-binding activity when assayed in the presence of reconstituted membranes containing photolyzed rhodopsin. Sodium dodecyl sulfate (SDS)-polyacrylamide gel electrophoresis (Fig. 6, inset) showed that the protein in this fraction consisted of three kinds of polypeptide chains which we designated T_α (39 kilodaltons), T_β (36 kilodaltons), and T_γ ($\sim$10 kilodaltons). The sedimentation profile in sucrose density gradients showed that T_α, T_β, and T_γ formed a complex in the absence of GNP (or GTP).

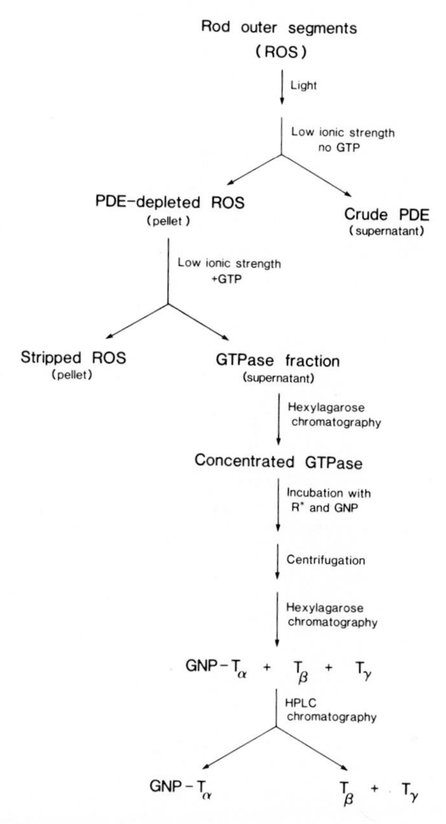

FIG. 5. Procedure for the isolation of transducin and of the Tα-GNP complex.

In contrast, T_α was dissociated from T_β and T_γ when this nucleotide was bound. As will be discussed shortly, this change in the state of association of the transducin complex turned out to be very useful in the purification of T_α.

V. AMPLIFIED FORMATION OF GNP-TRANSDUCIN IN A RECONSTITUTED MEMBRANE SYSTEM

The availability of purified transducin enables us to determine whether the amplified binding of GNP to transducin catalyzed by photolyzed rhodopsin could occur in the absence of phosphodiesterase. Purified transducin (containing less than 1% phosphodiesterase on a molar basis) was incubated in the dark with reconstituted membranes consisting of phosphatidylcholine and rhodopsin purified by chromatography on hydroxylapatite. These membranes contained 1

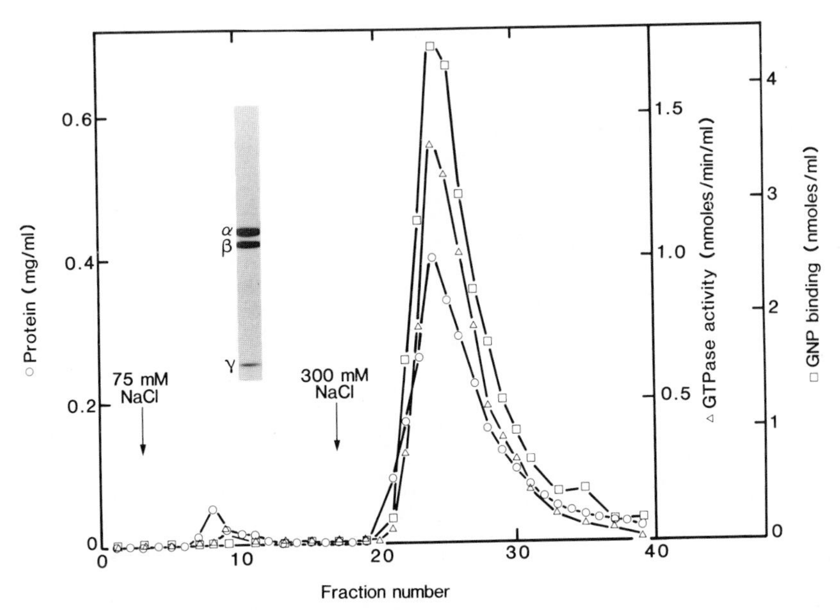

FIG. 6. Hexylagarose chromatography of transducin. The protein concentration (circles), GTPase activity (triangles), and GNP-binding activity (squares) of the column fractions are shown. The inset shows the SDS-polyacrylamide gel pattern of the peak fraction.

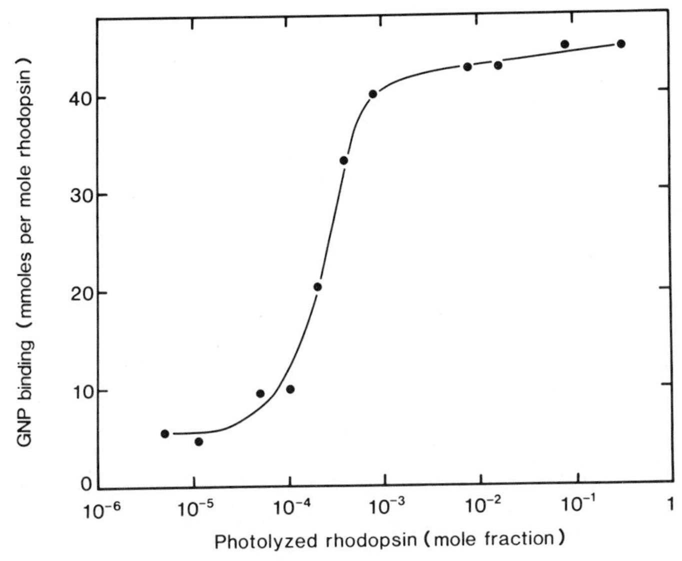

FIG. 7. Catalysis of the binding of GNP to purified transducin by photolyzed rhodopsin in reconstituted membranes.

molecule of transducin per 17 molecules of rhodopsin. They were then subjected to different degrees of illumination to determine the dependence of GNP uptake on the mole fraction of photolyzed rhodopsin (Fig. 7). The amount of nucleotide bound was half-maximal when the degree of photolysis was 0.027%. Thus *a single photolyzed rhodopsin molecule catalyzed the uptake of 71 GNP molecules by transducin molecules in this reconstituted membrane. It is evident that phosphodiesterase is not required for the catalyzed uptake of GNP by transducin.* The lower degree of amplification in this reconstituted system compared with native disk membranes (~500) is partly due to the fact that only half of the rhodopsin molecules are properly oriented in the reconstituted membranes. Also, the reconstituted vesicles are severalfold smaller than native ones and their lipid composition is different.

VI. ISOLATION OF GNP-α-TRANSDUCIN

The next step in elucidating the molecular mechanism of the cyclic-nucleotide cascade was to identify the subunit of transducin that contained the guanyl nucleotide-binding site. Transducin was incubated with radioactive GNP in the presence of photolyzed reconstituted membranes (Fig. 5). The T-GNP complex was separated from the membranes by centrifugation and purified by hexylagarose chromatography. High-pressure liquid chromatography then separated this complex into two fractions (Fig. 8). SDS-polyacrylamide gel electrophoresis showed that the first peak contained T_α, whereas the second contained T_β and T_γ. The radioactive GNP eluted together with T_α. The absence of detectable free nucleotide in fractions other than those containing T_α shows that the binding of GNP to T_α is very tight. Thus *T_α contains the binding site for GNP*. The separation of T_α from T_β and T_γ by high-pressure liquid chromatography was much better when GNP was bound than in the absence of added nucleotide. For this reason, it was not feasible to obtain T_α free of T_β and T_γ in the absence of GNP. The most likely explanation is that the affinity of T_α for T_β and T_γ is much lower when GNP is bound than in the absence of bound GNP.

VII. ACTIVATION OF PHOSPHODIESTERASE IN THE DARK BY THE ADDITION OF GNP-α-TRANSDUCIN

A key prediction of our working hypothesis is that phosphodiesterase can be activated by T-GNP in the absence of photolyzed rhodopsin. Indeed, this is so. Figure 8a shows that *phosphodiesterase bound to unilluminated membranes is activated by the addition of GNP-α-transducin (T_α-GNP)*. In contrast, phosphodiesterase is not activated by fractions containing only T_β and T_γ. The acti-

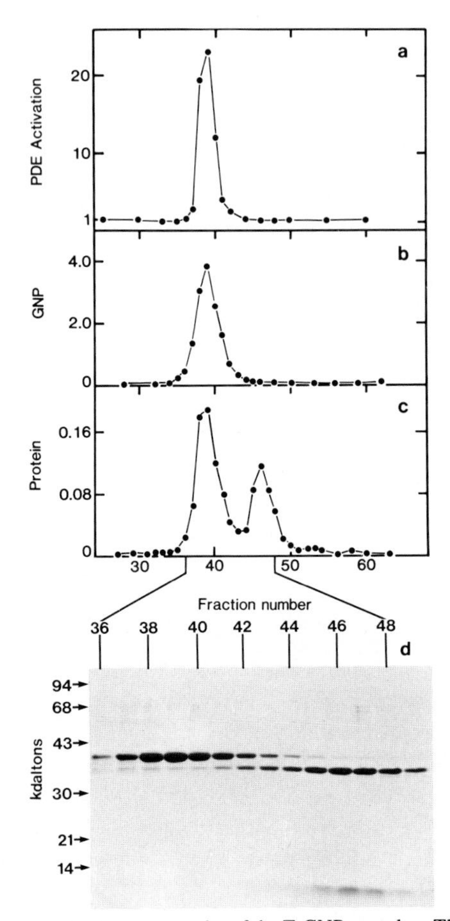

FIG. 8. High-pressure liquid chromatography of the T-GNP complex. The column fractions were assayed for activation of phosphodiesterase (PDE) bound to unilluminated ROS membranes(a), GNP binding (b), and protein content (c). SDS-polyacrylamide gel electrophoresis patterns of the peak column fractions are shown in (d).

vation of phosphodiesterase by T_α-GNP in this experiment strongly suggests that T_α-GTP is the information-carrying intermediate under physiological conditions.

The maximal degree of activation that can be obtained by the addition of T_α-GNP is of interest because it provides insight into the mechanism of activation of phosphodiesterase. It is known that phosphodiesterase can be activated by illuminating disk membranes in the presence of GTP (or GNP) or by adding a small amount of trypsin (Miki *et al.*, 1975). The catalytic activity of trypsin-activated phosphodiesterase is severalfold higher than that of light-activated phosphodiesterase. How does the catalytic activity of phosphodiesterase switched on by the addition of T_{α}-GNP compare with that obtained by illumination or treatment with trypsin? As shown in Fig. 9, the activity of phosphodiesterase bound to unilluminated ROS membranes increases nearly linearly with the amount of T_{α}-GNP added up to an asymptotic value. This limiting value is 2.2-fold as high as the light-activated level. Most striking, *the maximum catalytic activity obtained by the addition of T_{α}-GNP is nearly the same as that achieved by tryptic activation of phosphodiesterase*.

Why are these levels nearly the same? The simplest interpretation is that the addition of T_{α}-GNP removes the same inhibitory constraint as the proteolytic action of trypsin. An inhibitor of phosphodiesterase was recently isolated by heating the purified enzyme at pH 12 for 5 min at 100°C (Hurley, 1981). The pH was then lowered to 5.9 to precipitate the denatured phosphodiesterase, yielding a supernatant which exhibited a 13-kilodalton polypeptide on SDS-polyacrylamide gels. Highly active phosphodiesterase produced by tryptic digestion was rendered inactive by the addition of this supernatant fraction (Fig. 10). These experiments suggest that native phosphodiesterase is inhibited by its 13-

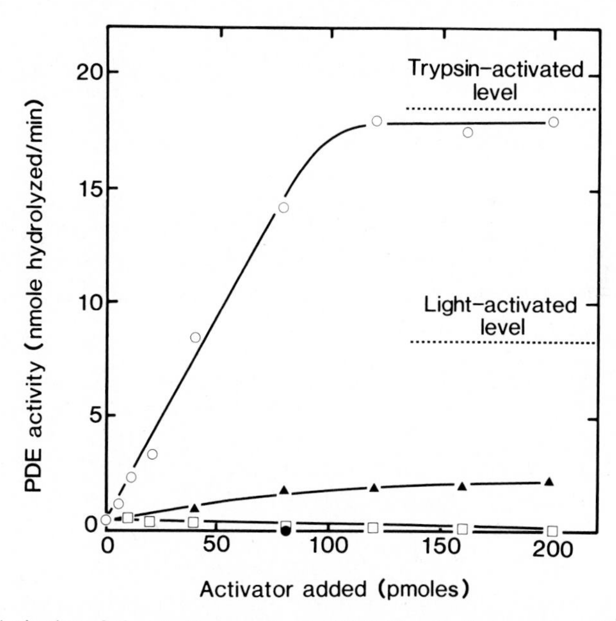

FIG. 9. Activation of phosphodiesterase by Tα-GNP. Phosphodiesterase in unilluminated ROS membranes was incubated with Tα-GNP (open circles), GNP alone (solid triangles), or T-GDP (open squares). Tα-GNP without added membranes (solid circles) contained almost no detectable phosphodiesterase activity. Maximal phosphodiesterase activity (upper dashed line) was determined by incubating unilluminated ROS membranes with trypsin for 5 min. The light-activated phosphodiesterase level (lower dashed line) was obtained by bleaching the membranes in the presence of 10 μM GNP.

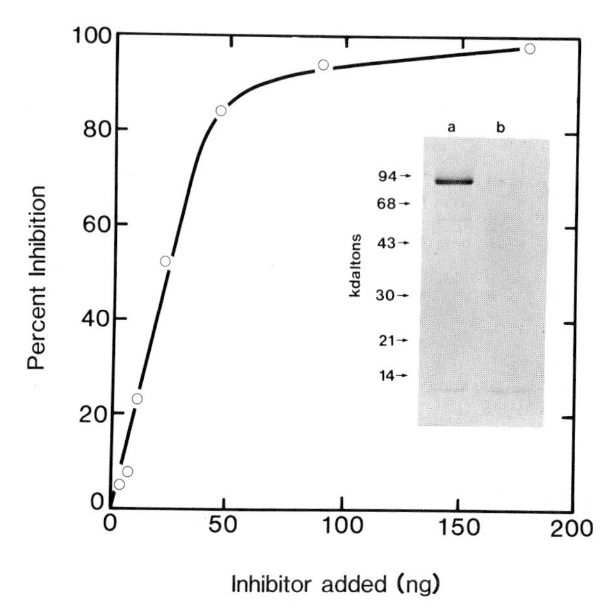

FIG. 10. Inhibition of trypsin-activated phosphodiesterase by the addition of a heat-stable protein fraction. The inset shows an SDS-polyacrylamide gel pattern of purified phosphodiesterase (a) and inhibitor isolated from the purified enzyme (b).

kilodalton subunit, and that trypsin activates the enzyme by cleaving this inhibitory subunit (I). How then does T-GNP or T-GTP activate phosphodiesterase? T-GTP could displace I from the enzyme or it could complete with phosphodiesterase for the binding of I. Alternatively, T-GTP could activate phosphodiesterase by altering the interaction of I with the catalytic subunit. We are carrying out experiments to determine how T-GTP reverses the inhibition imposed by an endogenous component of phosphodiesterase.

VIII. FLOW OF INFORMATION IN THE CYCLIC-NUCLEOTIDE CASCADE OF VISION

Our experiments show that a single photolyzed rhodopsin molecule activates several hundred phosphodiesterase molecules in two stages. First, T-GTP is formed in an exchange reaction catalyzed by photolyzed rhodopsin. Then T-GTP switches on phosphodiesterase. The first reaction can occur *in vitro* in the absence of phosphodiesterase, and the second can take place in the absence of photolyzed rhodopsin. The separability of these steps reveals that *information flows from photolyzed rhodopsin to T-GTP and then to phosphodiesterase*. The gain of 500 observed in the first step suggests that *transducin is the first amplified information-carrying intermediate in the cyclic-nucleotide cascade of vision.*

A reaction scheme for the cyclic-nucleotide cascade of vision that accounts for our experimental observations is shown in Fig. 11. The major features of this proposed light-activated amplification cycle are: (1) In the dark, nearly all the transducin is in the T-GDP form, which does not activate phosphodiesterase. On illumination, photolyzed rhodopsin (R*) acts catalytically to convert T-GDP into T-GTP, which switches on phosphodiesterase. (2) R* acts as a catalyst by forming an R*-T-GDP complex. GTP is then exchanged for bound GDP to give R*-T-GTP. (3) R* then dissociates from R*-T-GTP to form T-GTP. R* released from this complex binds another T-GDP. The rapid release and recycling of R* are critical for the highly amplified formation of T-GTP. (4) T-GTP activates phosphodiesterase (PDE) by relieving an inhibitory constraint. The PDE*-T-GTP complex rapidly hydrolyzes cyclic GMP. (5) The GTPase activity intrinsic to transducin converts T-GTP to T-GDP, which in turn deactivates phosphodiesterase. The PDE*-T-GDP complex then dissociates into T-GDP and inhibited PDE.

Some aspects of this proposed cycle deserve further comment:

1. A priori, R* could catalyze GTP–GDP exchange by altering the membrane potential. However, we found that the amplified exchange did not depend on ionic gradients nor was it affected by gramicidin A or alamethicin. Hence R* probably acts directly by forming an R*-T-GDP complex. The high fluidity of

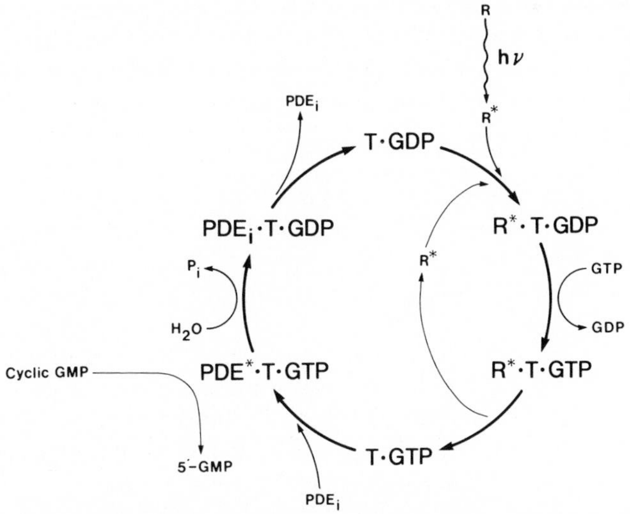

FIG. 11. Proposed light-activated amplification cycle. Information flows from photolyzed rhodopsin (R*) to transducin (T) and then to phosphodiesterase (PDE). The first stage of amplification is the formation of hundreds of T-GTP molecules per photolyzed rhodopsin molecule. Each T-GTP then activates a single phosphodiesterase molecule by relieving an inhibitory constraint.

the disk membrane (Poo and Cone, 1973; Liebman and Entine, 1974) may be important in enabling R* and T-GDP to encounter each other rapidly by diffusion.

2. The rapid and highly amplified formation of T-GTP requires that R* have a high affinity for T-GDP and a low affinity for T-GTP. Once R* encounters T-GDP by diffusion, it must bind it long enough to enable GTP to be exchanged for GDP. But once this exchange has occurred, R* must quickly leave T-GTP so that it can bind another T-GDP molecule and again catalyze a nucleotide exchange. At low ionic strength, T-GDP is bound to bleached disk membranes, whereas T-GTP is not (Kühn, 1980). This observation is consistent with the assumption that T-GDP under physiological conditions has a much higher affinity for R* than does T-GTP.

3. This amplification cycle is driven by the hydrolysis of GTP. Amplified activation of the phosphodiesterase does not require phosphoryl hydrolysis, as shown by the action of T-GNP. Rather, the hydrolysis of GTP serves to reset the system to the dark state. The absence of covalent bond changes in the steps leading to the formation of many activated phosphodiesterase molecules probably contributes to the rapidity of the activation phase of the cycle.

4. Is the formation of T-GTP sufficiently rapid to enable it to serve as an intermediate in visual excitation, which occurs in times of milliseconds to a few seconds (Penn and Hagins, 1972; Baylor *et al.*, 1979)? At 2 m*M* GTP, the physiological level (Robinson and Hagins, 1979), T-GTP is formed in less than 4 sec. An assay with higher temporal resolution is needed to determine whether T-GTP is kinetically competent to be on the excitation pathway.

5. We know that the α subunit of transducin is the information-carrying intermediate in the activation of phosphodiesterase. The roles of the β and γ subunits, on the other hand, are unclear. First, are they required for the catalyzed uptake of GTP by the α subunit of transducin? Second, do they contribute to the GTPase activity of the transducin complex? A direct way of answering these questions would be to isolate the complex of the α subunit with either GTP or GDP and determine the effects of adding the β and γ subunits on its properties. However, GNP bound to the α subunit cannot readily be displaced by GTP or GDP, and so this experiment is not yet feasible. It may be necessary to prepare T_α-GDP in a different way. Third, are the β and γ subunits required in addition to T_α-GTP for the activation of phosphodiesterase? The experiment shown in Fig. 9 shows that T_α-GTP is the information carrier, but it does not rule out the possibility that T_β and T_γ also participate in the activation process. A small amount of these subunits was present in the unilluminated membranes used to assay the activation of phosphodiesterase. This question could be answered by preparing a reconstituted system free of T_β and T_γ. It will also be very interesting to learn how T_α-GTP activates phosphodiesterase.

6. The hydrolysis of T-GTP to T-GDP is essential for the deactivation of phosphodiesterase and restoration of the dark state. However, hydrolysis of bound GTP is not enough to reset the system. The capacity of R* to catalyze

GTP–GDP exchange must also be turned off. There may be a more rapid deactivation event than isomerization of the retinal chromophore of R* back to the 11-cis form. Recent studies point to an adenosine 5′-triphosphate (ATP)-dependent inactivation of phosphodiesterase following illumination (Liebman and Pugh, 1979). It will be important to delineate the control elements in this amplification cycle.

The cycle proposed here is reminiscent of the elongation factor Tu-Ts cycle in protein synthesis, which also is driven by the hydrolysis of GTP (Kaziro, 1978). The affinities of transducin and elongation factor Tu for other proteins depend on whether GTP or GDP is bound to them. An even closer analogy, perhaps even a homology, is provided by the mechanism of activation of adenylate cyclase. A guanyl nucleotide-binding protein has recently been shown to be the information-carrying intermediate between a hormone receptor and adenylate cyclase (Cassel and Selinger, 1978; Pfeuffer, 1979; Ross and Gilman, 1980). Transducin may play a critical role in signal amplification and transduction in a wide variety of systems. Study of the cyclic-nucleotide cascade of vision affords a glimpse of what may be a recurring motif in nature in addition to being satisfying and rewarding for its own sake.

ACKNOWLEDGMENTS

We thank Ms. Cordula Atkinson for excellent technical assistance in the preparation of ROS membranes. We are indebted to Mr. Larry Weiss for helpful suggestions concerning high-pressure liquid chromatography. This work was supported by a grant from the National Eye Institute (EY-02005). J.B.H. is a fellow of the Helen Hay Whitney Foundation and B.K.-K.F. was a fellow of the Jane Coffin Childs Fund.

REFERENCES

Baehr, W., Devlin, M. J., and Applebury, M. L. (1979). Isolation and characterization of cGMP phosphodiesterase from bovine rod outer segments. *J. Biol. Chem.* **254,** 11669–11677.

Baylor, D. A., Lamb, T. D., and Yau, K.-W. (1979). Responses of retinal rods to single photons. *J. Physiol. (London)* **288,** 613–634.

Bitensky, M. W., Wheeler, G. L., Aloni, B., Vetury, S., and Matuo, Y. (1978). Light and GTP activated photoreceptor phosphodiesterase (PDE) regulation by a light-activated GTPase and identification of rhodopsin. *Adv. Cyclic Nucleotide Res.* **9,** 553–572.

Cassel, D., and Selinger, Z. (1978). Mechanism of adenylate cyclase activation through the β-adrenergic receptor: Catecholamine-induced displacement of bound GDP by GTP. *Proc. Natl. Acad. Sci. U.S.A.* **75,** 4155–4159.

Chader, G. J., Bensinger, R., Johnson, M., and Fletcher, R. T. (1973). Phosphodiesterase: An important role in cyclic nucleotide regulation in the retina. *Exp. Eye Res.* **17,** 483–486.

Fung, B. K.-K., and Stryer, L. (1980). Photolyzed rhodopsin catalyzes the exchange of GTP for bound GDP in retinal rod outer segments. *Proc. Natl. Acad. Sci. U.S.A.* **77,** 2500–2504.

Fung, B. K.-K., Hurley, J. B., and Stryer, L. (1981). Flow of information in the light-triggered cyclic nucleotide cascade of vision. *Proc. Natl. Acad. Sci. U.S.A.* **78,** 152–156.

Godchaux, W., III, and Zimmerman, W. F. (1979). Membrane-dependent guanine nucleotide binding and GTPase activities of soluble protein from bovine rod outer segments. *J. Biol. Chem.* **254,** 7874–7884.

Hurley, J. B. (1981). Isolation and assay of a phosphodiesterase inhibitor from retinal rod outer segments. *In* "Methods in Enzymology" (in press).

Kaziro, Y. (1978). The role of guanosine 5′-triphosphate in polypeptide chain elongation. *Biochim. Biophys. Acta* **505,** 95–127.

Kühn, H. (1980). Light- and GTP-regulated interaction of GTPase and other proteins with bovine photoreceptor membranes. *Nature (London)* **283,** 587–589.

Liebman, P. A., and Entine, G. (1974). Lateral diffusion of visual pigment in photoreceptor disk membranes. *Science* **185,** 642–644.

Liebman, P. A., and Pugh, E. N., Jr. (1979). The control of phosphodiesterase in rod disk membranes: Kinetics, possible mechanisms and significance for vision. *Vision Res.* **19,** 375–380.

Lipton, S. A., Rasmussen, H., and Dowling, J. E. (1977). Electrical and adaptive properties of rod photoreceptors in *Bufo marinus*. II. Effects of cyclic nucleotides and prostaglandins. *J. Gen. Physiol.* **70,** 771–791.

Miki, N., Keirns, J. J., Marcus, F. R., Freeman, J., and Bitensky, M. W. (1973). Regulation of cyclic nucleotide concentrations in photoreceptors: An ATP-dependent stimulation of cyclic nucleotide phosphodiesterase by light. *Proc. Natl. Acad. Sci. U.S.A.* **70,** 3820–3824.

Miki, N., Baraban, J. M., Keirns, J. J., Boyce, J. J., and Bitensky, M. W. (1975). Purification and properties of the light-activated cyclic nucleotide phosphodiesterase of rod outer segments. *J. Biol. Chem.* **250,** 6320–6327.

Miller, W. H., and Nicol, G. D. (1979). Evidence that cyclic GMP regulates membrane potential in rod photoreceptors. *Nature (London)* **280,** 64–66.

Miller, W. H., Gorman, R. E., and Bitensky, M. W. (1971). Cyclic adenosine monophosphate: Function in photoreceptors. *Science* **174,** 295–297.

Nicol, G. D., and Miller, W. H. (1978). Cyclic GMP injected into retinal rod outer segments increases latency and amplitude of response to illumination. *Proc. Natl. Acad. Sci. U.S.A.* **75,** 5217–5220.

Pannbacker, R. G., Fleischman, D. E., and Reed, D. W. (1972). Cyclic nucleotide phosphodiesterase: High activity in mammalian photoreceptor. *Science* **175,** 757–758.

Penn, R. D., and Hagins, W. A. (1972). Kinetics of the photocurrent of retinal rods. *Biophys. J.* **12,** 1073–1094.

Pfeuffer, T. (1979). Guanine nucleotide-controlled interactions between components of adenylate cyclase. *FEBS Lett.* **101,** 85–89.

Pober, J. S., and Bitensky, M. W. (1979). Light-regulated enzymes of vertebrate retinal rods. *Adv. Cyclic Nucleotide Res.* **11,** 265–301.

Poo, M.-M., and Cone, R. A. (1973). Lateral diffusion of rhodopsin in *Necturus* rods. *Exp. Eye Res.* **17,** 503–510.

Robinson, W. E., and Hagins, W. A. (1977). A light-activated GTPase in retinal rod outer segments. *Biophys. J.* **17,** 196a (Abstr.).

Robinson, W. E., and Hagins, W. A. (1979). GTP hydrolysis in intact rod outer segments and the transmitter cycle in visual excitation. *Nature (London)* **280,** 398–400.

Ross, E. M., and Gilman, A. G. (1980). Biochemical properties of hormone-sensitive adenylate cyclase. *Annu. Rev. Biochem.* **49,** 533–564.

Wheeler, G. L., and Bitensky, M. W. (1977). A light-activated GTPase in vertebrate photoreceptors: Regulation of light-activated cyclic-GMP phosphodiesterase. *Proc. Natl. Acad. Sci. U.S.A.* **74,** 4238–4242.

Woodruff, M. L., and Bownds, D. (1979). Amplitude, kinetics, and reversibility of a light-induced decrease in cyclic GMP in frog photoreceptor membranes. *J. Gen. Physiol.* **73,** 627–653.

Yee, R., and Liebman, P. A. (1978). Light-activated phosphodiesterase of the rod outer segments: Kinetics and parameters of activation and deactivation. *J. Biol. Chem.* **253,** 8902–8909.

Chapter 6

Rod Guanylate Cyclase Located in Axonemes

DARRELL FLEISCHMAN

Charles F. Kettering Research Laboratory
Yellow Springs, Ohio

I. DISTRIBUTION OF GUANYLATE CYCLASE IN THE RETINA

The involvement of cyclic nucleotides in photoreceptor metabolism was first revealed when Bitensky *et al.* (1971) found evidence of adenylate cyclase activity in frog retinal rod outer segment preparations. Soon afterward guanylate cyclase activity as well was found in bovine rod outer segments (Goridis *et al.*, 1973; Pannbacker and Fleischman, 1972; Pannbacker, 1973).

After isopycnic centrifugation of retinal homogenates on continuous sucrose density gradients, the profile of guanylate cyclase activity includes a peak that coincides approximately with the peak of rhodopsin optical density (Virmaux *et al.*, 1976; Zimmerman *et al.*, 1976), strongly suggesting that guanylate cyclase is indeed localized in rod outer segments. The guanylate cyclase activity is very slightly displaced toward densities lower than that of rhodopsin, however.

ISBN 0-12-153315-8

Whether this is related to the fact that outer segment fragments float at a lower buoyant density than osmotically intact outer segments (Zimmerman *et al.*, 1976) is not clear.

A study of the distribution of guanylate cyclase by microdissection of frozen retinas has confirmed the conclusion that its activity is greatest in the outer segments (Berger *et al.*, 1980). Smaller amounts of guanylate cyclase are found in other retinal layers. The highest level—about one-tenth that of the outer segments—is present in the inner plexiform layer of rabbit, monkey, and ground squirrel retinas. Very little guanylate cyclase is found in the inner segments.

Troyer *et al.* (1978) report that mouse retinas appear to contain three distinct forms of guanylate cyclase. Two of these—a soluble and a particulate form—display kinetic characteristics similar to those of the soluble and particulate forms of guanylate cyclase found in other mammalian tissues. A second particulate form with a low-K_m GTP (42 μM) is found in normal retinas but is absent in retinas of mice (C57BL) afflicted with photoreceptor dystrophy. These retinas are essentially devoid of photoreceptor cells. The low-K_m particulate cyclase therefore appears to be the form present in the photoreceptors and is presumably localized in the outer segments. The kinetic characteristics of this cyclase are unlike those of previously described forms of guanylate cyclase. It requires low concentrations (<5 mM) of Mg^{2+} or Mn^{2+} for maximal activity but displays little Ca^{2+} supported activity and is inhibited by high (>5 mM) divalent cation concentrations. These characteristics are similar to those of the guanylate cyclase found in purified rod outer segments (Krishnan *et al.*, 1978).

Analysis of sections of frozen retinas shows that there are high levels of cyclic guanosine 3′,5′-monophosphate (cyclic GMP) in the outer segments (about 150 μmoles kg^{-1} dry wt in dark-adapted rabbits). There are also rather high cyclic-GMP levels (about 60 μmoles kg^{-1} dry wt) in the retinal layers containing the inner portions of the photoreceptor cells—the inner segments and the outer nuclear and outer plexiform layers (Orr *et al.*, 1976). The cyclic-GMP concentration in each of these layers is diminished 30–50% in light-adapted retinas.

It seems reasonable to draw the following inferences from the experiments cited above. The proximal layers of the retina contain soluble and high-K_m particulate forms of guanylate cyclase resembling those found in other mammalian tissues. These are especially concentrated in the inner plexiform layer. The photoreceptor cells contain a distinctive low-K_m particulate form of guanylate cyclase, localized in the outer segment, which supplies large amounts of cyclic GMP to the entire photoreceptor cell.

Most workers report the specific activity of guanylate cyclase in gradient-purified or microdissected outer segments to be approximately 1.0 nmole mg^{-1} protein min^{-1} (Goridis *et al.*, 1973; Pannbacker, 1973; Fleischman and Denisevich, 1979; Berger *et al.*, 1980). A somewhat higher activity was found in bovine outer segments by Krishnan *et al.* (1978).

II. ASSOCIATION OF GUANYLATE CYCLASE WITH THE RETINAL ROD AXONEME

A. Attempts to Solubilize Guanylate Cyclase

In most tissues that have been studied, guanylate cyclase is either soluble or membrane-bound and can be solubilized by a nonionic detergent such as Triton X-100 (Goldberg and Haddox, 1977). However, Krishnan *et al.* (1978) were able to solubilize virtually no guanylate cyclase activity from bovine rod outer segments with 0.2% Triton X-100, and higher concentrations were inhibitory. The criterion of solubilization was the presence of activity in the supernatant after centrifugation of treated outer segments at 12,000 *g* for 10 min. Ammonyx LO (lauryldimethylamine *N*-oxide) at a concentration of 0.2% solubilized 10% of the activity but decreased specific activity by two-thirds. Activity was not released when outer segments were treated with acetone or *n*-butanol.

Goridis *et al.* (1973) were unable to solubilize guanylate cyclase with 0.1% emulphogene (centrifugation was at 10^5 *g* for 120 min). Some activity (probably about 25%) was released by 0.3% emulphogene when the outer segments were first illuminated for 20 min with a 75-W bulb. The light requirement seems especially interesting and should be examined in more detail, particularly since rhodopsin bleaching has been reported to influence the binding of rhodopsin kinase and other proteins to rod outer segments (Kühn, 1978).

B. Copurification of Guanylate Cyclase and Axonemes

1. Axoneme Isolation

Several years ago we began attempts to isolate the axoneme of bovine retinal rods with the hope of comparing its characteristics with those of the axonemes of motile cilia and flagella. The latter had been purified and studied extensively (see, for example, Linck, 1976), but there appeared to have been no biochemical studies on sensory receptor axenomes. Phototactic, mechanotactic, and chemotactic protozoa respond to appropriate stimuli by changing the orientation or pattern of beating of their cilia or flagella. Since several vertebrate sensory receptors have retained a ciliated structure, it has been suggested that axonemes might have a function in vertebrate sensory reception similar to the role they play in the tactic behavior of protozoa (Gray and Pumphrey, 1958; Eakin, 1965; Wiederhold, 1976). For recent discussions of the mechanism of phototaxis in *Euglena* and of the response of *Paramecium* to mechanical stimuli, see Doughty and Diehn (1979) and Eckert and Brehm (1979), respectively.

At an early stage in our efforts to isolate rod axonemes, R. G. Pannbacker suggested that we examine the preparations for cyclic-nucleotide phosphodiesterase, guanylate cyclase, and protein kinase activities. There have been reports that microtubule components can be phosphorylated in a cyclic adenosine 3′,5′-monophosphate (cyclic-AMP)-dependent reaction (reviewed by Mohri, 1976). Most of the guanylate cyclase of sea urchin sperm is localized in the flagellar membrane, while phosphodiesterase appears to be associated with both the membrane and the axoneme (Sano, 1976).

Our first experiments indicated that partially purified bovine rod axonemes contained guanylate cyclase, but little phosphodiesterase, activity (Pannbacker and Fleischman, 1972). Subsequently the yield and purity of the axoneme preparations was improved. A description of the isolation procedure (Fleischman and Denisevich, 1979) and a preliminary description of axenome morphology (Raveed and Fleischman, 1975) have been presented.

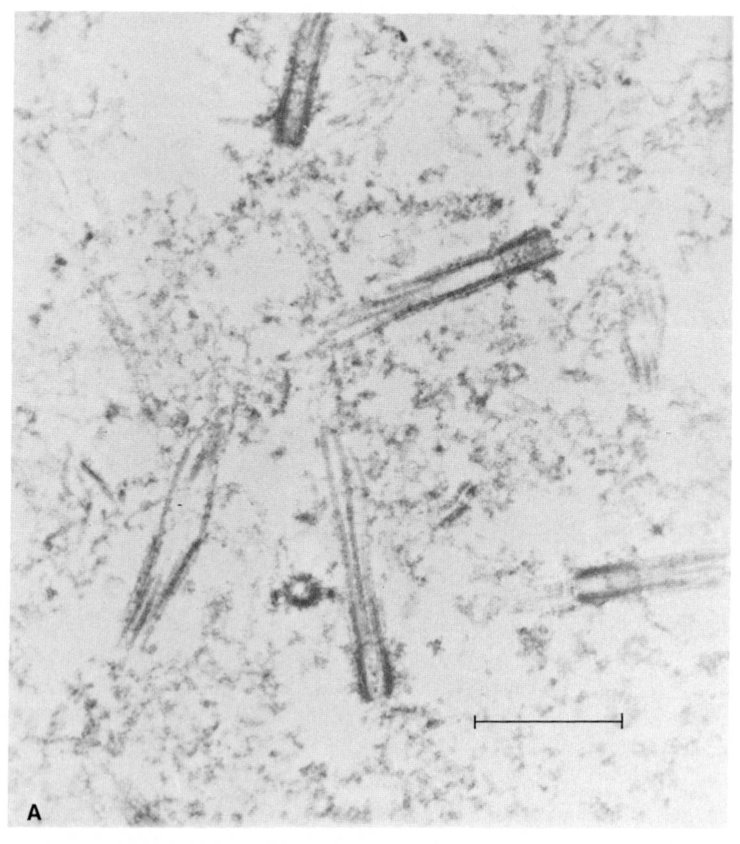

FIG. 1. Axonemes isolated from bovine retinal rods. The actinlike peptide has not been extracted. Scale bars: 1.0 μm. (A) Field. (B) Axoneme with rootlet and centriole.

Axoneme preparations include basal bodies and varying lengths of rod axonemes (Fig. 1). Portions of striated rootlet are sometimes present, and centrioles are often attached to the basal bodies by rootlet strands. The wine goblet-shaped arms connecting the microtubule doublets to the plasma membrane are retained with the axonemes. The best preparations are still contaminated with amorphous and filamentous material, some of which appears to originate from disintegrating microtubules and rootlets. Electrophoresis in sodium dodecyl sulfate–polyacrylamide gels indicates that a peptide with the molecular weight of tubulin (51,000) comprises at least one-third of the protein in the preparations (Fleischman *et al.*, 1980). There is an approximately equivalent amount of a peptide that comigrates with rabbit skeletal muscle actin. This peptide can be solubilized with a solution containing 2 m*M* Tris, 0.5 m*M* $CaCl_2$, 0.5 m*M* ATP, and 0.5 m*M* 2-mercaptoethanol at pH 8.0. This solution is commonly used to solubilize F-actin. Extraction of the actinlike peptide has no noticeable effect on the appearance of the axonemes but removes much of the amorphous and filamentous material. It is presumably derived from the rootlets or from inner segment microfilaments which remain attached to the basal bodies during the axoneme isolation.

The following procedure is the most effective that we have found for isolating rod axonemes. Retinas from freshly enucleated eyes are collected in a cold solution containing 10 m*M* PIPES, 5 m*M* $MgCl_2$, and 50% (w/w) sucrose at pH 7.0. The suspension is swirled vigorously, filtered through cheesecloth, and centrifuged for 1 hr at 13,000 *g*. The outer segment paste is removed from the top, layered on the bottom of a 26–38% (w/w) linear sucrose gradient in the same buffer, and centrifuged at 13,000 *g* for at least 1 hr. The outer segments are removed from the gradients, taking care to avoid the material floating beneath the outer segment bands. The outer segments are then dissolved (at a concentration of 5 mg protein ml^{-1} or less) in a solution containing 10 m*M* PIPES, 5 m*M* $MgCl_2$, 1 m*M* dithiothreitol, and 2% Triton X-100. Then 10 ml of the dissolved

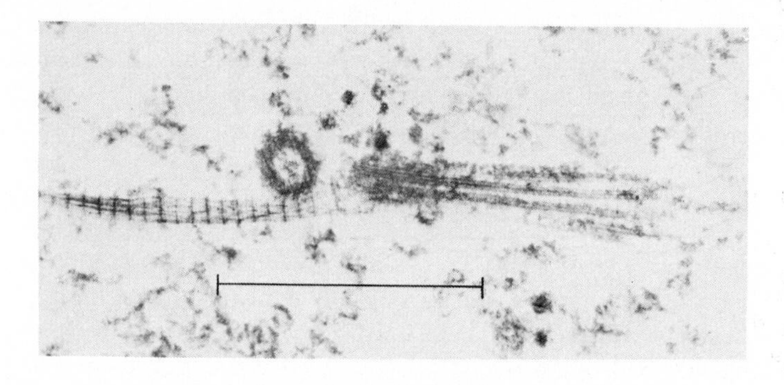

FIG. 1 (B)

outer segments is layered onto a 10-ml 45–65% (w/w) linear sucrose gradient made up in the Triton-containing buffer. After centrifugation at 13,000 *g* for at least 3 hr, the axonemes are found in a turbid band near the center of the gradient. They are removed with a syringe (in a volume of about 1 ml), diluted with 19 ml of the Triton-containing buffer, and collected by centrifugation at 13,000 *g* for 1 hr.

The axoneme yield is diminished if the isolations are performed at a higher pH, in the presence of lower concentrations of Mg^{2+}, or at higher ionic strength (e.g., in Ringer's solution). Nevertheless, under these conditions the peptide composition of the isolates, as indicated by gel electrophoresis, is almost unchanged. The only striking difference is the near absence of a 145,000-dalton peptide in such preparations. We suspect that this peptide is nexin, the protein forming the linkages that circumferentially connect the microtubule doublets (Stephens, 1970; Linck, 1976). In the absence of nexin, broken microtubules would be released from the axoneme and, because of their large surface/mass ratios, might sediment very slowly through the highly viscous gradients and fail to be collected.

2. Guanylate Cyclase Distribution in Gradient and Sepharose Fractions

The distribution of guanylate cyclase in a typical isolation gradient is presented in Fig. 2. The peak of cyclase activity is in the fraction containing the visibly turbid axoneme band. The distribution of cyclase activity between the axoneme fraction and the supernatant is variable; under conditions that lead to lower axoneme recovery, a larger proportion of cyclase is found near the bottom of the supernatant.

In order to determine whether the cyclase activity in the supernatant represents solubilized enzyme, a solution of Triton-dissolved outer segments was subjected to exclusion chromatography on a Sepharose 4B column. All the measurable guanylate cyclase activity was eluted in the void volume (Fleischman *et al.*, 1980). The exclusion limit of Sepharose 4B (for globular proteins) is about 2×10^7 daltons. Almost all the cyclase must therefore be clumped or attached to structures with large Stokes' radii such as microtubule fragments.

3. Attempts to Dissolve Guanylate Cyclase

The specific activity of guanylate cyclase in axoneme preparations averages about 30 nmoles mg^{-1} protein min^{-1}, a 30-fold enrichment over the activity in the outer segments (Fleischman and Denisevich, 1979). Under the best conditions, as much as 90% of the guanylate cyclase activity of the rod outer segments has been recovered in the axoneme preparation (Fleischman *et al.*, 1980).

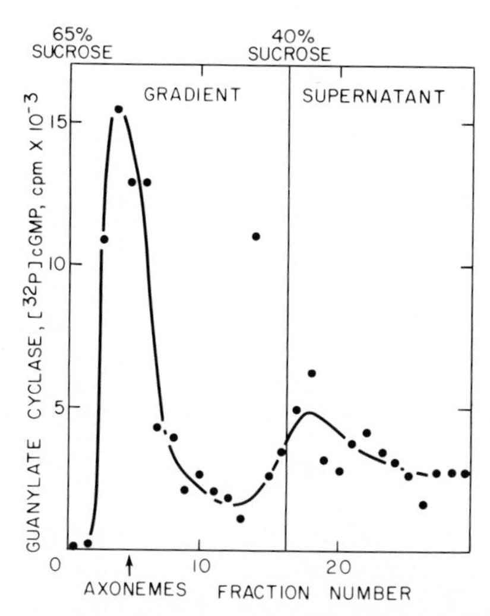

FIG. 2. Distribution of the guanylate cyclase activity of detergent-solubilized rod outer segments in a linear sucrose gradient. Fractions (0.6 ml) were collected from the bottom of an axoneme isolation gradient of the type described in the text. From Fleischman *et al.* (1980).

The results of our attempts to dissolve guanylate cyclase were similar to those of Krishnan *et al.* (1978). No more than 15% of the cyclase activity remained in the supernatant after centrifugation for 1 hr at 12,000 *g* of suspensions of axonemes that had been treated with the actin-solubilizing solution, with 5 m*M* EDTA containing a range of concentrations of NaCl, or with 1% 2-mercapto-ethanol. By the same criterion, about 25% of the cyclase was solubilized by 2% solutions of the chaotropic agents sodium azide and sodium thiocyanate, but in both cases the cyclase was partially inactivated and the tubulin was partially solubilized. The cyclase was inactivated by low concentrations (1% or less) of the anionic detergents sodium dodecyl sulfate, lauryldimethylamine N-oxide, and sodium lauryl sarcosinate (Fleischman *et al.*, 1980).

We have been exploring the methods employed to dissolve the axonemes of motile cilia and flagella, such as prolonged dialysis against EDTA at low ionic strength (Linck, 1976). After rod axonemes have been subjected to this treatment, about half of the tubulin and half of the guanylate cyclase activity remain in the supernatant after centrifugation for 1 hr at 12,000 *g*. However, all the cyclase still elutes from Sepharose 4B columns in the void volume, and after centrifugation in sucrose density gradients the cyclase is concentrated in fractions that are turbid and contain basal body–centriole pairs.

It should be stressed that insolubility in solutions of nonionic detergents does not constitute rigorous proof that a protein is not membrane-bound. We will be confident that guanylate cyclase is really attached to the axoneme only after it has been possible to employ more direct localization techniques such as the use of ferritin-labeled antibodies.

The question of where guanylate cyclase is located within the rod remains unanswered. The presence of cyclase activity in density gradient fractions containing basal bodies (after EDTA dialysis) suggests localization in the basal body. This possibility seems attractive for several reasons. The basal body lies near the mitochondria, where GTP is probably formed. The cyclase reaction has a very small standard free energy change, and it is thought that the reaction is "pulled" by the hydrolysis of pyrophosphate, the other reaction product. There is said to be much more pyrophosphatase in the inner segment than in the outer segment (Lowry, cited by Berger *et al.,* 1980). Finally, the cyclase apparently must supply cyclic GMP to the inner segment as well as to the outer segment (Section I), a task suitable for a basal body enzyme.

Calmodulin has been found in the ciliary basal bodies of a number of organisms, including *Tetrahymena* (Means and Dedman, 1980). *Tetrahymena* also contains a particulate guanylate cyclase that is stimulated by Ca^{2+} in the presence of an endogenous protein (Nakazawa *et al.,* 1979). There is evidence of calmodulin-like activity in vertebrate photoreceptors (Liu *et al.,* 1979), as well as an effect of low Ca^{2+} concentrations on retinal cyclic-GMP levels (Cohen *et al.,* 1978; Killbride, 1980). The presence of guanylate cyclase and calmodulin in rod basal bodies is therefore an appealing idea.

A strong argument against the presence of cyclase in rod basal bodies, however, is the observation that it seems to be located almost entirely in the outer segment (Berger *et al.,* 1980).

The failure of part of the cyclase to sediment through sucrose gradients and the elution of all the cyclase from Sepharose 4B columns in the void volume can be reconciled most easily if the enzyme is attached to a structure such as a microtubule which has a large surface/mass ratio as well as a large Stokes' radius. This argument suggests that cyclase may be attached to the microtubules extending into the outer segment. Most early observers reported that axoneme microtubules extended only a short distance into the outer segment (Cohen, 1972). However, more recent workers have observed microtubules near the distal ends of the outer segments of cat (Steinberg and Wood, 1975) and monkey (Young, 1971) rods and cones. After leaving the cilium, the axoneme separates into groups of one or two microtubules which follow the channels defined by the clefts in the disks. Thus microtubule-bound guanylate cyclase might be found throughout the outer segments.

III. OTHER ENZYMATIC ACTIVITIES IN THE AXONEME

Isolated axonemes contain traces of phophodiesterase and GTPase activities, probably as contaminants, and little if any pyrophosphatase. They do, however, display adenylate cyclase activity (Fleischman and Denisevich, 1979). The ratio of adenylate cyclase activity to guanylate cyclase activity is comparable to that found in gradient-purified outer segments (1:10–100) (Pannbacker, 1973; Miki *et al.*, 1974). It has been argued that the adenylate cyclase found in outer segments results from contamination with other retinal cells (Hendricks *et al.*, 1973; Manthorpe and McConnell, 1974), but the simultaneous enrichment in adenylate and guanylate cyclase activities in axonemes leads us to suggest that both cyclic nucleotides may be formed by the same enzyme. The cyclic-AMP concentration in microdissected rabbit outer segments is reported to be 5.5 μmoles kg^{-1} dry wt, whereas that in the inner segment is two- to fourfold higher (Orr *et al.*, 1976). Thus the ratio of cyclic AMP to cyclic GMP found in the outer segments is of the same order of magnitude as the reported ratio of adenylate cyclase to guanylate cyclase.

IV. GUANYLATE CYCLASE KINETICS

The kinetic behavior of axoneme guanylate cyclase (Fleischman and Denisevich, 1979) appears to be similar to that of cyclase in intact outer segments (Krishnan *et al.*, 1978). The substrate is the GTP complex of Mg^{2+} or Mn^{2+}, and additional free Mg^{2+} or Mn^{2+} is required. Plots of the reciprocal of the velocity against $[GTP \cdot Mg^{2-}]^{-1}$ at varying concentrations of free Mg^{2+} are linear and intersect at the $1/V$ axis. Plots of $1/V$ against $[Mg^{2-}]^{-1}$ at varying $[GTP \cdot Mg^{2-}]$ are linear and do not intersect at the $1/V$ axis. The enzyme behaves similarly in the presence of Mn^{2+} and $GTP \cdot Mn^{2-}$. Such behavior is consistent with a mechanism in which the enzyme must bind the free divalent cation before binding the substrate:

$$\mathrm{E + Mg^{2+} \overset{K_1}{\rightleftharpoons} EMg}$$

$$\mathrm{EMg + GTP \cdot Mg \overset{K_2}{\rightleftharpoons} EMgGTP \cdot Mg}$$

$$\mathrm{EMgGTP \cdot Mg \xrightarrow{k_P} E + cyclic\ GMP + PP_i + 2Mg^{2+}}$$

Appropriate analyses of the slopes and intercepts of the double reciprocal plots indicate that, for Mg^{2+}, $K_1 = 1 \times 10^{-3}$ *M*, $K_2 = 8 \times 10^{-4}$ *M*, and for Mn^{2+}, $K_1 = 1 \times 10^{-4}$ *M*, $K_2 = 1.4 \times 10^{-4}$ *M*. These dissociation constants suggest that the enzyme is less than half-saturated with substrate at physiological GTP concentrations (400 μM in dark-adapted frog rods, DeAzeredo and Lust, 1978). The rate of cyclic-GMP formation therefore could be influenced by light-induced changes in GTP concentration.

The maximum velocity is the same in the presence of Mg^{2+} or Mn^{2+}. The pH optimum is 7.0–9.0. Monovalent cations are stimulatory when the divalent cation concentration is less than optimal. High concentrations of Ca^{2+} (about 1 mM) are inhibitory; the binding of two Ca^{2+} ions to each enzyme molecule appears to be required for inhibition. Since it seems unlikely that such high Ca^{2+} concentrations are ever attained under physiological conditions, it is probable that the Ca^{2+} effect is simply a manifestation of the general inhibition of the cyclase by high concentrations of divalent cations (Krishnan *et al.*, 1978).

In a single experiment, we found that EGTA enhanced guanylate cyclase activity two- to threefold. In view of the unresolved question of how Ca^{2+} and cyclic GMP interact in photoreceptors, it is of the greatest importance to examine the effects of very low Ca^{2+} concentrations on guanylate cyclase activity using EGTA to buffer the Ca^{2+} concentration.

REFERENCES

Berger, S. J., DeVries, G. W., Carter, J. G., Schulz, D. W., Passonneau, P. N., Lowry, O. H., and Ferrendelli, J. A. (1980). The distribution of the components of the cyclic GMP cycle in the retina. *J. Biol. Chem.* **255,** 3128–3133.

Bitensky, M. W., Gorman, R. E., and Miller, W. H. (1971). Adenyl cyclase as a link between photon capture and changes in membrane permeability of frog photoreceptors. *Proc. Natl. Acad. Sci. U.S.A.* **68,** 561–562.

Cohen, A. I. (1972). Rods and cones. *In* "Handbook of Sensory Physiology" (M. G. F. Fuortes, ed.), vol. 7, Part 2, pp. 63–110. Springer-Verlag, Berlin and New York.

Cohen, A. I., Hall, I. A., and Ferrendelli, J. A. (1978). Calcium and nucleotide regulation in incubated mouse retinas. *J. Gen. Physiol.* **71,** 595–612.

DeAzeredo, F. A. M., and Lust, W. D. (1978). Guanine nucleotide concentrations *in vivo* in outer segments of dark and light adapted frog retina. *Biochem. Biophys. Res. Commun.* **85,** 293–300.

Doughty, M. J., and Diehn, B. (1979). Photosensory transduction in the flagellated alga *Euglena gracilis*. I. Action of divalent cations, Ca^{2+} antagonists and Ca^{2+} ionophore on motility and photobehavior. *Biochim. Biophys. Acta* **588,** 148–168.

Eakin, R. M. (1965). Evolution of photoreceptors. *Cold Spring Harbor Symp. Quant. Biol.* **30,** 363–370.

Eckert, R., and Brehm, P. (1979). Ionic mechanisms of excitation in *Paramecium*. *Annu. Rev. Biophys. Bioeng.* **8,** 353–383.

Fleischman, D., and Denisevich, M. (1979). Guanylate cyclase of isolated bovine retinal rod axonemes. *Biochemistry* **18,** 5060–5066.

Fleischman, D., Denisevich, M., Raveed, D., and Pannbacker, R. G. (1980). Association of guanylate cyclase with the axoneme of retinal rods. *Biochim. Biophys. Acta* **630,** 176–186.

Godchaux, W., III. and Zimmerman, W. F. (1979). Soluble proteins of intact bovine rod cell outer segments. *Exp. Eye Res.* **28,** 483–500.

Goldberg, N. D., and Haddox, M. K. (1977). Cyclic GMP metabolism and involvement in biological regulation. *Annu. Rev. Biochem.* **46,** 823–896.

Goridis, C., Virmaux, N., Urban, P. F., and Mandel, P. (1973). Guanyl cyclase in a mammalian photoreceptor. *FEBS Lett.* **30,** 163–166.

Gray, E. G., and Pumphrey, R. J. (1958). Ultrastructure of the insect ear. *Nature (London)* **181,** 618.

Hendriks, T., DePont, J. J. H. H. M., Daemen, F. J. M., and Bonting, S. L. (1973). Biochemical aspects of the visual process. XXIV. Adenylate cyclase and rod photoreceptor membranes: A critical appraisal. *Biochim. Biophys. Acta* **330,** 156–166.

Killbride, P. (1980). Calcium effects on frog retinal cyclic guanosine monophosphate levels and their light-initiated rate of decay. *J. Gen. Physiol.* **75,** 457–465.

Krishnan, N., Fletcher, R. T., Chader, G. J., and Krishna, G. (1978). Characterization of guanylate cyclase of rod outer segments of the bovine retina. *Biochim. Biophys. Acta* **523,** 506–515.

Kühn, H. (1978). Light-regulated binding of rhodopsin kinase and other proteins to cattle photoreceptor membranes. *Biochemistry* **17,** 4389–4395.

Linck, R. W. (1976). Flagellar doublet microtubules: Fractionation of minor components and α-tubulin from specific regions of the A-tubule. *J. Cell Sci.* **20,** 405–439.

Liu, Y. P., Krishna, G., Aguirre, G., and Chader, G. J. (1979). Involvement of cyclic GMP phosphodiesterase activator in an hereditary retinal degeneration. *Nature (London)* **280,** 62–64.

Manthorpe, M., and McConnell, D. G. (1974). Adenylate cyclase in vertebrate retina: Relationship to specific fractions and to rhodopsin. *J. Biol. Chem.* **249,** 4608–4613.

Means, A. R., and Dedman, J. R. (1980). Calmodulin—An intracellular calcium receptor. *Nature (London)* **285,** 73–77.

Miki, N., Keirns, J. J., Marcus, F. R., and Bitensky, M. W. (1974). Light regulation of adenosine 3′,5′-cyclic monophosphate levels in vertebrate photoreceptors. *Exp. Eye Res.* **18,** 281–297.

Mohri, H. (1976). The function of tubulin in motile systems. *Biochim. Biophys. Acta* **456,** 85–127.

Nakazawa, K., Shimonaka, H., Nagao, S., Kudo, S., and Nozawa, Y. (1979). Magnesium-sensitive guanylate cyclase and its endogeneous activating factor in *Tetrahymena pyriformis*. *J. Biochem. (Tokyo)* **86,** 321–324.

Orr, H. T., Lowry, O. H., Cohen, A. I., and Ferrendelli, J. A. (1976). Distribution of 3′:5′-cyclic AMP and 3′:5′-cyclic GMP in rabbit retina *in vivo:* Selective effects of dark and light adaptation and ischemia. *Proc. Natl. Acad. Sci. U.S.A.* **73,** 4442–4445.

Pannbacker, R. G. (1973). Control of guanylate cyclase activity in rod outer segments. *Science* **182,** 1138–1140.

Pannbacker, R. G., and Fleischman, D. E. (1972). Cyclic nucleotide synthesis in bovine rod outer segments. *Abstr. 16th Annu. Meet. Biophys. Soc.* SAPM H-11.

Raveed, D., and Fleischman, D. (1975). The basal apparatus of bovine retinal rods. *Proc.—Annu. Meet., Electron Microsc. Soc. Am.* **33,** 478–479.

Sano, M. (1976). Subcellular localizations of guanylate cyclase and 3′,5′-cyclic nucleotide phosphodiesterase in sea urchin sperm. *Biochim. Biophys. Acta* **428,** 525–531.

Steinberg, R. H., and Wood, I. (1975). Clefts and microtubules of photoreceptor outer segments in the retina of the domestic cat. *J. Ultrastruct. Res.* **51,** 397–403.

Stephens, R. E. (1970). Isolation of nexin—The linkage protein responsible for the maintenance of the nine-fold configuration of flagellar axonemes. *Biol. Bull. (Woods Hole, Mass.)* **139,** 438.

Troyer, E. W., Hall, I. A., and Ferrendelli, J. A. (1978). Guanylate cyclases in CNS: Enzymatic characteristics of soluble and particulate enzymes from mouse cerebellum and retina. *J. Neurochem.* **31,** 825–833.

Virmaux, N., Nullans, G., and Goridis, C. (1976). Guanylate cyclase in vertebrate retina: Evidence for specific association with rod outer segments. *J. Neurochem.* **26,** 233–235.

Wiederhold, M. L. (1976). Mechanosensory transduction in "sensory" and "motile" cilia. *Annu. Rev. Biophys. Bioeng.* **5,** 39–62.

Young, R. Y. (1971). The renewal of rod and cone outer segments in the rhesus monkey. *J. Cell Biol.* **49,** 303–318.

Zimmerman, W. F., Daemen, F. J. M., and Bonting, S. L. (1976). Distribution of enzyme activities in subcellular fractions of bovine retina. *J. Biol. Chem.* **251,** 4700–4705.

Chapter 7

Light Control of Cyclic-Nucleotide Concentration in the Retina

THOMAS G. EBREY,[1] PAUL KILBRIDE,[2] JAMES B. HURLEY,[3] ROGER CALHOON,[1] AND MOTOYUKI TSUDA[4]

I. INTRODUCTION

The role of cyclic guanosine 3′,5′-monophosphate (cyclic GMP) in visual physiology is not known. There are a number of reasons to think that it plays an important role in vision. First, there is a relatively high concentration of GMP in the retina, most of it being in the photoreceptor cells (Farber and Lolley, 1974; Orr *et al.*, 1976; Berger *et al.*, 1980). The concentration of cyclic GMP in the photoreceptor cells under normal conditions is about 50 μM. Second, light controls the concentration of cyclic GMP (Goridis *et al.*, 1974; Ferrendelli and Cohen, 1976; Kilbride and Ebrey, 1979; Woodruff and Bownds, 1979). Third, a number of enzymes have been isolated that are involved in the control of cyclic GMP (Pober and Bitensky, 1979; Hurley *et al.*, 1980). Finally, a number of drugs that affect cyclic-nucleotide metabolism have an effect on the electrical activity of the photoreceptor cells (Ebrey and Hood, 1973; Lipton *et al.*, 1977;

[1]Department of Physiology and Biophysics, University of Illinois, Urbana, Illinois.
[2]Pharmacology Department, Washington University Medical School, St. Louis, Missouri.
[3]Department of Structural Biology, Sherman Fairchild Center, Stanford University School of Medicine, Stanford, California.
[4]Department of Physics, Sapporo Medical College, Sapporo, Japan.

ISBN 0-12-153315-8

Nicol and Miller, 1978). None of these points directly to the exact function of cyclic GMP in the photoreceptors or what its physiological role might be—excitation, adaptation, the control of metabolism, providing a cellular energy supply, or involvement in some other physiological process.

We have investigated three aspects of light control of cyclic-nucleotide concentrations in the retina. First, how fast does the cyclic-nucleotide concentration change after light activation, and are these changes affected by calcium—another important factor that can control the physiological activity of photoreceptors. Second, what components might be involved at the enzymatic level in the regulation of cyclic nucleotide concentrations in the retina? Third, are the kinds of enzymes that control cyclic-nucleotide concentrations in vertebrate retinas also found in invertebrate retinas?

II. THE EFFECT OF LIGHT ON CYCLIC-GMP CONCENTRATION IN FROG RETINAS

How fast does light cause the cyclic-GMP concentration to fall in the frog retina (*Rana catesbeiana*)? To measure this, we have worked out a technique of rapidly freezing the retina soon after the initiation of illumination (Kilbride and Ebrey, 1979). The retina was carefully dissected away from the pigment epithelium in oxygenated Ringer's solution and placed in the bottom of a 30-ml Corex test tube. At appropriate intervals after the initiation of a steady light, an aluminum plunger previously immersed in liquid nitrogen was dropped on top of the retina (Fig. 1). The plunger lowered the temperature of the retina by about 50°C within 100 msec after contact, which presumably stopped all the light-initiated biochemical reactions in the retina. To prevent any further cyclic-GMP metabolism, perchloric acid and methanol were added, and the frozen retina, tube, and hammer assembly were slowly warmed in a dry ice–ethanol bath. Care was taken that the Ringer's solution was oxygenated during the dissection process; the Ringer's solution was buffered and also contained bicarbonate which in some cases seemed to improve the health of the retina. Electroretinograms of retinas handled exactly the same way as those used for the cyclic-GMP measurements seemed normal (Kilbride and Ebrey, 1979). Moreover, the cyclic-GMP concentration that we found, which is influenced by the health of the photoreceptor cells, was in the same range as those found by other workers (Goridis *et al.*, 1974; de Azeredo *et al.*, 1978). A flow diagram for the experimental procedure is shown in Fig. 2.

The results of the experiments on retinas in normal Ringer's solution are shown in Fig. 3. Three interesting points are: First, in the dark the concentration of cyclic GMP in the retina is about 170 pmoles cyclic GMP per retina. If we assume that about 50% of this is in the rod outer segments, then this gives a

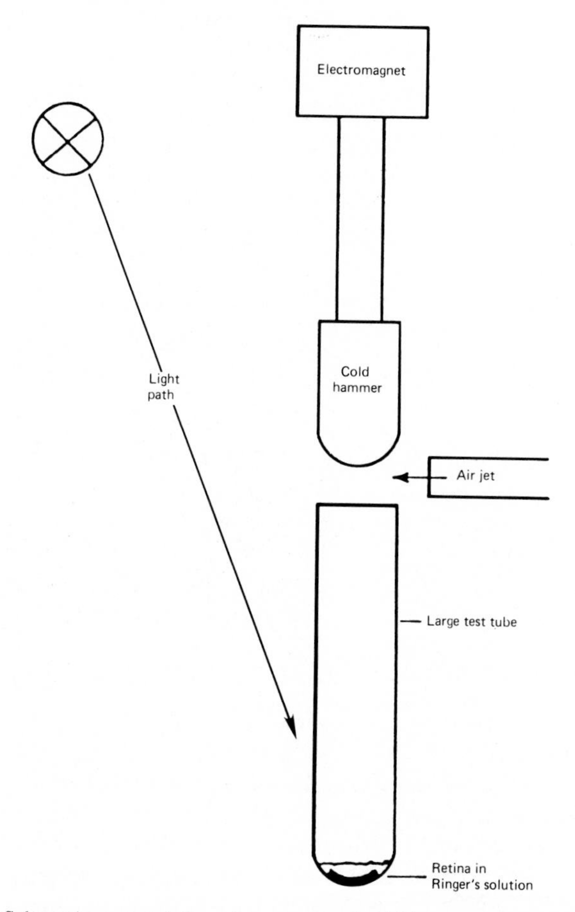

FIG. 1. Schematic representation of an automated cold hammer-dropping apparatus.

concentration in the rod outer segments of roughly 50 μM cyclic GMP. Second, when a light is turned on, even with the brightest light used, we can detect no change in the cyclic-GMP concentration in the retina until 3–5 sec after the light is turned on. There seems to be no change by 1 sec, the shortest time used in this set of experiments. The third point is that at lower light intensities, even if the light remains on, the cyclic-GMP concentration returns to its normal (dark) level. Thus after light activation of the enzymes involved in controlling cyclic-GMP levels, there is some further alteration in the enzyme activities.

These experiments could be explained by the following: (1) Either the cyclic GMP in the retina is divided into two or more pools that respond differently in the light so that the smaller pool changes its concentration quickly and the larger pool changes more slowly (or two pools of roughly equal size change their concen-

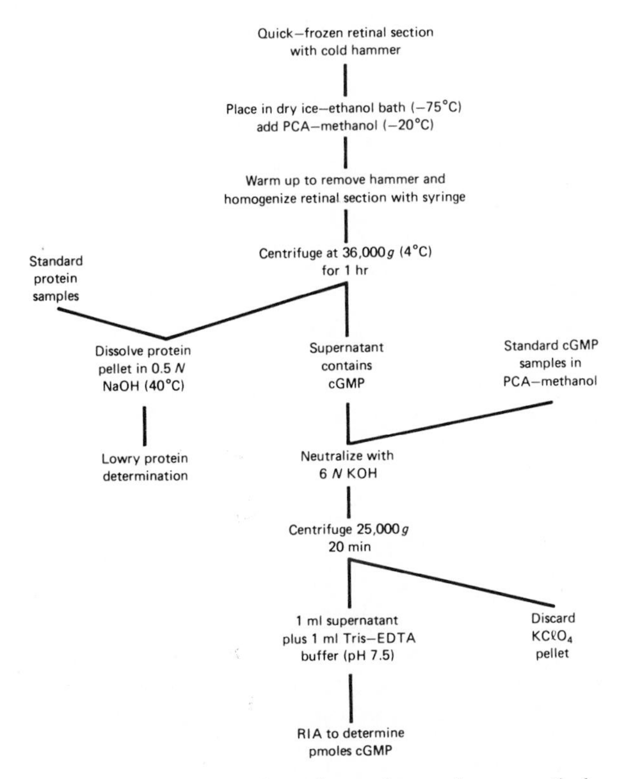

FIG. 2. Flow chart showing the experimental procedure used once a retinal section has been quick-frozen.

trations within the first second in opposite directions, one rising and the other falling rather quickly), or (2) if there is only one major pool of cyclic GMP in the photoreceptor cells, then the concentration of cyclic GMP in the photoreceptors does not change significantly 1 sec after excitation with a bright light. Although there is no evidence for multiple pools of cyclic GMP in the outer segments, we cannot exclude this possibility; and if a fraction of the cyclic GMP in the outer segments even as high as 10% were to change quickly, we might not be able to detect that change in our experiments.

The effect of changing calcium concentrations on both the level of cyclic nucleotides in the retina and the rate at which they are metabolized has been examined (Kilbride, 1980). Figure 4 illustrates the results of these experiments. When the calcium concentration is raised from 2 to 20 m*M*, there is only a slight decrease in the concentration of cyclic GMP in the retina—less than twofold (Fig. 4A). When the calcium concentration is lowered from 2 m*M* to approxi-

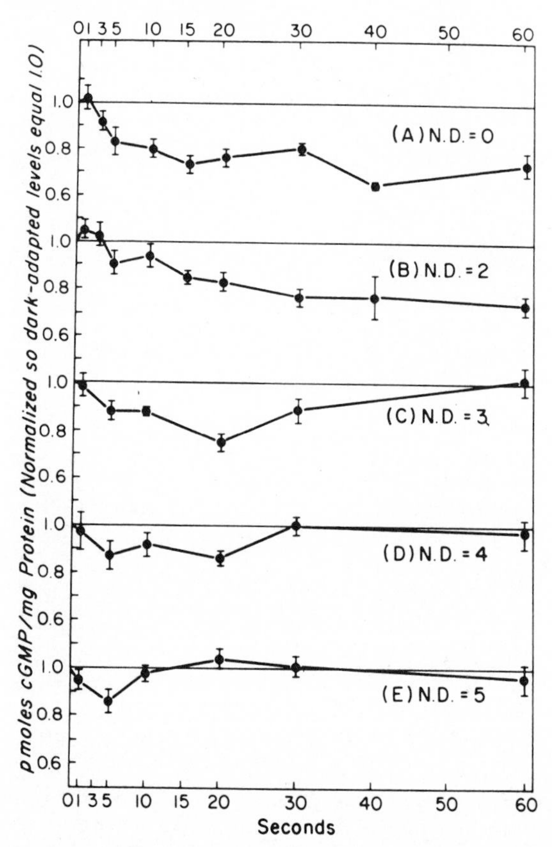

FIG. 3. Concentrations of cyclic GMP (picomoles of cyclic GMP per milligram of protein) as a function of light exposure (seconds) in retinal sections normalized so dark-adapted values were 1. The light intensity was attenuated by five different neutral density (N.D.) filters. The light without neutral density filter (N.D. = 0) corresponds to approximately 7×10^7 rhodopsin molecules bleached per second per rod outer segment. The points represent the mean ± SEM with an average of seven light-exposed retinal sections assayed per point (Kilbride and Ebrey, 1979).

mately 10^{-9} *M* with 3 m*M* EGTA, then the cyclic-GMP concentration rises from about 50 pmoles cyclic GMP mg^{-1} protein to 500 pmoles cyclic GMP mg^{-1} protein (Fig. 4B). Figure 5 shows that, after the cyclic-GMP concentration has been substantially raised in the low-calcium media, light initiates a much more rapid decrease in the GMP concentration so that within 1 sec there is a substantial decrease. Moreover, the maximal extent of the decrease in cyclic-GMP concentration is much larger, so that the final level of cyclic GMP in the retina after illumination with a bright light is about the same as that seen with normal calcium.

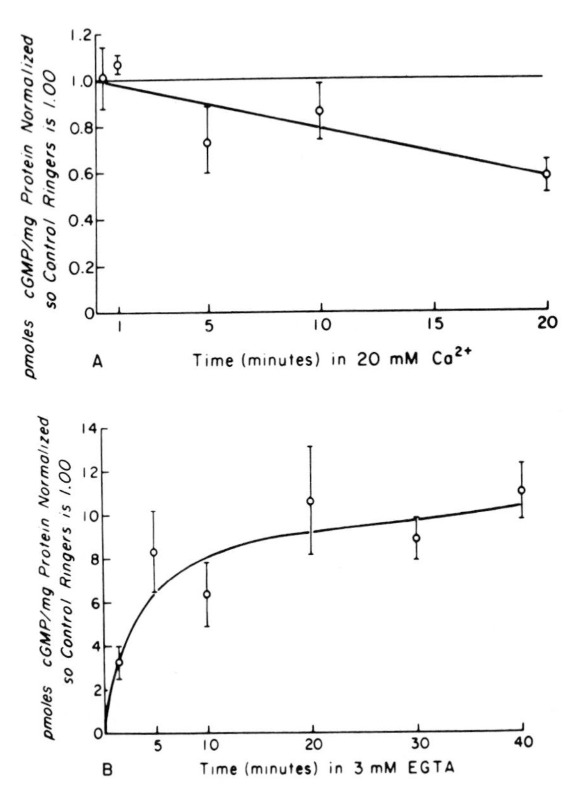

FIG. 4. Concentrations of cyclic GMP (picomoles of cyclic GMP per milligram of protein) as a function of dark-adapted retinal section incubation time (minutes) in a high-calcium (20 m*M* Ca^{2+}) (A) and a low-calcium (3 m*M* EGTA) (B) concentration of Ringer's solution normalized so the cyclic-GMP values of dark-adapted retinal sections incubated in normal Ringer's solution were 1. The points represent the mean ± SEM with an average of six (A) and four (B) retinal sections assayed per point (Kilbride, 1980).

III. LIGHT-ACTIVATABLE ENZYMES IN BOVINE ROD OUTER SEGMENTS

The second question we investigated concerned how light exerts its control on cyclic-nucleotide levels in rods. Two key enzymes have already been implicated in light control of cyclic-nucleotide concentrations—a light-activated phosphodiesterase and a light-activated GTPase. We were interested in what other components might be involved. Figure 6b shows the phosphodiesterase activity of a crude bovine rod outer segment extract as a function of protein concentration (Hurley *et al.*, 1981). As the concentration of the enzyme is increased, its

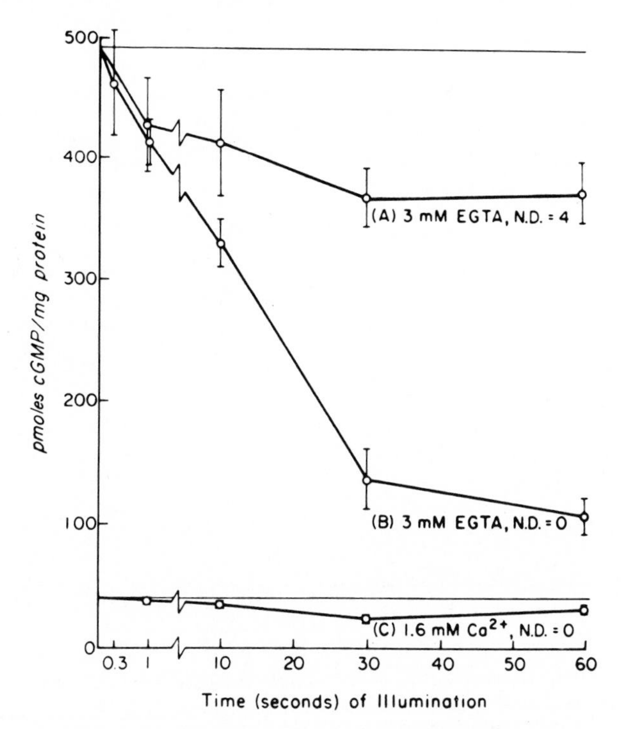

FIG. 5. Concentration of cyclic GMP (picomoles of cyclic GMP per milligram of protein) as a function of light exposure (seconds) in retinal sections normalized to the average cyclic-GMP value of dark-adapted, similarly calcium-treated retinal sections. The retinal sections were incubated in Ringer's solutions with two different calcium concentrations, and the light intensity was attenuated by two different neutral density filters. The points represent the mean ± SEM with an average in each case of four (A), eight (B), and seven (C) light-exposed retinal sections assayed per point (Kilbride, 1980).

specific activity decreases. This suggests that there is an inhibitor that is bound to the phosphodiesterase at high concentrations and inhibits its action. In order to show this we attempted to purify the phosphodiesterase and the GTPase activity from bovine rods. Figure 6a shows that the specific activity of phosphodiesterase, which is purified by DEAE-cellulose and Sephadex G-200 chromatography, is relatively concentration-independent, suggesting that an inhibitor has been at least partially removed. There is a fraction from the DEAE-cellulose column which, when added back to purified or partially purified phosphodiesterase, inhibits its activity. This inhibitor has a number of unusual properties such as stability to boiling at certain pH values (Hurley *et al.*, 1980). Bovine phosphodiesterase can be purified almost to homogeneity; and by sodium dodecyl sulfate (SDS) gel chromatography it consists of two peaks with molecular

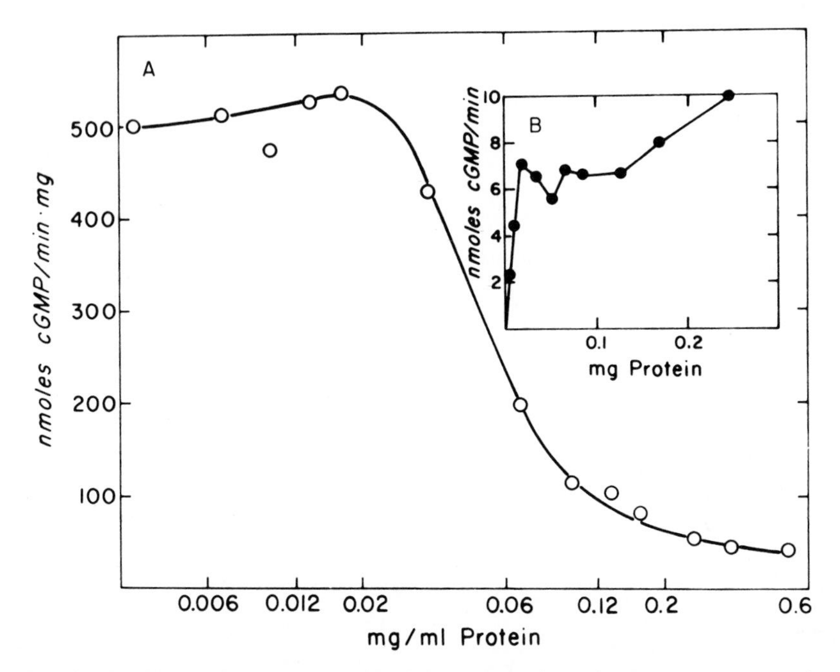

FIG. 6. Specific activity (A) and activity of the crude bovine rod outer segment extract phosphodiesterase (B) as a function of protein concentration.

weights of about 84,000 and 88,000. On a DEAE-cellulose column, the phosphodiesterase and guanosine triphosphatase (GTPase) run together, but they can be separated by gel chromatography. Although good yields of pure phosphodiesterase can be obtained with Sephadex G-200, the GTPase activity is not recovered. The GTPase can be separated from the phosphodiesterase on a Sepharose 6B column with the addition of 1 m*M* GMP to the buffers; the GMP seems to give added stability to the otherwise unstable GTPase. When purified, the GTPase activity appears as a doublet with molecular weights of 36,000 and 39,000. The maximal turnover number of the phosphodiesterase is quite high, and its maximum activity (using controlled trypsin digestion) is about 25 μmoles cyclic GMP min^{-1} mg^{-1}. In contrast, the activity of the GTPase is very low and has a turnover number of 0.05 sec^{-1}.

We wished to determine which of these enzymes were necessary for light activation of the phosphodiesterase. Reconstitution experiments were performed (Hurley, 1980) using the following components—purified rhodopsin in egg phosphatidylcholine vesicles, GTPase, phosphodiesterase, and the crude inhibitor fraction that had no GTPase or phosphodiesterase activity. Activation was achieved by adding bleached rhodopsin and the required cofactor, guanosine 5′-triphosphate (GTP). Table I shows that the purified rhodopsin plus phos-

TABLE I
RECOMBINATION OF PURIFIED ROD OUTER SEGMENT CYCLIC-GMP PHOSPHODIESTERASE WITH ITS REGULATORS[a]

Recombination mixture	Phosphodiesterase activity (nmoles cyclic GMP min^{-1})	
	Without GTP	With 100 μM GTP
Bleached rhodopsin + PDE	61	72
Bleached rhodopsin + PDE + inhibitor	35	31
Bleached rhodopsin + GTPase	14	19
Bleached rhodopsin + GTPase + inhibitor	2.6	3.8
Bleached rhodopsin + GTPase + PDE	68 (54)	74 (55)
Bleached rhodopsin + GTPase + PDE + inhibitor	29 (26)	59 (55)

[a] Where indicated, ~5 μg purified phosphodiesterase (PDE), ~6 μg purified GTPase (27 pmoles GTP min^{-1}), and ~30 μg DEAE-cellulose inhibitor were added. The numbers in parentheses are activities due to added PDE corrected for residual PDE activities in the isolated GTPase.

phodiesterase did not activate the phosphodiesterase. Addition of the GTPase still did not lead to significant activation of the phosphodiesterase by the addition of bleached rhodopsin and GTP. Only when the inhibitor fraction was added to the GTPase and the phosphodiesterase could bleached rhodopsin plus GTP cause activation of the phosphodiesterase. Thus we believe that the inhibitor is a required component in light regulation of the phosphodiesterase.

IV. LIGHT ACTIVATION IN OCTOPUS OUTER SEGMENTS

A third area of interest is the existence of light-activated enzymes in invertebrate retinas. We have made two approaches to the problem. First, we looked for light-activated GTPase activity in octopus photoreceptor cells (Calhoon *et al.*, 1980). Photoreceptor membranes consisting primarily of outer segments were prepared as previously described (Tsuda, 1979), and a homogenate was made of the crude membranes. We looked for both ATPase and GTPase activities and, although there was considerable ATPase activity, only the GTPase activity could be activated with light. We found that the maximal activation took place at ca. 1 μM GTP and that raising the salt concentration to a value close to physiological levels for octopus photoreceptors (roughly 400 mM NaCl or KCl) usually resulted in greater activation by light. The pH optimum was broad and centered at about 8. The amount of light activation we observed was variable. Occasionally it was as low as 1.5; the largest activation we found was ca. fivefold. This light activation does not appear to be dependent on enzyme concentration. When the crude photoreceptor membrane preparation was washed with a low-salt buffer

(10 m*M* Tris, pH 8, 2 m*M* dithiothreitol, 25 μM phenylmethylsulfonyl fluoride) or buffer containing 5 m*M* EDTA, most of the GTPase activity remained with the membrane fraction. These conditions allowed extraction of the light-activated GTPase of bovine photoreceptor membranes.

We have also attempted to find light-activated phosphodiesterase activity in octopus photoreceptor membranes but so far have not been successful. We do not know whether this is due to the absence of this enzyme, to our using an improper set of assay conditions, or to the condition of our samples. However, we did test for whether or not octopus rhodopsin could activate bovine phosphodiesterase. In this case partially purified octopus photoreceptor membranes were used. Bovine phosphodiesterase was prepared as previously described (Hurley *et al.*, 1980). One micromole of GTP was added to the mixture which was then irradiated. We found fivefold activation with octopus photoreceptor membranes; a similar activation was found when bovine rhodopsin was used (Ebrey *et al.*, 1980). This suggests that not only is there a fairly close homology between octopus and bovine rhodopsin, but since octopus rhodopsin can activate the bovine phosphodiesterase system, a light-activated phosphodiesterase may yet be found in the octopus system.

V. SUMMARY

The previous results suggest several tentative conclusions. First, the complex regulatory mechanism by which bleached rhodopsin controls phosphodiesterase activity reinforces the notion that cyclic GMP plays an important role in photoreceptor physiology. Moreover, the presence of a light-activated GTPase in octopus photoreceptor membranes and the ability of octopus photoreceptor membranes to activate bovine rod phosphodiesterase suggest that there may be light regulation of cyclic nucleotide levels in the octopus photoreceptor.

Finally, several observations suggest that cyclic GMP probably does not act as the excitatory internal transmitter linking bleached rhodopsin with the initial closing of the sodium permeability sites which signals visual excitation in vertebrate photoreceptors. First, noise problems of any internal transmitter signal activated by a photon and causing a reduction in the concentration of the internal transmitter, rather than an increase in its concentration, make cyclic GMP an unlikely candidate for the transmitter. Second, it appears that under physiological conditions the change in cyclic-GMP concentration is too slow to be involved in visual excitation (Fig. 3). Moreover, Fig. 4A shows that increasing the Ca^{2+} concentration has no significant effect on cyclic-GMP concentration, especially within a minute after the increase, even though such elevated Ca^{2+} levels turn off the sodium permeability sites within seconds. Finally, Ostroy and co-workers (Meyertholen *et al.*, 1980) have found recently that the effect of removing bicarbonate from toad ringers is similar to the effect of light in that it reduces cyclic-GMP levels in the toad rods by about 65%; however, it does this without greatly affecting the receptor potential. Thus, mimicking the effect of light on the cyclic-GMP levels in the rod does not mimic the effect of light on the sodium permeability sites. Taken together, these results suggest that cyclic GMP is not the excitatory internal transmitter.

REFERENCES

Berger, S. J., DeVries, G. W., Carter, J. G., Schulz, D. W., Passonneau, P. N., Lowry, O. H., and Ferrendelli, J. A. (1980). The distribution of the components of the cyclic GMP cycle in retina. *J. Biol. Chem.* **255,** 3128–3133.

Calhoon, R., Tsuda, M., and Ebrey, T. G. (1980). A light-activated GTPase from octopus photoreceptors. *Biochem. Biophys. Res. Commun.* **94,** 1452–1457.

de Azeredo, F. A. M., Lust, W. D., and Passonneau, J. V. (1978). Guanine nucleotide concentratione *in vivo* in outer segments of dark and light adapted frog retina. *Biochem. Biophys. Res. Commun.* **85,** 293–300.

Ebrey, T. G., and Hood, D. C. (1973). The effects of cyclic nucleotide phosphodiesterase inhibitors on the frog rod receptor potential. *In* "Biochemistry and Physiology of Visual Pigments" (H. Langer, ed.), pp. 341–350. Springer-Verlag, Berlin and New York.

Ebrey, T. G., Tsuda, M., Sassenrath, G., West, J. L., and Waddell, W. H. (1980). Light activation of bovine rod phosphodiesterase by non-physiological visual pigments. *FEBS Lett.* **116,** 217–219.

Farber, D. B., and Lolley, R. N. (1974). Cyclic guanosine monophosphate: Elevation in degenerating photoreceptor cells of the C3H mouse retina. *Science* **186,** 449–451.

Ferrendelli, J. A., and Cohen, A. I. (1976). The effects of light and dark adaptation on the levels of cyclic nucleotides in retinas of mice heterozygous for a gene for photoreceptor dystrophy. *Biochem. Biophys. Res. Commun.* **73,** 421–427.

Goridis, C., Virmaux, N., Cailla, H. L., and Delaage, M. A. (1974). Rapid light-induced changes of retinal cyclic GMP levels. *FEBS Lett.* **49,** 167–169.

Hurley, J. B. (1980). Isolation and recombination of bovine rod outer segment cGMP phosphodiesterase and its regulators. *Biochem. Biophys. Res. Commun.* **92,** 505–510.

Hurley, J. B., Barry, B., and Ebrey, T. G. (1981). Isolation of an inhibitory protein for the cGMP phosophodiesterase of bovine rod outer segment. *Biochim. Biophys. Acta* **675** (in press).

Kilbride, P. (1980). Calcium effects on frog retinal cyclic guanosine 3′,5′-monophosphate levels and their light initiated rate of decay. *J. Gen. Physiol.* **75,** 457–465.

Kilbride, P., and Ebrey, T. G. (1979). Light-initiated changes of cyclic guanosine monophosphate levels in the frog retina measured with quick-freezing techniques. *J. Gen. Physiol.* **74,** 415–426.

Lipton, S. A., Rasmussen, H., and Dowling, J. E. (1977). Electrical and adaptive properties in *Bufo marinus*. II. Effects of cyclic nucleotides and prostaglandins. *J. Gen. Physiol.* **70,** 771–791.

Meyertholen, I., Wilson, M., and Ostroy, S. (1980). Removing bicarbonate/CO_2 reduces the cGMP concentration of the vertebrate photoreceptor to the levels normally observed on illumination. *Biochem. Biophys. Res. Commun.* **96,** 785–792.

Nicol, G. D., and Miller, W. H. (1978). Cyclic GMP injected into retinal rod outer segments increases latency and amplitude of response to illumination. *Proc. Natl. Acad. Sci. U.S.A.* **75,** 5217–5220.

Orr, H. T., Lowry, O. H., Cohen, A. I., and Ferrendelli, J. A. (1976). Distribution of 3′:5′-cyclic

AMP and 3′:5′-cyclic GMP in rabbit retina *in vivo:* Selective effects of dark and light adaptation and ischemia. *Proc. Natl. Acad. Sci. USA.* **73,** 4442–4445.

Pober, J. S., and Bitensky, M. W. (1979) Light-regulated enzymes of vertebrate retinal rods. *Adv. Cyclic Nucleotide Res.* **11,** 265–708.

Tsuda, M. (1979). Transient spectra of intermediates in the photolytic sequence of octopus rhodopsin. *Biochim. Biophys. Acta* **545,** 537–546.

Woodruff, M. L., and Bownds, M. D. (1979). Amplitude, kinetics, and reversibility of a light-induced decrease in guanosine 3′,5′-cyclic monophosphate in frog photoreceptor membranes. *J. Gen. Physiol.* **73,** 629–653.

Chapter 8

Cyclic-GMP Phosphodiesterase and Calmodulin in Early-Onset Inherited Retinal Degenerations

G. J. CHADER,[1] Y. P. LIU,[1] R. T. FLETCHER,[1] G. AGUIRRE,[2] R. SANTOS-ANDERSON,[3] AND M. T'SO[3]

I. PHOSPHODIESTERASE: GENERAL CHARACTERISTICS

Cyclic 3′:5′-nucleotide phosphodiesterase (EC 3.1.4.17) catalyzes the hydrolysis of cyclic adenosine 3′,5′-monophosphate (cyclic AMP) or cyclic

[1]Laboratory of Vision Research, National Eye Institute, National Institutes of Health, U.S. Department of Health and Human Services, Bethesda, Maryland.
[2]Section Ophthalmology, School of Veterinary Medicine, University of Pennsylvania, Philadelphia, Pennsylvania.
[3]Department of Ophthalmology, University of Illinois Medical School, Chicago, Illinois.

ISBN 0-12-153315-8

guanosine 3′,5′-monophosphate (cyclic GMP) to the corresponding 5′-nucleotide (5′-AMP or 5′-GMP). The phosphodiesterase (PDE) enzyme is the only enzyme that can break down or metabolize cyclic nucleotides and therefore, along with the appropriate cyclase enzymes, is of pivotal importance in controlling the concentration of intracellular messengers. In the past, the cyclase enzymes were studied most extensively, particularly with respect to their response to hormonal stimulus, and were thought to play a predominant role in controlling cyclic-nucleotide concentration. More recently, PDE has also been shown to play an active role in the regulation of cyclic-nucleotide concentration. This is particularly true in the photoreceptor cell where it may very well have a central role in the visual process, as well as other yet undefined roles in photoreceptor cell function.

PDE was first described by Butcher and Sutherland (1962) and for some time was essentially thought to be a simple enzyme of high activity, little affected by cell control mechanisms. Also, the apparent K_m seemed to be several orders of magnitude higher than the tissue concentration of cyclic nucleotide, and thus the physiological role of the enzyme was suspect. It is now apparent that multiple forms of the enzyme are present in most tissues, with both low and high affinities for cyclic nucleotides. These have been shown to be separable into individual molecular forms by several techniques (Thompson and Appleman, 1971). In neural tissue, in particular, forms of different cyclic-nucleotide specificity (cyclic AMP versus cyclic GMP), molecular weight, kinetic properties, and subcellular localization have been described (Strada and Thompson, 1978). A specific activator has also been described. A decade ago, Kakiuichi and Yamazaki (1970) demonstrated the presence of Ca^{2+}-activated PDE activity in neural tissue. Earlier, Cheung (1967) had reported a protein activator of PDE called calmodulin. It is now clear that Ca^{2+} and calmodulin act in concert to activate "Ca^{2+}-dependent PDE." As seen in Fig. 1, calmodulin first binds Ca^{2+} and then combines with PDE to form an "active" ternary complex. All PDE types are not Ca^{2+}-dependent; several tissues exhibit Ca^{2+}-independent PDE activity.

MODEL FOR CALMODULIN ACTION

1. ACTIVATOR (inactive) + Ca^{2+} $\rightleftharpoons$ ACTIVATOR* — Ca^{2+} (active)

2. ACTIVATOR* — Ca^{2+} + PDE (low activity) $\rightleftharpoons$ ACTIVATOR* — Ca^{2+} | PDE* (high activity)

FIG. 1. Interaction of calcium, calmodulin, and PDE to produce activated PDE.

II. RETINAL PHOSPHODIESTERASE

A. Early Studies

Bitensky *et al.* (1971) first reported the presence of a light-sensitive cyclic-nucleotide system in photoreceptor outer segments. Initially, it appeared to take the form of a light-inhibited adenylate cyclase in the isolated outer segments. It soon became apparent, however, that other factors had to be taken into consideration as well. We were unable to confirm a light effect on this cyclase in our laboratory but did see a two- to threefold increase in 5′-nucleotide after incubation (Chader *et al.*, 1973). This indicated a light effect on PDE, and we concluded that this "differential effect in light and dark on cAMP degradation . . . can easily be misinterpreted as an inhibitory effect of light on the activity of the adenylate cyclase enzyme." A similar conclusion was reached by Bitensky and co-workers at the same time (Miki *et al.*, 1973).

Similarly it soon became clear that the nucleotide of greatest interest in the rod outer segment (ROS) was cyclic GMP rather than cyclic AMP. Pannbacker *et al.* (1972) were the first to report this, as well as the extraordinarily high activity of PDE in isolated outer segments. Both Miki *et al.* (1973) and our laboratory (Chader *et al.*, 1974a) confirmed this independently, and it was demonstrated that cyclic-GMP PDE was activated by light in the presence of nucleotide triphosphate. Guanosine 5′-triphosphate (GTP) was found to function in this capacity as well as adenosine 5′-triphosphate (ATP), although the exact role of the nucleotide is not yet clear.

B. PDE Specificity and Compartmentalization

Although it is probable that PDE is distributed in all layers of the retina, a specific, high-activity, cyclic-GMP PDE is compartmentalized in the photoreceptor outer segments. This is seen in the distribution of PDE activity after a conventional subcellular fractionation procedure (Chader *et al.*, 1974b). Table I shows that not only is the PDE activity highest in the outer segment fraction but that the preferred substrate is cyclic GMP. PDE sediments with ROS on both sucrose and Ficoll gradients (Fig. 2). The enzyme can probably best be classified as a "membrane-associated" protein, however, since it is easily removed from disk membrane fragments when the disks are ruptured by hypotonic shock. This could perhaps afford a high degree of flexibility of the enzyme in associating with other membrane and/or soluble cellular constituents and controlling cyclic-GMP concentration.

Evidence from ontogenic studies on the chick retina also indicates compartmentalization of the major portion of retinal cyclic-GMP PDE with the outer

TABLE I
DISTRIBUTION OF CYCLIC-NUCLEOTIDE PHOSPHODIESTERASE ACTIVITY IN SUBCELLULAR FRACTIONS OF THE BOVINE RETINA[a]

	PDE activity ($nmoles\ mg^{-1}\ min^{-1}$)	
Fraction	Cyclic AMP	Cyclic GMP
Homogenate	0.5	1.6
Nuclear	0.5	1.5
Mitochondrial	0.4	1.9
ROS	2.9	16.1
Microsomal	0.4	0.7
Supernatant	0.9	3.9

[a] A fresh homogenate diluted to 10% in 0.32 *M* sucrose and 40 m*M* $MgCl_2$, pH 7.6, was centrifuged at 900 *g* for 10 min at 4°C. The resultant pellet after washing was termed the nuclear fraction. The supernatant fluid was centrifuged at 11,500 *g* for 20 min, and the mitochondrial fraction obtained. The latter supernatant fluid was centrifuged at 100,000 *g* for 60 min; the pellet obtained was termed the microsomal fraction, and the soluble fluid was termed the supernatant. Outer segments were prepared by discontinuous sucrose gradient centrifugation. PDE activity was determined with a 5 μM substrate concentration (Chader *et al.*, 1974b).

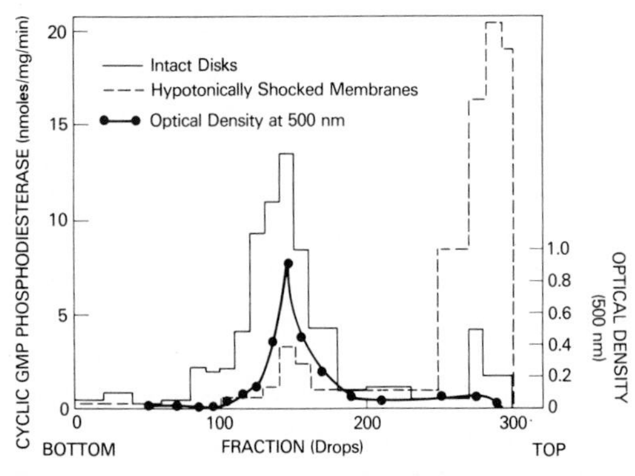

FIG. 2. Ficoll gradient (5–20%) of an isolated bovine rod outer segment preparation. Cyclic-GMP PDE activity was determined with an intact disk preparation (solid line) or after hypotonic disruption (dashed line) (Chader *et al.*, 1974b).

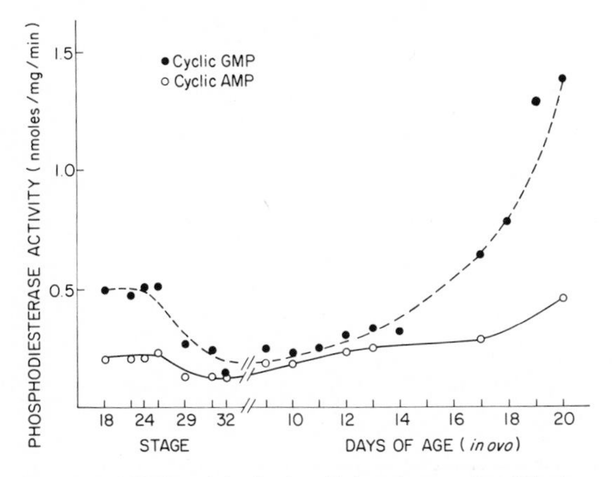

FIG. 3. Development of PDE activity in the chick embryo retina (Chader *et al.*, 1974c).

segments (Chader *et al.*, 1974c). Development of the inner retina (ganglion cells to photoreceptor inner segments) is virtually complete within the first 2 weeks of embryonic development (Coulombre, 1955). After day 14, a sharp rise in cyclic-GMP but not cyclic-AMP PDE activity occurs, continuing through the time of hatching on day 21 (Fig. 3). Morphologically, the outer segments first appear on days 14–15; elongation occurs until approximately day 19, when this and the other components of the retina are functionally complete. Although other organization changes occur during this period, outer segment development is by far the major event in the third embryonic week and correlates closely with cyclic-GMP PDE development. Microdissection studies on rat retinas during comparable stages of development also indicate that the major increase in PDE occurs at the time of photoreceptor outer segment development (Lolley and Farber, 1976).

C. PDE Properties and Control

Cyclic-AMP PDE types are present in complex K_m forms within cells, whereas cyclic-GMP kinetic forms are generally somewhat simpler (Strada and Thompson, 1978). In the retina, we found complex kinetics with cyclic-AMP PDE for soluble PDE (i.e., two K_m values), while the soluble cyclic-GMP PDE activity exhibited a single K_m value (Chader *et al.*, 1974b). In retinal homogenates (soluble plus particulate), two K_m types are observed (Farber and Lolley, 1976; Aguirre *et al.*, 1978). In ROS cyclic-GMP PDE exhibits apparent simple kinetics, although K_m values reported from various laboratories vary about 10-fold (Table II) in the general range of 10^{-4} *M*. Activity data vary widely in the literature but, as pointed out in this chapter and by Baehr *et al.* (1979), ‘‘many

TABLE II
K_m VALUES FOR ROD OUTER SEGMENT (ROS) CYCLIC-GMP PHOSPHODIESTERASE

Source of PDE	K_m value (μM)	Reference
Isolated bovine ROS	180	Pannbacker *et al.*, 1972
Isolated frog ROS	160	Miki *et al.*, 1973
Isolated bovine ROS	30	Chader *et al.*, 1974b
Isolated bovine ROS	220	Manthorpe and McConnell, 1975
Microdissected photoreceptor layer of rat retina	490	Lolley and Farber, 1975
Purified (frog)	70	Miki *et al.*, 1975
Purified (bovine)	[a]	Coquil *et al.*, 1975
Purified (bovine)	150	Baehr *et al.*, 1979

[a] Not determined because of "complex kinetics."

factors appear to modulate the activity of retinal ROS phosphodiesterase," making *in vitro* determinations unreliable. Moreover, the purified enzyme looses its ability to be modulated by light, indicating that calculation of turnover numbers, etc., may have little significance for the physiological state.

The purified PDE of frog ROS has a reported molecular weight of 240,000 (Miki *et al.*, 1975), and that of bovine ROS was reported originally as 105,000 (Coquil *et al.*, 1975) and more recently as 170,000 for a "core enzyme" (Baehr *et al.*, 1979). It is interesting to note that Baehr and co-workers have found a second ROS protein that copurifies with PDE. This protein inhibits PDE activity (50% for a 1:1 molar ratio) and has a molecular weight of about 80,000. If one assumes that it is in close association with PDE *in vivo,* even though they may not exist as a holoenzyme complex, a final molecular size close to that in frog ROS is probable. Similarly, Miki *et al.* report that limited proteolysis with trypsin activates PDE and that the trypsin-activated species has a molecular weight of 170,000. The possible relationship of the 80,000 unit to the 28,000-molecule-weight cyclic-AMP PDE inhibitor reported by Dumler and Etingof (1976) certainly should be investigated in the future.

Bitensky and co-workers have recently presented compelling evidence that other proteins are involved in the light-dependent activation of PDE. A light-sensitive guanosinetriphosphatase (GTPase) in isolated frog photoreceptors has an action spectrum similar to that of rhodopsin and PDE activation and appears to be necessary for maximal cyclic-GMP hydrolysis (Wheeler and Bitensky, 1977; Wheeler *et al.*, 1977). Other factors (an H-component) have been found in turn to regulate GTP hydrolysis, indicating the complexity of the PDE control mechanisms operative in the outer segment (Shinozawa *et al.*, 1980).

D. Cations and Calmodulin

Along with nucleotide specificity (cAMP versus cGMP), cellular compartmentalization, and diverse kinetic forms, several other potential mechanisms for controlling PDE activity are known. Enzyme activity, for example, is totally dependent on the presence of divalent cation, Mg^{2+} and Mn^{2+} being the most efficacious (Chader *et al.*, 1974d). In dialysis experiments (Fig. 4A), it can be shown that Ca^{2+} inhibits ROS PDE activity ostensibly by competing with Mg^{2+} for a divalent cation-binding site. EDTA effectively inhibits PDE activity, whereas dialysis against EGTA decreases PDE activity only to the level seen with dialysis against buffer alone. The addition of EGTA directly to isolated bovine ROS preparations does not significantly affect PDE activity (Fig. 4B). Likewise, the addition of Ca^{2+}, even in the presence of the ionophore X537A, has little effect on enzyme activity. These studies tend to indicate that ROS PDE is not markedly affected by fluctuations in Ca^{2+} concentration. Similarly, the data indicate that adult bovine ROS PDE is a calmodulin-independent type of enzyme. No stimulation with added Ca^{2+} was noted; EGTA had no inhibitory effect, although EDTA totally abolished activity. Calmodulin is present in bovine ROS, however (Liu and Schwartz, 1978), and may have a function other than PDE regulation in the adult retina. Interestingly, Cohen *et al.* (1978) have shown that Ca^{2+} has a marked effect on the level of cyclic GMP in intact,

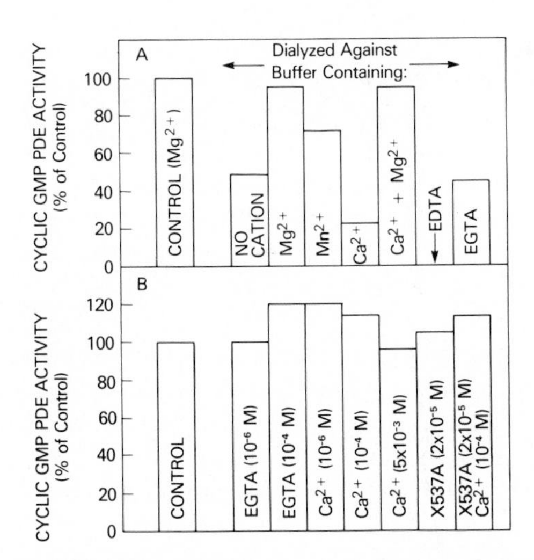

FIG. 4. PDE activity of isolated bovine rod outer segments under several conditions *in vitro*. (A) PDE values after dialysis of rod outer segments against 1.5 m*M* cation or chelating agent compared to undialyzed control rod outer segments analyzed with buffer containing 1.5 m*M* Mg^{2+}. (B) PDE activity in the absence (control) or presence of EGTA, calcium, and/or the ionophore X537A.

dark-adapted mouse retinas in culture, although this effect may not be mediated through PDE.

III. CYCLIC GMP AND PHOSPHODIESTERASE IN RETINAL DEGENERATIONS

A. RCS Rat

Retinal degeneration in the Royal College of Surgeons (RCS) rat has been studied for several years (Dowling and Sidman, 1962) as a possible model for the human disease retinitis pigmentosa (RP). In this disease, the primary lesion appears to be in the pigment epithelium (Mullen and LaVail, 1976), resulting in secondary retinal damage and an accumulation of outer segment debris between the retinal and pigment epithelial tissues (Bok and Hall, 1971). Lolley and Farber (1975) first reported that degradation of cyclic GMP in retinas of RCS rats was abnormal during the early postnatal period as the photoreceptor outer segment degenerated and debris accumulated. Subsequently (Lolley and Farber, 1976), it was demonstrated that both the cyclic-GMP content and PDE activity of affected retinas developed normally for the first 1–2 weeks after birth but soon thereafter fell below control levels. *In vitro* studies by Lolley and Farber (1976) indicate that the accumulated debris can inhibit retinal PDE activity. This and the temporal relationship between debris accumulation and abnormal PDE activity *in vivo* suggest that the retinal degeneration may at least be exacerbated by the inhibition of retinal PDE by outer segment debris. Biochemically and morphologically, however, this disease is quite distinct from those seen in the C3H mouse and the dog breeds discussed below.

B. C3H Mouse

An early-onset degeneration, also known as a ''dysplasia'' of the retinal photoreceptor cell, is seen in the C3H mouse (Sidman and Green, 1965). In this model, degeneration is extremely rapid and differs from that in the RCS rat in that no outer segment debris accumulates between the retinal and pigment epithelial cell layers. Schmidt and Lolley (1973) first described a defect in cyclic-AMP metabolism in C3H retinas early in the postnatal period. Subsequently, this was pinpointed as a specific deficiency in cyclic-GMP PDE activity in the retinal photoreceptor layer (Farber and Lolley, 1976). In normal mouse retinas during the first postnatal week, a single kinetic type of PDE ($K_m = 3.2 \times 10^{-5}\ M$) is observed. In the second postnatal week, as the photoreceptor outer segments appear, a second kinetic type of PDE becomes evident at $K_m = 2.8 \times 10^{-4}\ M$. In the affected C3H retina, the PDE type at $K_m = 3.2 \times 10^{-5}\ M$ is found, but the latter type ($K_m = 2.8 \times 10^{-4}\ M$) cannot be detected. Since this higher-K_m type of PDE is localized in the outer segment, cyclic-GMP concentrations quickly rise to levels considerably higher than in control, nonaffected retinas (Farber and Lolley, 1974). This abnormally high concentration of cyclic GMP early in outer segment development of the C3H mouse could therefore be causally related to subsequent photoreceptor cell degeneration. This thesis is supported by experiments by Lolley *et al.* (1977) on eye rudiments of *Xenopus laevis* in culture. They showed that isobutylmethylxanthine (IBMX), an inhibitor of PDE, added to such cultures increased the intracellular cyclic-GMP concentration and resulted in arrested development and degeneration of photoreceptor cells.

C. Irish Setter

Several dog species with inherited retinal degeneration have now been described. The Irish setter in particular has been well studied (Parry *et al.*, 1955). In this breed, the disease is inherited as an autosomal recessive trait manifested soon after birth (Aguirre and Rubin, 1975). Night blindness is clinically apparent within 6–8 weeks, and total blindness quickly thereafter. Figure 5 shows typical

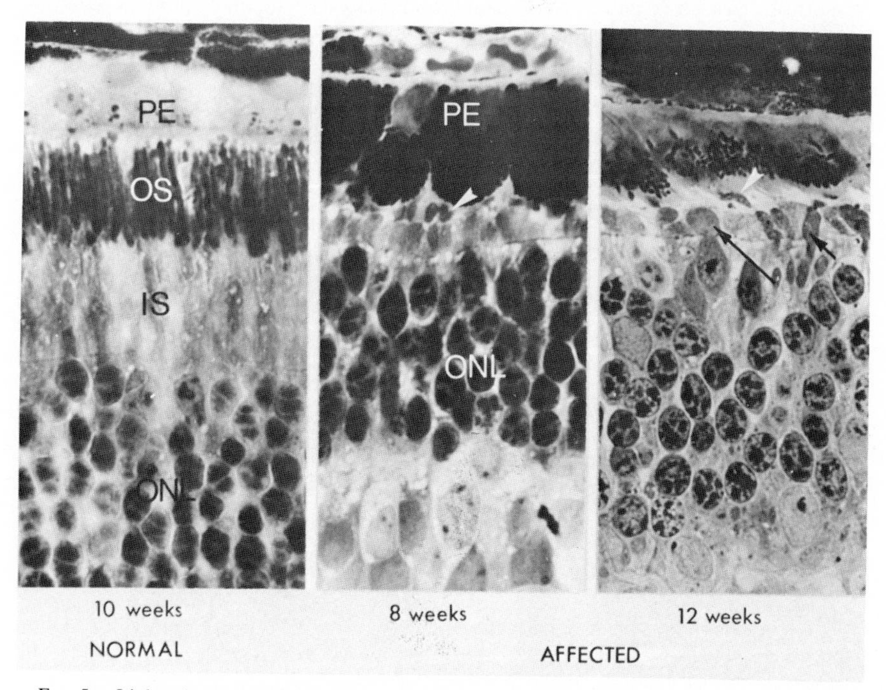

FIG. 5. Light micrographs from the retina of control (10-week-old) and affected (8- to 12-week-old) Irish setters. Tissues were fixed in 2.5% glutaraldehyde and 2% osmium tetroxide, dehydrated, and embedded in Epon. PE, Pigment epithelium; OS, outer segments; IS, inner segments; ONL, outer nuclear layer. White arrowheads indicate minimal remaining outer segment material; black arrows indicate remaining cones (Aguirre *et al.*, 1978).

light micrographs from retinas of a 10-week-old unaffected retina. Photoreceptor inner segments and outer segments are fully elongated and abut an area of nonpigmented pigment epithelium in the tapetal zone. In affected animals, even by 8 weeks of age, the photoreceptor layer is reduced in width and density. The inner segments are diminutive, and minimal outer segment material remains. Rod photoreceptor loss accentuates the prominence of the cones, and the outer nuclear layer is greatly reduced in width. By approximately 18 weeks (Fig. 6, 128 days), the photoreceptor layer is considerably narrowed and basically consists of broad, prominent cone inner segments with shortened outer segments. The loss of rod photoreceptors has reduced the outer nuclear layer by this time to 3 nuclei, whereas 10 are normal. The relentless progression of the disease is evident in Fig. 6 (348 days) which shows that the photoreceptor and outer nuclear layers have disappeared and the inner nuclear layer is now located adjacent to the pigment epithelium.

A defect in cyclic-GMP metabolism is present in affected retinas of the Irish setter, which is analogous to that seen in the C3H mouse. In early studies (Aguirre *et al.*, 1978), we found extremely high cyclic-GMP concentrations in retinas of affected animals within 2–3 months after birth (Table III). In general, levels were about 10-fold higher in affected neural retinas than in controls, while no significant differences were observed in other tissues including the pigment epithelium–choroid complex. Two apparent kinetic types of PDE activity are observed in retinal homogenates from control animals at this time: K_m-A at 1.2×10^{-4} *M* and K_m-B at 2.0×10^{-5} *M*. In affected retinas, only the lower-K_m-B type (1.8×10^{-5} *M*) is observed. By analogy with the situation in the mouse (Lolley and Farber, 1975), it is reasonable to assume that the higher-K_m kinetic type of PDE enzyme is compartmentalized in the outer segments in control retinas and is deficient in photoreceptor cells of affected retinas. At this

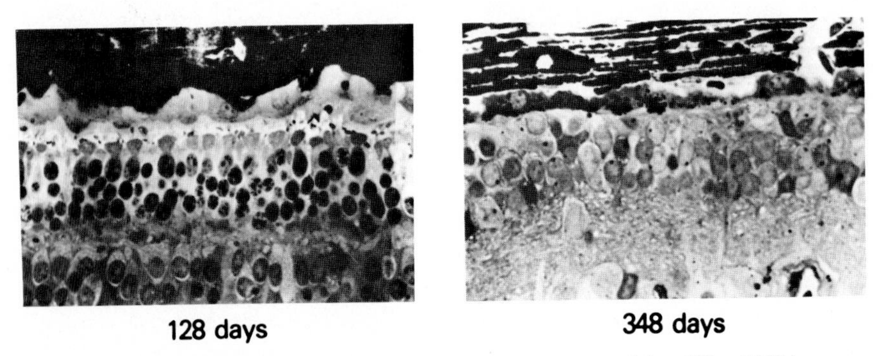

FIG. 6. Light micrographs from older affected Irish setters at postnatal days 128 and 348 (approximately 18 and 50 weeks of age, respectively).

TABLE III
CYCLIC-NUCLEOTIDE CONCENTRATIONS IN TISSUES OF NORMAL AND AFFECTED IRISH SETTERS[a]

Tissue	Cyclic-GMP concentration (pmoles mg^{-1} protein)	
	Control	Affected
Neural retina	12.1 ± 0.7	102.8 ± 8.6
Pigment. epithelium	9.2 ± 1.2	6.9 ± 0.9
Visual cortex	3.7 ± 0.2	5.0 ± 0.4
Liver	2.9 ± 0.1	5.1 ± 0.3

[a] Neural retinas were obtained from 10-week-old control animals (N = 8 eyes) and 8- and 12-week-old affected animals (N = 10 eyes). Visual cortex and liver samples were from 27-week-old control and 12-week-old affected animals. Pigment epithelium samples included choroid (Aguirre *et al.*, 1978).

time, however (8–12 weeks), few outer segments remain; the PDE deficiency thus could very well be secondary to the loss of ROS.

To examine this point, cyclic-GMP concentrations, PDE activity, and calmodulin levels were determined at times early in outer segment development, as well as when the disease was fully apparent (i.e., 8–12 weeks). Figure 7 demonstrates that the cyclic-GMP level rises about eightfold during the period of outer segment development (approximately postnatal days 10–40) in retinas of nonaffected control setters. In retinas of affected setters, cyclic-GMP levels are markedly higher than in control retinas, even at the earliest times of outer segment development (days 10–15). Cyclic GMP in affected retinas remains at extremely high levels; these levels fall after loss of the photoreceptor and outer nuclear layers (see the value at 50 weeks and the morphological characteristics at 348 days in Fig. 6).

Cyclic-GMP PDE activity rises sharply during outer segment development (Fig. 8A). The abnormally high cyclic-GMP content in affected retinas appears to reflect the low cyclic-GMP PDE activity in these retinas during the entire postnatal period (Fig. 8A). Even as the outer segments first begin to develop at about day 9, cyclic-GMP PDE activity is already lower in affected retinas (244 pmoles mg^{-1} min^{-1}) than in the heterozygous control retinas (309 pmoles mg^{-1} min^{-1}). Cyclic-AMP PDE activity is lower than that for cyclic GMP (Fig. 8B)

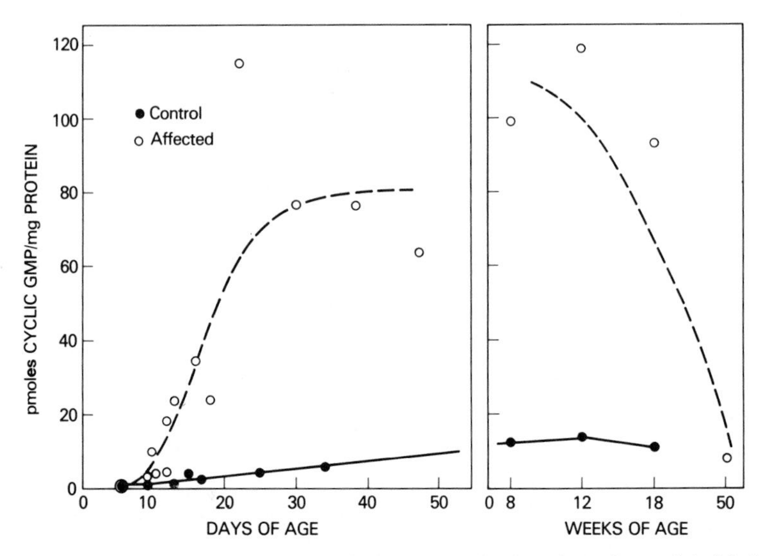

FIG. 7. Developmental pattern of cyclic-GMP concentration in retinas of control (solid circles) and affected (open circles) Irish setters. Each point represents a single separate determination (manuscript in preparation).

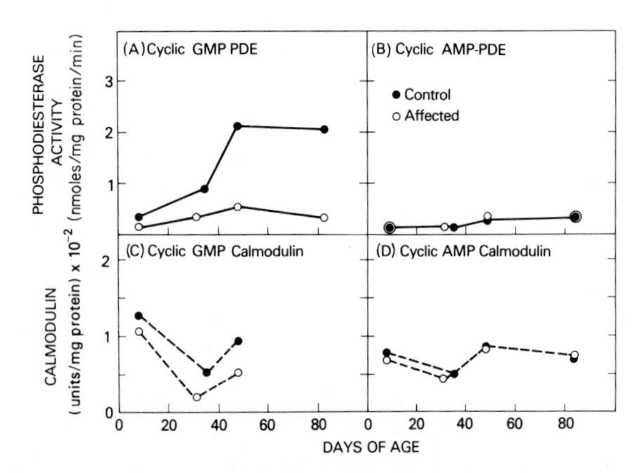

FIG. 8. Developmental pattern of cyclic-GMP PDE activity (A), cyclic-AMP PDE activity (B), and calmodulin levels using cyclic GMP (C) or cyclic AMP (D) as substrate in retinas of control (solid circles) and affected (open circles) Irish setters. PDE activity was determined in retinal homogenates using 1 μM cyclic [^{3}H]nucleotide. Calmodulin was determined using purified, activator-deficient bovine brain PDE as previously described (Liu *et al.*, 1979).

and is virtually identical in control and affected retinas. It does not exhibit an increase as the outer segments develop, as would be expected if one assumes it is mainly localized in the inner layers of the retina.

Although evidence indicated that ROS PDE in adult retina was of the Ca^{2+}-independent type (Fig. 4), it was of interest to compare the calmodulin levels in control and affected retinas during the critical period of retinal development occurring just after birth. Using 1 μM cyclic [^{3}H]GMP as substrate, calmodulin was found to be lower in affected retinas than in control retinas before (day 9), during (days 31–34), and after (day 48) ROS development (Fig. 8C). No difference was observed in calmodulin concentration using 1 μM cyclic [^{3}H]AMP under similar conditions. From these results it is clear that an abnormality in PDE activity is present in affected retinas only with regard to cyclic-GMP metabolism. Also, an apparent deficiency in calmodulin is observed in affected retinas with regard to cyclic GMP but not cyclic AMP.

Since cyclic-GMP PDE activity appeared low in affected retinas and calmodulin levels also appeared to be deficient, the *in vitro* activity of retinal cyclic-GMP PDE was investigated in the presence of purified calmodulin or EGTA, an active Ca^{2+} chelator. The addition of calmodulin increased PDE activity in 9-day-old retinas of both control and affected animals, whereas the addition of EGTA decreased PDE activity in both cases (Fig. 9). This is in contrast to the situation in isolated ROS of the adult bovine retina (Fig. 4) where EGTA had very little

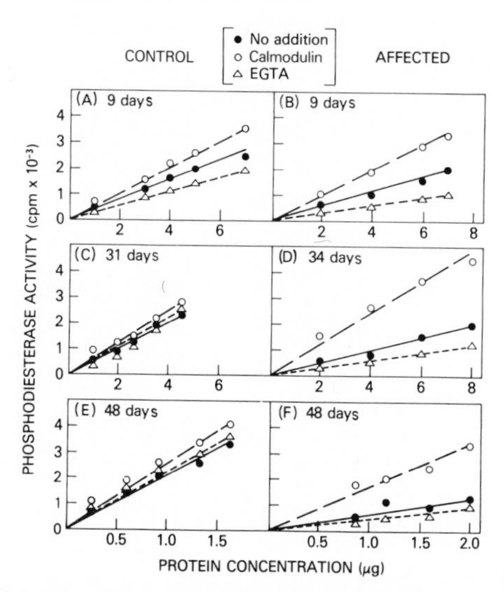

FIG. 9. Cyclic-GMP PDE activity in retinas of affected and control Irish setters of different ages. Solid circles, No additions to retinal homogenate incubation; open circles, addition of 10 μg purified ox brain calmodulin to incubation; open triangles, addition of 250 μM EGTA to incubation. PDE activity was determined using 1.0 μM cyclic [^{3}H]GMP (Liu *et al.*, 1979).

effect on PDE activity. In retinas of older control animals (31 days, 48 days of age), however, neither calmodulin nor EGTA had any significant effect on PDE activity. In affected retinas, calmodulin continued to stimulate activity and EGTA to inhibit activity in the older dogs (34 days, 48 days of age). These data are consistent with the hypothesis that, before ROS development, i.e., 9 days of age, the cyclic-GMP PDE present in the retina is of the calmodulin-dependent type. In control retinas, this appears to switch to a calmodulin-independent type as outer segments elongate and mature (day 31 and thereafter). In affected retinas, the PDE appears to remain as the calmodulin-dependent type.

Since use of 1 μM cyclic [^{3}H]GMP as substrate measures only part of the total retinal cyclic-GMP PDE activity, experiments were designed using 1 mM cyclic [^{3}H]GMP to measure total PDE activity. This is particularly important, since the kinetic type of PDE that was found to be deficient in dogs with advanced retinal degeneration (8–12 weeks) was of the higher-K_m type ($K_m A = 1.2 \times 10^{-4}$ M; Aguirre *et al.*, 1978). Table IV shows a comparison of the effects of added calmodulin on PDE activity using 1 μM and 1 mM cyclic [^{3}H]GMP as substrate. In the early stages of ROS development (day 12), calmodulin increased PDE activity by 20–25% at either micromolar or millimolar concentrations of cyclic [^{3}H]GMP. During the subsequent period of ROS elongation and maturation (days 17, 21, and 33), calmodulin activation of PDE activity was variable, but in each case tested with micromolar or millimolar concentrations of substrate, stimulation by the protein activator was evident. Since calmodulin is thought to occur normally in manyfold excess over the molar concentration of PDE in tissues,

TABLE IV

EFFECT OF CALMODULIN ON CYCLIC-GMP PHOSPHODIESTERASE ACTIVITY OF AFFECTED IRISH SETTER RETINAS USING 1 μM AND 1 mM CYCLIC[^{3}H]-GMP[a]

Animal age (postnatal days)	Cyclic-GMP PDE activity (pmoles mg^{-1} protein min^{-1}): 1 μM substrate, No calmodulin	1 μM substrate, Added calmodulin	1 μM substrate, Activation (%)	1 mM substrate, No calmodulin	1 mM substrate, Added calmodulin	1 mM substrate, Activation (%)
12	240	305	25	2650	3168	20
17	314	436	39	3290	4750	44
21	317	659	110	5188	7687	48
33	473	653	38	8540	9100	7

[a] PDE was assayed in the absence or presence of 10 μg purified ox brain calmodulin. Triplicate determinations were performed on retinal samples of each age.

even a small increase in PDE activity elicited by exogenous calmodulin indicates the probability of a significantly lower calmodulin level *in vivo* in affected retinas than in controls.

Several explanations of these data are possible. One hypothesis is outlined in schematic form in Fig. 10 which attempts to synthesize the biochemical and morphological findings. In this model, the cyclic-GMP PDE activity (or activities) of the normal retina are calmodulin-dependent prior to outer segment formation, but a biochemical switch occurs during ROS elongation, changing the enzyme to the calmodulin-independent type. It is well known that many types of enzymes (e.g., isozymes) undergo changes during the late fetal and early postnatal period of development. Specific evidence that the levels of PDE activity and PDE sensitivity to regulation by Ca^{2+}–calmodulin change during development has also been reported from several laboratories (Davis and Kuo, 1976; Singer *et al.*, 1978). In the affected retina, no such normal switch appears to occur during photoreceptor development. This, coupled with lowered calmodulin levels, leads to the abnormally high cyclic GMP characteristic of the disease. This scheme is testable *in vivo*. Theoretically at least, the injection of calmodulin should decrease retinal cyclic-GMP levels and delay or halt the progress of the disease. There is some hope that this in fact can occur *in vivo,* since the addition of purified calmodulin to retinal homogenates *in vitro* restores PDE activity to approximately that seen in control retinas (Fig. 9).

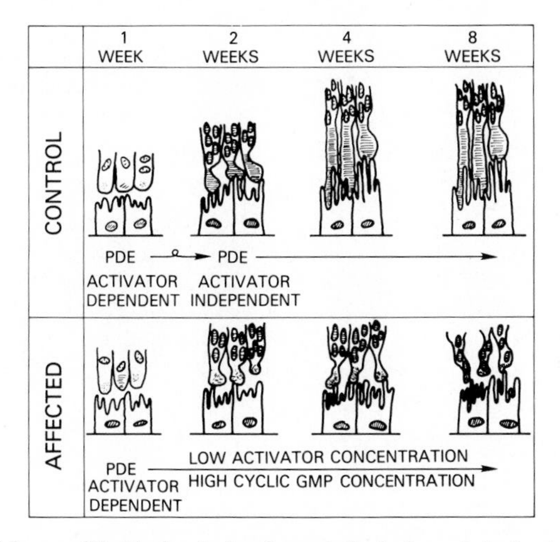

FIG. 10. Model for possible biochemical and morphological events in the development of outer segments in retinas of control and affected Irish setters. Note that the retinal orientation is inverted in comparison to that in Figs. 5 and 6.

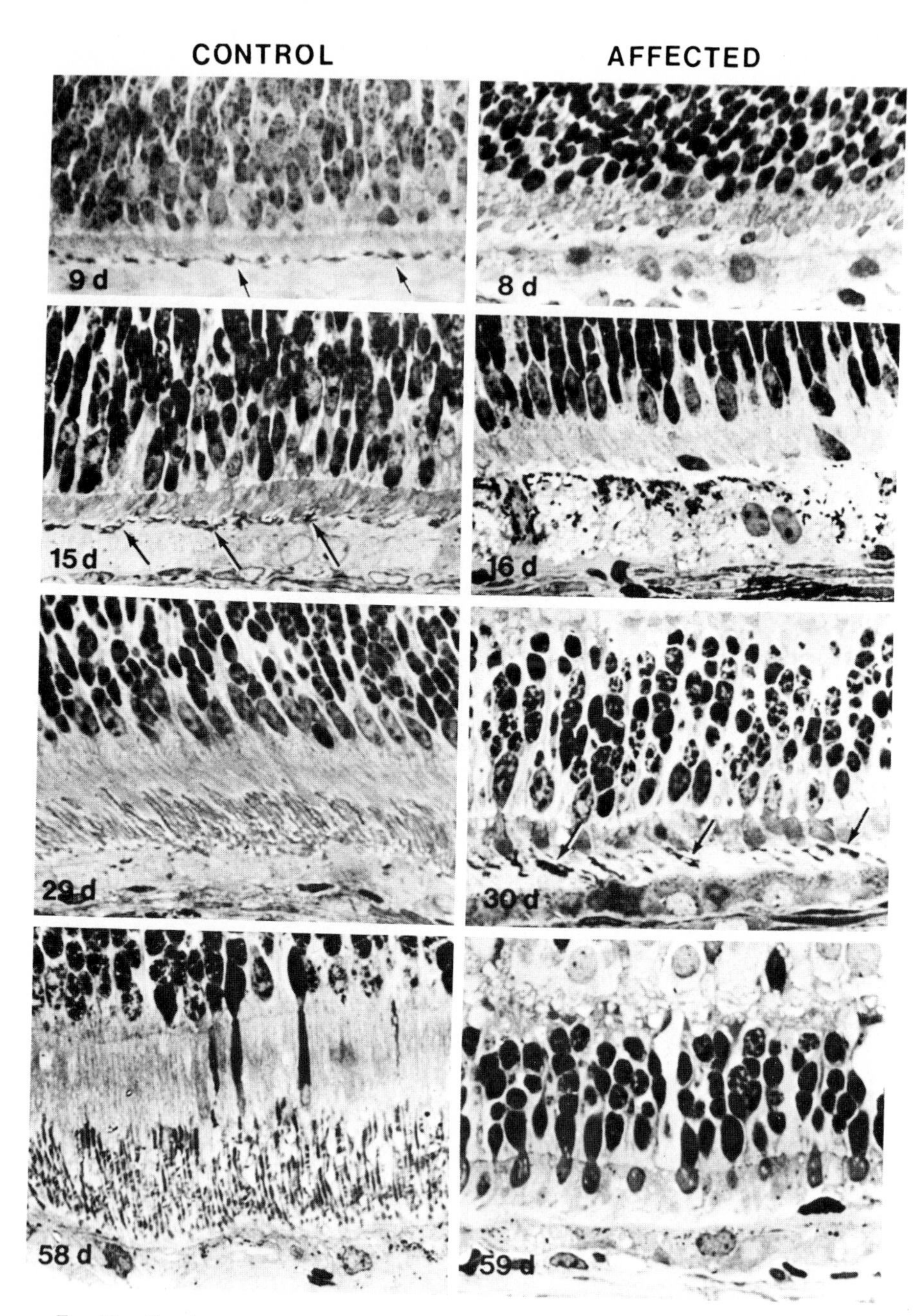

FIG. 11. Developmental pattern (light micrographs) of normal and affected collie retinas from approximately 1 to 8 postnatal weeks of age. Arrows indicate developing outer segments in control (9 and 15 days) retinas or abnormally developed outer segments in affected retinas at 30 days of age. Retinal orientation as in Fig. 10 (Santos-Anderson *et al.*, 1980).

D. Collie

Retinal degeneration in the collie is similar to that in the Irish setter in that it is an inherited disease of rapid onset first observed in the early postnatal period (Wolf *et al.*, 1978). Night blindness is observed by about 6 weeks and total blindness within 6–8 months. Figure 11 documents the course of the disease from 9 days after birth in affected animals in comparison to age-matched controls (Santos-Anderson *et al.*, 1980). At day 9, outer segments begin to form in the collie, as in the setter, whereas very little outer segment material is observed in the 8-day affected retina. By 2 weeks of age (15 days), outer segments are already well formed and regularly oriented in control retinas but are present only in occasional packets in affected retinas. The pigment epithelium is hypertrophic and vacuolated at this time, in contrast to the ''normal'' pigment epithelium observed in the setter and C3H mouse under comparable conditions. At 1 month of age (30 days), outer segments are evident in affected retinas but are sparse and of irregular form and length. By 2 months of age (58 days) the outer segments in control retinas have reached their final mature length, while affected retinas demonstrate a significant loss of rod outer and inner segments and cell bodies as well as the presence of macrophages in the subretinal space. As in the setter, the disease is thus perhaps best described as a rod–cone dysplasia, since the outer segments never fully mature before degeneration ensues. It is distinct from the Irish setter disease, however, since it shows marked pigment epithelium cell involvement early in the postnatal period; comparable cellular changes are not observed in the course of the disease in the setter (Santos-Anderson *et al.*, 1980). Similarly, the diseases are genetically distinct as is evident by the normal progeny that results from breeding affected animals of the two different breeds (Acland *et al.*, 1980).

Biochemically, the cyclic-GMP concentration rises in retinas of control animals at the time of early outer segment elongation (Fig. 12) and then levels off

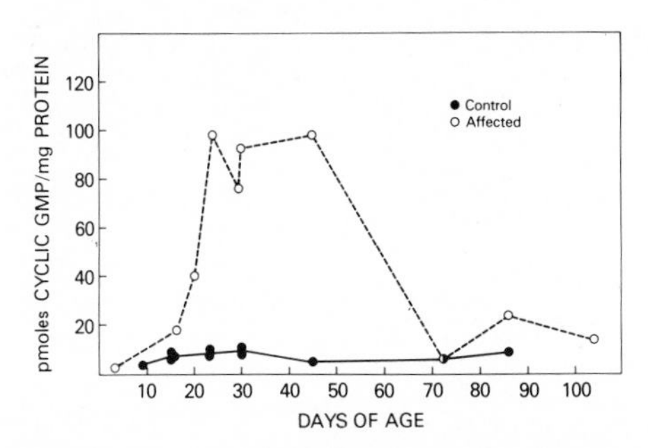

FIG. 12. Developmental pattern of cyclic-GMP concentration in retinas of control (solid circles) and affected (open circles) collies. Values are averages of at least three determinations and, when appropriate, are given ±SEM.

(Woodford *et al.*, 1980). In affected retinas, it is 10- to 20-fold higher than in controls even in earliest ROS development, a situation quite comparable to that seen in the Irish setter. It is interesting to see the levels drop again by 70 days of age. This correlates with a complete disappearance of the ROS and a breakdown of the inner segments (2–2½ months). Figure 13 compares the patterns of PDE and calmodulin development in control and affected retinas. Cyclic-GMP PDE activity measured at a 1 μM substrate concentration rises in control retinas but at a somewhat slower rate than that seen in the setter (Fig. 13A). PDE activity does not rise in affected retinas but rather declines with photoreceptor degeneration. Cyclic-AMP PDE activity is considerably higher in affected than in control retinas (Fig. 13B). This is interesting, since it occurs early in the postnatal developmental period and is too great simply to reflect a proportionally higher amount of nonphotoreceptor cell material in affected retinas. The calmodulin concentration (measured with 1 μM cyclic [^{3}H]GMP) rises rapidly in control retinas, reaching peak levels well before ROS fully elongate (Fig. 13C). The activator is late in appearing in affected retinas, peaks at about 40 days of age, and declines thereafter. This generally parallels the delayed appearance of ROS and their subsequent degeneration. No marked change in calmodulin (1 μM cyclic [^{3}H]AMP) was noted in control retinas over the period studied (Fig. 13D). As with the cyclic-AMP PDE activity, the measured calmodulin level was higher in affected than in control retinas until about 40 days. The decline thereafter again may reflect terminal photoreceptor cell degeneration.

Figure 14 compares the effects of calmodulin and EGTA on cyclic-GMP PDE activity in control and affected retinas at about 2 months of age. Calmodulin was not found to stimulate nor did EGTA inhibit PDE activity. The PDE at this time

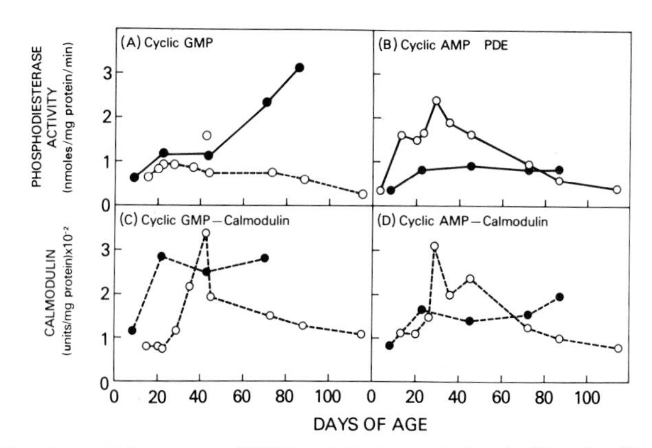

FIG. 13. Developmental patterns of PDE activity in control and affected collie retinas. Solid circles, control; open circles, affected. (A) Cyclic GMP; (B) cyclic AMP; (C) calmodulin concentration measured with cyclic GMP; (D) calmodulin concentration measured with cyclic AMP. The concentration of cyclic [^{3}H]nucleotide used was 1 μM (Woodford *et al.*, 1981).

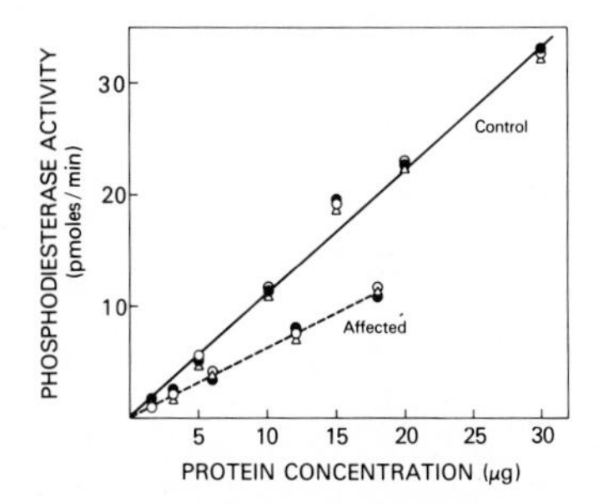

FIG. 14. Cyclic-GMP PDE activity in control and affected retinas of collie dogs at 45 days of age assayed in the presence (open circles) or absence (solid circles) of 10 μg purified ox brain calmodulin or of 250 μM EGTA (open triangles).

in affected retinas as well as in control retinas appears to be of the calmodulin-independent type. It seems that, although there is a substantial calmodulin level in the collie retina, it probably has effects other than on the PDE. It is now well known that calmodulin affects several enzyme systems in tissues [e.g., adenosinetriphosphatase (ATPase), adenylate cyclase]. Although it is not known what systems might be specifically affected in the retina, the collie data are in sharp contrast to those for the setter where calmodulin markedly stimulates PDE activity in affected retinas under identical conditions.

IV. CONCLUSIONS

The role of cyclic GMP and light-activated PDE in the normal visual process is not clear. Similarly, a definitive link between abnormal PDE activity, high cyclic-GMP levels, and retinal dysplasia or degeneration has yet to be established.

Much circumstantial evidence, however, points to the importance of cyclic GMP and PDE in photoreceptor function—high concentration, nucleotide specificity, compartmentalization, etc. Light modulation of PDE activity (Miki *et al.*, 1973; Chader *et al.*, 1974a) and, more importantly, of actual cyclic-GMP levels in the photoreceptor outer segment (Fletcher and Chader, 1976; Orr *et al.*, 1976) further implicates cyclic-GMP changes in light-regulated physiological processes *in vivo*. Woodruff *et al.* (1977) have correlated cyclic-GMP concentration with physiologically functioning of the frog retina, while Yee and Liebman (1978) have demonstrated that the rapid activation and turnover of cyclic-GMP PDE is consistent with a role for the enzyme in photoreception. Moreover, Miller and Nichol (1979) have direct evidence that cyclic GMP causes depolarization of the rod plasma membrane to the sodium equilibrium potential and that hyperpolarization is dependent on cyclic-GMP hydrolysis. Figure 15A gives a model outlining

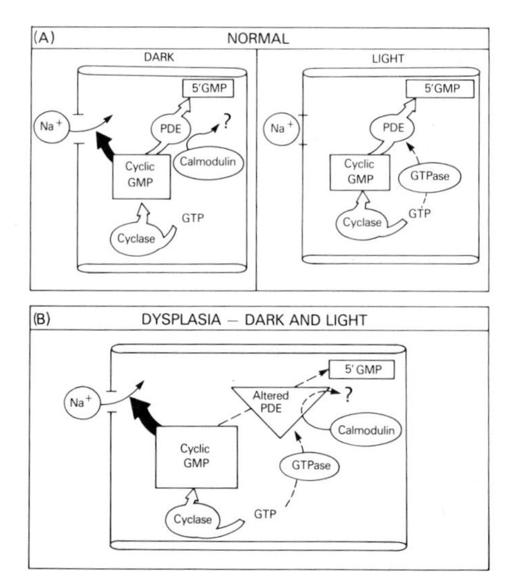

FIG. 15. Model for control of PDE activity and cyclic-GMP action in normal (A) and dysplastic (B) rod photoreceptor outer segments.

some of the possible factors that control cyclic-GMP concentration in light- and dark-adapted ROS and the subsequent action of cyclic-GMP. The ''activation'' of PDE is not simply a light-triggered event but is undoubtedly a complex series of regulatory steps involving (1) nucleoside concentration, (2) other enzymes such as GTPase, (3) light-evoked opsin–membrane conformational charges, and (4) calcium, calmodulin or ''helper'' factors such as those recently described by Shinozawa *et al.* (1980) for GTPase. Much work has yet to be done in sorting out these factors; defining the precise role of cyclic GMP (e.g., maintaining open sodium channels, controlling dark adaptation) will be a continuing challenge.

In retinal degeneration, particularly the early-onset (dysplasia) type as exemplified in the setter, it is clear that impaired PDE activity in affected retinas leads to abnormally high cyclic-GMP levels. This lesion is specific, in that general membrane protein synthesis does not differ in affected and control retinas in the early postnatal stages of the disease (Chader *et al.*, 1980). Possible consequences of such altered PDE activity are schematically shown in Fig. 15B. Simplistically, the ''altered PDE'' activity may result from a drastic reduction in the actual amount of high-K_m enzyme compartmentalized in the photoreceptor outer segments. As yet no direct evidence equates lowered PDE activity with an actual decrease in the number of PDE protein molecules. Alternatively, developmental abnormalities of enzyme production (isoenzymes, etc.) are well known and could be responsible for the problem in the Irish setter. This is clearly not the same situation in the collie though, since PDE activity in this breed is refractory to added calmodulin. Genetic probing at the nuclear or ribosomal level, as well as immunological investigation, may shed light on this problem and define the amount as well as the type of PDE present in affected retinas.

Since the control of PDE activity in the photoreceptor is known to be complex, it is possible that abnormalities in such control mechanisms could also contribute to the problem (Fig. 15B). Both activators (calmodulin) and inhibitors (Dumler and Etingof, 1976; Baehr *et al.*, 1979) may be involved in normal or abnormal photoreceptor functioning. Nucleoside triphosphate concentration and GTPase changes may also be involved. Since purified soluble PDE looses its ability to be light-regulated, a simple change in the membrane-binding properties of PDE in affected retinas could dramatically alter enzyme activity.

A second question fundamental to the disease process is: Does the high cyclic-GMP concentration itself precipitate retinal dysplasia, or is this just an early parallel or contributing event? Work by Lolley *et al.* (1977) and more recently by Ulshafer *et al.* (1980) indicates that cyclic nucleotides and PDE inhibitors can indeed damage and destroy retinal rod photoreceptors, although the mechanism is obscure. If high cyclic-GMP levels (as in dark adaptation) maintain or contribute to a depolarized state with decreased cyclic GMP, resulting in blocking of the sodium channels, one would predict that the high cyclic-GMP levels in photoreceptor cell dysplasia could lead to a constant state of depolarization. Such a continuous condition could result in a severe metabolic imbalance (e.g., energy deprivation) and photoreceptor degeneration. It will be interesting to see if an analogous continuous dark-adapted state is common in humans, particularly under early-onset conditions. It will be particularly important in the future to define a method for decreasing the cyclic-GMP concentration, possibly through stimulation of residual PDE activity in the remaining photoreceptor cells. Calcium (and calmodulin) are attractive candidates for this purpose, since PDE might be stimulated and outer segment guanylate cyclase is also known to be inhibited by calcium (Krishnan *et al.*, 1978). Since the outer segment is in a constant state of renewal (Young, 1976), it is at least theoretically possible that, if a proper regimen of treatment could be initiated at an early enough time, the disease process might not only be halted but actually reversed.

REFERENCES

Acland, G., Aguirre, G., Chader, G., Fletcher, R., and Farber, D. (1980). Canine early onset hereditary retinal degeneration: Genetic and biochemical distinction of these diseases. *Invest. Ophthalmol. Visual Sci., Suppl.* **1,** 250.

Aguirre, G., and Rubin, L. (1975). Rod-cone dysplasia (progressive retinal atrophy) in irish setters. *J. Am. Vet. Med. Assoc.* **166,** 157–164.

Aguirre, G., Farber, D., Lolley, R., Fletcher, R., and Chader, G. (1978). Rod-cone dysplasia in irish setters: A defect in cyclic GMP metabolism in visual cells. *Science* **201,** 1133–1134.

Baehr, W., Devlin, M., and Applebury, M. (1979). Isolation and characterization of the cGMP phosphodiesterase from bovine rod outer segments. *J. Biol. Chem.* **254,** 11669-11677.

Bitensky, M., Gorman, R., and Miller, W. (1971). Adenyl cyclase as a link between photon capture and changes in membrane permability of frog photoreceptors. *Proc. Natl. Acad. Sci. U.S.A.* **68,** 561-562.

Bok, D., and Hall, M. (1971). The role of the pigment epithelium in the etiology of inherited retinal dystrophy in the rat. *J. Cell Biol.* **49,** 664-682.

Butcher, R., and Sutherland, E. (1962). Adenosine 3′,5′-phosphate in biological materials. I. Purification and properties of cyclic 3′,5′-nucleotide phosphodiesterase and use of this enzyme to characterize adenosine 3′,5′-phosphate in human urine. *J. Biol. Chem.* **237,** 1244-1250.

Chader, G., Bensinger, R., Johnson, M., and Fletcher, R. (1973). Phosphodiesterase: An important role in cyclic nucleotide regulation in retina. *Exp. Eye Res.* **17,** 483-486.

Chader, G., Herz, L., and Fletcher, R. (1974a). Light activation of phosphodiesterase activity in retinal rod outer segments. *Biochim. Biophys. Acta* **347,** 491-493.

Chader, G., Johnson, M., Fletcher, R., and Bensinger, R., (1974b). Cyclic nucleotide phosphodiesterase of the bovine retina: Activity, subcellular distribution and kinetic parameters. *J. Neurochem.* **22,** 93-99.

Chader, G., Fletcher, R., and Newsome, D. (1974c). Development phosphodiesterase activity in the chick retina. *Dev. Biol.* **40,** 378-380.

Chader, G., Fletcher, R., Johnson, M., and Bensinger, R. (1974d). Rod outer segment phosphodiesterase: Factors affecting the hydrolysis of cyclic AMP and cyclic GMP. *Exp. Eye Res.* **18,** 509-515.

Chader, G., Liu, Y., O'Brien, P., Fletcher, R., Krishna, G., Aguirre, G., Farber, D., and Lolley, R. (1980). Cyclic GMP phosphodiesterase activator: Involvement in a hereditary retinal degeneration. *Neurochemistry* **1,** 441-458.

Cheung, W. (1967). Cyclic 3′,5′-nucleotide phosphodiesterase: Pronounced stimulation by snake venom. *Biochem. Biophys. Res. Commun.* **29,** 479-482.

Cohen, A., Hall, I., and Ferrendelli, J. (1978). Calcium and cyclic nucleotide regulation in incubated mouse retinas. *J. Gen. Physiol.* **71,** 595-612.

Coquil, J. F., Vermaux, N., and Goridis, C. (1975). Cyclic nucleotide phosphodiesterase of retinal photoreceptors: Partial purification and some properties of the enzyme. *Biochim. Biophys. Acta* **403,** 425-437.

Coulombre, A. (1955). Correlation of structure and biochemical changes in the developing retina of the chick. *Am. J. Anat.* **96,** 153-193.

Davis, C., and Kuo, J. (1976). Ontogenic changes in levels of phosphodiesterase for adenosine 3′:5′-monophosphate and guanosine 3′:5′-monophosphate in the lung, liver, brain and heart from guinea pigs. *Biochim. Biophys. Acta* **444,** 554-562.

Dowling, J., and Sidman, R. (1962). Inherited retinal dystrophy in the rat. *J. Cell Biol.* **14,** 73-109.

Dumler, I., and Etingof, R. (1976). Protein inhibitor of cyclic adenosine 3′:5′-monophosphate phosphodiesterase in retina. *Biochim. Biophys. Acta* **429,** 474-484.

Farber, D., and Lolley, (1974). Cyclic guanosine monophosphate: Elevation in degenerating photoreceptor cells of the C3H mouse retina. *Science* **186,** 449-451.

Farber, D., and Lolley, R. (1976). Enzymatic basis for cyclic GMP accumulation in degenerative photoreceptor cells of mouse retina. *J. Cyclic Nucleotide Res.* **2,** 139-148.

Fletcher, R., and Chader, G. (1976). Cyclic GMP: Control of concentration by light in retinal photoreceptors. *Biochem. Biophys. Res. Commun.* **70,** 1297-1302.

Kakiuichi, S., and Yamazaki, R. (1970). Calcium-dependent phosphodiesterase activity and its activating factor (PAF) from brain. *Biochem. Biophys. Res. Commun.* **41,** 1104-1110.

Krishnan, N., Fletcher, R., Chader, G., and Krishna, G. (1978). Characterization of guanylate cyclase of rod outer segments of the bovine retina. *Biochim. Biophys. Acta* **523,** 506-515.

Liu, Y. P., and Schwartz, H. (1978). Protein activator of cyclic AMP phosphodiesterase and cyclic nucleotide phosphodiesterase in bovine retina and bovine lens. *Biochim. Biophys. Acta* **526,** 186–193.

Liu, Y. P., Krishna, G., Aguirre, G., and Chader, G. (1979). Involvement of cyclic GMP phosphodiesterase activator in an hereditary retinal degeneration. *Nature (London)* **280,** 62–64.

Lolley, R., and Farber, D. (1975). Cyclic nucleotide phosphodiesterases in dystrophic rat retinas: Guanosine 3′,5′-cyclic monophosphate anomalies during photoreceptor cell degeneration. *Exp. Eye Res.* **20,** 585–597.

Lolley, R., and Farber, D. (1976). A proposed link between debris accumulation, guanosine 3′-5′ cyclic monophosphate changes and photoreceptor cell degeneration in retina of RCS rats. *Exp. Eye Res.* **22,** 477–486.

Lolley, R., Farber, D., Rayborn, M., and Hollyfield, J. (1977). Cyclic GMP accumulation causes degeneration of photoreceptor cells: Simulation of an inherited disease. *Science* **196,** 664–666.

Manthorpe, M., and McConnell, D. (1975). Cyclic nucleotide phosphodiesterase associated with bovine retinal outer segment fragments. *Biochim. Biophys. Acta* **403,** 438–445.

Miki, N., Keirns, J., Marcus, F., Freeman, J., and Bitensky, M. (1973). Regulation of cyclic nucleotide concentrations in photoreceptors: An ATP-dependent stimulation of cyclic nucleotide phosphodiesterase by light. *Proc. Natl. Acad. Sci. U.S.A.* **76,** 3820–3824.

Miki, N., Baraban, J., Keirns, J., Boyce, J., and Bitensky, M. (1975). Purification and properties of the light-activated cyclic nucleotide phosphodiesterase of rod outer segments. *J. Biol. Chem.* **250,** 6320–6327.

Miller, W., and Nichol, G. (1979). Evidence that cyclic GMP regulates membrane potential in rod photoreceptors. *Nature (London)* **280,** 64–66.

Mullen, R., and LaVail, M. (1976). Inherited retinal dystrophy: Primary defect in pigment epithelium determined with experimental rat chimeras. *Science* **192,** 799–801.

Orr, H., Lowry, O., Cohen, A., and Ferrendelli, J. (1976). Distribution of 3′:5′-cyclic AMP and 3′:5′-cyclic GMP in rabbit retina *in vivo:* Selective effects of dark and light adaptation and ischemia. *Proc. Natl. Acad. Sci. U.S.A.* **73,** 4442–4445.

Pannbacker, R., Fleishman, D., and Reed, D. (1972). Cyclic nucleotide phosphodiesterase: High activity in a mammalian photoreceptor. *Science* **175,** 757–758.

Parry, H., Tansley, K., and Thomson, L. (1955). Electroretinogram during development of hereditary retinal degeneration in the dog. *Br. J. Ophthalmol.* **39,** 349–352.

Santos-Anderson R., T'so, M., and Wolf, D. (1980). An inherited retinopathy in collies: A light and electron microscopic study. *Invest. Ophthalmol. Visual Sci.* **19,** 1281–1294.

Schmidt, S. and Lolley, R. (1973). Cyclic nucleotide phosphodiesterase: An early defect in inherited retinal degeneration of C3H mice. *J. Cell Biol.* **57,** 117–123.

Shinozawa, T., Uchida, S., Martin, E., Cafiso, D., Hubbell, W., and Bitensky, M. (1980). Additional component required for activity and reconstitution of light-activated vertebrate photoreceptor GTPase. *Proc. Natl. Acad. Sci. U.S.A.* **77,** 1408–1411.

Sidman, R., and Green, M. (1965). Retinal degeneration in the mouse. Localization of the *rd* locus in linkage group XVII. *J. Hered.* **56,** 23–42.

Singer, A., Dunn, A., and Appleman, M. (1978). Cyclic nucleotide phosphodiesterase and protein activator in fetal rabbit tissue. *Arch. Biochem. Biophys.* **187,** 406–413.

Strada, S., and Thompson, W. (1978). Multiple forms of cyclic nucletoide phosphodiesterase: Anomalies or biologic regulators? *Adv. Cyclic Nucleotide Res.* **9,** 265–283.

Thompson, W., and Appleman, M. (1971). Multiple cyclic nucleotide phosphodiesterase activities from rat brain. *Biochemistry* **313,** 311–316.

Ulshafer, R., Garcia, C., and Hollyfield, J. (1980). Elevated levels of cGMP destroy rod photoreceptors in the human retina. *Invest. Ophthalmol. Visual Sic., Suppl.* p. 38.

Wheeler, G., and Bitensky, M. (1977). A light-activated GTPase in vertebrate photoreceptors:

Regulation of light-activated cyclic GMP phosphodiesterase. *Proc. Natl. Acad. Sci. U.S.A.* **74,** 4238–4242.

Wheeler, G., Matuo, Y., and Bitensky, M. (1977). Light-activated GTPase in vertebrate photoreceptors. *Nature (London)* **269,** 822–824.

Wolf, E., Vainisi, S., and Santos-Anderson, R. (1978). Inherited rod-cone dysplasia in the collie. *J. Am. Vet. Med. Assoc.* **173,** 1131–1133.

Woodford, B., Chader, G., Farber, D., Liu, L., Fletcher, R., Santos-Anderson, R., and T'so, M. (1981). Cyclic nucleotide metabolism in inherited retinopathy in collies: A biochemical and histochemical study. *Invest. Ophthalmol.* (submitted for publication).

Woodruff, M., Bownds, D., Green, S., Morrisey, J., and Shedlovsky, A. (1977). Guanosine 3′,5′-cyclic monophosphate and the *in vitro* physiology of frog photoreceptor membranes. *J. Gen. Physiol.* **69,** 667–669.

Yee, R., and Liebman, P. (1978). Light-activated phosphodiesterase of the rod outer segment: Kinetics and parameters of activation and deactivation. *J. Biol. Chem.* **253,** 8902–8908.

Young, R. (1976). Visual cells and the concept of renewal. *Invest. Ophthalmol.* **15,** 700–725.

Chapter 9

Control of Rod Disk Membrane Phosphodiesterase and a Model for Visual Transduction

P. A. LIEBMAN

Department of Anatomy
School of Medicine
University of Pennsylvania
Philadelphia, Pennsylvania

AND

E. N. PUGH, JR.

Department of Psychology
University of Pennsylvania
Philadelphia, Pennsylvania

I. PROPERTIES OF TRANSDUCTION

The transduction of single-photon absorptions into the closure of Na^+ channels and a change in membrane potential that serves vision in vertebrate rods has a number of characteristic properties that must ultimately be understood in terms of their underlying biochemistry and biophysics. The important properties and their implications that concern us are

1. Rods respond to single photons (Yau *et al.*, 1977).
2. The semisaturation intensity is 30 photons (1 photon changes outer segment membrane conductance by 3%) (Penn and Hagins, 1972; Baylor *et al.*, 1979a).
3. The production of a reliable signal requires that a rod outer segment (ROS)

ISBN 0-12-153315-8

contains at least 3600 modulatable conductance channels (Yoshikami and Hagins, 1973).

4. Points 2 and 3 imply that a single photon closes at least 100 channels of the ROS plasma membrane (overall gain $\geqslant 10^2$).

5. Communication between one photon absorption and 100 or more plasma membrane channels requires a cytoplasmic transmitter, since absorber and channels occupy separate membranes. There must be 100 or more transmitter particles released per photon to deal with the 100 or more Na^+ channels.

6. The membrane conductance decreases after single-photon activations is slow and kinetically complex, embodying a fourth- or higher-order delay (Penn and Hagins, 1972; Baylor *et al.*, 1979a). The delay is longer in larger rods.

7. The decrease in conductance in response to single photons occurs smoothly and with near-identical amplitude, delay, and subsequent time course from trial to trial (Baylor *et al.*, 1979b).

8. Adaptation mechanisms are associated with an exchange of gain for speed (bandwidth) (Hagins, 1979). Photoreceptors thus behave like electronic operational amplifiers with a constant gain–bandwidth product.

The Ca^{2+} hypothesis of visual transduction (Hagins, 1972) has provided ideas and in some cases has led to experimental data consistent with several of the points on this list (Hagins, 1979). It has not attempted to address questions concerning the detailed coupling mechanisms between photon absorption and Ca^{2+} release from the rod disk membrane (RDM). In particular, no published version of it or simple idea traceable to it has attempted to explain points 6–8 dealing with speed, statistical reproducibility, or adaptation. As will be explained later, the elementary notion that conversion of a photon-hit rhodopsin molecule to, for example, metarhodopsin II (Meta II) might provide a temporary channel for the efflux of Ca^{2+} stored in the disks is inconsistent with points 6 and 7. Point 8 has not been addressed.

In this chapter, we summarize kinetic evidence from our laboratory that leads to a rather different model of visual transduction in which the diffusional interaction of rhodopsin with other enzymes on the surface of the RDM provides the earliest and rate-limiting steps of this process (Yee and Liebman, 1978; Liebman and Pugh, 1979, 1980a). Quantitative aspects of this model are consistent with or provide ready insight into all eight of the above features. The final stage of the model involves Ca^{2+}, but all of the control of gain, speed, and sensitivity originates in modulation of the cytoplasmic cyclic guanosine 3′,5′-monophosphate (cGMP) concentration.

II. METHOD

RDM were isolated in darkness from cow, rat, frog, or toad retinas by sucrose floatation in media containing 100 m*M* KC1, 2 m*M* $MgCl_2$, 1 m*M* dithiothreitol,

100 μM EDTA, and 20 mM pH 8 buffer (HEPES, MOPS, PIPES, or Tris, depending on a later pH adjustment). In cuvette suspensions containing 1–10 μM rhodopsin, guanosine 5′-triphosphate (GTP), and cGMP, bleaching results in rapid acidification of the medium, which we have shown to be caused by hydrolysis of cGMP in the reaction (Liebman and Evanczuk, 1981).

$$\text{cGMP}^- + \text{H}_2\text{O} \xrightleftharpoons{\text{PDE}} \text{GMP}^- \rightleftharpoons \text{GMP}^{-2} + \text{H}^+$$

This very useful property allows the activity of the light- and GTP-dependent enzyme, cGMP phosphodiesterase (PDE), to be measured in real time where controller effects may be easily studied (Liebman and Evanczuk, 1981). Several auxiliary enzymes appear to be involved. Like PDE, they are largely membrane peripheral proteins carried on the RDM (Kühn, 1980; Pober and Bitensky, 1979).

III. MODEL FOR CONTROL OF PDE ACTIVITY

Using the above assay, we quickly observed a preferential need for a GTP cofactor and that the system was much more sensitive to weak light than had been previously realized (Yee and Liebman, 1978). Fast measurements using pH-sensitive dyes showed a <50-msec activation delay after a 10% flash bleach. Subsequent work with J. H. Parkes, Jr., has shown that the delay tracks the thermally sensitive formation kinetics of Meta II (Parkes and Liebman, 1980). Our related work suggests that the duration of activation does not exceed the lifetime of Meta II (Liebman and Pugh, 1980b). Thus Meta II appears to be the activator of PDE. (Meta II and R* are identical in the subsequent discussion.)

The maximum velocity of our preparations was 15–30 moles of cGMP hydrolyzed per second per mole of rhodopsin, quite remarkably higher than had been previously suggested. More remarkable is the behavior of this velocity with light titration, as shown in Fig. 1. These curves show:

1. A bleach of 1 in 10^4 rhodopsin molecules produces nearly the same maximal activation that 1 in 6 gives, but the weak activation suffers a significant time delay.

2. Weak bleaches (10^{-4}–5×10^{-6}) cause the velocity to increase after a time delay whose duration is inversely proportional to the velocity.

3. Very weak bleaches ($<5 \times 10^{-6}$) cause a constant time delay, though activity continues to diminish with bleach.

4. The weakest bleach shown (10^{-6}) causes a velocity increase of 3 μmoles sec^{-1} in a sample of 3 μM rhodopsin. This implies a hydrolysis rate of 10^6 cGMP molecules sec^{-1} per R* molecule. If this result means that one R* molecule activates one PDE molecule, the turnover number (k_{cat}) for PDE is 10^6 sec^{-1}. This result can be compared to that for another very fast enzyme, acetylcholin-

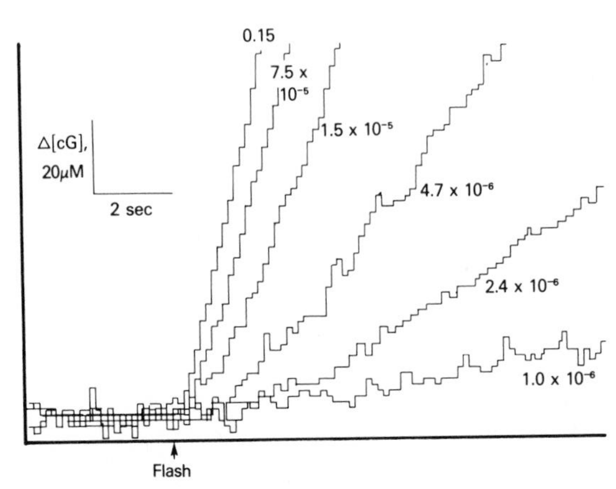

FIG. 1. Proton production (cGMP hydrolysis) as a function of time before and after flash bleaching the fraction of contained rhodopsin labeling the curves. *Bufo* RDM suspension (3 μM rhodopsin) in medium containing 8 mM cGMP and 500 μM GTP at 25°C; 1-msec flash duration.

esterase, whose k_{cat} is 10^4 with a K_m of 100 μM (the K_m of PDE for cGMP is also ~ 100 μM). In fact, the ratio k_{cat}/K_m for all of the fastest enzymes is only a little above 10^8 and is limited to this value by the speed of aqueous diffusion of substrate to the enzyme and orientation at the active site (Fersht, 1977). Clearly, PDE cannot exceed this limit, and the ratio $k_{cat}/K_m = 10^{10}$ that we have found must mean that *no fewer than* 50–100 PDE molecules, each operating at or below the diffusion limit, $k_{cat}/K_m = 10^8$, are required to explain the result. This result implies that 1 R* molecule must be able to communicate activation to at least 100 PDE molecules. Since this calculation was published (Yee and Liebman, 1978; Liebman and Pugh, 1979), a k_{cat} measurement of 2100 sec^{-1} has been obtained for purified PDE (Baehr *et al.*, 1979). Using this number, we found that in fact nearly 500 PDE molecules were activated by a single bleached rhodopsin molecule. It is interesting that the velocity increase associated with the weakest bleaches in Fig. 1 followed the mathematical form $V(t) = V_0(1 - e^{-t/\tau})$; i.e., it was first-order, with a time constant τ of 2.5 sec. We will return to this point later.

The data in Fig. 1 are used to plot the light titration curve in Fig. 2 (open circles). The curve is a saturating function, linear with bleach at the weakest intensity and showing half-saturation at about 10^{-5} bleach. However, the shape of the curve is not hyperbolic. It is rather like a Poisson sum, as was found by Liebman and Pugh (1979) for bovine RDM. This is consistent with the view that an individual R* molecule can activate all the PDE molecules within a large region or domain containing nearly 10^5 unbleached rhodopsin molecules. Two or more bleached rhodopsin molecules in the same domain can do no more than

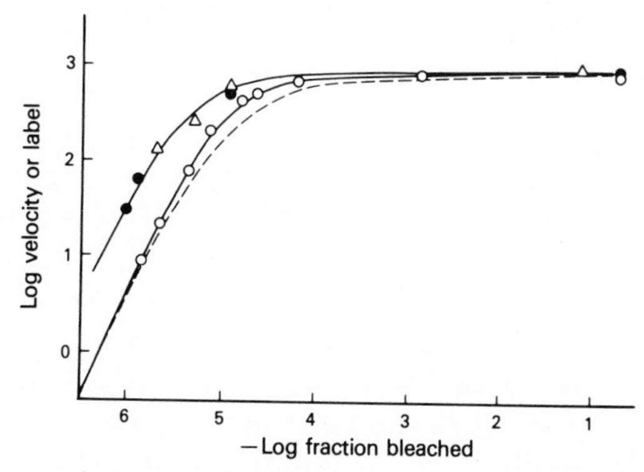

FIG. 2. Light titration of PDE velocity. Maximum slope of each curve of Fig. 1 plotted versus bleach in GTP (open circles). Similar data using GMPP(NH)P in place of GTP (solid circles) give same domain size as light titrations where the amount of [$\alpha-^{32}$P]GTP incorporation (open triangles) is measured. Dashed line is a hyperbolic saturation function fitted to linear and saturation regions of data. Solid lines are Poisson sums fitted to the data.

one, though the job of activation may be achieved with less delay if several R* molecules are present. Thus careful light titration shows activation to occur in domains of about 80,000 rhodopsin molecules. Within these domains *any* bleached rhodopsin molecule can fully activate the 500 PDE molecules suggested above.

How could such a one-on-many activation occur? Since both rhodopsin and PDE molecules are bound to the membranes, it seems natural to suggest that they must come into contact with each other directly on these membranes after bleaching to mediate the activation. A possible vehicle for this contact is the lateral diffusion of the constituents of the membranes (Liebman and Entine, 1974). If both rhodopsin and PDE molecules moved through Brownian agitation with about the same speed on the membrane surface, their relative motion would be governed by a diffusion coefficient (Liebman and Entine, 1974) D of about 1 $\mu m^2\ sec^{-1}$. Recent work has shown that molecules moving thus about a surface of area A will "sweep out" this area in a characteristic time τ_D of about A/D (Szabo *et al.*, 1980). A domain of 80,000 rhodopsin molecules occupies an area of about 2.5 μm^2 (30,000 μm^{-2}) and would be swept out with a τ_D of about 2.5 sec. It is significant that the characteristic time for activation of PDE at the weakest bleaches in the frog or toad RDM is just of this magnitude, 2.5 sec.

Furthermore, in collaboration with Charles Noback in this laboratory, we have computer-simulated such a diffusion-limited activation and have shown the build-up to be roughly first-order, the time constant A/D being about 2.5 sec for the present case. It thus seems plausible that activation is a lateral diffusion rate-limited process wherein one bleached rhodopsin molecule comes into contact with the PDE molecules one by one and activates them on the membrane.

Though plausible, the above scheme is thermodynamically repugnant. How could each PDE molecule "remember" to stay active after the R* molecule has separated and while the other PDE molecules are being recruited? This may be the role of GTP, the obligatory cofactor of activation. Each time a R*-PDE contact is made, R* might mediate (catalyze) the binding of a GTP molecule from solution to the PDE molecule or to a GTP-binding protein (GBP) which in turn activates or disinhibits the dormant PDE molecule. The PDE molecule would then remain active in the absence of R* as long as the GBP and GTP molecules stayed attached. The R* molecule would move on to "find" another GBP and/or PDE molecule while the activity of those previously activated continued (Yee and Liebman, 1978; Liebman and Pugh, 1979). This would produce a linear buildup of activated enzyme with time, and the cGMP hydrolyzed or H^+ formed would be the integral of this (velocity) ramp. The ramp would diminish in slope as the PDE content of a domain became exhausted, thus becoming first-order.

Several features of our measured kinetics are consistent with this picture. It is possible to calculate the approximate lifetime of the encounter complex between a rhodopsin and a PDE molecule, constrained only by the membrane viscosity. This is only about 5–10 μsec; i.e., a rhodopsin molecule might stay in contact with a PDE molecule for only 5–10 μsec before it departs on its diffusional path for elsewhere (Liebman and Pugh, 1980c). Our hypothesis clearly demands that a GTP molecule, if it is to be effective, must collide with the R*-PDE complex during its lifetime. GTP titration at a constant weak bleach shows that 25 μM GTP is required for half-maximal activation (Liebman and Pugh, 1980c). The calculated intercollision interval for GTP collisions with a binding site of the same size is about 10 μsec at 25 μM GTP. Thus theory predicts that about half the collisions might be effective at 25 μM GTP, in agreement with experimental data. With a strong bleach, where many R* molecules provide R*-PDE encounter complexes very much more frequently, theory predicts that the GTP requirement for activation should fall. Experiment shows this to be the case, half-maximal activation being achieved at ~0.5 μM GTP with a 10% bleach. Finally, work using radioactively labeled GTP or GTP analogues shows directly that hundreds of GTP molecules do in fact become bound for each R* molecule produced with weak bleaches (Liebman and Pugh, 1980c) (Fig. 2).

It thus seems clear that light converts rhodopsin to an unusual kind of enzyme that mediates the activation of hundreds of PDE molecules by catalyzing the binding of the allosteric activator, GTP. The approximately 500 activated PDE molecules in turn each hydrolyze cGMP molecules at the rate of about 2000 sec^{-1} for a total of as much as 10^6 sec^{-1} per R* molecule. Thus a two-stage amplifier is at work to lower cytoplasmic cGMP (Liebman and Pugh, 1979). Its first stage of gain can act only after a significant diffusion delay at single-photon levels.

What is it that determines the domain size of 80,000 rhodopsin molecules in the frog and toad and why is the domain size only 25,000 rhodopsin molecules or fewer for bovine RDM? How can the enormous amplification unleashed by a photon be arrested once again? At first the domain size appeared to be compatible with whole-disk vesicle activation (whole disks in cows or lobes of disks for amphibians), and we obtained preliminary verification of this by electron microscopy (Liebman and Pugh, 1979). However, numerous exceptions became obvious, and it also became clear that the light-activated guanosinetriphosphatase (GTPase) activity that had to be included in our model would have an important effect on domain size. Presence of the light-activated GTPase activity is shown in Fig. 3 where the activation cofactored by 200 nM GTP vanishes with a time constant of 10 sec and successive additions of GTP cause the activation to recover and then decay after an identical period. If instead of GTP, the nonhydrolyzable analogue GMPP(NH)P is used, the characteristic deactivation phase of

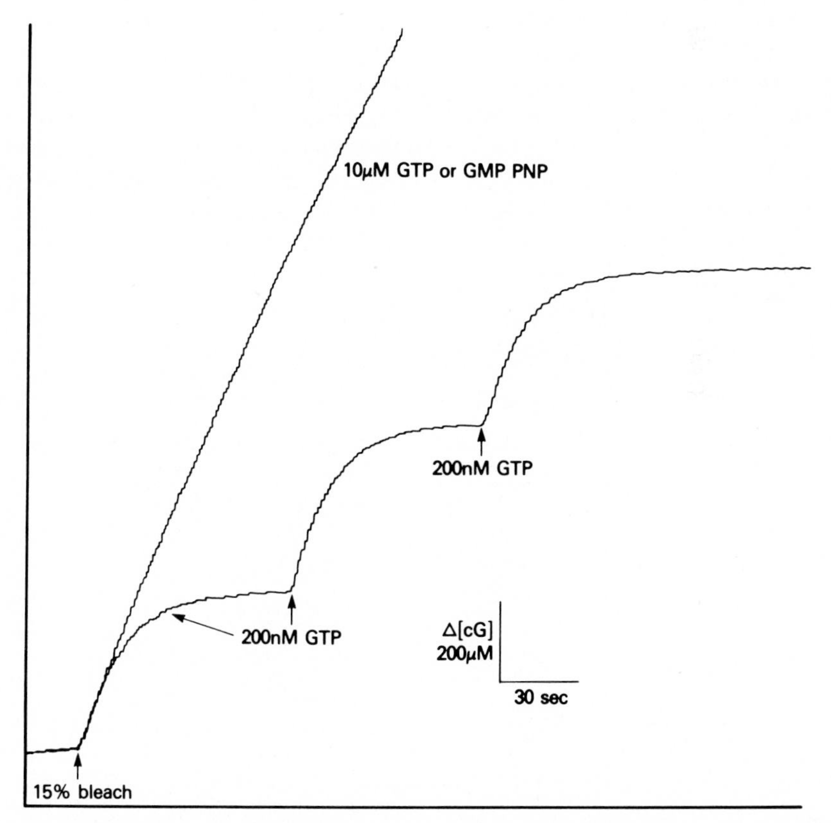

FIG. 3. The hydrolysis of cGMP in response to 15% bleach undergoes repetitive activation–decay cycles upon successive additions of 200 nM GTP. The decay phase seems to be missing at higher concentrations of GTP or GMPP(NH)P.

GTP disappears and activity continues, confirming that GTP hydrolysis can terminate activity and simultaneously showing that activation is not mediated through membrane protein phosphorylation. As expected, activity *continues* for long periods at a higher GTP concentration (Figs. 3 and 4). On the other hand, with weak bleaches, high concentrations of GTP cause *termination* of activity after several seconds rather than the continued activity found with strong bleaches (Liebman and Pugh, 1980a). (Fig. 4). What is the origin of this paradoxical behavior? Our model of PDE activation makes it clear that activity can be terminated only when both R* and GTP-activated PDE are inactivated. Thus GTPase activity should not by itself be sufficient, since R* will continue to cause further activation or reactivation of PDE through repeated catalysis of GTP binding. Figure 5 suggests that it is adenosine 5′-triphosphate (ATP) that was missing from the inactivation ingredients and that GTP can weakly mimic its activity-quenching effect (Liebman and Pugh, 1980a). There is little chance that ATP produces its effect by competing with GTP for the activator site, since a higher GTP level cannot compensate for the effects of even micromolar amounts of ATP. Instead, a high GTP level has the same effect as ATP, leading to early termination of activity (Liebman and Pugh, 1980a).

ATP can quench activity at any time after activation and can do so even in the absence of GTP (Liebman and Pugh, 1980a). The possibility of allosteric activation of a quencher without phosphate transfer, in analogy with the activator role

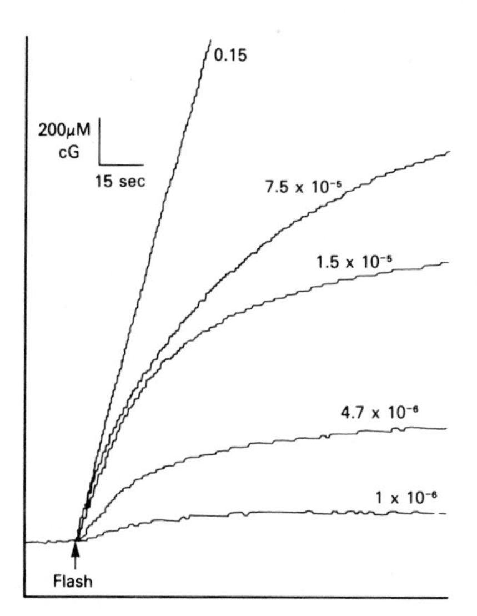

FIG. 4. Same data as in Fig. 1 on a slower time scale showing spontaneous decay of PDE light-induced activity with weak bleaches in the presence of excess (500 μM) GTP. These activation–decay cycles can be induced repeatedly on successive light flashes and are *not* due to GTP depletion of the surrounding medium.

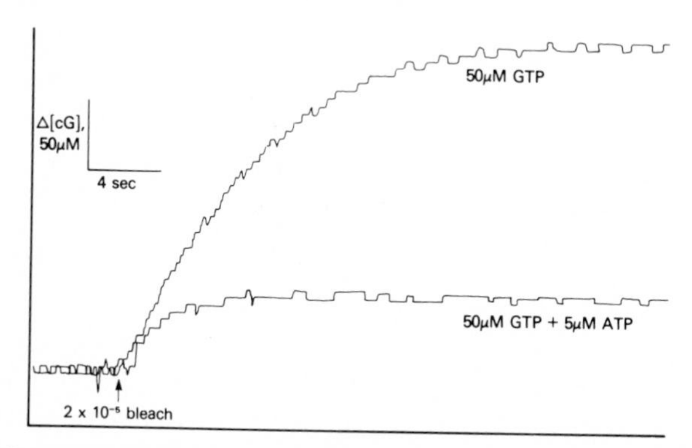

FIG. 5. PDE activity in response to weak bleach is terminated much more quickly when 5 μM ATP (lower curve) is added to the 50 μM GTP (upper curve). The initial velocity is very little affected by ATP.

of GTP, is not supported, for AMPP(NH)P leads to inhibition of the quench rather than to its support (Liebman and Pugh, 1980a). The K_m for quenching effectiveness of ATP is compatible with a role in rhodopsin phosphorylation. We tentatively conclude that it is the (multiple) phosphorylation of rhodopsin that quenches the ability to activate PDE (Liebman and Pugh, 1980a). We believe this reaction is very fast at the level of a single bleached rhodopsin molecule per domain, probably occurring in about 100–300 msec (Liebman and Pugh, 1980c). We also find the GTPase does its job in 1–3 sec at millimolar ambient GTP levels.

Under physiological conditions therefore R* activates PDE via catalysis of GTP binding, at the same time making itself available as a phosphorylatable substrate for opsin kinase. It continues to play its activator role until completely phosphorylated. We anticipate that the coupling efficiency of MII with PDE (GDP) diminishes gradually to zero with successive phosphorylation. Active PDE molecules independently deactivate with a rate constant characteristic of the membrane-bound GTPase. This enzyme appears to be light-activated but probably is so only insofar as light provides it with a substrate in the form of the GTP that becomes bound to GBP or PDE. Once GTP is split by this enzyme to give guanosine 5′-diphosphate (GDP) plus P_i, the activity of both the PDE and GTPase stops. When all PDE molecules have had their GTP hydrolyzed, the cycle of activation–deactivation is complete (Liebman and Pugh, 1980a).

A similar cycle of activation–deactivation seems to occur in hormonal control of adenylate cyclase by the hormone–receptor complex and GTP. Deactivation occurs upon both loss of hormone and GTP hydrolysis (Cassel *et al.*, 1977). The GDP resulting from hydrolysis remains bound in this case (Cassel and Selinger, 1978). GDP binding has been confirmed in RDM PDE-GBP (Fung and Stryer, 1980). The binding of ring-labeled or [α-^{32}P]GTP–GDP to RDM upon activation therefore becomes a valuable tool for testing our multiple PDE activation and control hypothesis. Thus we predict that labeled GTP will be bound upon light activation and remain bound upon deactivation after weak bleaches, while repeated nucleotide exchanges should occur with strong bleaches. Light titration of the amount of label incorporated should yield domains rather larger than those of the GTP-mediated enzyme velocity titration where a steady state allows only a fraction of the total PDE labeled to be active at any moment. In fact, domain sizes determined via labeling should match closely those obtained from light titration of PDE velocity using nonhydrolyzable GTP analogues, since each PDE molecule activated should remain active in the absence of GTP hydrolysis, just as the label remains attached. Our experimental results confirm this prediction; velocity and radiolabel domain sizes become comparable for nonhydrolyzable cofactors (Liebman and Pugh, 1980d) (Fig. 2). Furthermore, we find that the effect of ATP is to reduce the number of PDE molecules labeled as expected if the kinase is able to terminate quickly the activator role of R* by phosphorylation (Liebman and Pugh, 1980d).

The chemical equations and a summary diagram of our model for the control of PDE and its possible role in visual excitation are shown in Fig. 6. Our own studies have not carried us beyond the rapid loss of cGMP, but we can speculate on the role played by it. First, it seems clear that the reactions we are studying are among the fastest known in biochemistry, cGMP hydrolysis proceeding near the aqueous diffusion limit on light activation and the quenching reactions also being very rapid. Second, the power released by activation of a single enzyme molecule does not seem to be enough. The system was designed to amplify the

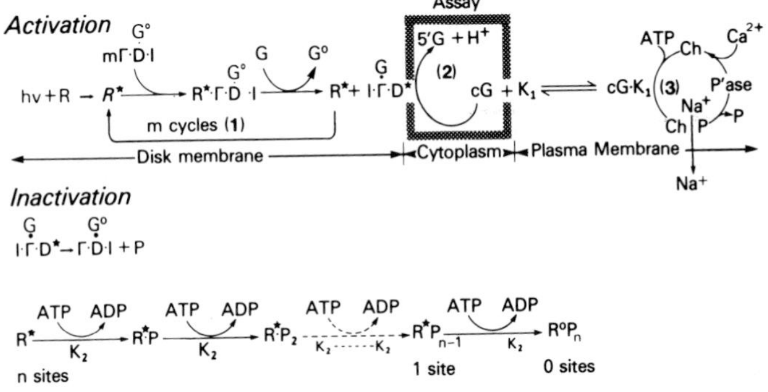

FIG. 6. Liebman–Pugh economy model of visual transduction by enzyme cascade: an abbreviated accounting of events discussed in this chapter. Γ, GTP-binding protein–GTPase complex: D, PDE; I, PDE inhibitor, G, GTP, K_1, cGMP-dependent protein kinase; G^0, GDP; K_2, rhodopsin (opsin) kinase. (1)–(3) indicate delays and gain stages. Stage (1) embodies two delays because its effect on cGMP concentration appears as the integral of the (increasing) number of PDE molecules diffusionally activated.

monomolecular, single-photon absorption event in rhodopsin into an effect on hundreds of enzyme molecules. Finally, this amplifier can be shut off again very quickly. At first not fully appreciative of the strength and speed of the ATP-mediated turnoff, we determined that a single photon could cause about 1% of cytoplasmic cGMP to be hydrolyzed during the interval between bleaching and receptor potential response peak (Liebman and Pugh, 1979). This suggested that cGMP depletion might actually be the sought-after cytoplasmic transmitter instead of Ca^{2+}. In the presence of the full ATP effect, however, we found that only 0.01% or 10^4 molecules of cGMP were hydrolyzed per photon with weak bleaches (Liebman and Pugh, 1980a). This is still a considerable amplification of one photon, and cGMP may still be a transmitter, but its effects are likely to be quite local, since diffusion of cGMP from other regions of the ROS would cause the concentration change to be attenuated and to recover quickly.

Moreover, the chemical mechanisms discussed here deal only with two stages of delay, though probably the first and longest ones. Electrophysiology tells us that at least two further delays must follow the cGMP depletion step. The role of cyclic nucelotides in other biochemical systems is to control the phosphorylation of proteins (Greengard, 1978). The control of cyclic nucleotide-dependent protein kinases by cyclic adenosine 3′,5′-monophosphate (cAMP) or cGMP is mediated with a stoichiometry of two or one nucleotides, respectively, per enzyme (Gill *et al.*, 1977; Lincoln *et al.*, 1977). This would be a no-amplification step, as would the reverse reaction of cGMP dissociation, but could embody another delay. The subsequent action of the kinase on its target proteins and the reverse phosphatase reaction would provide a fourth delay, probably with enzymatic gain. Finally, Ca^{2+} might bind to the target if it were a Na^+ channel, or might be released from binding sites upon phosphorylation. This kind of system might accommodate Ca^{2+} as an essential but passive (unmodulated) element needed to block a Na^+ channel. Availability of channels for blocking might thus be controlled entirely by cyclic nucleotides, Ca^{2+} playing a permissive role in a manner consistent with all the physiological results of Ca^{2+} manipulation.

Two further points need to be raised. The first is the question of recovery of the lost cGMP after the opsin kinase and GTPase arrest the depletion process. If the depletion caused by a single photon is as small as we believe, local concentration would recover quickly at the expense of distant regions serving as a reservoir for their neighbors. This would cause a pancytoplasmic loss of cGMP concentration, attenuated by 100- to 1000-fold relative to the maximum instantaneous local effect (the ratio of local PDE-depleted volume to the total). For flashes containing 100 or more photons, local depletion would begin to make a dent in the total concentration after diffusional equilibrium, and with 10^3-10^4 photons, longer-lasting depletion would occur as all regions of the ROS cytoplasm become "local" to absorbed photons. But what of the active synthetic source of cGMP, guanylate cyclase? What little evidence there is suggests that this ROS enzyme is neither directly modulated by light nor indirectly by physiological amounts of Ca^{2+} (Fleischman and Denisevich, 1979; Bitensky *et al.*, 1975). Quantitative calculations (P. A. Liebman, unpublished calculations based on data of Fleischman *et al.*, 1980) based on published turnover measurements (Fleischman and Denisevich, 1979; Fleischman *et al.*, 1980) and other information (D. Fleischman, personal communication) suggest that the fully activated PDE can deplete ROS cGMP about 1000 times faster than cyclase can replace it. Since at physiological light levels rods probably use only $^{1}/_{1000}$ of their total PDE, cyclase activity would be enough to keep pace with depletion and simple product inhibition might be adequate to control this activity. The distribution of cyclase along the axonemal spine of ROS (Fleischman *et al.*, 1980) seems well positioned for a rapid response to local cGMP concentration changes. Its separation from the rhodopsin-bearing disk membranes, however, militates against a light activation mechanism similar to that of PDE and in favor of the need for a cytoplasmic transmitter if a light effect is to occur.

Finally, it is of interest to consider an additional virtue of the lateral diffusion-mediated, multiple PDE activation mechanism. A virtue previously alluded to was the way in which a fundamental limit on gain or turnover number of $k_{cat}/K_m = 10^8$ can be overcome by turning on many copies of the enzyme without at the same time needing to increase K_m. The additional virtue relates to the need for reliability in rod signaling of single-photon absorptions (point 7 in Section I) and can be appreciated by considering the manner in which signals might be transduced were they instead mediated by a single molecular effect of light absorption. Consider the formation of conducting channels in black lipid films or at neuromuscular junctions (Bean *et al.*, 1969; Gordon and Haydon, 1972; Hladky and Haydon, 1972; Neher *et al.*, 1978). Here a single channel exists in one of two states, on or off. In the presence of a signal, the channel is turned on after a delay t_1 needed to pass through a transition state and into the conducting state. Upon removal of the signal, it is turned off after a delay t_2 required for dissociation. In a large number of single trials, both t_1 and t_2, are found to vary in an exponentially distributed manner about the mean times τ_1 and τ_2 (Neher *et al.*, 1978). These are, respectively, the mean delay and lifetime observed when a large number of channels are stimulated simultaneously by an impulse. Both turning on and turning off are chemical first-order processes. Similarly, if rhodopsin were a unimolecular Ca^{2+} channel formed upon bleaching with some mean lifetime, the delay until Ca^{2+} was released and the duration of Ca^{2+} release would vary from trial to trial in an exponentially distributed manner associated with variable amounts of Ca^{2+} release.

Were the activation of a single PDE molecule by a single bleached rhodopsin molecule responsible for transduction, there would be a similarly great variation in delay and in duration. The reason for this is simple. There are only a few PDE molecules among many rhodopsin molecules. Some rhodopsin molecules will be near a PDE molecule when bleached, and others far away. Nearby molecules will come into contact sooner than distant ones. Since any rhodopsin molecule can bleach on any single trial, the delay to PDE activation will sometimes be long and sometimes be short, while the PDE active state itself will sometimes be long and sometimes short, since the action of GTPase is also first-order, individual PDE molecule lifetimes becoming exponentially distributed. Such a mechanism could not account for the tightly controlled delay, amplitude, and time integral of ion flux of individual photon responses of rods (Baylor *et al.*, 1979).

The mechanism we describe here *could* account for these features. The activation of many PDE molecules and the multiple phosphorylations of bleached rhodopsin mediated by opsin kinase could produce the necessary averaging of monomolecular events necessary to constrain both delay and lifetime (and therefore integral activity) in a manner consistent with properties of the transduction process. Whether these events are used for this purpose remains to be seen. The work presented by Miller and Nicol (Chapter 24, this volume) (Miller Nicol, 1979; Nicol and Miller, 1978) suggests that they are indeed part of the transduction mechanism.

ACKNOWLEDGMENTS

This work was supported by NIH grants EY00012, EY01583, and EY00102.

REFERENCES

Baehr, W., Devlin, M. J., and Applebury, M. L. (1979). *J. Biol. Chem.* **254,** 11669.
Baylor, D. A., Lamb, T. D., and Yau, K.-W. (1979a). *J. Physiol. (London)* **288,** 589.
Baylor, D. A., Lamb, T. D., and Yau, K.-W. (1979b). *J. Physiol. (London)* **288,** 613.
Bean, R. C., Shepherd, W. C., Chan, H., and Eichner, J. T. (1969). *J. Gen. Physiol.* **53,** 741.
Bitensky, M. W., Miki, N., Kierns, J. J., Kierns, M., Baraban, J. M., Freeman, J., Wheeler, M. A., Lacy, J., and Marcus, F. R. (1975). *Adv. Cyclic Nucleotide Res.* **5,** 215.
Cassel, D., and Selinger, Z. (1978). *Proc. Natl. Acad. Sci. U.S.A.* **75,** 4155.
Cassel, D., Levkovitz, H., and Selinger, Z. (1977). *J. Cyclic Nucleotide Res.* **3,** 393.
Fersht, A. (1977). "Enzyme Structure and Mechanism." Freeman, San Francisco, California.
Fleischman, D., and Denisevich, M. (1979). *Biochemistry* **18,** 5060.
Fleischman, D., Denisevich, M., Raveed, D., and Pannbacker, R. G. (1980). *Biochim. Biophys. Acta* **630,** 176.
Fung, B. K.-K., and Stryer, L. (1980). *Proc. Natl. Acad. Sci. U.S.A.* **77,** 2500.
Gill, G. N., Walton, G. M., and Sperry, P. J. (1977). *J. Biol. Chem.* **252,** 6443.
Gordon, L. G. M., and Haydon, D. A. (1972). *Biochim. Biophys. Acta* **255,** 1014.
Greengard, P. (1978). *Science* **199,** 146.
Hagins, W. A. (1972). *Annu. Rev. Biophys. Bioeng.* **1,** 131.
Hagins, W. A. (1979). *In* "The Neurosciences: Fourth Study Program" (F. O. Schmitt and F. G. Worden, eds.). pp. 183–191. MIT Press, Cambridge, Massachusetts.
Hladky, S. B., and Haydon, D. A. (1972). *Biochim. Biophys. Acta* **274,** 294.
Kühn, H. (1980). *Nature (London)* **283,** 587.
Liebman, P. A., and Entine, G. (1974). *Science* **185,** 457.
Liebman, P. A., and Evanczuk, A. T. (1981). *In* "Methods in Enzymology" (in press).

Liebman, P. A., and Pugh, E. N., Jr. (1979). *Vision Res.* **19,** 375.
Liebman, P. A., and Pugh, E. N., Jr. (1980a). *Nature (London)* **287,** 734.
Liebman, P. A., and Pugh, E. N., Jr. (1980b). In preparation.
Liebman, P. A., and Pugh, E. N., Jr. (1980c). In preparation.
Liebman, P. A., and Pugh, E. N., Jr. (1980d). In preparation.
Lincoln, T. M., Dills, W. L., and Corbin, J. D. (1977). *J. Biol. Chem.* **252,** 4289.
Miller, W. H., and Nicol, G. D. (1979). *Nature (London)* **280,** 64.
Neher, R., Sakmann, B., and Steinbach, J. H. (1978). *Pfluegers Arch.* **375,** 219.
Nicol, G. D., and Miller, W. H. (1978). *Proc. Natl. Acad. Sci. U.S.A.* **75,** 5217.
Parkes, J. H., Jr., and Liebman, P. A. (1980). In preparation.
Penn, R. D., and Hagins, W. A. (1972). *Biophys. J.* **12,** 1073.
Pober, J. S., and Bitensky, M. W. (1979). *Adv. Cyclic Nucleotide Res.* **11,** 265.
Szabo, A., Schulten, K., and Schulten, Z. (1980). *J. Chem. Phys.* **72,** 4350.
Yau, K.-W., Lamb, T. D., and Baylor, D. A. (1977). *Nature (London)* **269,** 78.
Yee, R., and Liebman, P. A. (1978). *J. Biol. Chem.* **253,** 8902.
Yoshikami, S., and Hagins, W. A. (1973). *In* "Biochemistry of Visual Pigments" (H. Langer, ed.), pp. 245–255. Springer-Verlag, Berlin and New York.

CURRENT TOPICS IN MEMBRANES AND TRANSPORT, VOLUME 15

Chapter 10

Interactions of Rod Cell Proteins with the Disk Membrane: Influence of Light, Ionic Strength, and Nucleotides

HERMANN KÜHN

Institut für Neurobiologie der Kernforschungsanlage Jülich
5170 Jülich, Federal Republic of Germany

ISBN 0-12-153315-8

I. INTRODUCTION

An increasing number of light-activated enzymatic reactions in ROS[1] have been detected in the past decade, mostly involved in guanosine and adenosine nucleotide metabolism. Among them are PDE which needs both light and GTP for its activation (Pannbacker *et al.*, 1972; Miki *et al.*, 1973; Goridis and Virmaux, 1974; Baehr *et al.*, 1979; Liebman and Pugh, 1979); one or several GTPase(s) (Wheeler *et al.*, 1977; Biernbaum and Bownds, 1979; Caretta *et al.*, 1979; Robinson and Hagins, 1979), one or several ATPase(s) (Thacher, 1978; Uhl *et al.*, 1979a), and light-activated phosphorylation of rhodopsin by a kinase using both ATP and GTP as substrates (Bownds *et al.*, 1972; Kühn and Dreyer, 1972; Chader *et al.*, 1976). All these enzymes are expected to play a role in visual transduction and/or adaptation, but their precise roles have not yet been established.

A common feature of these reactions is that they are activated by *visible light,* i.e., by light that is not absorbed by the enzyme proteins themselves but by rhodopsin. In all cases where information about the enzyme proteins is available (Kühn *et al.*, 1973; Miki *et al.*, 1975; Baehr *et al.*, 1979; Kühn, 1980b), they are reported to be colorless proteins which cannot absorb visible light; and in all cases where the action spectrum of enzyme activation has been measured (Bownds *et al.*, 1972; Keirns *et al.*, 1975; Wheeler *et al.*, 1977), it matches the absorption spectrum of rhodopsin. This implies that some type of molecular interaction between bleached rhodopsin and the enzymes, direct or mediated by other molecules, must be involved in enzyme activation. The recent observation (Kühn, 1978, 1980a,b) that certain ROS proteins undergo a strong light-induced binding to the disk membrane may reflect this interaction involved in enzyme activation. Before describing this binding reaction, a short review of the protein inventory in ROS will be given.

II. METHODS

A. Preparation of Rod Outer Segments

Bovine ROS were purified from freshly dissected retinas by centrifugation on stepwise sucrose gradients as described earlier (McDowell and Kühn, 1977). The

[1]The following abbreviations are used in this chapter. ROS, rod outer segment(s); cGMP, cyclic guanosine 3′,5′-monophosphate; PDE, cyclic-GMP phosphodiesterase; PAGE, polyacrylamide gel electrophoresis; SDS, sodium dodecyl sulfate; DTT, dithiothreitol; EDTA, ethylenediaminetetra-acetic acid; K, thousand (e.g., 37K = 37,000); GTP, guanosine 5′-triphosphate; GTPase, guanosinetriphosphatase; ATP, adenosine 5′-triphosphate; ATPase, adenosine triphosphatase; PAS, periodic acid–Schiff.

buffer used throughout contained 70 m*M* sodium or potassium phosphate (pH 7), 1 m*M* $MgCl_2$, 0.1 m*M* EDTA, and 1 m*M* DDT. Electron microscopy of the purified ROS showed fragments 2–20 μm in length (mostly about 5 μm) with regularly packed disks, surrounded by an apparently intact plasma membrane (Krebs and Kühn, 1977). The A_{280}/A_{500} ratio of the unlysed ROS was between 2.0 and 2.4. The ROS were stored as pellets frozen at −70°C under argon. All experiments described in this chapter were done with bovine ROS.

B. Extraction of Proteins in Darkness and Light

ROS pellets were suspended under dim red light in the appropriate buffer with gentle homogenization. All buffers contained 1–2 m*M* DTT (the only exception was the 1 *M* NH_4Cl solution—see Section IV,B and Fig. 1g–i—which contained no DTT). Aliquots of the suspension were pipetted into transparent 0.5-ml centrifuge tubes and prewarmed to 20°C. The "light" samples were illuminated for 2–3 min at 20°C from the side through a water bath and, normally, through an orange Plexiglas filter ($\lambda > 540$ nm), bleaching about 50–80% of their rhodopsin. At the end of the illumination period, both the bleached and the dark-kept samples were cooled in ice water and then centrifuged for 7–30 min at 50,000 *g*. The supernatants were carefully removed and centrifuged again for 30–60 min at 50,000 *g* to make certain that they contained no membranous material. Normally, the supernatants were already clear and free of rhodopsin after the first centrifugation. Occasionally, an ultracentrifuge was used (220,000 *g*), leading to the same results. The pellets were resuspended in the dark for subsequent extractions with various buffers, using gentle homogenization. When the release of proteins from their light-induced binding to the membranes was to be studied (see, for example, Fig. 4), the suspensions were not cooled after illumination but further incubated at 20°C in the dark for various lengths of time before centrifugation. (For details, see Kühn, 1978, 1980a,b, 1981.) The supernatants were assayed for kinase or GTPase activity and for their polypeptide composition using SDS-PAGE.

C. Assay of Rhodopsin Kinase Activity

The kinase activity of extracts was assayed as described (Kühn, 1978), using washed and alum-treated ROS membranes as substrates in measuring the light-induced and kinase-catalyzed phosphate incorporation into rhodopsin from added [γ-^{32}P]ATP. The alum-treated ROS membranes were devoid of intrinsic kinase activity. The incubation conditions were 70 m*M* phosphate (pH 7), 3–4 m*M* [^{32}P]ATP, 4m*M* $MgCl_2$, and 5 m*M* DTT, with continuous white light at 30°C for 30 or 60 min.

D. GTPase Assay

The ^{32}P-labeled inorganic phosphate liberated from [γ-^{32}P]GTP (initial concentration 2–3 μM) was measured using the method of Neufeld and Levy (1969) with modifications (Kühn, 1980a,b; Kühn and Hargrave, 1981). The extracted GTPase showed no GTP-hydrolyzing activity in the absence of rhodopsin but became active when alum-treated ROS membranes were added with illumination. The GTPase activity of such mixtures was nearly independent of the amount of rhodopsin above a certain amount (4 μg rhodopsin per sample) but was strongly dependent (approximately linearly) on the amount of extracted GTPase added, thus providing a direct measure of the amount of GTPase present in extracts. The alum-treated ROS membranes alone were free of GTPase activity. When whole ROS were assayed for their intrinsic GTPase activity (e.g., Table I and Fig. 1a and b), the addition of alum-treated ROS was omitted. The buffer normally used contained either 50 mM Tris–Cl (pH 7.4), 2 mM $MgCl_2$, and 1 mM DTT, or 20 mM Tris–Cl, 5 mM $MgCl_2$, and 1 mM DTT, leading to the same results. GTPase activity was linear for at least 5 min, with a standard deviation from linearity normally below 5%. GTPase activities were expressed as picomoles of P_i liberated per minute at 30°C per amount of extract or ROS assayed (as specified in each case).

E. Gel Electrophoresis

In earlier studies (Kühn, 1978), the continuous buffer system of Weber and Osborn (1969) was used; electrophoresis was carried out for twice the length of time necessary to elute the tracking dye from the bottom of the gels, in order to achieve separation of the light-dependent bands at 68K and 48K from their light-independent neighbors (Fig. 3). The discontinuous gel system of Laemmli (1970) was found to give better resolution, particularly in the molecular-weight range of 30K–40K. This system was therefore mostly used in the studies presented here. It should be noted that both systems yield rather different apparent molecular weights for some of the ROS polypeptides by comparison of their mobilities with those of standard proteins. Furthermore, the sequence of certain polypeptide bands on the discontinuous gels is reversed as compared to continuous gels (Kühn, 1980b). For example, opsin (38K) migrates on top of the two GTPase polypeptides (37K and 35K) on continuous gels, but it migrates below them on discontinuous gels. Similarly, the light-dependent 48K polypeptide migrates faster than its light-independent neighbor on continuous gels (Fig. 3), but slower on discontinuous gels (Fig. 1). Furthermore, it was found that the separation and apparent molecular weight on discontinuous gels, particularly in the 30K–40K range, could be strongly influenced by slight modifications of the gel and buffer composition. Molecular weights were therefore always estimated from the continuous (Weber and Osborn, 1969) gels, using fluorescein-labeled chymotrypsinogen as an R_f marker. The gel system described by Swank and Munkres (1971) was used for the separation of small polypeptides.

III. PROTEINS IN ROD OUTER SEGMENTS

It has long been known that rhodopsin is the predominant protein in bovine and frog ROS (Bownds *et al.*, 1971; Heitzmann, 1972; Papermaster and Dreyer, 1974). Studies on the presence and characterization of other proteins have started only recently (Godchaux and Zimmerman, 1979a; Kühn, 1980b); exceptions are the PDE that was purified early and extensively studied (Miki *et al.*, 1975; Baehr *et al.*, 1979) and a high-molecular-weight membrane-bound protein of unknown function (Papermaster *et al.*, 1976).

A major difficulty in the investigation of the ROS protein inventory is discriminating between true ROS proteins and contaminants derived from other cell fragments. However, it seems reasonable to assume that enzymes (proteins) that undergo *light-induced* reactions are intrinsic to ROS, particularly if it is demonstrated that the reactions are related to rod physiology (for the cGMP system see, for example, Miller and Nicol, 1979) or that they occur *in vivo* (e.g., phosphorylation of rhodopsin, Kühn, 1974). Copurification of soluble proteins with rhodopsin on continuous density gradients has been used in a systematic study (Godchaux and Zimmerman, 1979a) as a criterion for discriminating between ROS proteins and contaminants; the major extractable proteins present in purified ROS preparations have thereby been designated intrinsic ROS proteins. However, for a number of polypeptide bands seen on SDS-PAGE of ROS preparations, particularly for the minor bands, it remains to be shown by more rigorous and specific methods (e.g., immunological methods) whether they are really intrinsic to ROS.

The polypeptide composition of bovine ROS purified by our routine method (Section II,A) is shown in Fig. 1a and b. It is for the most part similar to the composition published by Godchaux and Zimmerman (1979a) for their purified "intact ROS." The predominant protein is rhodopsin (opsin), but a number of additional polypeptides are also present. Staining for carbohydrate using PAS stain revealed two of the bands to be glycoproteins (Kühn, 1980b), namely, opsin and a high-molecular-weight polypeptide (>200K), in agreement with other reports (Molday and Molday, 1979; Bridges and Fong, 1980). Most of the proteins are soluble in aqueous buffers at low ionic strength (Fig. 1d, extractable proteins), whereas rhodopsin and the high-molecular-weight glycoprotein remain tightly associated with the membranes (Fig. 1c, intrinsic membrane proteins).

A. Intrinsic ROS Membrane Proteins

In thoroughly washed ROS membranes, the opsin band is by far the predominant polypeptide band (Fig. 1c). The second prominent band, the high-molecular-weight glycoprotein, comprises only 1–2% of the total material according to Coomassie staining (Papermaster and Dreyer, 1974; Krebs and Kühn, 1977). It has been shown that rhodopsin is located not only in the disk membrane but also in the plasma membrane surrounding the stack of disks (Jan and Revel, 1974). Because of their large number, the disks constitute about 97–99% of the total membrane area in ROS.

The polypeptide composition of washed disk membranes may, however, be more complex than it appears from Fig. 1c. Frog opsin has recently been separated on SDS-PAGE into three components of slightly different molecular size, each being a glycoprotein and binding a retinal chromophore (Molday and Molday, 1979). Such multiple forms have not been observed for bovine rhodopsin (Molday and Molday, 1979); instead, it has recently been suggested that the "opsin peak" of washed bovine disk membranes on SDS-PAGE contains a considerable amount of a nonopsin protein (Uhl *et al.*, 1979b).

B. Extractable ROS Proteins

In this section, the solubility of proteins of dark-kept bovine ROS will be discussed. Most of the proteins present in whole ROS are extractable at low ionic strength (Fig. 1d). The major extractable polypeptides are a doublet band at about 95K which has been identified as cGMP-PDE (Baehr *et al.*, 1979), a doublet near 50K, and a prominent doublet at 37K and 35K which binds GTP and GDP and carries GTPase activity (Godchaux and Zimmerman, 1979b; Kühn, 1980a,b). The apparent molecular weights vary somewhat among different reports and techniques used; Godchaux and Zimmerman (1979a,b) reported 41K and 37K for the GTPase doublet, and Baehr *et al.* (1979) reported 88K and 84K for the PDE doublet. (See also Section II,E which discusses the reliability of molecular-weight determinations using SDS-PAGE.)

A number of extractable ROS proteins are soluble regardless of the ionic strength of the buffer. In the following discussion these will be termed *soluble proteins*. Others are membrane-bound at moderate ionic strength and need extremes of ionic strength to be solubilized; they will be termed *peripherally bound membrane proteins*. This distinction serves practical purposes and does not exclude the possibility that some of the soluble proteins may be more-or-less loosely membrane-associated under physiological conditions.

1. Soluble Proteins

To extract the soluble proteins from ROS at moderate ionic strength (e.g., 70 or 100 m*M* phosphate or 100 m*M* Tris buffer), the surrounding plasma mem-

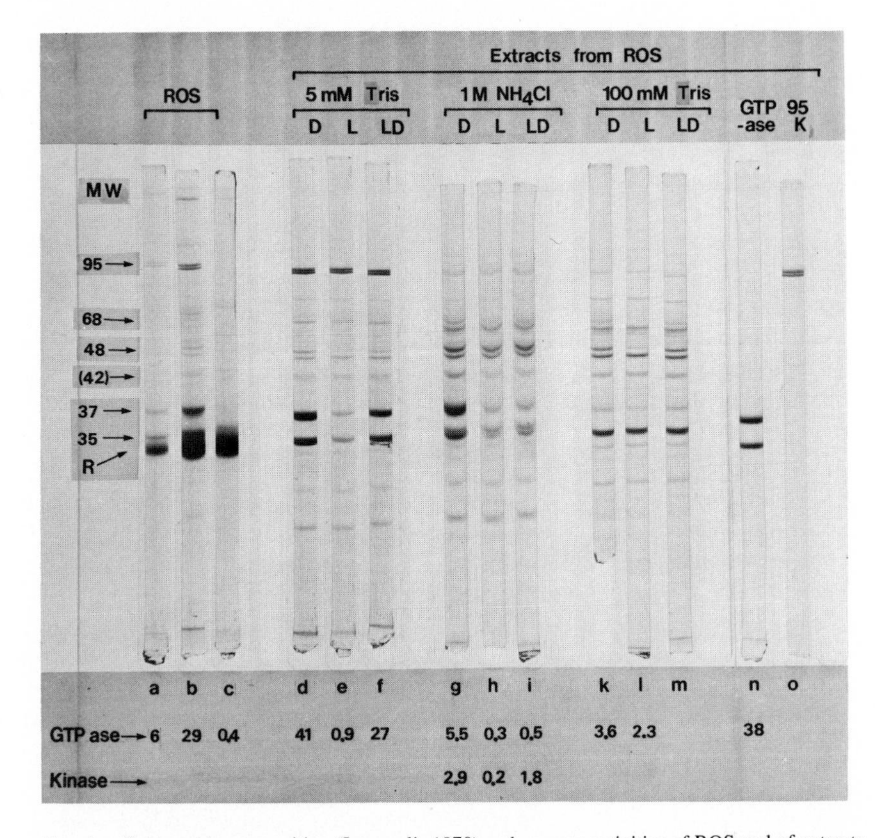

FIG. 1. Polypeptide composition (Laemmli, 1970) and enzyme activities of ROS and of extracts obtained from ROS in various buffers. The numbers on the left indicate molecular weights in thousands. The position of the rhodopsin (opsin) band (R) was identified by PAS staining of separate gels. Samples a b represent whole ROS containing 2.4 and 12 μg rhodopsin, respectively. Sample c ROS membranes after extensive hypotonic extraction; 12 μg rhodopsin. All other samples represent extracts of soluble proteins obtained in the buffers indicated at the top of the gels under the following illumination conditions: (D), dark extracts from unbleached ROS (D); extracts from bleached ROS centrifuged immediately after illumination (L); extracts from bleached ROS centrifuged after a 70-min dark incubation period at 20°C following illumination (LD). Extracts d–f are in 5 mM Tris-Cl and 0.5 mM $MgCl_2$; extracts g–i in 1 M NH_4Cl; and extracts k–m in 100 mM Tris-Cl and 0.5 mM $MgCl_2$ (see footnote 2). Sample n represents purified GTPase (2.8 μg protein), and sample o purified phosphodiesterase (Section VI).

GTPase and rhodopsin kinase assays were performed in each case with the same amount of extract or ROS, which was also applied to the corresponding gel. For details see Section II. GTPase activity is expressed as picomoles of P_i liberated per minute, and kinase activity as moles of phosphate incorporated per mole of rhodopsin (of alum-treated ROS) per 30 min at 30°C. (Reproduced with kind permission from Pergamon Press; Kühn, 1980b.)

brane must be disrupted. This is easily achieved by freezing and thawing (which, however, partially denatures some enzymes) or by homogenization. For quantitative extraction of soluble proteins, the stack of disks should also be mechanically disrupted, which is more difficult to achieve; it requires either strong homogenization or transient hypoosmotic shock (which has the disadvantage that peripheral proteins are also transiently solubilized).

The polypeptides extracted from dark-kept ROS after homogenization in 100 m*M* Tris-Cl are shown in Fig. 1k. The most prominent polypeptides are a doublet near 50K, and a band at about 35K. Both the doublets corresponding to PDE (95K) and to GTPase (37K and 35K) are absent. (The soluble 35K polypeptide seen in Fig. 1k differs from the GTPase 35K polypeptide seen in Fig. 1d, f, and n in its solubility at moderate ionic strength, in its slightly different mobility on modified Laemmli gels (Fig. 8), and in the fact that it does not undergo light-induced binding (Section IV,C). Note that much more extract was applied to the gel in Fig. 1k than to that in Fig. 1d in order to compensate for the lower extracting efficiency of the moderate-ionic-strength buffer.[2])

One of the minor proteins which is, for the most part, extractable at moderate ionic strength is *rhodopsin kinase,* the enzyme that catalyzes the light-induced phosphorylation of rhodopsin (Kühn, 1978). The active enzyme has a molecular weight of 69K ± 4K and consists most probably of one polypeptide chain of the same molecular weight (Kühn, 1978, 1980b).[3]

The presence of rhodopsin kinase and other soluble proteins in these purified ROS preparations indicates that these ROS are, on the average, sufficiently sealed to retain soluble proteins. On the other hand, no attempt was made to separate intact ROS from leaky ROS. If quantitative determination of soluble ROS proteins is intended, purification on isotonic density gradients leading to intact, sealed ROS (e.g., Schnetkamp *et al.,* 1979) should be used.

[2]Since 5 m*M* Tris buffer is more effective an extractant than 100 m*M* Tris or 1 *M* NH_4Cl, different amounts of extract were applied to gels in order to give approximately equal staining intensity. The amounts of extract applied to the gels correspond to the following amounts of rhodopsin present in the original ROS suspension before extraction: 35 μg rhodopsin for each of the samples applied to gels d-f, 80 μg rhodopsin for gels g-i, and 120 μg rhodopsin for gels k-m.

[3]A molecular weight of 50K-53K has recently been reported for "purified" rhodopsin kinase (Shichi and Somers, 1978). I find, however, that the active enzyme is eluted from Sephadex G-100 columns in 70 m*M* phosphate buffer with an apparent molecular weight of 69K ± 4K (Kühn, 1978). Analysis of the column fractions by SDS-PAGE showed that two polypeptides cochromatographed with kinase activity: a polypeptide of 50K-53K present in large amounts (corresponding to the light-independent band next to the light-dependent 48K band seen in crude extracts), and a 68K polypeptide present in much smaller amounts. We assume that the 68K polypeptide represents kinase, since it undergoes light-induced binding like kinase, and in a large number of extracts obtained under a variety of conditions, it was the only polypeptide whose presence always paralleled the presence of kinase activity.

2. Peripherally Membrane-Bound Proteins

Both PDE and GTPase are membrane-bound at moderate ionic strength (e.g., Ringer's solution, or 100 m*M* Tris buffer, phosphate buffer, NaCl, or KCl) but are soluble at low ionic strength (e.g., 5–15 m*M* Tris–Cl), in agreement with other reports (Miki *et al.*, 1975; Wheeler *et al.*, 1977; Bignetti *et al.*, 1978; Baehr *et al.*, 1979). This suggests that they are also membrane-bound under physiological ionic strength conditions. After extensive washing of ROS membranes with a buffer of moderate ionic strength, these two proteins are the only major extractable proteins that remain bound to the washed membranes (Section VI; Baehr *et al.*, 1979).

Losses of these two proteins during ROS purification are not likely to occur, even if the ROS are leaky, provided the ionic strength is sufficiently high to keep the proteins membrane-bound. The amount of these proteins has therefore been estimated by densitometry of Coomassie-stained gels. As compared to the amount of rhodopsin, about 12% of the GTPase polypeptide (37K) and about 4% of the 95K doublet are present in our bovine ROS preparations. If Coomassie blue stains the different proteins, including rhodopsin, with equal intensity, it can be estimated from this that 1 molecule of GTPase is present per about 8 rhodopsin molecules, and 1 PDE molecule per about 120 rhodopsin molecules. After purification of the GTPase, we found somewhat less GTPase (about 1 GTPase molecule per 15 rhodopsin molecules; Kühn, 1980a). These estimates are in the same range as recently published values: Godchaux and Zimmerman (1979a) estimated the relative amount of each of the two GTPase polypeptides to be 7% of the total ROS proteins; and Baehr *et al.* (1979) estimated that between 40 and 140 rhodopsin molecules were present for each PDE molecule.

IV. LIGHT-INDUCED BINDING OF PROTEINS

Several polypeptides that are extractable from dark-kept ROS become membrane-bound if the ROS are illuminated. These polypeptides sediment with the bleached membranes upon centrifugation and are therefore absent, or present in reduced amounts, in the supernatants. They can be extracted from the membranous pellets by various subsequent treatments. At least five polypeptides have been found to undergo this light-induced binding; their apparent molecular weights are 68K, 48K, 37K, 35K, and ~6K. The 68K polypeptide is the most likely candidate for rhodopsin kinase (Kühn, 1978) (see footnote 3). The three polypeptides of 37K, 35K, and ~6K together are associated with GTPase activity (Kühn, 1980a). PDE (95K) does *not* undergo this light-induced binding. Various aspects of the light-induced binding are exhibited under different ionic conditions, as will be described in the following sections.

A. Light-Induced Binding at Low Ionic Strength

Comparison of low-ionic-strength extracts obtained from *unbleached* ROS (Fig. 1d) with extracts from *bleached* ROS (Fig. 1e) clearly shows the light-dependent presence of four polypeptides (48K, 37K, 35K, ~6K).[4] The ~6K polypeptide, which migrates with the front band on the gels in Fig. 1, can be resolved using another type of gel (Swank and Munkres, 1971) where it migrates at an R_f of 0.61, indicating that its molecular weight is below 10K (6K is an approximate value).

The light-induced binding of the four polypeptides is slowly reversed in the dark: When ROS are illuminated and then further incubated for about 1 hr in the dark before centrifugation, the bound polypeptides are found to be partly released again into the soluble supernatant (Fig. 1f). This release is strongly temperature-dependent and does not take place at 0°C (Kühn, 1980b). The extent to which the bound polypeptides are released in the dark at 20°C was found to vary considerably among different ROS preparations; the reason for these variations is not clear at the present time.

The GTPase activities of the extracts are shown below the gels in Fig. 1d–f. The "dark" extract (Fig. 1d) contains about 40 times more GTPase activity than the corresponding "light" extract (Fig. 1e), and two-thirds of the bound GTPase is found subsequently to be released from the light-induced binding (Fig. 1f). It should be noted that the extracted GTPase is inactive in its soluble form and requires the addition of bleached membrane-bound rhodopsin for activation (Section II,D).

The light-induced binding of the extractable polypeptides is observed not only when suspensions of "whole" ROS are illuminated (Section II,B and Fig. 1), but also when previously separated components (i.e., purified disk membranes and extract containing soluble proteins) are mixed under illumination (Kühn, 1978; Kühn and Hargrave, 1981). Binding of the 37K and 35K polypeptides (GTPase) to bleached ROS membranes has also been observed under somewhat similar conditions by Godchaux and Zimmerman (1979b) (but has not been interpreted as binding due to previous illumination). These authors mixed hypotonic ROS extract with a preparation of previously bleached, washed ROS membranes (stored frozen) and found that the two GTPase polypeptides became membrane-bound. Virmaux (1975) has mentioned in an abstract that two

[4]Light–dark differences in the presence of the 68K polypeptide (kinase) are not easily seen under these low-ionic-strength conditions. This may be partly due to the fact that kinase is rapidly denatured at low ionic strength (H. Kühn, unpublished observations). It seems likely that the denatured enzyme does not undergo light-induced binding. Under some conditions, two light-dependent polypeptide bands were discerned in the 60K–80K region in hypotonic extracts (Kühn, 1980b).

"opsin-like proteins," which are presumably identical to the GTPase polypeptides, are extractable from dark-kept but not from bleached ROS.

To demonstrate light-induced binding of proteins, the membranes containing the bound proteins are normally separated from soluble proteins by sedimentation (Fig. 1 and Section II,B). When the disk membranes are subjected to flotation rather than sedimentation, the light-dependent polypeptides are similarly found to be bound to the floated disks as shown in Fig. 2. Smith *et al.* (1975) have shown that the majority of the disks remain osmotically intact when ROS are lysed in water; only the osmotically intact (i.e., swollen) disks float upon centrifugation in 5% Ficoll and can thus be separated from broken disks and other membranous material sedimenting under the same conditions. With the use of this technique, the *same yield* of floating (i.e., osmotically intact) disks is obtained from ROS in which 80% of the rhodopsin has been previously bleached (L in Fig. 2), as from unbleached ROS (D in Fig. 2). The bleached disks contain the light-dependent polypeptides (48K, 37K, 35K, ~6K), which can be subsequently extracted (extract L′ in Fig. 2), whereas the unbleached disks contain no such extractable polypeptides. This demonstrates unequivocally that the light-induced binding of the polypeptides occurs to the disk membrane.

B. Light-Induced Binding at High Ionic Strength

Both the GTPase polypeptides and the 68K polypeptide are soluble in 1 *M* NH_4Cl in the dark and become membrane-bound upon illumination of ROS. Under the conditions used in Fig. 1g–i, GTPase was bound irreversibly, whereas kinase (activities shown below the gels) was bound reversibly.

C. Light-Induced Binding at Moderate Ionic Strength

At moderate ionic strength, the two polypeptide bands of 68K (kinase; see footnote 3) and 48K are the most prominent light-dependent bands. Both are soluble in the dark and become reversibly membrane-bound upon illumination (Figs. 1k–m and 3a–c). The amount of kinase found in extracts from dark-kept ROS was up to 20 times greater than in corresponding extracts from bleached ROS (Kühn, 1978).

The rate at which the light-induced binding occurs is unknown. The time course of the release of the kinase, following its light-induced binding, is shown in Fig. 4. It is a slow reaction, taking about 1–2 hr at room temperature. The time course of this release does not correlate with the decay or formation of any of the slow photoproducts of bovine rhodopsin; it occurs more slowly than the decay of metarhodopsin II into metarhodopsin III but faster than the decay of metarhodopsin III (Kühn, 1978). The 48K protein is released, following its light-

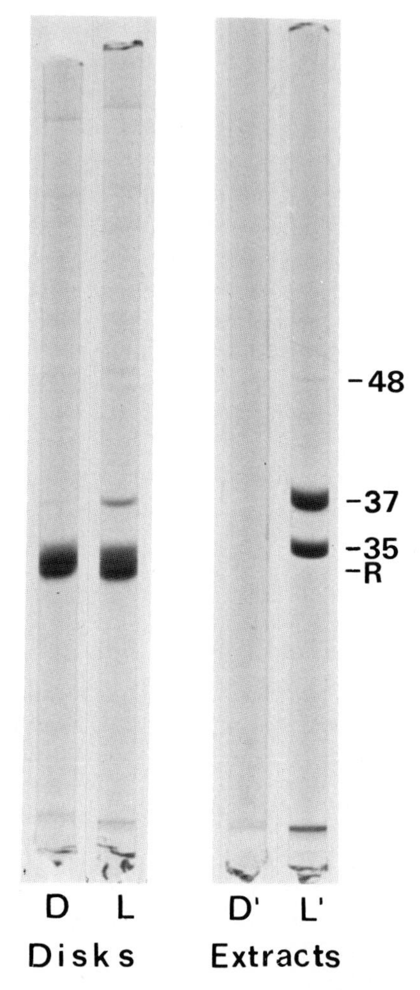

FIG. 2. Polypeptides present in disks purified from dark-kept (D) and bleached (L) ROS. Bovine ROS suspended in 100 m*M* Tris–Cl were bleached or kept unbleached, respectively. After sedimentation, the membranous pellets were lysed with water containing 1 m*M* DTT and 5% Ficoll. The osmotically intact disks were separated from other material by flotation (Smith *et al.*, 1975), and portions of them were then treated with 40 μM GTP in 5 m*M* Tris–Cl and 1 m*M* DTT to extract soluble proteins. The bleached disks contained, in addition to opsin, the polypeptides of 48K, 37K, 35K, and ~6K, which could be subsequently extracted (extract L′), whereas the unbleached disks contained no such extractable polypeptides (extract D′). The two gels on the left each contained 7 μg of rhodopsin (opsin), and each of the gels on the right contained the extract corresponding to 58 μg rhodopsin (opsin) in the disk suspensions before extraction.

induced binding, at about the same rate as the kinase (Kühn, 1980b). The extent to which both proteins are released in the dark in their soluble form was found to vary among different preparations. The occasional finding that more kinase was found in supernatants obtained at long times in the dark after bleaching than in supernatants of unbleached ROS (Fig. 4) may be explained by the assumption that the ROS preparations used were not completely dark-adapted.

It should be noted that not only kinase and the 48K protein but also GTPase undergoes light-induced binding at moderate ionic strength. This is not very obvious from the 100 m*M* Tris–Cl supernatants shown in Fig. 1k and 1 since most of the GTPase is membrane-bound at this ionic strength in both light and darkness. Only a small fraction (see footnote 2) of the 37K polypeptide (and GTPase activity) is seen to be soluble and to undergo light-induced binding. (Any light–dark difference in the amount of 35K polypeptide is masked by the presence of a light-independent, soluble polypeptide of approximately the same molecular weight.) However, the major fraction of the GTPase, which is membrane-bound because of the ionic strength, also undergoes light-induced binding, superimposed on the ionic strength-dependent binding, as will be shown in Section V,C.

V. ADDITIONAL FACTORS THAT INFLUENCE THE BINDING OF PROTEINS

A. Mg^{2+} Ions

$MgCl_2$ significantly enhances the binding of GTPase to disks in both darkness and light (Kühn, 1980b). This is demonstrated, for example, in Fig. 5 for the control samples [(D) and (L)] to which no GTP was added. More GTPase is extracted from unbleached ROS in the presence of 1 m*M* EDTA [Fig. 5B, (D) control] than in the presence of 1 m*M* $MgCl_2$ [Fig. 5A, (D) control]. Similarly, more GTPase is also extracted from bleached ROS in the presence of EDTA than in the presence of $MgCl_2$ [(L) controls in Fig. 5B and A, respectively]. Thus the *ratio* of GTPase extractability in darkness versus light is greater in the presence of 1 m*M* $MgCl_2$, whereas the absolute *amounts* of extracted GTPase are higher in the presence of EDTA. Relatively high concentrations of $MgCl_2$ (>0.2 m*M*) are required to enhance the binding of GTPase significantly. $CaCl_2$ was found to have a similar but somewhat stronger effect (H. Kühn, unpublished observations). The binding of kinase and of the 48K protein seems to be less influenced by Mg^{2+} (Kühn, 1978) than the binding of GTPase.

B. Nucleotides

When ROS suspensions at low ionic strength are bleached in the *presence* of GTP, the light-induced binding of GTPase does not occur, i.e., GTPase remains

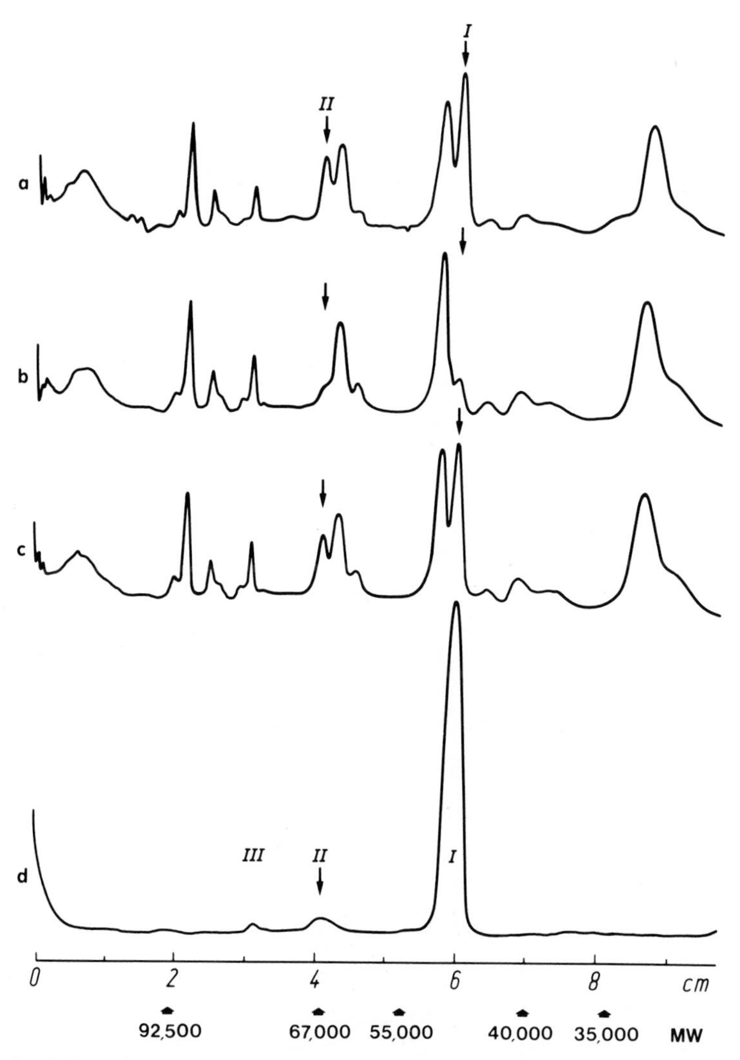

FIG. 3. Polypeptide composition of crude and purified extracts obtained from ROS at moderate ionic strength (70–74 m*M* phosphate). SDS-PAGE was performed in the continuous system described by Weber and Osborn (1969) with modifications. The ordinate (not shown) of the densitograms represents the relative absorbance at 580 nm of Coomassie-stained gels. (a) Extract from dark-kept ROS; (b) extract from ROS centrifuged immediately after bleaching; (c) extract from bleached ROS centrifuged 90 min after bleaching; (d) proteins purified by specific transient adsorption to bleached disk membranes (Section VI,C). Band I represents the 48K protein, and band II kinase (67K–68K). [Reproduced with kind permission from *Biochemistry* (Kühn, 1978, p. 4393).]

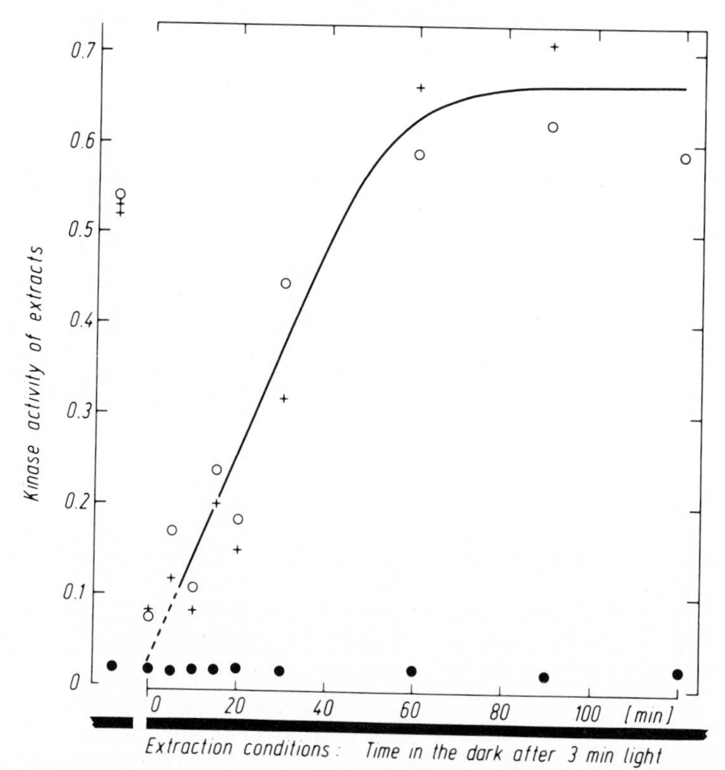

FIG. 4. Kinase extractability as a function of the time elapsed between bleaching and centrifugation. Each of 10 identical ROS suspensions in 70 m*M* phosphate (pH 7.0), 1 m*M* $MgCl_2$, and 0.1 m*M* EDTA was centrifuged without bleaching (which is the dark sample before time zero), immediately after bleaching (which is time zero), or at various times after bleaching as indicated on the abscissa. Kinase activity of the clear supernatants is expressed as moles of phosphate incorporated per mole of rhodopsin (of alum-treated ROS; see Section II,C) per hour. The curve was drawn by hand. [Reproduced with kind permission from *Biochemistry* (Kühn, 1978, p. 4391).]

soluble as in the unbleached control sample (Kühn, 1980a). When ROS are bleached in the *absence* of GTP such that the light-induced binding of GTPase occurs, subsequent addition of GTP causes rapid dissociation of the enzyme from the bleached membranes. This is demonstrated in Fig. 5 for various concentrations of GTP. Under the experimental conditions used, the light-induced binding is completely reversed by GTP concentrations higher than ca. 0.2 m*M*, both in the presence of $MgCl_2$ (Fig. 5A) and in the presence of EDTA (Fig. 5B). Lower concentrations of GTP cause a partial release of bound GTPase, and even a concentration as low as 1-3 μM GTP causes a small but significant release of GTPase in the presence of both Mg^{2+} and EDTA [compare with the (L) controls in Fig. 5A and B, respectively]. Obviously, Mg^{2+} is not needed for this solubiliz-

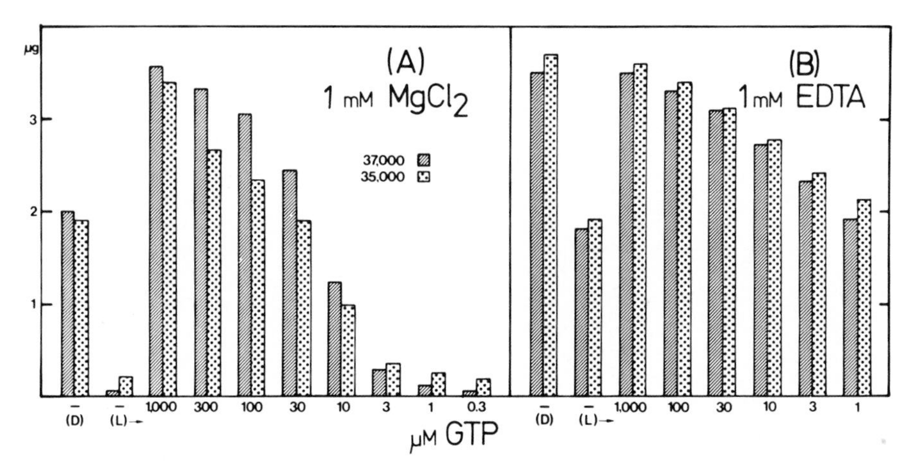

FIG. 5. Solubilization of 37K and 35K polypeptides from bleached ROS by various concentrations of GTP. All the 18 samples are aliquots of the same ROS suspension in 5 m*M* Tris–Cl. Either $MgCl_2$ (A) or EDTA (B) was added before bleaching. All the samples except the (D) controls were bleached at 20°C in the absence of GTP to induce the binding of GTPase to the membranes, and GTP was then added in the dark in the final concentrations indicated on the abscissa. Immediately after mixing, the samples were cooled to 0°C and centrifuged. The amount of 37K and 35K polypeptides present in the supernatants was determined from densitograms of stained SDS-PAGE gels and is shown on the ordinate. (Reproduced with kind permission from Pergamon Press; Kühn, 1980b.)

ing effect of GTP; it should be noted, on the other hand, that Mg^{2+} is required for the hydrolysis of GTP by GTPase, which is inhibited by EDTA (H. Kühn, unpublished experiments).

This solubilizing effect was found to be relatively, but not quite absolutely, specific for GTP (1980b). At a 200 μM concentration, GDP was found to be almost as effective as GTP, and the γ-blocked GTP analogue guanylyl imidodiphosphate was partially (about 30%) effective. ATP was effective in the presence of $MgCl_2$ but ineffective in the presence of EDTA. Many nucleotides tested, including cGMP, were completely ineffective (Kühn, 1980b).

In the cases of GDP and ATP, it cannot be excluded that small amounts of GTP, perhaps present as impurity or formed during the incubation in the presence of ROS, may in fact be responsible for the solubilization of the GTPase at these relatively high nucleotide concentrations (200 μM). Godchaux and Zimmerman (1979b) reported that at lower concentrations (20 μM), both GDP and ATP are much less effective than GTP, suggesting that the solubilizing effect may in fact be specific for GTP.

One conclusion that can be drawn from these specificity studies in the presence and absence of $MgCl_2$ is that hydrolysis of the γ-phosphate group of GTP is not a necessary step in solubilizing GTPase following its light-induced binding.

C. Ionic Strength

As has been seen in previous sections, ionic strength is an important factor in determining the solubility of PDE and GTPase, both being peripherally bound disk membrane proteins at moderate ionic strength in darkness as well as in light. It should be emphasized that the ionic strength and not the osmolality of the solution is important; uncharged solutes cannot replace salts. GTPase is, for instance, equally extractable from dark-kept bovine ROS in 5 m*M* Tris–Cl as in 5 m*M* Tris–Cl containing 300 or 600 m*M* sucrose (H. Kühn, unpublished experiments). Similarly, it has been shown by Bignetti *et al*. (1978) that both PDE and GTPase can be extracted from frog ROS in 200 m*M* mannitol solution but not in Ringer's solution.

The ionic strength-dependent binding of PDE and GTPase is *reversible*. If extract in 5–10 m*M* Tris buffer, containing all of the extractable ROS proteins, is added to washed disk membranes and the buffer concentration raised to 100 m*M*, both GTPase and PDE become preferentially membrane-bound, whereas the other proteins remain soluble (Kühn and Hargrave, 1981; see also Section VI,B).

Even though GTPase is already membrane-bound at moderate ionic strength in darkness, illumination causes an additional binding which is superimposed on the ionic strength-dependent binding. In the experiment shown in Fig. 6, ROS were suspended in 100 m*M* Tris–Cl, divided into two equal portions, and the L suspension was bleached. Both of the supernatants (D1 and L1) obtained after centrifugation contained only low amounts of GTPase activity and polypeptides because of the ionic strength-dependent binding of GTPase to the membranes. Subsequent treatment of the membranous pellets in the dark at low ionic strength caused solubilization of the PDE (95K doublet) from both pellets (Fig. 6, D2 and L2) but solubilization of GTPase only from the unbleached pellet. Bleaching in 100 m*M* Tris–Cl thus prevented GTPase from being extracted by the subsequent hypotonic treatment in the dark, indicating that light-induced binding had taken place at moderate ionic strength and persisted during the hypotonic treatment. The hypotonic treatment suffices to break the ionic strength-dependent type of binding but not the light-induced binding.

The membranous pellets were subsequently treated with 40 μM GTP at low ionic strength to extract the residual GTPase still bound to the membranes. As expected, much more GTPase (activity and polypeptides) could be extracted from the bleached pellet (extract L3 in Fig. 6) than from the unbleached pellet (D3) from which most of the GTPase had been previously extracted. The total amount of GTPase in the combined light extracts (L1–L3) was approximately the same as in the combined dark extracts (D1–D3).

Another example of light-induced binding at moderate ionic strength has been given in the disk flotation experiment shown in Fig. 2. Bleaching in 100 m*M* Tris–Cl (supernatants not shown) prevented GTPase from being eluted by the

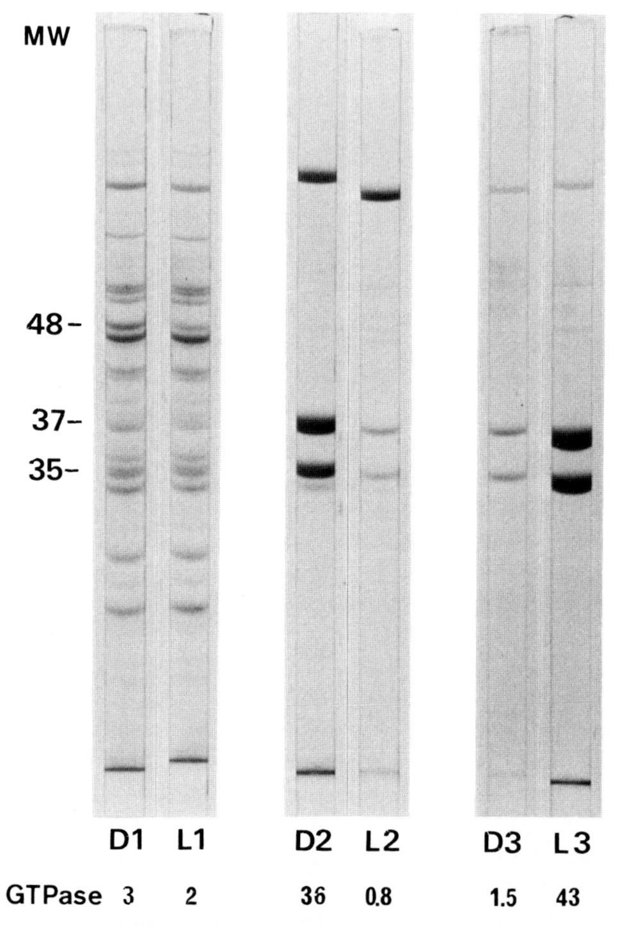

FIG. 6. Light-induced binding superimposed on ionic strength-dependent binding. The first two gels show the polypeptide composition of supernatants obtained after centrifugation of unbleached (D1) and bleached (L1) ROS suspended in 100 m*M* Tris–Cl. Most of the GTPase was absent from both supernatants because of the ionic strength-dependent binding to the membranes. Both the unbleached and bleached membranous pellets were then reextracted in the dark with 5 m*M* Tris–Cl, breaking the ionic strength-dependent binding and leading to supernatants D2 and L2, respectively. The final treatment, using 40 μM GTP and 5 m*M* Tris–Cl, broke the light-induced binding of the GTPase (supernatant L3). The GTPase activities of the extracts, corresponding to the same volumes applied to the gels, are shown below the gels.

subsequent hypotonic treatment with water–Ficoll such that it was still attached to the bleached disks after their purification by flotation, whereas the unbleached disks were free of GTPase (Fig. 2, L and D, respectively). Subsequent treatment of the disks with GTP eluted GTPase from the bleached disks, whereas no

GTPase could be eluted from the unbleached disks (Fig. 2, L′ and D′, respectively).

The results concerning the binding of GTPase under various conditions of light, ionic strength, and presence of GTP are schematically summarized in Fig. 7. GTPase is membrane-bound in 100 m*M* Tris buffer in darkness as well as in light because of the ionic strength-dependent interaction (x in Fig. 7). Light causes an additional interaction (y in Fig. 7). This superimposed light-induced interaction is not directly obvious upon centrifugation, since GTPase sediments with both the bleached and the unbleached membranes in 100 m*M* Tris–Cl. Subsequent treatment of the membranous pellets with low-ionic-strength buffer breaks the ionic strength-dependent binding but not the light-induced binding, such that the light–dark difference in the two samples becomes obvious upon centrifugation (Fig. 7, middle). GTP finally breaks the light-induced binding such that the GTPase is eluted from the bleached membranes.

The scheme shown in Fig. 7 poses a number of questions: Which one of the GTPase subunits interacts with the membrane? Are both types of binding (ionic strength-dependent and light-induced) located in the same subunit, or in different subunits as depicted in Fig. 7? What is the geometrical array of the GTPase subunits, and what is the stoichiometry and arrangement of the ~6K polypeptide with respect to the two large subunits? Where are the binding sites on the membrane? Recent experiments (Kühn and Hargrave, 1981) indicate that the light-induced binding site is located in the rhodopsin molecule. Purified disks

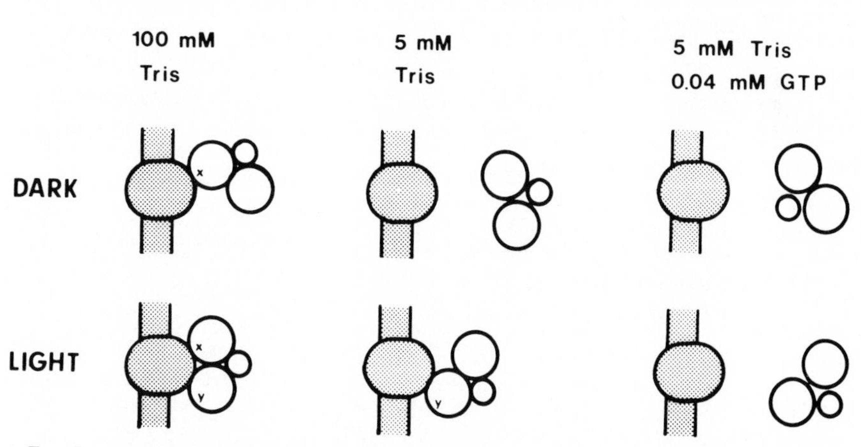

FIG. 7. Schematic presentation of GTPase binding as a function of ionic strength, light, and GTP. The membrane containing rhodopsin is represented by the shaded symbols, and the GTPase by the three open circles corresponding to the polypeptides of 37K, 35K, and ~6K. At moderate ionic strength, GTPase is membrane-bound (symbolized by x). Light induces an additional binding (symbolized by y) which is not broken by lowering the ionic strength (middle) but is broken by GTP (right side). It is not known whether the two types of binding (x and y) are located on the same subunit or on different subunits of the GTPase, nor is it known if both types of binding sites are located on the rhodopsin molecule as depicted.

were modified by limited proteolysis with thermolysin, leading to cleavage of rhodopsin into two large, noncovalently associated, membrane-bound fragments (Pober and Stryer, 1975) without changing the absorption spectrum of rhodopsin. GTPase extract was added to such modified disks and to untreated control disks for binding experiments in light and darkness. The light-induced binding (i.e., the light–dark difference in binding) of GTPase was found to be greatly diminished after this proteolytic modification of rhodopsin. Similarly, the solubilizing effect of GTP on the GTPase was also greatly diminished. Intactness of the major part of rhodopsin's polypeptide chain is thus required for both the light-induced binding of the GTPase, and its reversal by GTP, to take place. On the other hand, a more gentle modification of rhodopsin, removing only 12 amino acids from its carboxyl terminus, has no influence, neither on the light-induced binding nor its reversal by GTP (Kühn and Hargrave, 1981).

Table I shows that moderate ionic strength is as important as light activation in obtaining optimum GTPase activity. The ROS preparation used, which was obtained from cattle dark-adapted prior to slaughter, exhibited GTPase activities seven times higher in light than in darkness under standard assay conditions (50 m*M* Tris–Cl, 2 m*M* $MgCl_2$). This shows that, even with bovine ROS and working under dim red light, reasonably dark-adapted preparations can be obtained that show the same extent of light activation (a factor of 6–7) as previously reported for dark-adapted frog ROS prepared under infrared illumination (Wheeler *et al.*, 1977). Lowering the $MgCl_2$ concentration to 0.2 m*M* in the presence of 100 m*M* Tris buffer does not cause a strong change in GTPase activities in light or in darkness (Table I, second row). At low ionic strength (5 m*M* Tris–Cl, 0.2 m*M* $MgCl_2$), however, GTPase is nearly inactive even in light. This indicates that conditions under which the light-induced binding takes place (5 m*M* Tris, 0.2 m*M* $MgCl_2$, light) are not sufficient for GTPase to be active;

TABLE I

GTPase Activity as a Function of Ionic Strength and Light[a]

Ion concentrations during assay (m*M*)		GTPase activity (pmoles min^{-1})	
Tris–Cl	$MgCl_2$	In light	In darkness
50	2	29.3 ± 0.4	4.4 ± 0.3
100	0.2	26.8 ± 0.4	3.7 ± 0.3
5	0.2	3.2 ± 0.2	1.8 ± 0.2

[a] ROS suspensions prepared from dark-adapted bovine eyes were assayed for their intrinsic GTPase activity (i.e., without extraction and without the addition of alum-treated ROS). Each sample contained 11.1 μg rhodopsin (ROS) in a final volume of 200 μl. The [^{32}P]GTP concentration was 3 μM. Values are given as mean ± SE of three determinations.

conditions favoring the ionic strength-dependent binding are additionally required. It thus appears that both types of binding shown in Fig. 7, the light-induced binding *and* the ionic strength-dependent binding, are equally important for GTPase activity.

The observation, made at low ionic strength, that GTP reverses the light-induced binding of the GTPase (Figs. 2 and 5–7), raises the question whether this solubilizing effect of GTP is sufficiently strong to solubilize the GTPase even at moderate ionic strength, i.e., to break the ionic strength-dependent binding. Godchaux and Zimmerman (1979b) have shown that the GTPase can in fact be solubilized from previously bleached membranes using 20 μM GTP in 100 mM NaCl/buffer, whereas we failed to solubilize the GTPase from freshly bleached membranes under only slightly different conditions (Kühn, 1980b; H. Kühn, unpublished). It thus appears that slight variations in the experimental conditions can affect the solubility of the GTPase at moderate strength, and the question remains whether *in situ* the GTPase is membrane-bound or soluble.

Figure 8 shows a binding experiment in darkness and light in an attempt to approach intracellular conditions of nucleotide concentrations (1 mM GTP, 2 mM ATP; see Robinson and Hagins, 1979) and ionic strength. It is obvious that the GTPase is membrane-bound in darkness but becomes partially soluble upon illumination at these high nucleotide concentrations (supernatants D1 and L1, respectively). Interestingly, the two large subunits of the GTPase are partially separated under these conditions, the 37K subunit becoming more soluble than the 35K subunit (supernatant L1). This may be taken as a preliminary piece of evidence that the 37K subunit is the one that undergoes the light-induced binding followed by solubilization with GTP, whereas the 35K subunit is responsible for the ionic strength-dependent binding (see Fig. 7). Further extraction of the pellets with GTP in 50 mM Tris–Cl again shows that the GTPase is more easily extractable from the bleached than from the unbleached pellet (supernatants L2 and D2, respectively). The final treatment at low ionic strength solubilizes most of the residual GTPase still membrane-bound which is, of course, more from the dark pellet than from the bleached pellet (D3 and L3, respectively).

Even this experiment (Fig. 8) does not unambiguously answer the question whether the GTPase *in situ* is membrane-bound or soluble. It suggests, however, that the GTPase is membrane-bound at least in the dark; the light-induced increase in extractability of the 37K subunit, observed in the presence of GTP, may as well reflect an increase of its mobility within the membrane after it has reacted with photolyzed rhodopsin and GTP.

VI. PURIFICATION OF ROS PROTEINS USING TRANSIENT LIGHT-INDUCED BINDING

The different solubilities of some ROS proteins in light and darkness can be used to purify them in a few simple centrifugation steps carried out under the

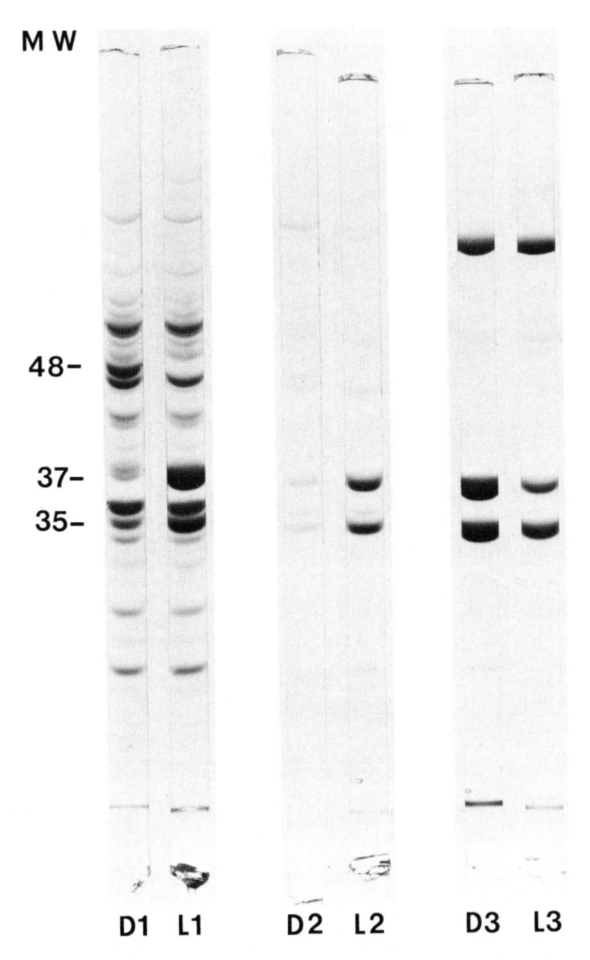

FIG. 8. Partial solubilization of GTPase at moderate ionic strength by light and high concentrations of GTP and ATP. The first two gels show supernatants obtained after centrifugation of unbleached (D1) and bleached (L1) ROS, suspended in KCl–Ringer's solution containing 130 m*M* KCl/20 m*M* NaCl/5 m*M* $MgCl_2$/8 m*M* Tris–Cl, pH 7.5/2 m*M* ATP/1 m*M* GTP. The 37K subunit of the GTPase is preferentially solubilized upon illumination at this high GTP concentration. In contrast, the 48K polypeptide undergoes quantitative light-induced binding, probably enhanced by the presence of the nucleotides (Kühn, 1980a,b). Both the unbleached and the bleached pellet were reextracted in the dark with 50 m*M* Tris–Cl/0.5 m*M* GTP (supernatants D2 and L2) and finally with 5 m*M* Tris–Cl/ca. 0.05 m*M* GTP (supernatants D3 and L3). The amount of supernatant applied to each gel corresponds to 92 μg of rhodopsin present in the original suspension before extraction.

appropriate conditions of ionic strength and illumination, using the disk membranes as an "affinity chromatography support" material for specific adsorption and desorption of the proteins (Kühn, 1981).

A. GTPase

The purification of GTPase is easy to perform because of its specific elution with GTP following light-induced binding to the membranes (Godchaux and Zimmerman, 1979b; Kühn, 1980a,b). Several methods are possible and nearly equally suitable. A way that is somewhat more convenient than a previously used method (Kühn, 1980a,b) is described in the following. The ROS are first washed thoroughly in the dark with 100 m*M* Tris buffer to remove the soluble proteins including the 48K protein (for details, see the next section). Subsequent bleaching in 5 m*M* Tris buffer induces the binding of GTPase to the disks, whereas PDE is eluted. The bleached pellet is washed once more with 5 m*M* Tris buffer, and GTPase is then eluted, using 40 μM GTP in 5 m*M* Tris buffer. DTT (1–2 m*M*) must be present in all the buffers. Purified GTPase can be separated from excess GTP using molecular sieve chromatography on a short column of Sephadex G-25 at 0°–4°C. The yield is normally about 70 μg of each of the two large polypeptides (37K, 35K) per milligram of rhodopsin.

The polypeptide composition and activity of purified GTPase is depicted in Fig. 1n. Even when 10 times more protein was applied to the gels than that shown in Fig. 1n, no polypeptides other than the 37K, 35K, and ~6K polypeptides were detectable. The ~6K polypeptide migrates in the front on the gels in Fig. 1 but is resolved on another type of gel (Swank and Munkres, 1971), as mentioned in Section IV,A. The 37K and 35K polypeptides are always present in approximately equal amounts, whereas the stoichiometry of the ~6K polypeptide with respect to them is not yet defined. The ~6K polypeptide accompanies the 37K and 35K polypeptides under all conditions tested: It is peripherally membrane-bound in 100 m*M* buffer and soluble in 5 m*M* buffer; it undergoes light-induced binding and is eluted from bleached membranes with the same nucleotides that elute GTPase. It cochromatographs with GTPase on molecular sieve columns. The following evidence suggests that it is not a product due to proteolytic cleavage of larger polypeptides: Its presence and quantity are independent of the presence of protease inhibitors (phenylmethylsulfonyl fluoride and Leupeptin, Sigma) during the preparation and extraction of ROS; and its quantity is not increased if ROS are incubated for 20 hr at room temperature in the absence of protease inhibitors.

B. Phosphodiesterase

The procedure for purifying PDE is largely similar to that described above for GTPase; in fact, both enzymes can be purified at the same time from the same

ROS preparation. The soluble ROS proteins must first be quantitatively extracted in the dark at moderate ionic strength where PDE and GTPase remain membrane-bound. It is important for quantitative extraction that not only the plasma membrane be broken but also that the disks be separated from each other. This is most conveniently achieved by hypotonic lysis of the ROS. The ROS are gently homogenized in 5 m*M* Tris buffer which dissociates all the extractable proteins from the membranes. Subsequent addition of Tris–Cl (100 m*M*) and $MgCl_2$ (1 m*M*) leads to reattachment of PDE and GTPase to the membranes, whereas the other proteins remain soluble and are removed by centrifugation. After two washings performed under these conditions, the ROS pellet is suspended in 5 m*M* Tris–Cl and 1 m*M* $MgCl_2$, and the suspension is bleached to induce the binding of GTPase. The resulting supernatant (Fig. 1o) contains PDE in nearly pure form. Following quantitative extraction of PDE by repeated washing of the bleached pellet, GTPase can be extracted in pure form using GTP, as described in the previous section.

The major source of impurity in the PDE preparation is residual GTPase, if the light-induced binding was incomplete. It is quantitatively separated from the PDE by column chromatography on Sephadex G-100 Superfine (Kühn, 1981; see also Baehr *et al.*, 1979). Purified PDE contains the high-molecular-weight (ca. 95K) polypeptide doublet and a third polypeptide (ca. 14K) which is not seen in Fig. 1o) but which is resolved using the gel system of Swank and Munkres (1971). This is in agreement with the polypeptide composition (88K, 84K, 13K) recently published by Baehr *et al.* for extensively purified bovine ROS PDE.

C. The 48K Protein

A similar but somewhat more elaborate procedure has been used to purify the 48K protein partially (Kühn, 1978). Previously purified washed disk membranes were mixed with extract containing the soluble proteins in 70 m*M* phosphate buffer and illuminated. The 48K protein and the kinase were preferentially bound to the bleached disks. The other proteins were removed by centrifugation, and the disks were then repeatedly incubated in the dark at 20°C for time periods of about 1 hr to allow for release of the bound proteins. The resulting extracts (Fig. 3d) contained mainly the 48K protein together with some kinase (68K) and a higher-molecular-weight polypeptide. The 48K protein was obtained in about 95% purity. Its function is as yet unknown; it may be a light-regulated enzyme or an inhibitor or an activator of an enzyme.

VII. CONCLUDING REMARKS

The data presented pose a number of questions, some of which will be discussed in this section.

A. Some Properties of GTPase and Its Possible Function

It has been shown that GTPase is a peripheral disk membrane protein at moderate ionic strength and is reversibly dissociated from the membrane by lowering the ionic strength. The molecular weight of the isolated GTPase is the same (80K–90K) in a solution of 10 m*M* or 100 m*M* Tris–Cl (H. Kühn, unpublished experiments); i.e., low-ionic-strength conditions leading to dissociation from the membrane do not lead to dissociation of the enzyme into its subunits (37K, 35K, ~6K). The solubilized enzyme is inactive in the absence of ROS membranes but is fully reactivated if rhodopsin-containing membranes are added at moderate ionic strength with illumination, i.e., under conditions where both ionic strength-dependent and light-induced binding to the membranes takes place.

Quantitative estimates show that the turnover of GTP hydrolysis by this GTPase is extremely slow. The molar ratio of GTPase versus rhodopsin is estimated to be about 1 GTPase per 10 rhodopsins; this estimate is an average obtained from gel scans of whole ROS (12 μg of 37K polypeptide per 100 μg of rhodopsin) and from the yield of purified GTPase (7 μg of each of the 37K and 35K polypeptides per 100 μg rhodopsin) (see also Section III,B,2). The maximum GTPase activity shown in Table I (Section V,C), as measured under standard assay conditions, is 2.6 pmoles of GTP hydrolyzed per microgram of rhodopsin per minute, corresponding to a turnover rate of 1 mole GTP hydrolyzed per mole of GTPase per minute at 30°C. In a number of different bovine ROS preparations, this turnover rate was always found to be between 0.8 and 1.4 moles $mole^{-1}$ $min.^{-1}$. Wheeler *et al.* (1977), and Caretta *et al.* (1979) report similarly low GTPase activities for frog ROS (500 pmoles GTP min^{-1} mg^{-1} ROS protein at 0.05% bleaching, and 2.5 nmoles min^{-1} mg^{-1} rhodopsin at 4% bleaching, respectively).

Normally, only micromolar concentrations of GTP are used in GTPase assays, i.e., three orders of magnitude lower than the GTP concentrations reported for intact frog ROS (ca. 0.7–2.0 m*M*; Biernbaum and Bownds, 1979; Robinson and Hagins, 1979). However, the K_m of the light-activated GTPase of frog ROS is only 0.5 μM GTP (Wheeler *et al.*, 1977) such that higher GTP concentrations would not lead to a much higher turnover. Extracted GTPase from bovine ROS, assayed in the presence of GTPase-depleted ROS membranes (Section II,D), has a similarly high affinity for GTP (H. Kühn, unpublished experiments); the apparent K_m was found to be 0.2 μM GTP, but straightforward Michaelis kinetics could not be applied since the enzyme was present in concentrations (ca. 0.1 μM) near the lower substrate concentrations. (The same may apply to the K_m measurements by Wheeler *et al.*, 1977.)

In any event, the turnover of this GTPase is extremely slow and is not strongly increased even at higher substrate concentrations because of the high affinity of the enzyme for GTP. This low turnover makes it very unlikely that GTPase serves as an ion pump in establishing ionic gradients. It rather suggests that GTP binding and hydrolysis by this GTPase serves to regulate other enzymatic activities in the rod, for instance, activation and deactivation of PDE as proposed by Wheeler and Bitensky (1977) and by Shinozawa and Bitensky (1980). The term "GTPase" should therefore be regarded as preliminary until its precise function is established. Baehr *et al.* (1979) have reported that this protein (which they termed 80K protein) inhibits the activity of purified PDE as assayed in the absence of ROS membranes. Godchaux and Zimmerman (1979b) have carefully studied the binding of GTP and GDP, and the hydrolysis of GTP, by this protein which they term "GTP binding protein." Fung and Stryer (1980) have shown that bleaching of one rhodopsin leads to up to 500 GTP/GDP exchanges. Very recently, Fung *et al.* (1981) have in fact demonstrated that the 37K subunit of GTPase in its GTP-binding form activates PDE.

It has been reported that light induces a relatively rapid (half-time in the range of 7 sec) hydrolysis of intrinsic GTP in intact frog ROS (Biernbaum and Bownds, 1979; Robinson and Hagins, 1979). It is not clear at present whether this GTP hydrolysis reflects the activity of the GTPase reported here and elsewhere (e.g., Wheeler and Bitensky, 1977) or whether a different enzyme is involved. Biernbaum and Bownds (1979) report a strong Ca^{2+} dependence of GTP hydrolysis, whereas isolated GTPase under our assay conditions is found to be independent of Ca^{2+} and to have the same activity in 0.5 m*M* Ca^{2+} and 2 m*M* Mg^{2+} as in 1 m*M* EGTA and 3 m*M* Mg^{2+} (H. Kühn, unpublished). However, the system of intact ROS is much more complex, such that a direct comparison is difficult.

It should be kept in mind that extractable GTPase is present in ROS in such a large amount that, if fully activated, it would be able to deplete the GTP pool fully within a few minutes in spite of its slow turnover rate. If it is assumed that in bovine ROS, as reported for frog ROS (Biernbaum and Bownds, 1979), ca. 0.3 mole GTP is present per mole of rhodopsin, this would imply that only three molecules of GTP are present for each molecule of GTPase. It should also be kept in mind that the turnover number of 0.8–1.4 min^{-1} (see above) was measured at 30°C rather than at 37°C, that the ROS preparations had been frozen before use, and that all three of the components (rhodopsin, GTPase, and GTP) in the assay mixture were diluted by a factor of about 1000 as compared to the conditions in native ROS. It is therefore likely that the rate of GTP hydrolysis, extrapolated to native conditions, is higher than that observed under our assay conditions, possibly approaching the rates of intrinsic GTP hydrolysis reported by Biernbaum and Bownds (1979) and by Robinson and Hagins (1979).

B. Possible Role of Light-Induced Protein Binding

The data presented here and elsewhere (Kühn, 1980b) suggest that light induces a profound change in the arrangement and interaction of proteins in the disk membrane, a change that is reversible in the dark and is strongly influenced by certain nucleotides, particularly GTP. Nucleotides influence not only the light-induced binding of GTPase but also that of kinase and of the 48K protein (Kühn, 1980b). Most of the phenomena observed are not yet understood in detail, but a number of experiments are suggested by these observations.

As far as kinase is involved, some hints are available for interpreting the role of its light-induced binding. A priori, it is necessary for kinase to interact with its substrate, bleached rhodopsin, in order to phosphorylate it; the only surprise is that the interaction is so strong that kinase sediments with bleached rhodopsin for some time after bleaching. Light induces in rhodopsin both the capacity to bind kinase and to be phosphorylated. Both light-induced capacities decay in the dark within the same amount of time (half-time 20–30 min; see Fig. 3 and Kühn, 1978); this suggests that both are connected with each other and that the phosphorylation reaction may be regulated in the following way: Upon bleaching, rhodopsin (or the membrane) undergoes structural changes which transiently expose a site for kinase binding. Only as long as kinase is bound to this site can it phosphorylate rhodopsin. Based on this model, the light-induced binding of kinase and its subsequent release serve as a mechanism for regulating the phosphorylation activity. ATP and Mg^{2+} needed for phosphorylation are not needed for the binding of kinase (Kühn, 1978).

The GTPase system appears to be much more complex. A sophisticated sequence of reactions seems to be involved, including the binding and exchange of GTP and GDP, respectively (Godchaux and Zimmerman, 1979b; Fung and Stryer, 1980), the light-induced interaction of GTPase with rhodopsin, and the interference of GTP in this interaction, and, of particular importance, potential interactions of GTPase with other proteins (e.g., phosphodiesterase) on the surface of the disk membrane. It seems that the function of this GTPase can be understood only in connection with other proteins, as part of a light-regulated multienzyme complex, in analogy to other systems such as hormone-activated adenylate cyclase (Shinozawa *et al.*, 1979; Rodbell, 1980). Moderate ionic strength, which causes GTPase to be bound as a peripheral membrane protein, is as necessary for its activity as illumination of rhodopsin (Table I).

It is clear that some type of light-induced interaction between GTPase and rhodopsin, directly or indirectly, must be involved in GTPase activation, since the presence of bleached rhodopsin is obligatory for GTPase to be active. It has been proposed (Kühn, 1980a,b) that the light-induced binding, as observed in the centrifugation experiments, reflects this interaction needed for enzyme activation. It is interesting to note in this context that PDE does not undergo light-induced binding, in contrast to GTPase (Fig. 1). The finding that light-activated GTPase serves to activate PDE (Fung *et al.*, 1981) implies that PDE needs to interact with GTPase but not necessarily with bleached rhodopsin directly; the lack of light-induced binding of PDE may therefore reflect the lack of a direct contact between PDE and bleached rhodopsin, in contrast to the case of GTPase, where a direct interaction is indicated by its light-induced binding.

Recent experiments have provided insights into the stoichiometry and kinetics of light-induced GTPase interactions. Flash light- (500 nm) induced changes in light scattering (at 700 nm) of ROS and disk membrane suspensions were used to monitor structural changes involved in GTPase association/dissociation reactions. Two light-induced signals (i.e., changes in light scattering) could be attributed specifically to interactions of the GTPase with disks. Both signals depend on the presence of GTPase; they occur even if purified GTPase (see Fig. 1,n) is added back to thoroughly washed membranes (Fig. 1,c). The membranes alone, or in combination with PDE or various other preparations of ROS proteins except GTPase, do not show any of these signals.

The first GTPase signal (the ''binding signal'') is observed in the absence of GTP, as a light-induced increase in turbidity (OD_{700}). It saturates if equimolar amounts of rhodopsin with respect to GTPase present are bleached, i.e., in the case of ROS, at about 10% bleaching. If more GTPase is added to the membranes, proportionally higher bleaching levels are necessary to reach saturation. This indicates a 1:1 stoichiometric interaction between photolyzed rhodopsin and GTPase and most probably reflects the light-induced binding observed in the centrifugation experiments described in this paper.

The second signal (''dissociation signal'') occurs in the presence of GTP. It is of opposite sign (decrease in OD_{700}), and it saturates if less than 0.5% of the rhodopsin is bleached, indicating a highly nonstoichiometric interaction. Its kinetics strongly depend on the flash light intensity and GTP concentration, being faster at high intensity and high GTP concentration. The dissociation signal starts with some delay after the flash and after the onset of the binding signal, and both signals take place within a few hundred milliseconds at 20°. Following saturating illumination, both signals recover slowly with increasing time between the saturating flash and a subsequent flash (half-time in the 10-minute range).

These and other data are interpreted in the following model (Kühn *et al.*, 1981). Photolyzed rhodopsin (R*) has a high affinity for the dark-adapted form of the GTPase [which normally contains tightly bound GDP (Godchaux and Zimmerman, 1979b; Fung and Stryer, 1980)]. This leads to the binding of one GTPase per R* in a complex (R*-GTPase-GDP) which, in the absence of GTP, is stable for tens of minutes (binding signal). If GTP is present, on the other hand, the complex R*-GTPase-GDP is not stable: It exchanges bound GDP for GTP, rapidly followed by dissociation into R* and GTPase-GTP. This is observed as the light-scattering dissociation signal, and as the reversal of the light-induced binding in centrifugation experiments (Section V,B,C). After its dissociation from the complex, R* participates in further cycles of GTPase binding, GDP/GTP exchange, and dissociation. This explains the highly nonstoichiometric saturation behavior of this signal: One photolyzed rhodopsin catalyzes many cycles of GDP/GTP exchange, in agreement with the data of Fung and Stryer (1980), until all of the GTPase is in the GTP-binding form which is unreactive with rhodopsin photolyzed by further flashes (observed as saturation). Preliminary experiments (N. Bennett *et al.*) have shown that the dissociation signal, and thus the formation of GTPase–GTP, immediately precedes PDE activation.

All of the effects observed, i.e., the light-induced binding of GTPase (binding signal), its reaction with GTP (dissociation signal), and the activation of the PDE (in agreement with the data of Liebman and Pugh, 1979) occur sufficiently fast that their participation in the mechanism of visual excitation cannot be excluded.

ACKNOWLEDGMENTS

I wish to thank Ms. O. Mommertz for expert technical assistance, R. Esser for obtaining bovine eyes (particularly dark-adapted ones) from the slaughterhouse, Ms. B. Goebbels for typing the manuscript, and Dr. M. Chabre for stimulating discussions. This work was supported by SFB 160 of Deutsche Forschungsgemeinschaft.

REFERENCES

Baehr, W., Devlin, M. J., and Applebury, M. L. (1979). Isolation and characterization of cGMP phosphodiesterase from bovine rod outer segments. *J. Biol. Chem.* **254,** 11669–11677.

Biernbaum, M. S., and Bownds, M. D. (1979). Influence of light and calcium on guanosine 5′-triphosphate in isolated frog rod outer segments. *J. Gen. Physiol.* **74,** 649–669.

Bignetti, E., Cavaggioni, A., and Sorbi, R. T. (1978). Light-activated hydrolysis of GTP and cyclic GMP in the rod outer segments. *J. Physiol. (London)* **279,** 55–69.

Bownds, D., Gordon-Walker, A., Gaide-Huguenin, A., and Robinson, W. (1971). Characterization and analysis of frog photoreceptor membranes. *J. Gen. Physiol.* **58,** 225–237.

Bownds, D., Dawes, J., Miller, J., and Stahlman, M. (1972). Phosphorylation of frog photoreceptor membranes induced by light. *Nature (London), New Biol.* **237,** 125–127.

Bridges, C. D. B., and Fong, S.-L. (1980). Lectins as probes of glycoprotein and glycolipid oligosaccharides in rods and cones. *Neurochem. Int.* **1,** 255–267.

Caretta, A., Cavaggioni, A., and Sorbi, R. T. (1979). Phosphodiesterase and GTPase in rod outer segments: Kinetics *in vitro*. *Biochim. Biophys. Acta* **583,** 1–13.

Chader, G. J., Fletcher, R. T., O'Brien, P. J., and Krishna, G. (1976). Differential phosphorylation by GTP and ATP in isolated rod outer segments of the retina. *Biochemistry* **15,** 1615–1620.

Fung, B. K., and Stryer, L. (1980). Photolyzed rhodopsin catalyzes the exchange of GTP for bound GDP in retinal rod outer segments. *Proc. Natl. Acad. Sci. U.S.A.* **77,** 2500–2504.

Fung, B. K., Hurley, J. B., and Stryer, L. (1981). Flow of information in the light-triggered cyclic nucleotide cascade of vision. *Proc. Natl. Acad. Sci. U.S.A.* **78,** 152–156.

Godchaux, W., III, and Zimmerman, W. F. (1979a). Soluble proteins of intact bovine rod cell outer segments. *Exp. Eye Res.* **28,** 483–500.

Godchaux, W., III, and Zimmerman, W. F. (1979b). Membrane-dependent guanine nucleotide binding and GTPase activities of soluble protein from bovine rod cell outer segments. *J. Biol. Chem.* **254,** 7874–7884.

Goridis, C., and Virmaux, N. (1974). Light-regulated guanosine 3′,5′-monophosphate phosphodiesterase of bovine retina. *Nature (London)* **248,** 57–58.

Heitzmann, H. (1972). Rhodopsin is the predominant protein in rod outer segment membranes. *Nature (London), New. Biol.* **235,** 114.

Jan, L. Y., and Revel, J. P. (1974). Ultrastructural localization of rhodopsin in the vertebrate retina. *J. Cell Biol.* **62,** 257-273.

Keirns, J. J., Miki, N., Bitensky, M. W., and Keirns, M. (1975). A link between rhodopsin and disc membrane cyclic nucleotide phosphodiesterase: Action spectrum and sensitivity to illumination. *Biochemistry* **14,** 2760-2766.

Krebs, W., and Kühn, H. (1977). Structure of isolated bovine rod outer segment membranes. *Exp. Eye Res.* **25,** 511-526.

Kühn, H. (1974). Light-dependent phosphorylation of rhodopsin in living frogs. *Nature (London)* **250,** 588-590.

Kühn, H. (1978). Light-regulated binding of rhodopsin kinase and other proteins to cattle photoreceptor membranes. *Biochemistry* **17,** 4389-4395.

Kühn, H. (1980a). Light- and GTP-regulated interaction of GTPase and other proteins with bovine photoreceptor membranes. *Nature (London)* **283,** 587-589.

Kühn, H. (1980b). Light-induced, reversible binding of proteins to bovine photoreceptor membranes: Influence of nucleotides. *Neurochem. Int.* **1,** 269-285.

Kühn, H. (1981). Light regulated binding of proteins to photoreceptor membranes, and its use for the purification of several rod cell proteins. *In* "Methods in Enzymology." Academic Press, New York (in press).

Kühn, H., and Dreyer, W. J. (1972). Light dependent phosphorylation of rhodopsin by ATP. *FEBS Lett* **20,** 1-6.

Kühn, H., and Hargrave, P. A. (1981). Light-induced binding of GTPase to bovine photoreceptor membranes: Effect of limited proteolysis of the membranes. *Biochemistry* **20,** 2410-2417.

Kühn, H., Cook, J. H., and Dreyer, W. J. (1973). Phosphorylation of rhodopsin in bovine photoreceptor membranes: A dark reaction after illumination. *Biochemistry* **12,** 2495-2502.

Kühn, H., Bennett, N., Michel-Villaz, M., and Chabre, M. (1981). Light- and GTP-dependent changes in the interaction of GTPase with bovine photoreceptor membranes. *Hoppe-Seyler's Z. Physiol. Chem.* **362,** 217.

Laemmli, U. K. (1970). Cleavage of structural proteins during the assembly of the head of bacteriophage T4. *Nature (London)* **227,** 680-685.

Liebman, P. A., and Pugh, E. N. (1979). The control of phosphodiesterase in rod disk membranes: Kinetics, possible mechanisms and significance for vision. *Vision Res.* **19,** 375-380.

McDowell, J. H., and Kühn, H. (1977). Light-induced phosphorylation of rhodopsin: Substrate activation and inactivation. *Biochemistry* **16,** 4054-4060.

Miki, N., Keirns, J. J., Marcus, F. R., Freeman, J., and Bitensky, M. W. (1973). Regulation of cyclic nucleotide concentrations in photoreceptors: An ATP-dependent stimulation of cyclic nucleotide phosphodiesterase by light. *Proc. Natl. Acad. Sci. U.S.A.* **70,** 3820-3824.

Miki, N., Baraban, J. M., Keirns, J. J., Boyce, J. J., and Bitensky, M. W. (1975). Purification and properties of the light-activated cyclic nucleotide phosphodiesterase of rod outer segments. *J. Biol. Chem.* **250,** 6320-6327.

Miller, W. H., and Nicol, G. D. (1979). Evidence that cyclic GMP regulates membrane potential in rod photoreceptors. *Nature (London)* **280,** 64-66.

Molday, R. S., and Molday, L. L. (1979). Identification and characterization of multiple forms of rhodopsin and minor proteins in frog and bovine rod outer segment disc membranes. *J. Biol. Chem.* **254,** 4653-4660.

Neufeld, A. H., and Levy, H. M. (1969). A second ouabain-sensitive sodium-dependent adenosine triphosphatase in brain microsomes. *J. Biol. Chem.* **244,** 6493-6497.

Pannbacker, R. G., Fleischman, D. E., and Reed, D. W. (1972). Cyclic nucleotide phosphodiesterase: High activity in a mammalian photoreceptor. *Science* **175,** 757-759.

Papermaster, D. S., and Dreyer, W. J. (1974). Rhodopsin content in the outer segment membranes of bovine and frog retinal rods. *Biochemistry* **13,** 2438–2444.

Papermaster, D. S., Converse, C. A., and Zorn, M. (1976). Biosynthetic and immunochemical characterization of a large protein in frog and cattle rod outer segment membranes. *Exp. Eye Res.* **23,** 105–115.

Pober, J. S., and Stryer, L. (1975). Light dissociates enzymatically cleaved rhodopsin into two different fragments. *J. Mol. Biol.* **95,** 477–481.

Robinson, W. E., and Hagins, W. A. (1979). GTP hydrolysis in intact rod outer segments and the transmitter cycle in visual excitation. *Nature (London)* **280,** 398–400.

Rodbell, M. (1980). The role of hormone receptors and GTP-regulatory proteins in membrane transduction. *Nature (London)* **284,** 17–21.

Schnetkamp, P. P. M., Klopmakers, A. A., and Daemen, F. J. M. (1979). The isolation of stable cattle rod outer segments with an intact plasma membrane. *Biochim. Biophys. Acta* **552,** 379–389.

Shichi, H., and Somers, R. L. (1978). Light-dependent phosphorylation of rhodopsin: Purification and properties of rhodopsin kinase. *J. Biol. Chem.* **253,** 7040–7046.

Shinozawa, T., and Bitensky, M. W. (1980). Co-operation of peripheral and integral membrane proteins in the light dependent activation of rod GTPase and phosphodiesterase. *Photochem. Photobiol.* **32,** 497–502.

Shinozawa, T., Sen, I., Wheeler, G. L., and Bitensky, M. W. (1979). Predictive value of the analogy between hormone-sensitive adenylate cyclase and light-sensitive photoreceptor cyclic-GMP phosphodiesterase: A specific role for a light-sensitive GTPase as a component in the activation sequence. *J. Supramol. Struct.* **10,** 185–190.

Smith, H. G., Stubbs, G. W., and Litman, B. J. (1975). The isolation and purification of osmotically intact discs from retinal rod outer segments. *Exp. Eye Res.* **20,** 211–217.

Swank, R. T., and Munkres, K. D. (1971). Molecular weight analysis of oligopeptides by electrophoresis in polyacrylamide gel with sodium dodecyl sulfate. *Anal. Biochem.* **39,** 462–477.

Thacher, S. M. (1978). Light-stimulated, magnesium-dependent ATPase in toad retinal rod outer segments. *Biochemistry* **17,** 3005–3011.

Uhl, R., Borys, T., and Abrahamson, E. W. (1979a). Evidence for structural changes in the photoreceptor disk membrane, enabled by magnesium ATPase activity and triggered by light. *FEBS Lett.* **107,** 317–322.

Uhl, R., Borys, T., Semple, N., Pasternak, J., and Abrahamson, E. W. (1979b). The presence of two major protein components in the bovine photoreceptor disc membrane. *Biochem. Biophys. Res. Commun.* **90,** 58–64.

Virmaux, N. (1975). Interdisc compartment and disc membranes of bovine retinal outer segments: Separation by a mild chemical method and protein analysis. *Exp. Eye Res.* **20,** 185.

Weber, K., and Osborn, M. (1969). The reliability of molecular weight determinations by dodecyl sulfate-polyacrylamide gel electrophoresis. *J. Biol. Chem.* **244,** 4406–4412.

Wheeler, G. L., and Bitensky, M. W. (1977). A light-activated GTPase in vertebrate photoreceptors: Regulation of light-activated cyclic-GMP phosphodiesterase. *Proc. Natl. Acad. Sci. U.S.A.* **74,** 4238–4242.

Wheeler, G. L., Matuo, Y., and Bitensky, M. W. (1977). Light-activated GTPase in vertebrate photoreceptors. *Nature (London)* **269,** 822–824.

CURRENT TOPICS IN MEMBRANES AND TRANSPORT, VOLUME 15

Chapter 11

Biochemical Pathways Regulating Transduction in Frog Photoreceptor Membranes

M. DERIC BOWNDS

Laboratory of Molecular Biology
Department of Zoology and Neurosciences Training Program
University of Wisconsin–Madison
Madison, Wisconsin

I. INTRODUCTION

Studies on visual transduction in vertebrate rod photoreceptor membranes are attracting increasing interest because a number of reactions that may intervene between input (photon absorption) and output (a conductance decrease) are being specified, and our knowledge of this system is becoming much more complete than for other receptor systems. This is in part due to the fact that the vertebrate rod outer segment is a specialized structure, obtained easily and in quantity, whose enzymatic machinery is dedicated mainly to the transduction process. It is a modified cilium of cylindrical shape, which consists of a plasma membrane enclosing a stacked series of 1000–2000 disk membranes containing the visual pigment rhodopsin as a major intrinsic membrane protein. The molecules of the outer segment are separate from the major sites of protein synthesis and energy

ISBN 0-12-153315-8

metabolism housed in the inner segment portion of the receptor cell. This is in contrast to other systems, such as hormone-sensitive cells, in which the molecules involved in reception and intracellular signaling are less easily distinguished from the vegetative apparatus of the cell.

Two major aspects of visual transduction in frog rod outer segments must be explained in biochemical terms: excitation and adaptation. Excitation refers to the process whereby a single photon absorption by one of the 10^7–10^9 rhodopsin molecules in a rod outer segment can cause the closing of many sodium channels in the plasma membrane. It is generally accepted that this process involves the mobilization of an internal transmitter substance which communicates between the site of photon absorption in the disk membrane system and the permeability mechanism of the plasma membrane, calcium and cyclic guanosine 3′,5′-monophosphate (cyclic GMP) being the most frequently mentioned putative transmitters (for review, see Hubbell and Bownds, 1979). Adaptation involves a desensitization of this process, which occurs after a very small number of photons have been absorbed. Thus activation of the first few rhodopsin molecules enhances reactions that make further rhodopsin bleaching less effective in closing sodium channels.

Experiments in this laboratory over the past several years have put emphasis on searching for light-dependent chemistry that might be relevant to excitation and adaptation. Attention has been restricted to studying, in fresh and minimally perturbed frog rod outer segment preparations, rapid chemical changes triggered by the low levels of illumination at which rod cells show their optimal function. An effort has been made to define the physiologically relevant expression and controls of the reactions studied, as a prelude to more mechanistic studies involving dissociation and reconstruction of the important controlling elements.

Several limitations in working with isolated rod outer segments, either intact or broken, should be mentioned. Although intact outer segments preserve some light-dependent chemistry as well as the ability to undergo a permeability change, they would be expected to decay in their responsiveness to illumination because the normal dark current, ion gradients, and any metabolites provided by the inner segment are absent. Disruption of the outer segment's plasma membrane, which is required to assay the light-sensitive enzymology, introduces the further problem that the reactions observed may be expressed differently than in the living cell. In spite of these reservations, *in vitro* studies have revealed a number of interesting reactions which almost certainly contribute to the normal light response of the rod cell, and such studies must be pursued to define the chemistry that ultimately must be monitored in single living rod cells.

This chapter briefly reviews our work on a number of chemical reactions which may be relevant to the transduction process in frog rod outer segments: (1) activation of a cyclic-GMP phosphodiesterase (PDE) which is sensitive to a number of regulators including calcium ions and adenosine 5′-triphosphate

(ATP); (2) a resulting rapid drop in cyclic-GMP levels, which has a stoichiometry and time course appropriate for the internal transmitter presumed to mediate between photon absorption in the disk membrane system and the permeability decrease in the plasma membrane; (3) a dephosphorylation of two small proteins whose phosphorylation is controlled by cyclic-GMP and calcium levels; (4) a slower hydrolysis of guanosine 5′-triphosphate (GTP) which may drive the efflux of calcium from the outer segment; and (5) a phosphorylation of rhodopsin, which is sensitive to cyclic-GMP and calcium levels. The purpose of this summary is not to provide a detailed review of each of these reactions but rather to suggest how they might be related to each other in the formulation of one possible scheme for the pathways that regulate excitation and adaptation. These studies have demonstrated that cyclic-nucleotide and calcium pathways are likely to interact, and that the system is more complicated than any of the models suggested thus far.

II. THE LIGHT-INDUCED DECREASE IN CYCLIC GMP

Following initial work on light-sensitive enzymes of cyclic-nucleotide metabolism in outer segments (for reviews, see Pober and Bitensky, 1979; Hubbell and Bownds, 1979) Bownds and Brodie noted in 1975 that the addition of cyclic-nucleotide PDE inhibitors to dark-adapted rod outer segments enhanced their permeability. A link between permeability and internal cyclic-GMP levels was shown in subsequent studies which demonstrated, in isolated outer segments, that light suppression of both cyclic-GMP levels and permeability occurred over the same range of light intensities. Further, several pharmacological perturbations were found to have corresponding effects on permeability and cyclic-GMP levels (Brodie and Bownds, 1976; Woodruff *et al.*, 1977). Electrophysiological studies (Lipton *et al.*, 1977b; Miller and Nicol, 1979) also have supported a correlation between cyclic GMP and permeability.

Several laboratories have now observed in retinas a light-induced drop in cyclic GMP mediated by a light-activated PDE (Fletcher and Chader, 1976; Orr *et al.*, 1976). Studies in our laboratory have permitted a detailed description of the amplitude, kinetics, and stoichiometry of the light-induced decrease in cyclic GMP in intact outer segments (Woodruff *et al.*, 1977; Woodruff and Bownds, 1979). The work has shown that cyclic GMP meets several of the requirements for the internal transmitter presumed to mediate between photon absorption by rhodopsin in the disk membrane system and the permeability decrease in the plasma membrane. The cyclic-GMP decrease can be rapid ($t\frac{1}{2} \sim 125$ msec) and reversible, an observation compatible with the finding of Yee and Liebman (1978) that activation of the PDE occurs within milliseconds of light absorption. At very low levels of illumination at least 10^4–10^5 molecules of cyclic GMP are hydrolyzed for each rhodopsin molecule bleached; as the light intensity increases, the magnitude of the cyclic-GMP decrease becomes larger, with saturation occurring at light levels that bleach 5×10^4 rhodopsin molecules per outer segment per second.

It is important to emphasize that these data establish that the isolated outer segment is competent to effect very rapid changes in cyclic GMP, but they do not reveal what happens in a living receptor cell. All the reactions discussed in this chapter ideally should be monitored in living rod outer segments, a technically difficult task. Kilbride and Ebrey (1979) and Kilbride (1980) have approached this issue by making rapid measurements of the light-initiated cyclic-GMP decrease in intact frog retinal sections. No cyclic-GMP decrease was detected 1 sec after illumination in Ringer's solution containing 1.6 m*M* calcium, but a decrease was observed when the calcium concentration was lowered by the addition of 3 m*M* EGTA. In both cases the decrease was slower than that reported by Woodruff and Bownds (1979). It was suggested that the rapid light-sensitive decrease in cyclic GMP observed by Woodruff *et al.* (1979) in isolated rod outer segments might occur only at low calcium concentrations and be much slower at millimolar levels. Polans *et al.* (1981), however, have recently found that the cyclic-GMP decrease in isolated outer segments is substantially complete within 1 sec at calcium concentrations varying from 10^{-9} to 10^{-3} *M*.

The issue of the kinetics of the light-induced decrease in cyclic GMP *in vivo* remains unresolved. The slower kinetic data obtained by Kilbride and Ebrey do not permit one to conclude that the cyclic-GMP decrease is slower than the conductance changes occurring in the outer segments of the retinal sections used, since the relevant conductance changes were not measured. This is an important point, because extracellular current recordings from sections of toad retinas have revealed that the time to peak of the outer segment photoresponse can be much slower (0.6–4 sec) than previous intracellular voltage recordings have indicated (Baylor *et al.*, 1979). Measurements using frog retinal sections further assume that the light-sensitive changes in the levels of cyclic GMP occur only in the rod outer segments. This assumption is based on work with other species (Farber and Lolley, 1974; Orr *et al.*, 1976). If this were not the case in the frog, the time course of the cyclic-GMP decrease in the retina would represent a composite of changes rather than just the outer segments response.

III. LIGHT-INDUCED PROTEIN DEPHOSPHORYLATION

One possible role for the cyclic-GMP decrease is indicated by the recent finding of cyclic-GMP-sensitive protein phosphorylation reactions (Polans *et al.*, 1979). Two small extrinsic membrane proteins of frog rod outer segments (designated components I and II, with molecular weights of 13,000 and 12,000) are phosphorylated in the dark and dephosphorylated upon illumination. Half-maximal dephosphorylation is observed at the same light intensity that causes half-maximal changes in cyclic-GMP levels and permeability. Further, a number of pharmacological agents that perturb cyclic-nucleotide levels, the membrane potential, and presumably the permeability of living rod cells (Lipton *et al.*, 1977a,b) have similar effects on the phosphorylation of components I and II. More recent studies (Hermolin and Bownds, 1981) have shown that the phosphorylation of components I and II is specifically cyclic-GMP-dependent and is inhibited by calcium ions. The calcium effect becomes maximum when the calcium concentration reaches 10^{-4} *M* and causes at least a 50% decrease in the maximal phosphorylation observed. A puzzling aspect of the phosphorylation–dephosphorylation sequence is that the dephosphorylation is observed only in outer segments that remain attached to the retina. Isolation of outer segments apparently lesions the dephosphorylation step (see Polans *et al.*, 1979). It is possible that a phosphatase enzyme is either lost or altered in activity. Phosphatases, as a class of enzymes, are much less understood than kinases (cf. Krebs and Beavo, 1979).

These observations suggest that the central pathway shown in Fig. 1 must be considered seriously. In dark outer segments, cyclic-GMP levels, the phosphorylation of components I and II, and sodium permeability are high. Light absorption by rhodopsin causes it to activate the PDE that hydrolyzes cyclic GMP.

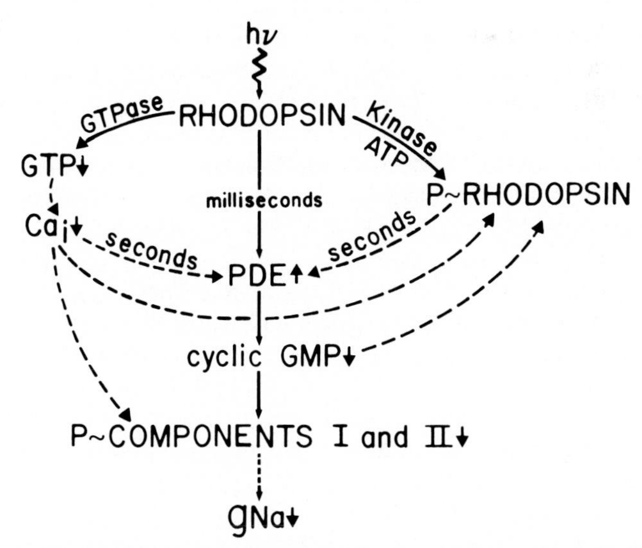

FIG. 1. Pathways that may regulate transduction in frog photoreceptor membranes. Solid lines indicate pathways that appear to be firmly established; dashed lines indicate pathways suggested but not proven by recent experiments.

[This activation process involves several additional protein factors (Wheeler and Bitensky, 1977; Robinson *et al.*, 1980; and see below).] It is suggested that, as cyclic-GMP levels fall, the activity of the cyclic-GMP-dependent protein kinase decreases and the level of phosphorylated components I and II is lowered as the phosphate groups are removed by phosphatase activity. Components I and II are then presumed to be either part of the plasma membrane permeability mechanism or involved in its control, so that their dephosphorylation leads to the closing of sodium channels. This last assumption needs to be tested by studies on both the localization of components I and II and the kinetics of their light-induced dephosphorylation.

In the text that follows, Fig. 1 is developed as one way of organizing data on the reactions studied by this and other laboratories. Solid lines indicate pathways that are firmly established, and dashed lines indicate pathways suggested but not proved by available data. The "calcium hypothesis" (see Hagins and Yoshikami, 1974) has been omitted to reduce confusion and also because rapid intracellular increases in calcium induced by illumination have not yet been observed.

IV. THE LIGHT-INDUCED DECREASE IN GTP

A search for small molecules other than cyclic GMP showing light-sensitive chemistry has led to the discovery and characterization of a light-induced decrease in GTP in outer segments (Biernbaum and Bownds, 1979; see also Robinson and Hagins, 1979). Only a small fraction of the GTP decrease can be accounted for by cyclic-nucleotide metabolism or protein phosphorylation reactions. The decrease has a half-time of 7 sec and is observed as continuous light intensity is increased over 5 log units. An interesting feature of the data is that GTP levels inside the intact outer segment are linked to external calcium levels and can be depleted to the same level caused by illumination if the external calcium concentrations is reduced to 10^{-8} M. Restoration of external calcium causes net synthesis of GTP. Such behavior is characteristic of a reversible energy-requiring calcium transport system. One possibility is that light activates a calcium pump which *extrudes* calcium from outer segments. Gold and Korenbrot (1980), and more recently Yoshikami *et al.* (1980), have measured a calcium extrusion from the outer segments of living retinas, which might reflect a decrease in cytoplasmic calcium levels. The suggested lowering of cytoplasmic calcium caused by light activation of a GTP-dependent extrusion mechanism is shown on the left in Fig. 1.

It should be pointed out that the light-induced drop in GTP probably does not reflect solely activation of the GTPase associated with PDE activation (Wheeler and Bitensky, 1977) because this GTPase is activated over a narrower range of light intensities than the light-induced GTP decrease and does not show the same calcium sensitivity.

V. THE CYCLIC-GMP PHOSPHODIESTERASE COMPLEX

Recent studies in this laboratory (Robinson *et al.*, 1980; Kawamura and Bownds, 1981) have revealed a number of controls of the light-activated PDE that is central in controlling cyclic-GMP levels, controls in addition to the GTPase and inhibitor components already noted (see Pober and Bitensky, 1979, for review). In outer segments that have not been extensively washed or purified, a light-induced increase in the K_m of the enzyme is observed (from 100 to 900 μM). The component responsible for this shift is removed by the sucrose flotation procedures normally employed in preparing outer segments. A further labile component makes the enzyme sensitive to calcium concentration if ATP is present. At high calcium levels ($\sim 10^{-3}$ M), PDE activation is very sensitive to illumination, as reported by several laboratories (cf. Yee and Liebman, 1978), but as the calcium concentration is lowered, the enzyme becomes less effectively activated by light. The intensity response function for the enzyme moves to higher light intensities, partially mimicking the changes that occur during light adaptation of the rod photoreceptor (Fain, 1976). The presence of ATP desensitizes the enzyme. The ATP-dependent desensitization is most pronounced at 10^{-9}–10^{-8} M Ca^{2+}, is intermediate at 10^{-7}–10^{-6} M Ca^{2+}, and is smallest at 10^{-4}–10^{-3} M Ca^{2+}.

These data indicate one possible physiological role for the light-induced calcium movements mentioned above. The suggestion is that in dark-adapted outer segments calcium levels are relatively high ($\geq 10^{-6}$ M) and PDE has maximal light sensitivity (Fig. 1, central pathway). On illumination the GTP-dependent calcium pump lowers internal calcium levels, and the PDE is desensitized (Fig. 1, left pathway). Both PDE activation and the conductance decrease that it indirectly controls then shift to a higher range of light intensity. This could be responsible for at least part of the shift of the intensity response curve of the rod photoreceptor that occurs during light adaptation. This proposal could be taken as the converse of the calcium hypothesis (see Hagins and Yoshikami, 1974), in which light causes an increase in cytoplasmic calcium levels. It may be appropriate to consider the resting (dark) state of the vertebrate photoreceptors as analogous to the excited state of hormone-sensitive cells or muscle cells. In the latter, excitation most frequently involves an increase in plasma membrane conductance and an increase in intracellular calcium, whereas here photoexcitation might cause suppression of conductance and a lowering of cytoplasmic calcium.

As PDE has come to be appreciated more as a multienzyme complex, further controls are being found. Liebman and Pugh (1979) recently have made the interesting observation that PDE activated by a flash of light is rapidly inactivated if ATP is present and suggest that this ATP may be used in rhodopsin phosphorylation. A role for rhodopsin phosphorylation in sensitivity control in photoreceptors had been suggested earlier by both Miller *et al.* (1975) and Kühn *et al.* (1977). Current studies in our laboratory on the kinetics of PDE inactivation and rhodopsin phosphorylation at low light intensities suggest a correlation between the two, and thus the right portion of Fig. 1 includes a second possible pathway for sensitivity control: an inactivation of PDE caused by the interaction of phosphorylated rhodopsin (or opsin) with PDE. This is a very tentative suggestion, for it has been found that under certain conditions rhodopsin phosphorylation can be suppressed with no apparent effect on the ATP-dependent PDE desensitization (J. H. Hermolin, M. A. Karell, and D. Bownds, unpublished observation).[1]

VI. MULTIPLE SITES OF ACTION FOR CALCIUM AND CYCLIC GMP

Recent experiments suggest that even more arrows must be added to the possible controls shown in Fig. 1. Hermolin and Bownds (1981) (see above) have found that the *in vitro* phosphorylation of components I and II is regulated by calcium ions. Increasing calcium inhibits this phosphorylation by at least 50%, with maximum inhibition observed as calcium levels approach 10^{-4} *M*. If, as suggested above, calcium levels fall in the period of seconds after light absorption, calcium inhibition of the kinase might be released, and rephosphorylation and the return of dark permeability might occur more rapidly than recovery of the dark cyclic-GMP levels or resensitization of PDE. This chemistry might underlie the behavior of the dark current noted recently by Albani and Yoshikami (1980). This current, and presumably the sodium permeability of the outer segment, recover more rapidly than the light sensitivity after a strong rhodopsin bleach.

Finally, it has been found that in gently disrupted outer segments the rhodopsin phosphorylation occurring at low light intensities can be inhibited by both cyclic GMP and calcium ions (Hermolin and Bownds, 1981). This had not been observed previously in experiments using higher light intensities. Cyclic-GMP inhibition becomes maximal as cyclic-GMP levels are increased from 10^{-5} to 10^{-4} *M*.[2] Calcium inhibition of rhodopsin phosphorylation becomes maximal when the calcium concentration reaches 10^{-4}–10^{-3} *M*. These influences can also

[1]Further work now has failed to demonstrate a linkage of rhodopsin phosphorylation with the rapid ATP-dependent inactivation of PDE.

[2]More recent experiments have shown that GMP also inhibits rhodopsin phosphorylation. Thus the cyclic-GMP effect noted may actually be due to this hydrolysis product.

be rationalized, for after illumination stops, and as cyclic-GMP levels and calcium concentrations increase to their original dark values, it seems appropriate that rhodopsin phosphorylation cease.

It seems likely that many of these reactions are being observed because one is examining minimally disrupted, unwashed outer segment preparations which retain easily eluted controls (cf. Robinson *et al.*, 1980). Current experiments are directed toward understanding the locus of the calcium and cyclic-GMP effects and evaluating the role of rhodopsin phosphorylation and other protein phosphorylation in PDE control.

VII. OUTSTANDING PROBLEMS

Regardless of the details and the correctness of the scheme in Fig. 1, experiments show that the system is likely to be more complicated than any of the models proposed thus far (cf. Hagins and Yoshikami, 1974; Farber *et al.*, 1978; Liebman and Pugh, 1979). Cyclic-nucleotide and calcium pathways interact. Effects of calcium said to provide indirect evidence for the calcium hypothesis (Hagins and Yoshikami, 1974, 1977) could be explained by effects of calcium on cyclic-nucleotide chemistry. Advocacy of either cyclic GMP or calcium as a primary internal transmitter is becoming increasingly irrelevant because sufficient data are lacking. Cyclic GMP can change with a stoichiometry and time course appropriate for the internal transmitter in isolated outer segments, but the kinetics of the light-induced decrease *in vivo* are not known. The sodium conductance of the plasma membrane can be rapidly inhibited by increasing calcium levels (Yoshikami and Hagins, 1980), but there are as yet no data showing how or whether the cell uses this mechanism. There is now considerable confusion over the possible roles for calcium in this system. Several recent electrophysiological studies have provided evidence that calcium does not function as an internal transmitter in generating the light response (Arden and Low, 1978; Bertrand *et al.*, 1978), and conflicting opinions are held concerning its possible role in adaptation processes (cf. Bastian and Fain, 1979; Flaming and Brown, 1979).

The construction of the summary scheme in Fig. 1 was made possible by chemical data obtained in just the past 2 years, data that provide components in excess of the number needed to generate the four rate constants that Baylor *et al.* (1979) use in their model of the photoreceptor response to light. A further point is that the data suggest that parallel pathways may at first proceed independently from rhodopsin bleaching and then subsequently interact to control sensitivity changes occurring during light adaptation. This is an alternative to the linear models most commonly used to describe the photoreceptor response (Baylor *et al.*, 1974). The complexity emerging also leads one to be cautious in interpreting studies that emphasize fitting the kinetics of single reactions (such as the rhodopsin bleaching sequence and the activation–inactivation of PDE) to the kinetics of the light-induced conductance change. It seems most likely that rate-limiting steps are provided by several reactions of the type shown in Fig. 1, and that appropriate modeling must consider them all together. It is possible that further important components of the system remain to be specified.

Figure 1 is only one way of organizing the data, and it would be very surprising if parts of the summary were not incorrect. The scheme may be fundamentally incorrect in emphasizing the cyclic-GMP pathway as being more rapid than calcium changes. One could view calcium as the primary regulator of plasma membrane conductance, desensitization and feedback pathways being a function of the cyclic-GMP-dependent chemistry. The pathways shown, however, seem most appropriate for the information presently existing, given that no very rapid light-induced increases in cytoplasmic calcium levels have yet been found. The recent data on calcium extrusion from living rods (see above) is compatible with a light-induced lowering, rather than an increase, in cytoplasmic calcium levels.

The pathways likely to regulate transduction are sufficiently complex to make interpretation of even simple electrophysiological and pharmacological experiments quite difficult, and this raises a further bothersome point. As complexity increases, a scheme like that in Fig. 1 becomes an all-encompassing and untestable hypothesis. There always seems to be a way to explain away data that seem to disagree with a favored hypothesis. What one must await is more information on the physical location of the reactions shown (plasma versus disk membrane, etc.), as well as measurements of their kinetics *in vivo*. Determining which of the pathways are most relevant, and also whether further reactions should be considered, will take a great deal of patient work. Because the chemistry is potentially overwhelming, studies emphasizing aspects likely to be relevant to the *in vivo* physiology will probably be most productive.

ACKNOWLEDGMENTS

Work summarized in this chapter was supported by NIH grant EY 00463, training grant GM-01874, and a grant from Fight for Sight, New York, New York.

REFERENCES

Albani, C., and Yoshikami, S. (1980). Two different mechanisms control the amplitude-intensity response and the dark current in retinal rods. *Fed. Proc., Fed. Am. Soc. Exp. Biol.* **39,** 2066.

Arden, G. B., and Low, J. C. (1978). Changes in pigeon cone photocurrent caused by reduction in extracellular calcium activity. *J. Physiol. (London)* **280,** 55–76.

Bastian, B. L., and Fain, G. L. (1979). Light adaptation in toad rods: Requirement for an internal messenger which is not calcium. *J. Physiol. (London)* **297,** 493–520.

Baylor, D. A., Hodgkin, A. L., and Lamb, T. D. (1974). The electrical response of turtle cones to flashes and steps of light. *J. Physiol. (London)* **242,** 685–727.

Baylor, D. A., Lamb, T. D., and Yau, K.-W. (1979). The membrane current of single rod outer segments. *J. Physiol. (London)* **288,** 589–611.

Bertrand, D., Fuortes, M. G. G., and Pochobradsky, J. (1978). Actions of EGTA and high calcium on the cones in the turtle retina. *J. Physiol. (London)* **275,** 419–437.

Biernbaum, M. S., and Bownds, M. D. (1979). Influence of light and calcium on guanosine 5′-triphosphate in isolated frog rod outer segments. *J. Gen. Physiol.* **74,** 649–669.

Bownds, D., and Brodie, A. E. (1975). Light-sensitive swelling of isolated frog rod outer segments as an *in vitro* assay for visual transduction and dark adaptation. *J. Gen. Physiol.* **66,** 407–425.

Brodie, A. E., and Bownds, D. (1976). Biochemical correlates of adaptation processes in isolated frog photoreceptor membranes. *J. Gen. Physiol.* **68,** 1–11.

Fain, G. L. (1976). Sensitivity of toad rods: Dependence on wave-length and background illumination. *J. Physiol. (London)* **261,** 71–101.

Farber, D. B., and Lolley, R. N. (1974). Cyclic guanosine monophosphate: Elevation in degenerating photoreceptor cells of the C3H mouse retina. *Science* **186,** 449–451.

Farber, D. B., Brown, B. M., and Lolley, R. N. (1978). Cyclic GMP: Proposed role in visual cell function. *Vision Res.* **18,** 479–499.

Flaming, D. G., and Brown, K. T. (1979). Effects of calcium on the intensity-response curve of toad rods. *Nature (London)* **278,** 852–853.

Fletcher, R. T., and Chader, G. J. (1976). Cyclic GMP: Control of concentration by light in retinal photoreceptors. *Biochem. Biophys. Res. Commun.* **70,** 1297–1302.

Gold, G. H., and Korenbrot, J. I. (1980). Light-induced Ca efflux from intact rod cells in living retinas. *Fed. Proc., Fed. Am. Soc. Exp. Biol.* **39,** 1814.

Hagins, W. A., and Yoshikami, S. (1974). A role for Ca^{2+} in excitation of retinal rods and cones. *Exp. Eye Res.* **18,** 299–305.

Hagins, W. A., and Yoshikami, S. (1977). Intracellular transmission of visual excitation in photoreceptors: Electrical effects of chelating agents introduced into rods by vesicle fusion. *In* "Vertebrate Photoreception" (H. B. Barlow and P. Fatt, eds.), pp. 97–139. Academic Press, New York.

Hermolin, J. H., and Bownds, D. (1981). Cyclic GMP and calcium regulation of protein phosphorylation in frog outer segments. *J. Gen. Physiol.* (submitted for publication).

Hubbell, W. L., and Bownds, M. D. (1979). Visual transduction in vertebrate photoreceptors. *Annu. Rev. Neurosci.* **2,** 17–34.

Kawamura, S., and Bownds, D. (1981). Light adaptation of the phosphodiesterase of frog photoreceptor membranes mediated by ATP and calcium ions. *J. Gen. Physiol.* **77,** 571–591.

Kilbride, P. (1980). Calcium effects on frog retinal cGMP levels and their light initiated rate of decay. *J. Gen. Physiol.* **75,** 457–465.

Kilbride, P., and Ebrey, T. G. (1979). Light initiated changes of cyclic GMP levels in the frog retina measured with quick freezing techniques. *J. Gen. Physiol.* **14,** 415–426.

Krebs, E. G., and Beavo, J. A. (1979). Phosphorylation-dephosphorylation of enzymes. *Annu. Rev. Biochem.* **48,** 923–959.

Kühn, H., McDowell, J. H., Leser, K. H., and Bader, S. (1977). Phosphorylation of rhodopsin as a possible mechanism of adaptation. *Biophys. Struct. Mech.* **3,** 175–180.

Liebman, P. A., and Pugh, E. N. (1979). The control of phosphodiesterase in rod disk membranes: Kinetics, possible mechanisms and significance for vision. *Vision Res.* **19,** 375–380.

Lipton, S. A., Ostroy, S. E., and Dowling, J. E. (1977a). Electrical and adaptive properties of rod photoreceptors in *Bufo marinus.* I. Effects of altered extracellular Ca^{2+} levels. *J. Gen. Physiol.* **70,** 747–770.

Lipton, S. A., Rasmussen, H., and Dowling, J. E. (1977b). Electrical and adaptive properties of rod photoreceptors in *Bufo marinus.* II. Effects of cyclic nucleotides and prostaglandins. *J. Gen. Physiol.* **70,** 771–791.

Miller, H. A., Brodie, A. E., and Bownds, D. (1975). Light-activated rhodospin phosphorylation may control light sensitivity in isolated rod outer segments. *FEBS Lett.* **59,** 20–23.

Miller, W. H., and Nicol, G. D. (1979). Evidence that cyclic GMP regulates membrane potential in rod photoreceptors. *Nature (London)* **280,** 64–66.

Orr, H. T., Lowry, O. H., Cohen, A. I., and Ferrendelli, J. A. (1976). Distribution of 3′:5′-cyclic AMP and 3′:5′-cyclic GMP in rabbit retina *in vivo*: Selective effects of dark and light adaptation and ischemia. *Proc. Natl. Acad. Sci. U.S.A.* **73,** 4442–4445.

Pober, J. S., and Bitensky, M. W. (1979). Light-regulated enzymes of vertebrate retinal rods. *Adv. Cyclic Nucleotide Res.* **11,** 265–301.

Polans, A. S., Hermolin, J., and Bownds, M. D. (1979). Light-induced dephosphorylation of two proteins in frog rod outer segments—Influence of cyclic nucleotides and calcium. *J. Gen. Physiol.* **74,** 595–613.

Polans, A. S., Kawamura, S., and Bownds, D. (1981). Influence of calcium on cyclic GMP levels in frog rod outer segments. *J. Gen. Physiol.* **77,** 41–48.

Robinson, P. R., Kawamura, K. S., Abramson, B., and Bownds, M. D. (1980). Control of the cyclic GMP phosphodiesterase of frog photoreceptor membranes. *J. Gen. Physiol.* **76,** 631–645.

Robinson, W. E., and Hagins, W. A. (1979). GTP hydrolysis in intact rod outer segments and the transmitter cycle in visual excitation. *Nature (London)* **280,** 398–400.

Wheeler, G. L., and Bitensky, M. W. (1977). A light-activated GTPase in vertebrate photoreceptors: Regulation of light-activated cyclic GMP phosphodiesterase. *Proc. Natl. Acad. Sci. U.S.A.* **74,** 4238–4242.

Woodruff, M. L., and Bownds, M. D. (1979). Amplitude, kinetics and reversibility of a light-induced decrease in guanosine 3′,5′-cyclic monophosphate in frog photoreceptor membranes. *J. Gen. Physiol.* **73,** 629–653.

Woodruff, M. L., Bownds, D., Green, S. H., Morrisey, J. L., and Shedlovsky, A. (1977). Guanosine 3′,5′-cyclic monophosphate and the *in vitro* physiology of frog photoreceptor membranes. *J. Gen. Physiol.* **69,** 667–679.

Yee, R., and Liebman, P. A. (1978). Light-activated phosphodiesterase of the rod outer segment: Kinetics and parameters of activation and deactivation. *J. Biol. Chem.* **253,** 8902–8909.

Yoshikami, S., and Hagins, W. A. (1980). Kinetics of control of the dark current of retinal rods by Ca^{++} and by light. *Fed. Proc., Fed. Am. Soc. Exp. Biol.* **39,** 1814.

Yoshikami, S., George, J., and Hagins, W. A. (1980). Light causes large, fast Ca^{++} efflux from outer segments of live retinal rods. *Fed. Proc., Fed. Am. Soc. Exp. Biol.* **39,** 2066.

CURRENT TOPICS IN MEMBRANES AND TRANSPORT, VOLUME 15

Chapter 12

The Use of Incubated Retinas in Investigating the Effects of Calcium and Other Ions on Cyclic-Nucleotide Levels in Photoreceptors

ADOLPH I. COHEN

Departments of Anatomy-Neurobiology, and Ophthalmology
Washington University School of Medicine
St. Louis, Missouri

I. INTRODUCTION

Biochemical investigations of rod outer segments as carried out in a number of laboratories over the last decade have revealed that, although photopigment is the predominant protein of rod disks, these organelles contain an assemblage of enzymes whose activities (indirectly) are initiated or notably modulated by the action of light. Recent attempts to model responses of rods and cones have suggested that a considerable number of steps must intervene between the capture of a photon and alterations in membrane potential (Baylor *et al.*, 1974; Hagins and Yoshikami, 1975), suggesting that transductive and adaptive physiologies in

ISBN 0-12-153315-8

photoreceptors are rather complex. One is led therefore to consider whether the above light-modulated enzymes may be elements in a complex metabolism in the outer segment which is related to transduction or to an adaptation process. A detailed examination of these enzymes and their cofactors reveals a prodigious possibility for complex interactions. Various nucleotides are the key molecular elements in these interactions. The light-modulated enzymes thus far discovered are a guanosinetriphosphatase (Robinson and Hagins, 1977, 1979; Wheeler and Bitensky, 1977; Bitensky *et al.,* 1978; Shinozawa *et al.,* 1980), an adenosine-triphosphatase (Thacher, 1978; Dacko *et al.,* 1980), guanosine 5′-triphosphate (GTP)-utilizing protein kinases (Chader *et al.,* 1980), and a moderately specific cyclic-GMP phosphodiesterase (cGMP-PDE) (Bitensky *et al.,* 1975; Fletcher and Chader, 1976) which rapidly hydrolyzes both cyclic guanosine 3′,5′-monophosphate (cyclic GMP, cGMP) (Chader *et al.,* 1974; Goridis and Virmaux, 1974) and adenosine 3′,5′-monophosphate (cyclic AMP, cAMP). Triphosphate nucleotides are substrates for membrane ion pumps and also for phosphokinases which are modulated by levels of specific cyclic nucleotides. Nucleotide triphosphates are also substrates for the highly specific cyclase enzymes that form cGMP and cAMP. Another complexity is that particular nucleotide triphosphates may serve as necessary, nonsubstrate cofactors for enzymes acting on other nucleotides and related molecules. Thus GTP is an activating cofactor for the cGMP-PDE of the outer segment and for adenylate cyclase. Finally, the behavior of this entire group of enzymes is often markedly affected by the levels of divalent cations—such as calcium, magnesium, and manganese. Rodnight (1979) provides many examples of the calcium activation of certain cyclase and PDE activities in neural tissue. Clearly, it is reasonable to expect that a high degree of physical segregation and multiple chemical control factors are required to bring order to a system with the above cross-reacting potentials.

Recently we have studied certain aspects of these systems. Using quick-frozen dark- or light-adapted eyes from which freeze-dried sections were obtained (Lowry and Passonneau, 1972), we measured cAMP and cGMP levels, adenylate and guanylate cyclases, and PDE (Orr *et al.,* 1976; De Vries *et al.,* 1979), as well as some other enzymes and substrates relating to cyclic-nucleotide metabolism (Berger *et al.,* 1980), in particular photoreceptor and other retinal strata. In rabbit (Orr *et al.,* 1976), monkey (Berger *et al.,* 1980), and 13-line ground squirrel retinas (De Vries *et al.,* 1979) cGMP seems to be concentrated in photoreceptors, and the activities of a cGMP-PDE and a particulate guanylate cyclase are markedly concentrated in the photoreceptor outer segment layer. In the monkey, guanylate kinase, nucleoside diphosphokinase, and inorganic pyrophosphatase are low in outer segments but high in inner segments (Berger *et al.,* 1980). Light *adaptation* diminishes cGMP in rabbit photoreceptor layers but not in ground squirrels. (Current methods of freezing whole eyes are too slow to permit changes to be followed on the time scale of transduction.)

The data for cAMP are more difficult to assess. Although never as concentrated as cGMP in the outer segment layer, cAMP is found in significant concentrations at all retinal levels (Lolley *et al.*, 1974; Orr *et al.*, 1976; De Vries *et al.*, 1979). Light-adapted mouse retinas have lower levels of cAMP (Ferrendelli and Cohen, 1976), and Orr *et al.* (1976) reported a sharp loss in cAMP at the outer plexiform level on light-adapting the rabbit retina. Farber and Lolley (1978) have reported a sharp light-induced loss of cAMP in the outer retina of the ground squirrel, although this finding is at variance with that of De Vries *et al.* (1979).

One problem of the layer analysis technique as applied to retinas of frozen intact eyes is that processes of pigment epithelial cells may interdigitate with photoreceptor outer segments and, in some species, to some extent with inner segments as well. This interdigitation is modest in mice, rabbits, and monkeys but extensive in the ground squirrel. In addition, processes of glial cells of Müller interdigitate with photoreceptor inner segments, somata, axons, and terminals in all retinas. Thus, in such sections, one cannot be certain as to the proportion of a molecular concentration that is actually receptoral. Another problem with the utilization of whole, quick-frozen eyes is that the variables that can be applied prior to isolation and freezing are restricted to light exposure, degree of anoxia, and the few pharmacological agents that cross the blood–retina barrier.

The use of isolated outer segments (e.g., Fletcher and Chader, 1976) if clean, i.e., free of pigment epithelial cells, their processes, and other cell debris, reduces the contamination problem for this one photoreceptoral portion, which can then be exposed to a variety of environments, but all the isolated outer segments are necessarily damaged in order to achieve their isolation. In careful preparations many isolated segments apparently heal, as judged by osmotic or dye penetration tests, but others remain leaky. The molecule being assayed may be lost or, as a result of the loss of metabolic cofactors, its concentration may shift to an abnormal level. Furthermore, any steady-state metabolic traffic of outer segments with the inner segments and pigment epithelium is lost and with time may be reflected in abnormal concentrations of some molecules within the outer segments. Each class of preparation therefore has both advantages and liabilities.

In 1976, we found that retinas isolated under dim red light from dark-adapted mice had higher concentrations of *both* cGMP and cAMP than those isolated from light-adapted mice under moderate white light. When we compared these to retinas from coisogenic animals, which were receptorless because of homozygosity for the gene *rd,* the light effect was missing, as expected, and the receptorless retinas had much less cGMP but slightly more cAMP. Assuming no transneuronal changes, quantitation suggested that the missing cell type, i.e., the photoreceptors, would have to contain well over 90% of the normal retina's cGMP but be poorer (per unit protein) in cAMP than other retinal levels. This assessment agrees well with our early and current layer analysis data from quick-frozen normal eyes of other species (Orr *et al.*, 1976; De Vries *et al.*, 1979) and with an observation of Lolley *et al.* (1974) that the outer half of the freeze-dried mouse retina was rich in cGMP.

II. THE INCUBATED RETINA PREPARATION

These findings motivated us to develop an incubated mouse retina preparation which would be physiologically intact and permit exposure of the retina to a wide range of ionic and pharmacological variables.

A further opportunity provided by the incubated retina preparation is the employment of light effects combined with ionic and pharmacological manipulations to extract information on the behavior of cyclic nucleotides in photoreceptors. For example, if transmission from the photoreceptors to all other cells could be totally blocked, a light-induced change in the retinal level of some molecule would have to occur in the photoreceptors—according to the dogma that photopigment is totally confined to these cells. A possible concern is an indirect, nonsynaptic influence of photoreceptors on other cells via the transient or sustained activity of photoreceptors modifying the environment of these cells—by changing extracellular potassium levels, for example. However, when known or suspected, suitable controls can be devised for such factors.

This chapter largely reviews work published elsewhere utilizing this incubated retina preparation (Cohen *et al.*, 1978; Mitzel *et al.*, 1978; Cohen and Ferrendelli, 1979).

A. Effects of Isolation Conditions

We first examined the effects of the isolation procedure. Mice were decapitated, and both of their readily proptosed eyes evulsed into ice-cold Earle's medium within 20 sec. Earle's medium contains 115 m*M* NaCl, 5 m*M* KCl, 1.8 m*M* $CaCl_2$, 0.8 m*M* $MgSO_4$, 2.6 m*M* $NaHCO_3$, 0.9 m*M* NaH_2PO_4, and 5.5 m*M* dextrose equilibrated with 95% O_2 and 5% CO_2 at pH 7.4. Both retinas were removed with an average speed of 5.5 min under dim red light and then frozen in liquid nitrogen. The freezing rate is too slow relative to a light flash to permit studies in the time frame of transduction (ca. 100 msec). The levels of cGMP (ca. 68.5 pmoles mg^{-1} protein) did not differ significantly when the procedure was performed under infrared illumination, however, dim red light produced a significant decline in cAMP (23.2 pmoles mg^{-1} protein in infrared versus 17.1 pmoles mg^{-1} protein in dim red), thus suggesting that the retina's cAMP is more sensitive to weak bleaching than cGMP. However, the cGMP level might have reacted and recovered. Because the cold isolating bath could cause cessation of the functioning of membrane ion pumps, isolations under dim red light were also compared when the isolation medium was at 4° and 37°C, but no changes were seen in the level of either cyclic nucleotide.

B. Morphological Effects of Incubation

Incubation of the retinas beyond 5 min began to produce swelling of Müller cells (Mitzel *et al.*, 1978), and incubations beyond 15 min exaggerated this swelling and began to cause the swelling of synaptic processes in the inner plexiform layer. For this reason in most experiments incubations were limited to 5 min. The reasons for the swelling are not clear, but the swelling is potassium-dependent since it can be retarded for 20 min (but not for 30 min) if the potassium level is reduced to 2 m*M* from the 5 m*M* level of Earle's medium.

III. BEHAVIOR OF cGMP LEVELS IN INCUBATED RETINAS

The findings reported in this section are summarized in Table I. On transferring dark-adapted retinas to incubation flasks at 37°C a rise in the cGMP level from about 70 pmoles mg^{-1} protein to 100–125 pmoles mg^{-1} protein was seen within 2 min, followed by a long, slow rise over 30 min. When at 2 min the retinas were exposed to a strong white bleaching light, the cGMP level fell by about 50% within the next 3 min. The same effect occurred in the presence of

TABLE I
FACTORS INCREASING DARK RETINAL cGMP[a]

Factor	Increase[b]
Incubation alone	Slight
Incubation, 100 m*M* aspartate (replacing Cl^-)	Slight
Incubation, 100 m*M* K^+ (replacing Na^+)	Slight
Incubation, 50 μM–5 m*M* ouabain	Slight
IBMX (1 m*M*)	Slight
IBMX, 10–100 m*M* aspartate	Slight
Millimolar Mg^{2+} with micromolar Ca^{2+}	Moderate
Millimolar Mg^{2+} with micromolar Ca^{2+}, IBMX	Large
Chelating Ca^{2+}, 0.1–3 m*M* EGTA, no Ca^{2+} salt	Large
Chelating Ca^{2+}, IBMX	Large
Chelating Ca^{2+}, IBMX, 10 m*M* aspartate	Large

[a] Significant light-induced declines in cGMP were seen in all the media.
[b] Slight, <50%; moderate, up to 3-fold; large, 4- to 70-fold.

10–100 mM aspartate; thus, if aspartate blocked light signaling by postreceptoral neurons (Sillman *et al.*, 1969; Dowling and Ripps, 1972), at least 50-60 pmoles cGMP mg^{-1} protein were in the incubated photoreceptors. The failure of light to reduce the cGMP level below 40 pmoles mg^{-1} protein may be attributable to not all of the photoreceptor's cGMP being in the outer segment, as suggested by layer analyses in other species; in part, to not all of the cGMP being in photoreceptors; and perhaps to a bound cGMP fraction not being sensitive to hydrolytic enzyme. A experiment related to aspartate exposure can be performed by altering the medium to promote a marked depolarization of the photoreceptors and thus the virtual elimination of any photovoltage. For this purpose 50 μM–5 mM ouabain was added to the incubation medium (Cohen *et al.*, 1978; Cohen, unpublished). Although a 10–15% reduction in the dark level of cGMP was sometimes observed in such media, light still caused about a 50% decrease in the cGMP level when applied for 3 min after 2–5 min of dark incubation in ouabain-containing media. When 100 mM Na^+ was replaced by equimolar potassium, there was likewise no effect on the dark cGMP level and light still produced a 50% fall in cGMP. If these depolarizing situations prevent a photovoltage, then it appears that light produces these changes in photoreceptors and that this fall is independent of the membrane potential.

Experiments revealed that the cGMP level was influenced by general metabolism, but glycolysis had to be compromised to see this result. Thus, while the cGMP dark level was only slightly reduced by replacing 95% O_2–5% CO_2 by 95% N_2–5% CO_2, it was reduced to half its usual dark value in 5 min when 1 mM iodoacetate was also present—with almost total elimination of light sensitivity of the cGMP level. Penn and Hagins (1969) showed that 10 mM KCN totally abolished the dark voltage and photovoltage of slices of rat retinas in 2 min. Mitzel *et al.* (1978) found that mouse retinas in Earle's medium with 10 mM KCN for 5 min exhibited ATP and phosphocreatine losses of 60–75%. The reaction of the dark level of cGMP to KCN was somewhat dependent on the extent of preincubation in cyanide-free medium, the cGMP *increasing* in the absence of preincubation but slightly decreasing when cyanide exposure followed 15 min of preincubation. Accordingly, if cyanide was abolishing "dark" voltage and photovoltage, it was doing so without a fall in cGMP levels. Replacing 4 mM Cl^- by F^- diminished dark levels of cGMP by about 25% and reduced the usual light-induced diminution in cGMP from 50% to about 30%. This may also involve interference with glycolysis.

In the case of divalent ions, when retinas were incubated in Earle's medium lacking calcium salt and containing in addition various concentrations of EGTA, a 4-fold increase in the dark level of cGMP was seen after 2 min of exposure to medium with 0.1 mM EGTA, and this increased to a 10- to 20-fold increase when 1–3 mM EGTA was added. Adding equimolar calcium and EGTA had no effect, so calcium chelation rather than a specific effect of EGTA seems to be the agency of the action. Important, however, is the observation that in media both lacking calcium salt and processing 3 mM EGTA, light readily reduced the cGMP concentration from its dark, calcium-chelation-enhanced level to far below its normal dark level (Cohen *et al.*, 1978). Since a profound calcium chelation should effectively block synaptic transmission, this light-induced loss of over 90% of the elevated dark cGMP level must occur in the photoreceptors; thus these cells must be the locus of the chelation-induced cGMP increase.

Restoring normal calcium levels in the dark brought the cGMP concentration back to its usual dark value. However, adding calcium, up to 40 mM, to dark-adapted, dark-incubated retinas in otherwise normal media did not reduce the normal dark level of cGMP. (This addition of high calcium was done in HEPES-buffered rather than bicarbonate–CO_2-buffered medium.) The same negative result was seen with high Ca^{2+} in the presence of high K^+. Nor did adding 20 μM of the calcium ionophore A23187 to an incubation medium with 20 mM calcium reduce the normal dark cGMP value. Retinas were also preincubated in 10^{-6} and 10^{-7} M Ca^{2+}-EGTA-buffered media with 20 μM A23187 and only subsequently exposed to 20 mM Ca^{2+} levels. (Dissociation constants for Ca^{2+}-EGTA and Mg^{2+}-EGTA at 37°C were not available and had to be estimated.) However, these maneuvers were likewise ineffectual in reducing cGMP below its usual dark level. In view of all the above maneuvers, the effect of light in reducing the cGMP level does not appear to be significantly influenced by calcium even if light causes a sustained increase in the free calcium level.

Under certain circumstances increasing concentrations of Mg^{2+} ion could also cause an increase in cGMP (Cohen and Ferrendelli, 1979). When the calcium concentration in the incubation medium was 50 μM or less, progressively increasing the Mg^{2+} concentration from 0.5 to 40 mM progressively increased the 2-min dark cGMP level from 100 to 230 pmoles mg^{-1} protein. When the calcium level was 500 μM or more, the same levels of Mg^{2+} were without effect on the cGMP level.

IV. BEHAVIOR OF cAMP LEVELS IN INCUBATED RETINAS

The studies of Lipton *et al.* (1977) on the electrophysiological responses of toad rods exposed to agents putatively modifying levels of cyclic nucleotides contained some suggestions that cAMP and cGMP might have antagonistic actions. That photoreceptors contain cGMP is well established, but that they contain cAMP is less certain because of the earlier mentioned analytical problems involving mixed cell types in freeze-dried retinal layers and the possible contamination of isolated outer segments, although analyses of both isolated outer segment preparations (Fletcher and Chader, 1976) and freeze-dried retinal slices including photoreceptors (Orr *et al.*, 1976; De Vries *et al.*, 1979) yield finite amounts of cAMP which are somewhat less than those at deeper retinal levels.

However, because outer segments are membrane-rich, data referred to mass or to protein tend to minimize cytosol concentrations in outer segments. One way of establishing that photoreceptors actually contain cAMP is to block transmission from these neurons to other neurons and show that light can modify the retinal cAMP level despite the block. Such attempts were encouraged by the lower cAMP levels in retinas from light-adapted mice (Ferrendelli and Cohen, 1976), despite our expectation at the time that much if not all of this light-induced diminution was postreceptoral.

Unfortunately, in the incubated mouse retina, partly but not entirely because of isolation in dim red light, white light proved ineffective in further reducing cAMP. Sometimes prolonged dark incubation caused a modest but significant increase in the cAMP of retinas isolated in dim red light and some light sensitivity returned. However, such prolonged incubations produced considerable morphological damage (Mitzel *et al.*, 1978), and it was not practical to carry out experiments requiring large numbers of retinas under infrared because many infrared-isolated retinas prove to be contaminated by nonretinal tissue which similarly reflects this radiation. These must be discarded.

It proved possible to circumvent the above problem, and the findings reported in this section are summarized in Table II.

Adding 1 m*M* 3-isobutylmethylxanthine (IBMX) to media caused in 2 min a sixfold dark increase in cAMP in retinas incubated in HCO_3^--CO_2-buffered media and a fourfold increase in cAMP in HEPES-buffered media. The level of cGMP was increased by 30% in the former and not at all in the latter. Both dark cyclic-nucleotide levels in incubated retinas then fell by about 50% on exposure to white light. Moreover, when in addition to IBMX the dark incubation medium also contained 10–100 m*M* aspartate, light was equally effective in reducing both nucleotide levels. Magnesium ion (35 m*M*) in IBMX-containing HEPES-

TABLE II
FACTORS INCREASING DARK RETINAL cAMP[a]

Factor	Increase[b]
Prolonged dark incubation	Slight
Co^{2+} (1–20 m*M*)	Moderate
IBMX (1 m*M*)	Large
IBMX, Co^{2+} (2 m*M*), normal or low Ca^{2+}	Large
IBMX, 10–100 m*M* aspartate	Large
IBMX, 5 μ*M*–5 m*M* ouabain	Large
IBMX, 50 m*M* K^+ (replacing Na^+)	Large
IBMX, 50 m*M* $Tris^+$ (replacing Na^+)	Large

[a] Significant light-induced declines in cAMP were seen in all the media.
[b] Slight, <50%; moderate, up to 3-fold; large, 4- to 10-fold.

buffered media also permitted light-induced reductions, as did 2 or 5 m*M* cobaltous ion. Moreover, cobaltous ion was found to elevate cAMP levels in retinas isolated in dim red light whether IBMX was present or not. After a 5-min dark incubation in IBMX-free HEPES-buffered Earle's medium, a control level of cAMP was typically about 15 pmoles mg^{-1} protein. On adding cobaltous ion at 1–5 m*M*, the 5-min dark cAMP level progressively increased for these cobalt levels to about 48 pmoles mg^{-1} protein. Such heightened cAMP levels produced by Co^{2+} alone were likewise markedly light-sensitive ($>50\%$ loss).

The literature contains many references to the effectiveness of aspartate in blocking postreceptoral light responses (e.g., Sillman *et al.*, 1969; Dowling and Ripps, 1972), and to the efficacy of cobaltous and magnesium ions in blocking receptor transmission presynaptically (e.g., Winkler, 1972; Dowling and Ripps, 1973; Cervetto and Piccolino, 1975; Dacheux and Miller, 1976; Evans *et al.*, 1978) when applied at appropriate concentrations. To sum up, IBMX and/or Co^{2+} ion caused the accumulation of cAMP in the retina. A substantial part of this pharmacologically elevated pool of cAMP was light-sensitive in the presence of aspartate, magnesium, or cobalt. If postreceptoral neurons cannot sense light onset, this light-sensitive fraction must be in the photoreceptors.

To further explore the influence of Co^{2+} and Mg^{2+} on cyclic nucleotides in background calcium levels, dark-adapted retinas were dark-incubated in HEPES-buffered Earle's medium where 20 m*M* Co^{2+}, Mg^{2+}, or $Tris^{+}$ replaced isoosmotic levels of Na^{+} in the rinsing and incubation solutions. The second variable was that all solutions either contained 1.8 m*M* Ca^{2+} or calcium salt was omitted from the medium used to dissect the eyes and rinse and incubate the retinas.

This experiment revealed no effect on the dark levels of the cyclic nucleotides or on their previous light sensitivity as related to replacing Na^{+} by $Tris^{+}$. The dark cAMP level was indifferent to 20 m*M* Mg^{2+} but elevated by 20 m*M* Co^{2+} whether calcium salt was present or not, and still exhibited a sharp light-induced fall in the presence of 20 m*M* Co^{2+} ion. The dark cGMP level was indifferent to the presence of cobaltous ion whether calcium salt was present or not, and was increased only by 20 m*M* Mg^{2+} when calcium salt was omitted. The dark cGMP level could always be reduced by light irrespective of the presence of Mg^{2+}, Co^{2+}, Ca^{2+} or $Tris^{+}$. Thus Co^{2+} and Mg^{2+} have different effects on cyclic nucleotides. If they cause the blocking of synaptic transmission by a common mechanism, they nevertheless do not have similar effects on cyclic-nucleotide systems. A comparison of the heightened dark levels of cAMP or cGMP following the use of these ions to the residual nucleotide levels after light exposure strongly suggests that much of their action is on the photoreceptors. Evans and Erulkar (1980) have recently reported a differential action of cobalt and magnesium at the neuromuscular synapse.

To explore further the effects of calcium on cAMP levels, retinas were isolated in cold solutions containing IBMX and then incubated in solutions containing IBMX, 1–3 m*M* EGTA, and no calcium salt. In such solutions a transient rise in cAMP was seen at 2 min of dark incubation, but this rise was about 40% less than that seen with normal calcium levels and IBMX. After 5 min in the dark, the cAMP level had fallen by 40% from the 2-min level. In the continued presence of the chelator and IBMX, light applied after 2 min of darkness did not accelerate the fall in cAMP. These data suggest that the maintenance of or increase in cAMP levels in the light-sensitive compartment is blocked by chelating calcium.

V. CONTRIBUTION OF GLIA TO CYCLIC-NUCLEOTIDE LEVELS

Another aspect to be considered is the possible existence and behavior of a cyclic-nucleotide pool in glia, such as in the glial cell of Müller, the principal glial species in the retina. Some evidence exists that this cell behaves like a potassium electrode (Miller, 1973). The question is raised as to whether photoreceptor activity in incubation media with IBMX (including synaptic blockers) can modify extracellular potassium levels and thus influence putative cyclic-nucleotide levels in Müller cells. Experiments were therefore designed to study the effects of light on dark levels of cyclic nucleotides as a function of K^+ concentrations, with 10 m*M* aspartate present to block any postphotoreceptoral light response and 1 m*M* IBMX present to help keep cAMP at a level where light sensitivity was most readily detected. On incubating dark-adapted retinas in the dark in high-potassium media with 10 m*M* aspartate and 1 m*M* IBMX, the cAMP level rose fairly rapidly over 1 min and then exhibited a slower steady rise for at least 10 min. Dark-incubated retinas showed little response to increasing K^+. When 5-min dark levels of cAMP were studied as a function of K^+ in the above medium, over a K^+ range from 10 to 120 m*M*, from an initial dark level of 160 pmoles cAMP mg^{-1} protein, there was *at most* an increase in *dark* cAMP levels of 20–40 pmoles mg^{-1} protein. Over most of this range, when 2 min of darkness was followed by 3 min of light, the cAMP level fell, but there was a progressive reduction in the net extent of this light-induced decline with increasing K^+. The cAMP level *after light exposure* was 60 pmoles mg^{-1} protein for a solution containing 10 m*M* K^+, a decline of 100 pmoles. The postillumination level of cAMP increased linearly for increasing K^+ concentrations, reaching 160 pmoles mg^{-1} protein at a 120 m*M* K^+ level, thus yielding a light-induced decline of, at most, 20 pmoles cAMP mg^{-1} protein. Light-adapted, light-incubated mouse retinas exhibited twofold increases in cAMP in shifting to high concentrations of potassium (*without IBMX or aspartate present*), while dark-adapted retinas proved inert to K^+ changes under dark incubation (Ferrendelli *et al.*, 1980). Thus, in light-adapted retinas (*without IBMX and aspartate*), cAMP can increase with increasing K^+. Wassenaar and Korf (1976) found an essentially linear increase in light-adapted rat retinas from 10 to 50 m*M* K^+, the maximal K^+ level tested. No manipulations of K^+ levels such as might be encountered physiologically mimicked the effect of light on cyclic AMP, thus arguing that a cyclic AMP system in glia, modulated by external potassium levels, is unlikely to be the basis of the light effect on cyclic AMP. However, it appears that the diminishing extent of the light-induced decline in cyclic AMP as a function of increasing potassium is a complex net result. One element is a progressive increase with potassium of a light-insensitive (partly because of aspartate) but Haloperidol-preventable accumulation of cyclic AMP in the inner retina. Other elements are persistently light-sensitive pools of cyclic AMP in the outer retina, one of which is potassium-sensitive. The dark levels and light-induced falls in cGMP were not influenced by the potassium level.

VI. THE USE OF 3-ISOBUTYLMETHYLXANTHINE

The phosphodiesterase inhibitor IBMX was required to observe a number of phenomena. By itself, it elevated cAMP to a level where it was clearly light-sensitive. Whereas 1 m*M* IBMX increased the 5-min dark retinal level of cAMP about 6-fold, with 50 μM or 5 m*M* ouabain present the increase was 15-fold even though these ouabain levels by themselves produced no increase in cAMP. One potential problem with IBMX is that it, like methylxanthines in general, has a potential for increasing free calcium levels in cells (Friedman *et al.*, 1974). Attempts were made to use nonmethylxanthine PDE inhibitors. Papaverine and Ro 20-1724 (the latter a kind gift of Drs. W. E. Scott and H. Sheppard of Hoffmann-La Roche, Inc., Nutley, New Jersey) were employed and, while both caused increases in cAMP but not cGMP, the increases were far below (at most 2 times) those achieved with IBMX, and the elevated cAMP levels were not clearly light-sensitive. As employed, 1 m*M* IBMX only partially decreased PDE activity, since 3 m*M* IBMX yielded about 2.5 times more cAMP than 1 m*M* IBMX in 5 min of dark incubation. This is in contrast to its effects on retinal homogenates where PDE inhibition by 1 m*M* IBMX was virtually total. The data suggest that any cyclic nucleotide effects related to a calcium increase attributable to the use of IBMX were swamped by the results of its PDE inhibition.

VII. CONCLUSIONS

Photoreceptors contain both cAMP and cGMP. How these nucleotides respond in incubated retinas is summarized in Tables I and II. The major synthetic site of photoreceptor cGMP is probably in the outer segment. Fleishman and Denisevich (1979) have shown that the particulate guanylate cyclase co-isolates with the ciliary apparatus of the outer segments. These authors, Krishnan *et al.* (1977)

and Troyer *et al.* (1978) have shown that this enzyme is unique among cyclases in being inhibitable by calcium over a certain concentration range. Thus one hypothesis explaining the effect of calcium chelation on cGMP levels is disinhibition of the guanylate cyclase of the outer segment. Most soluble guanylate cyclases are activated by calcium. Particulate guanylate cyclases are generally either indifferent to or activated by calcium. That calcium chelation increases cGMP is thus far unique to the retina among nervous tissues. In addition to the particulate Ca^{2+}-inhibitable guanylate cyclase of the outer segment, the ground squirrel retina may have a soluble, Ca^{2+}-activated guanylate cyclase in its cone inner segments (De Vries and Ferrendelli, 1980). The isolated PDE of the outer segment requires Mg^{2+}, GTP, and opsin for activation and is said to be indifferent to calcium (Bitensky *et al.*, 1975). However, P. R. Robinson *et al.* (1980) have recently reported that the V_{max} of the PDE of outer segments of frog rods is reduced by low calcium levels. This also causes cGMP to accumulate. Carter *et al.* (1979) compared the outer segment level in freeze-dried slices of the light-adapted monkey retina for PDE activity, with either cGMP or cAMP as substrate. PDE affinity was 50-fold greater for cGMP, but V_{max} was similar for both substrates. With cAMP as substrate they checked the effects of calcium and found a greater than 2-fold *increase* when calcium was chelated. If the cAMP-PDE of the outer segment is the same enzyme as cGMP-PDE, then this suggests some calcium sensitivity for PDE in a direction that does not fit the accumulation of cGMP when calcium is chelated. If it is a different enzyme, and if mouse rods resemble monkey rods, increasing hydrolysis of cAMP with decreasing calcium would be one among many possible mechanisms explaining how chelating calcium eliminates the light-sensitive cAMP level.

Since in the current study a light-sensitive cAMP pool in incubated retinas was accumulated only with cobalt or IBMX, it is probable that the levels generated by these agents were unphysiological. Some cAMP from this artificially heightened pool may have diffused into the outer segments where the light activation of PDE produced its destruction. The evidence strongly supports the presence of a cAMP metabolism in photoreceptors, but no claim is made to demonstration of a normal metabolism including the light-activated destruction of cAMP, although the existence of such is possible. Light-adapted mouse retinas have less cAMP than dark-adapted retinas. The fact that light produced significant destruction of cAMP in the presence of IBMX and levels of ouabain that should have eliminated photovoltage and thus confined light effects to outer segments shows that a diffusion mechanism is possible. At the moment there is no evidence that photoreceptors are postsynaptic to any cell using a chemical transmitter said to activate adenylate cyclase in postsynaptic membranes. However, cAMP might be associated with glycogen metabolism, microtubule assembly, the modulation of protein synthesis, or photovoltage-controlled activities.

The behavior of cobalt, magnesium, and calcium levels in manipulating cyclic-nucleotide levels suggests that, when ionic agents are employed for physiological manipulations, one should be aware that it is quite possible that their effects will not be confined to membrane channels or to a single metabolic system of any cell.

No claim is made in this report linking cyclic nucleotides to either transduction or adaptation. The time scale of our studies is wrong for the former. The general assessment following these studies is that a complex and partly unique nucleotide system exists in the outer segment and elsewhere in the photoreceptor, but that not all the components have been identified. Our laboratory has only begun to study a subset of the reactive elements. For already identified components, their precise location and interactions are still incompletely understood. Until we know the actors and their position on stage, understanding the play will be difficult.

ACKNOWLEDGMENT

This work was supported in part by grants EY-00258 and EY-02294 from the National Eye Institute.

REFERENCES

Baylor, D. A., Hodgkin, A. L., and Lamb, T. D. (1974). Reconstruction of the electrical responses of turtle cones to flashes and steps of light. *J. Physiol. (London)* **242,** 759–791.

Berger, S. J., De Vries, G. W., Carter, J. G., Schultz, D. W., Passonneau, P. N., Lowry, O. H., and Ferrendelli, J. A. (1980). The distribution of the components of the cyclic GMP system in retina. *J. Biol. Chem.* **255,** 3128–3133.

Bitensky, M. W., Miki, N., Keirns, J. J., Baraban, J. M., Freeman, J., Wheeler, M. A., Lucy, J., and Marcus, F. R. (1975). Activation of photoreceptor disc membrane phosphodiesterase by light and ATP. *Adv. Cyclic Nucleotide Res.* **5,** 213–240.

Bitensky, M. W., Wheeler, G. L., Aloni, B., Vetury, S., and Matuo, Y. (1978). Light- and GTP-activated photoreceptor phosphodiesterase: Regulation by a light-activated GTPase and identification of rhodopsin as the phosphodiesterase binding site. *Adv. Cyclic Nucleotide Res.* **9,** 553–572.

Carter, J. G., Berger, S. J., and Lowry, O. H. (1979). The measurement of cyclic GMP and cyclic AMP phosphodiesterase. *Anal. Biochem.* **100,** 244–253.

Cervetto, L., and Piccolino, M. (1975). Mechanisms of synaptic transmission in the vertebrate retina. *In* "Golgi Centennial Symposium Proceedings" (M. Santini, ed.), pp. 577–581. Raven, New York.

Chader, G., Fletcher, R., Johnson, M., and Bensinger, R. (1974). Rod outer segment phosphodiesterase: Factors affecting the hydrolysis of cyclic AMP and cyclic GMP. *Exp. Eye Res.* **18,** 509–515.

Chader, G. J., Fletcher, R. T., and Krishna, G. (1980). Guanine nucleotides: Importance in visual processes of the rod outer segment. *In* "Future Directions in Ophthalmic Research" (M. Sears and Y. Pouliquen, eds.) (in press).

Cohen, A. I., and J. A. Ferrendelli (1979). The effect of light and putative synaptic blockers on the cAMP level of the incubated mouse retina. *Invest. Ophthalmol. Visual Sci.* **18** (Suppl.), 21.

Cohen, A. I., Hall, I. A., and Ferrendelli, J. A. (1978). Calcium and cyclic nucleotide regulation in incubated mouse retinas. *J. Gen. Physiol.* **71,** 595–612.

Dacheux, R. F., and Miller, R. F. (1976). Photoreceptor-bipolar cell transmission in the perfused retina eyecup of the mudpuppy. *Science* **191,** 963–964.

Dacko, S. M., Zuckerman, R., and Weiter, J. J. (1980). Evidence for ATP-dependent internalization of transmitter in ROS disc. *Invest. Ophthalmol. Visual Sci.* **19,** ARVO Suppl., 131.

De Vries, G. W., and Ferrendelli, J. A. (1980). Evidence for multiple guanylate cyclases in a vertebrate photoreceptor. *Invest. Ophthalmol. Visual Sci.* **19,** ARVO Suppl., 21.

De Vries, G. W., Cohen, A. I., Lowry, O. H., and Ferrendelli, J. A. (1979). Cyclic nucleotides in the cone-dominant ground squirrel retina. *Exp. Eye Res.* **29,** 315–322.

Dowling, J. E., and Ripps, H. (1972). Adaptation in skate photoreceptors. *J. Gen. Physiol.* **60,** 698–719.

Dowling, J. E., and Ripps, H. (1973). Effect of magnesium on horizontal cell activity in the skate retina. *Nature (London)* **242,** 101–103.

Evans, G. J., and Erulkar, S. D. (1980). Comparison of the effect of magnesium and cobalt on transmitter release at frog neuromuscular junction in low calcium. *Fed. Proc., Fed. Am. Soc. Exp. Res.* **39,** p. 2463.

Evans, J. A., Hood, D. C., and Holtzman, E. (1978). Differential effects of cobalt ions on rod and cone synaptic activity in the isolated frog retina. *Vision Res.* **18,** 145–151.

Farber, D. B., and Lolley, R. N. (1978). cAMP and cGMP content of cone-dominant retinas of ground squirrel. *Assoc. Res. Vision Ophthalmol.* Abstracts, p. 255.

Ferrendelli, J. A., and Cohen, A. I. (1976). The effects of light and dark adaptation on the levels of cyclic nucleotides in retinas of mice heterozygous for a gene for photoreceptor dystrophy. *Biochem. Biophys. Res. Commun.* **73,** 421–427.

Ferrendelli, J. A., De Vries, G. W., Cohen, A. I., and Lowry, O. H. (1980). Localization and roles of cyclic nucleotide systems in retina. *Neurochemistry* **1,** 311–326.

Fleishman, D., and Denisevich, M. (1979). Guanylate cyclase of isolated bovine retinal rod axonemes. *Biochemistry* **18,** 5060–5066.

Fletcher, R. T., and Chader, G. J. (1976). Cyclic GMP: Control of concentration by light in retinal photoreceptors. *Biochem. Biophys. Res. Commun.* **70,** 1297–1302.

Friedman, H. H., Bianchi, C. P., and Weiss, S. J. (1974). Structural aspects of the effects of ethylaminobenzoates on caffeine contracture. *J. Pharmacol. Exp. Ther.* **189,** 423–433.

Goridis, C., and Virmaux, N. (1974). Light-regulated guanosine 3′,5′-monophosphate phosphodiesterase of bovine retina. *Nature (London)* **248,** 57–58.

Hagins, W. A., and Yoshikami, S. (1975). Ionic mechanisms in excitation of photoreceptors. *Ann. N.Y. Acad. Sci.* **264,** 314–325.

Krishnan, N., Fletcher, R. T., Chader, G. J., and Krishna, G. (1977). Characterization of guanylate cyclase of rod outer segments of the bovine retina. *Biochim. Biophys. Acta* **523,** 506–515.

Lipton, S. A., Rasmussen, H., and Dowling, J. E. (1977). Electrical and adaptive properties of rod photoreceptors in *Bufo marinus.* II. Effects of cyclic nucleotides and prostaglandins. *J. Gen. Physiol.* **70,** 771–791.

Lolley, R. N., Schmidt, S. Y., and Farber, D. B. (1974). Alterations in cyclic AMP metabolism associated with photoreceptor cell degeneration in the C3H mouse. *J. Neurochem.* **22,** 701–707.

Lowry, O. H., and Passonneau, J. V. (1972). "A Flexible System of Enzymatic Analysis." Academic Press, New York.

Miller, R. F. (1973). Role of K^+ in generation of b-wave of electroretinogram. *J. Neurophysiol.* **36,** 1–38.

Mitzel, D. L., Hall, I. A., De Vries, G. W., Cohen, A. I., and Ferrendelli, J. A. (1978). Comparison of cyclic nucleotide and energy metabolism of intact mouse retina *in situ* and *in vitro*. *Exp. Eye Res.* **27,** 27–37.

Orr, H. T., Lowry, O. H., Cohen, A. I., and Ferrendelli, J. A. (1976). Distribution of 3′:5′-cyclic AMP and 3′:5′-cyclic GMP in rabbit retina *in vivo*: Selective effects of dark and light adaptation and ischemia. *Proc. Natl. Acad. Sci. U.S.A.* **73,** 4442–4445.

Penn, R. D., and Hagins, W. A. (1969). Signal transmission along retinal rods and the origin of the electroretinographic a-wave. *Nature (London)* **223,** 201–205.

Robinson, P. R., Kawamura, S., and Bownds, D. (1980). Control of the cyclic GMP phosphodiesterase of frog photoreceptor membranes by light and calcium ions. *Fed. Proc., Fed. Am. Soc. Exp. Res.* **39,** 2138.

Robinson, W. E., and Hagins, W. A. (1977). A light-activated GTPase in retinal rod outer segments. *Biophys. J.* **17,** 196a.

Robinson, W. E., and Hagins, W. A. (1979). GTP hydrolysis in intact rod outer segments and the transmitter cycle in visual excitation. *Nature (London)* **280,** 398–400.

Rodnight, R. (1979). Functional roles for cyclic nucleotides in post-synaptic events. *Int. Rev. Biochem.* **26,** 2–80.

Shinozawa, T., Uchida, S., Martin, E., Cafiso, D., Hubbell, W., and Bitensky, M. (1980). Additional component required for activity and reconstitution of light-activated vertebrate photoreceptor GTPase. *Proc. Natl. Acad. Sci. U.S.A.* **77,** 1408–1411.

Sillman, A. J., Itu, H., and Tomita, T. (1969). Studies on the mass receptor potential of the isolated frog retina. I. General properties of the response. *Vision Res.* **9,** 1435–1442.

Thacher, S. M. (1978). Light-stimulated, magnesium-dependent ATPase in toad retinal rod outer segments. *Biochemistry* **17,** 3005–3011.

Troyer, E. W., Hall, I. A., and Ferrendelli, J. A. (1978). Guanylate cyclases in CNS: Enzymatic characteristics of soluble and particulate enzymes from mouse cerebellum and retina. *J. Neurochem.* **31,** 825–833.

Wassenaar, J. S., and Korf, J. (1976). Characterization of catecholamine receptors in rat retina. *In* "Transmitters in the Visual Process" (S. L. Bonting, ed.), pp. 199–218. Pergamon, Oxford.

Wheeler, G. L., and Bitensky, M. W. (1977). A light-activated GTPase in vertebrate photoreceptors: Regulation of light-activated cyclic GMP phosphodiesterase. *Proc. Natl. Acad. Sci. U.S.A.* **74,** 4238–4242.

Winkler, B. S. (1972). The electroretinogram of the isolated rat retina. *Vision Res.* **12,** 1183–1198.

Chapter 13

Cyclic AMP: Enrichment in Retinal Cones

DEBORA B. FARBER

Jules Stein Eye Institute
University of California School of Medicine
Los Angeles, California and
Developmental Neurology Laboratory
Veterans Administration Medical Center
Sepulveda, California

I. INTRODUCTION

The last decade has witnessed extraordinary progress in the understanding of molecular mechanisms that take place in the retina. For many researchers, the involvement of cyclic nucleotides as regulators of the metabolism and function of photoreceptor cells became a focal point of interest. Data on this subject began to accumulate and to share importance with the results of studies on the biochemistry of rhodopsin.

The early work on cyclic nucleotides in visual cells revolved mainly around cyclic adenosine 3′,5′-monophosphate (cyclic AMP), the factors that controlled its synthesis and hydrolysis, and its possible physiological significance. But with the discovery of high levels of cyclic guanosine 3′,5′-monophosphate (cyclic GMP) in rod photoreceptors—levels much higher than those of cyclic AMP—and a rod outer segment-specific cyclic-GMP phosphodiesterase with a higher

ISBN 0-12-153315-8

affinity for cyclic GMP than for cyclic AMP, there was a change in focus, and much of the attention turned to cyclic GMP. Furthermore, abnormalities in the metabolism of cyclic GMP, specifically, were shown to be associated with inherited retinal degenerations affecting primarily the rod visual cells.

Relatively little work had been done on the biochemistry of cone photoreceptors when we began to study the cyclic-nucleotide metabolism of retinas dominated by cones. It was exciting to learn that, whereas rods were enriched in cyclic GMP, cyclic AMP was the predominant cyclic nucleotide of cones (Farber and Lolley, 1978). Rod and cone visual cells have several other characteristics distinguishing one from the other, including outer segment and synaptic morphology, diurnal time of shedding, visual pigments, and the intensity of light to which each responds.

II. CONE- VERSUS ROD-DOMINANT RETINAS: CYCLIC-NUCLEOTIDE CONTENT AND EFFECT OF LIGHT ADAPTATION

We have studied two animal species that have retinas dominated by cones: the 13-line ground squirrel, *Citellus tridecemlineatus,* and the western fence lizard, *Sceloporus occidentalis*. In both retinas, the levels of cyclic AMP are severalfold higher than those of cyclic GMP. The cyclic AMP/cyclic GMP ratios are 8.2 for the ground squirrel and 2.2 for the western fence lizard. These cyclic AMP/cyclic GMP ratios contrast strongly with those of rod-dominant retinas (e.g., those of rats, mice, and toads), which in general range between 0.2 and 0.3 (Farber *et al.*, 1981).

Rod- and cone-dominant retinas differ not only in their relative levels of cyclic AMP and cyclic GMP but also in their response to illumination; that is, light modulates the amount of the opposite cyclic nucleotide. For example, the cyclic-GMP concentration is reduced after exposure to light in dark-adapted, rod-dominant retinas *in vitro* and in isolated rod outer segments (Goridis *et al.*, 1974; Fletcher and Chader, 1976; Farber and Lolley, 1977; DeVries *et al.*, 1978), and the cyclic-AMP content of dark-adapted, cone-dominant retinas is decreased after light adaptation. Cyclic AMP in the outer segments of rods (Orr *et al.*, 1976) and cyclic GMP in cone-dominant retinas remain unchanged following exposure to light (Farber *et al.*, 1981).

Our data on the cyclic-nucleotide levels in cone-dominant retinas were obtained using freshly dissected tissue. It was suggested that the lower levels of cyclic GMP, as compared to those of cyclic AMP that we observed in the ground squirrel retina, possibly resulted from the loss of outer segment material during dissection of the tissue (DeVries *et al.*, 1979). It is true that cone outer segments break very easily at the outer limiting membrane and that the detached receptors adhere firmly to the apical villi of the pigment epithelial cells. To avoid this problem and, at the same time, take advantage of the shorter elapsed time between enucleation and extraction of the cyclic nucleotides, we measured the cyclic-AMP and cyclic-GMP content of the intact eyes of dark- and light-adapted ground squirrels and lizards. The intact eyes, like the dissected retinas, contained more cyclic AMP than cyclic GMP (cyclic AMP/cyclic GMP ratio = 8.0 for ground squirrel and 3.6 for lizard). We estimated that about 65% of the cyclic-AMP and 80–90% of the cyclic-GMP content of the intact eye were localized in the retina. Furthermore, the effect of light on the cyclic-nucleotide content of the intact eyes was similar to that for the dissected retinas: Light reduced by about 50% the cyclic-AMP concentration of dark-adapted eyes, whereas the levels of cyclic GMP were not significantly affected (Farber *et al.*, 1981).

III. EFFECT OF FREEZING ON THE CYCLIC-NUCLEOTIDE CONTENT OF GROUND SQUIRREL OCULAR TISSUES

The temperature at which the ocular tissues are maintained prior to assay is critical in the measurement of cyclic-AMP levels. We found that freezing either the retina or the whole eye reduced the cyclic-AMP content of freshly dissected tissues considerably, via a mechanism that has not been clarified as yet. In contrast, the levels of cyclic GMP were minimally affected. We should point out that light decreases the levels of cyclic AMP in fresh as well as frozen dark-adapted retinas and whole eyes. Furthermore, once the dark- or light-adapted tissue is frozen, freeze-drying under normal lighting conditions causes an additional reduction in cyclic-AMP levels (Farber *et al.*, 1981). We have estimated that dark-adapted as well as illuminated retinas or whole eyes contain, after freeze-drying in the light, about 16–18% of the cyclic-AMP and 80% of the cyclic-GMP content of freshly dissected, dark-adapted ocular tissues. Thus caution is recommended when measuring cyclic-AMP levels in retinal samples of cone-dominant retinas. This caution is particularly relevant for microdissected retinal layers prepared using freeze-drying techniques. We and others (DeVries *et al.*, 1979) have found that cyclic-AMP and cyclic-GMP levels are similar in freeze-dried retinal sections and that there is no difference in cyclic-AMP content between light- and dark-adapted freeze-dried tissues.

IV. LOCALIZATION OF CYCLIC NUCLEOTIDES IN THE GROUND SQUIRREL RETINA

Approximately one-half of the cyclic-AMP and almost all of the cyclic-GMP content of the retina of the ground squirrel are localized in the photoreceptors.

This was determined during studies in which we selectively destroyed the cone visual cells of ground squirrels and then measured the cyclic-nucleotide levels of the remaining inner retinal layers. The relative difference in cyclic-AMP and cyclic-GMP concentrations between controls and photoreceptorless retinas was attributed to the cyclic-nucleotide content of the cones.

In order to destroy the photoreceptors, we injected ground squirrels intracardially with iodoacetic acid (Farber *et al.*, 1981). Schubert and Bornstein (1951) and Nöell (1952) had observed earlier that intravenous injections of sodium iodoacetate caused a characteristic degeneration of rods in several species of animals; all other cells of the retina remained undamaged morphologically unless high doses of the drug were used. With the ground squirrel, the effect of iodoacetic acid was also selective for the cones which, 11 days after treatment, were replaced by a layer of large macrophages containing dense material. Under the electron microscope, these masses were identified as partially degraded photoreceptors. In some patches of the retina, though, macrophages and debris were absent altogether, and the pigment epithelium lay atop the inner nuclear layers which remained unchanged. Biochemical analysis of the same retina from which pieces were taken for morphological study showed that cyclic AMP and cyclic GMP both were decreased by degeneration of the cone visual cells. The iodoacetic acid treatment reduced the levels of cyclic AMP from 91.0 pmoles mg^{-1} protein in the control dark-adapted retina to 49.2 pmoles mg^{-1} protein in the experimental dark- or light-adapted retina. The levels of cyclic GMP decreased from 11.1 to 0.7 pmole mg^{-1} protein.

Thus cyclic AMP is quite evenly distributed between the cones and inner retinal layers, whereas the low levels of cyclic GMP present in the ground squirrel retina are almost exclusively localized in the visual cells. The relative loss of cyclic AMP after iodoacetic acid treatment is about four times greater than that of cyclic GMP (Farber *et al.*, 1981). Therefore we estimate that cone photoreceptors of the ground squirrel contain four times more cyclic AMP than cyclic GMP. The precise localization of cyclic AMP within the visual cells still remains to be determined.

V. CONCLUSION

Although there is very limited information on the role that cyclic AMP may play in cone metabolism or function, it is tempting to extend analogies between the rod and cone systems. While rods have high levels of cyclic GMP which are decreased by light, cones have high levels of cyclic AMP which are modulated by light. As described in other chapters of this book, cyclic GMP appears to be an important component of normal rod photoreceptors, possibly acting in the visual transduction mechanism or in other homeostatic processes. It may be that cyclic AMP has a similar function in the physiological events that take place in cone photoreceptors.

ACKNOWLEDGMENTS

This work was supported by the National Eye Institute (grants EY 2651, EY 331 and RCDA KO4-EY 144). My appreciation is extended to Dr. Richard N. Lolley for his constructive criticism of the manuscript and also to Dr. David G. Chase for his help with the electron microscopy. Thanks are due also to Dennis W. Souza for technical assistance and to Louise V. Eaton, Ethel Mason, and Andrea Williams for help in manuscript preparation.

REFERENCES

DeVries, G. W., Cohen, A. I., Hall, I. A., and Ferrendelli, J. A. (1978). Cyclic nucleotide levels in normal and biologically fractionated mouse retina: Effects of light and dark adaptation. *J. Neurochem.* **31,** 1345–1351.

DeVries, G. W., Cohen, A. I., Lowry, O. H., and Ferrendelli, J. A. (1979). Cyclic nucleotides in the cone-dominant ground squirrel retina. *Exp. Eye Res.* **29,** 315–321.

Farber, D. B., and Lolley, R. N. (1977). Light-induced reduction in cyclic GMP of retinal photoreceptor cells *in vivo*: Abnormalities in the degenerative diseases of RCS rats and *rd* mice. *J. Neurochem.* **28,** 1089–1095.

Farber, D. B., and Lolley, R. N. (1978). cAMP and cGMP of cone-dominant retinas of ground squirrel. *ARVO Abstr.* p. 255.

Farber, D. B., Souza, D. W., Chase, D. G., and Lolley, R. N. (1981). Cyclic nucleotides of cone-dominant retinas: Reduction of cyclic AMP levels by light and by cone degeneration. *Invest. Ophthalmol. Visual Sci.* **20,** 24–31.

Fletcher, R. T., and Chader, G. J. (1976). Cyclic GMP: Control of concentration by light in retinal photoreceptors. *Biochem. Biophys. Res. Commun.* **70,** 1297–1302.

Goridis, C., Virmaux, N., Cailla, H. L., and Delaage, M. A. (1974). Rapid, light-induced changes of retinal cGMP levels. *FEBS Lett.* **49,** 167–169.

Nöell, W. K. (1952). The impairment of visual cell structure by iodoacetate. *J. Cell. Comp. Physiol.* **40,** 25–55.

Orr, H. T., Lowry, O. H., Cohen, A. I., and Ferrendelli, J. A. (1976). Distribution of 3′,5′-cyclic AMP and 3′,5′-cyclic GMP in rabbit retina *in vivo*: Selective effects of dark and light adaptation and ischemia. *Proc. Natl. Acad. Sci. U.S.A.* **73,** 4442–4445.

Schubert, G., and Bornstein, H. (1951). Specific damage to retinal elements by iodine acetate. *Experientia* **7,** 461–462.

Chapter 14

Cyclic-Nucleotide Metabolism in Vertebrate Photoreceptors: A Remarkable Analogy and an Unraveling Enigma

M. W. BITENSKY,[1,2] *G. L. WHEELER,*[3] *A. YAMAZAKI,*[1]
M. M. RASENICK,[1] *AND P. J. STEIN*[4]

[1]Department of Pathology, Yale University School of Medicine, New Haven, Connecticut.
[2]Present address: Division of Life Sciences, Los Alamos Scientific Laboratory, Los Alamos, New Mexico.
[3]Department of Chemistry, University of New Haven, West Haven, Connecticut.
[4]Department of Biological Sciences, Purdue University, West Lafayette, Indiana. Present address: same as footnote 1.

ISBN 0-12-153315-8

I. INTRODUCTION: THE ROD OUTER SEGMENT PHOSPHODIESTERASE PUZZLE[5]

A. Early Musings

The possibility that cyclic nucleotides could be involved in the light-sensitive functions of vertebrate photoreceptors was first investigated in 1970 by M. W. Bitensky and W. H. Miller (Bitensky *et al.*, 1971, 1972; Miller *et al.*, 1971). These studies began with a single simple question: Can light regulate the levels of cyclic nucleotides in frog ROS? Fully dark-adapted frog ROS were isolated by sucrose flotation and disrupted mechanically in hypotonic buffers. The ROS disk membrane suspensions were incubated with [^{14}C]ATP in the presence or absence of light. It was expected that light might act like a hormone [by analogy with hormonal regulation of adenylate cyclase (AC)] in a wide variety of cell types (Perkins, 1974). In this first experiment, we were only prepared to assay the effect of light on the amount of ^{14}C-labeled cyclic AMP formed by AC. When the data for this experiment emerged from the scintillation counter, we found that the illuminated samples contained about 300 cpm of ^{14}C-labeled cyclic AMP formed, while in contrast the samples assayed in the dark contained more than 1800 cpm of cyclic AMP formed (Bitensky *et al.*, 1971). This demonstrated an unequivocal light-dependent modulation of cyclic-nucleotide levels, which was of a magnitude similar to that encountered in the classical activation of hepatic AC by the hormone glucagon (Rodbell *et al.*, 1971). Although light did indeed regulate cyclic-nucleotide metabolism in ROS, we later found that it actually activated a cyclic-GMP phosphodiesterase (PDE) and did not inhibit AC. It appears, nevertheless, that this first appraisal was quite useful, although a great deal of experimentation was needed to identify correctly and align the com-

[5] *Abbreviations:* AMP, adenosine 5′-monophosphate; ATP, adenosine 5′-triphosphate; Cyclic AMP, cyclic adenosine 3′,5′-monophosphate; GMP, guanosine 5′-monophosphate; GTP, guanosine 5′-triphosphate; Cyclic GMP, cyclic guanosine 3′,5′-monophosphate; GTPase, guanosinetriphosphatase; IMP, inosine 5′-monophosphate; ROS, rod outer segments; AC, adenylate cyclase;Rho, rhodopsin; App(NH)p, adenosine imidodiphosphate, Gpp(NH)p, guanosine imidodiphosphate; PDE, phosphodiesterase; SDS, sodium dodecyl sulfate; EGTA, ethylene glycol-bis(β-aminoethyl ether)-*N*,*N*′-tetraacetic acid; EDTA, ethylenediaminetetraacetic acid; H, helper; G, GTP-binding protein or G protein; G unit, GTP-binding protein plus the helper protein and putative cofactors.

ponents of the rod cyclic-nucleotide system. The initial light–hormone analogy confirmed by the earliest experiments has developed into a predictive tool which has been used in the design of subsequent studies with both PDE and AC (Shinozawa *et al.*, 1979; also see below).

In this chapter, we review some ideas and problems that have emerged in the course of elucidating the functional components of light-sensitive ROS. We wish also to list some more recent findings and ideas about future experimental directions. In the course of these ruminations, we will consider and emphasize the remarkable analogy between hormone-activated AC and light-activated PDE (Shinozawa *et al.*, 1979). In this context, we will describe reconstitution experiments that we have carried out with extensively and/or partially purified components of the disk membrane PDE system, and some of the crosses or combinations that have been arranged between components of the PDE and AC systems. Finally, encouraged by some recent "proofs of opportunity"[6] (Yee and Liebman, 1978; Woodruff and Bownds, 1979) and newer findings with R. Sorbi and P. Stein (see below), we will speculate about the possible physiological significance of ROS PDE.

B. Successive Approximations to a More Correct View

It was clear from the very outset that light caused a profound effect on cyclic-nucleotide metabolism in ROS disk membrane preparations. There were, however, several false starts that required clarification. An early and costly misconception was the assignment of the light effect to AC. This mistaken assignment was fostered by the dogma that cyclic AMP is the primary cyclic nucleotide in eukaryotic cells and that hormones regulate the production of cyclic AMP via the enzyme AC. In this early period, PDE was customarily viewed as a passive, albeit necessary, degradative enzyme which carried out the uninspired task of hydrolyzing free cyclic nucleotides. PDE was thus universally portrayed in the dreary, unimaginative role of returning cyclic-nucleotide concentrations to the "unstimulated" level. It was not yet the season to view PDE as a signal-controlled enzyme. Finally, by assaying both reactant and product concentrations, we arrived at the heretical realization that the observed effects of light on cyclic-nucleotide metabolism were indeed a consequence of activating a cyclic-GMP PDE. It is ironic that, had our first experiments sought a light-activated PDE, we would never have made the observation for lack of the requisite nucleoside triphosphate cofactor (Bitensky *et al.*, 1975). In attempting to study the effects of light on disk membrane AC activity, however, we reflexly

[6]The phrase "proofs of opportunity" is used to describe experiments that purport to show that light-mediated declines in cyclic GMP can occur with sufficient rapidity or are of sufficient magnitude to support transduction. Such data may indicate that cyclic GMP can participate in transduction; they do not prove participation.

incorporated appropriate substrate concentrations of ATP inadvertently contaminated with micromolar amounts of the real cofactor GTP (Kimura and Nagata, 1976).

After much additional study, we came to our present understanding that the light effects on disk membrane cyclic-nucleotide levels were actually a result of light-activated, GTP-dependent, PDE activity (Miki *et al.*, 1973; Bitensky *et al.*, 1973). It should be added that Pannbacker (1973) (who did not investigate the question of light modulation of PDE activity) was the first to conclude correctly that rod PDE activity showed an explicit preference for the hydrolysis of cyclic GMP rather than cyclic AMP.

C. Rhodopsin: A New Lyric for a Familiar Melody

Photoreceptor photopigments have fascinated vision researchers from the earliest moments in the history of retinal science. Systematic study of the light-induced bleaching of rhodopsin (Rho) in vertebrate retinas began in the 1800s with the pioneering work of Kühne (1878). It soon became apparent to us that we needed to confront the question of whether Rho was responsible for the light-dependent activation of PDE. Since Rho was clearly the photopigment mediating visual excitation in vertebrate rods, it was a preferred candidate for the role of primary activator for the by now eagerly studied light-sensitive ROS PDE. There were, however, certain innuendos suggesting that Rho might function otherwise, that is, the idea that Rho was the mediator of rod visual excitation via the exclusive domain of a light-gated fluctuation in the concentration of Ca^{2+} or some other cationic species (Hagins *et al.*, 1970). If illuminated Rho gated a calcium (or other cation) channel, the idea that it could also, in some independent (or even related) way, modulate a cyclic-nucleotide regulatory mechanism seemed at first both contrived and improbable. However, the action spectra obtained with bovine and amphibian Rho clearly established that the λ_{max} for the light activation of PDE corresponded perfectly to the λ_{max} for the absorption spectrum of Rho (Keirns *et al.*, 1975). The idea that Rho was responsible for light activation of PDE was subsequently verified in experiments that utilized purified bovine Rho reconstituted in phosphatidycholine vesicles as the only integral membrane protein supporting PDE activation (Shinozawa *et al.*, 1980).

D. Light Sensitivity of the Photoreceptor Phosphodiesterase: First Impressions

Early experiments revealed a striking light sensitivity of the ROS PDE system: We found that bleaching fewer than 1 in 1000 Rho molecules provided *maximal* activation of this enzyme (Bitensky *et al.*, 1972). Moreover, these data may only have indicated a lower limit for the sensitivity of the system. Subsequent studies have suggested that the sensitivity might be orders of magnitude higher (Yee and Liebman, 1978). Since the purification schedules for frog PDE indicate about 1 PDE molecule per 1000 Rho molecules, and sensitivity studies suggested that bleaching 1 in 1000 Rho molecules caused maximal PDE activation, it appeared that each bleached Rho could activate a single PDE. The intriguing possibility that one bleached Rho could activate many PDE molecules appears strongly supported by more recent estimates of sensitivity (Yee and Liebman, 1978; Woodruff and Bownds, 1979), which depend in part on earlier measurements of the catalytic constant for purified frog PDE (Miki *et al.*, 1975).

E. Activation of Phosphodiesterase by Small Admixtures of Bleached Rhodopsin

In addition to activation by a flash of light, it was also possible to achieve activation of PDE in another fashion. Fully dark-adapted ROS membrane suspensions (prepared under infrared light), which had been disrupted by syringing through a no. 26 needle, displayed almost no PDE activity even in the presence of added nucleoside triphosphate. The PDE in the reservoir could be fully activated with less than a 1% of admixture of illuminated Rho. The illuminated Rho could be heated or exposed to urea and then washed with EDTA to make certain that it did not contain any intrinsic residual PDE activity. This "purged," bleached disk membrane preparation, when added to a 100-fold excess of unilluminated disk membrane suspension, fully activated the associated PDE (Keirns *et al.*, 1975). Activation by mixing indicated that there must be a way in which the effects of illumination were communicated from the bleached Rho minority to all the PDE molecules associated with the unbleached Rho majority. The mechanism by which this activation proceeded clearly did not depend on the release of PDE from the "dark" membranes. We found that PDE was firmly bound to the disk membrane in the presence of 2 m*M* Mg^{2+}, an ionic environment in which activation by "mixing" readily occurred. The solution to this mixing puzzle required new insights into the activation mechanism and is discussed below.

F. Earlier Views of the Physiological Implications of Light-Activated Phosphodiesterase

What was the physiological significance of this most dramatic light-mediated hydrolysis of cyclic GMP? The possibility that this reaction could be a critical event in the generation of the receptor potential, i.e., that the fall in cyclic GMP functioned as a transducer in visual excitation, was strongly considered. Indeed the possibility was suggested by earlier electroexperiments (Bitensky *et al.*,

1972; Miller *et al.*, 1971). There was, however, some concern for the conceptual problem of a negative signal and the problem of hydrolysis rates. The catalytic constant established for PDE was 48,000 moles of cyclic GMP hydrolyzed per minute per mole of PDE (Miki *et al.*, 1975). The light sensitivity found in disk suspensions suggested that a single bleached Rho molecule could activate only one PDE molecule (Bitensky *et al.*, 1975). With this turnover number and one PDE molecule activated per bleached Rho molecule, it was felt that the resulting hydrolysis of cyclic GMP would not adequately impact on the free pool in a time period brief enough for visual excitation. (Bitensky *et al.*, 1975). There was also the serious objection that it was easier for a cell to perceive the appearance of 10 units of a transmitter when the appearance was superimposed on a background of zero than to perceive the disappearance of 10 or even 100 units of a "negative" transmitter from a population of 1000 (Bitensky *et al.*, 1975). These concerns become less compelling with the appearance of data that supported ratios of >100 PDE molecule activated per bleached Rho molecule (Yee and Liebman, 1978). The important possibility that light-mediated changes in cyclic GMP might be involved in other light-dependent rod functions including the regulation of rod sensitivity is not excluded by any of the aforementioned ideas.

Whatever one's aesthetic preference, it was clear that the light- and nucleoside triphosphate-dependent hydrolysis of cyclic GMP in vertebrate rods was a profoundly important reaction. This was true not only because of the striking sensitivity and high rates of the reaction but also because of the fact that the photopigment Rho was itself clearly committed to modulation of this cascade. The vertebrate rod has obviously made a major commitment in genetic information and gene product organization to the light-regulated hydrolysis of cyclic GMP. It was (and is) a compelling idea that this reaction must be of profound significance for the normal functioning of the rod.

II. FINDING THE PIECES: ELUCIDATION OF THE CASCADE FROM PHOTON CAPTURE TO ACTIVATION OF PHOSPHODIESTERASE

A. Phosphodiesterase

1. Purification and Characterization of Rod Outer Segment Phosphodiesterase

In order to understand the mechanism involved, it was clearly important to catalogue and characterize the protein components that participated in the light- and nucleoside triphosphate-mediated activation of rod disk PDE. This task was begun with extensive purification of PDE itself. The purification was facilitated by the fact that PDE is not an integral disk membrane protein but is released from the disk by incubation in hypotonic buffers and Mg^{2+}-chelating agents. When

PDE was associated with the disk membrane in the form of a magnesium-stabilized complex, both light and a nucleoside triphosphate were indispensable for its activation. However, activation of PDE by a polycation such as protamine or polyhistidine and activation of PDE by limited trypsin proteolysis were independent of membrane affiliation, light, and/or the nucleoside triphosphate (Miki *et al.*, 1973, 1974, 1975).

Purification of frog disk membrane PDE to homogeneity revealed that the enzyme was composed of two nonidentical but similarly sized (MW 115,000 and 125,000) subunits linked by noncovalent, hydrophobic interaction with an aggregate molecular weight of 240,000. The amino acid composition of the holoenzyme (i.e., both subunits) was determined, as was its isoelectric point (5.7). The K_m for cyclic GMP was 70 μM, and that for cyclic AMP was 3 mM. The turnover number for the pure protein with cyclic GMP as substrate was 48,000 moles of substrate hydrolyzed per mole of enzyme per minute. ROS PDE is without peer when ranked with other PDE by catalytic constant. While the ROS PDE preference for cyclic GMP/cyclic AMP was 23:1 at micromolar substrate concentrations, this preference was only 2½:1 with cyclic-nucleotide concentrations in the millimolar range (Miki *et al.*, 1973, 1975).

2. The Nucleoside Triphosphate Requirement

In the earliest studies on PDE we neglected to add a nucleoside triphosphate and were thus unable to detect light effects on ROS PDE activity. The decision to measure PDE activity under conditions identical to those used for the measurement of AC, even to the point of including a nucleoside triphosphate, made us immediately and profoundly aware that light could activate PDE in ROS providing a nucleoside triphosphate was present. These effects were quite dramatic and soon generated the conviction that the observed "light inhibition" of AC was in fact due to a previously undetected light activation of ROS PDE. In many subsequent experiments in which great care was taken to account for or inhibit PDE activity, we did not observe the effects of light on the ROS guanylate cyclase associated with rod outer segment membrane suspensions (Bitensky *et al.*, 1975).

The effects of a series of nucleoside triphosphates including ATP, ITP, and GTP were next examined. While such compounds provided comparable activation, their analysis was restricted to millimolar amounts. Under these conditions there was little to distinguish among the compounds, and this produced the incorrect impression that the nucleotide cofactor requirement for ROS PDE lacked specificity (Bitensky *et al.*, 1975). We also examined some nonhydrolyzable analogues of ATP including App(NH)p and β,γ-methylene-ATP (Bitensky *et al.*, 1975). Since neither of these analogues supported light-mediated PDE activation, even at millimolar concentrations, we arrived at the conclusion that an intact β,γ-phosphate configuration was necessary for PDE activation to take place. These earlier impressions about the nucleoside triphosphate requirement diverted us for some time from a more useful approach to analysis of the activation mechanism.

3. The Mechanism of Phosphodiesterase Activation

When we began to study the mechanism of activation of ROS PDE, the assembled information was both meager and misleading. It was clear that activation required the presence of bleached Rho (bovine and amphibian were equally effective in activating the eluted frog PDE) and the presence of a nucleoside triphosphate. Although we had some insight into the sensitivity of the system (i.e., its action spectrum and the properties of purified PDE), there was no understanding of how light-dependent activation took place. How was a signal communicated from bleached Rho to PDE? In a real sense, this question is a perfect analogue of the now classical problem of how photon capture by Rho is transduced into hyperpolarization of the rod plasma membrane. It is ironic that the solution of this enzymatic transduction problem may also answer the larger question of physiological transduction of photon capture. Indeed, since photon capture by Rho mediates both the activation of PDE and hyperpolarization of the plasma membrane, a more complete understanding of the relationship between the two effects is critical. The enigma alluded to in the title of this chapter refers both to the classical question of what mediates transduction and to the putative role of PDE activation in the transduction process.

4. Calmodulin

Because of perennial excitement about calcium as a transducer in rods and as a signal participant in other excitable tissues, we were concerned that calcium could itself be a regulator of photoreceptor PDE. This idea was attractive not only because of the ''calcium hypothesis'' (Hagins, 1972) but also because of the seminal work of Shiro Kakiuchi who had shown that calcium in the presence of a heat-stable modulator protein (calmodulin) had a profound effect on the PDE activity found in mammalian brain (Kakiuchi *et al.*, 1973). With a generous gift of highly purified brain calmodulin from Dr. Kakiuchi came the opportunity to evaluate the effects of calcium on rod PDE activities in the presence and absence of light and calmodulin. These experiments failed to show any effects of calcium or calmodulin on the light activation of PDE. Activation took place with the same speed and to the same extent whether or not EGTA, calcium, or calmodulin, alone or in combination, were present during the PDE activation and assay steps (Bitensky *et al.*, 1975).

5. Search for a Phosphorylated Intermediate

The view that a variety of hydrolyzable nucleoside triphosphates could support activation of PDE soon provoked a determined search for a phosphorylated intermediate. We expected to observe, in the presence of illuminated Rho and a nucleoside triphosphate, the generation of a phosphorylated protein that was indispensable for the activation of PDE. We anticipated that the phosphorylation step would involve Rho, PDE, or some as yet unknown intermediate which participated in the activation sequence. The search for the phosphorylated intermediate as a necessary component of the light-dependent activation of PDE became an *idée fixe*. This passionate belief in the inevitability of a phosphorylation mechanism was fueled not only by the earlier impression that a hydrolyzable nucleoside triphosphate such as ATP was required for activation but also by the consensus that cyclic-nucleotide action was properly expressed via protein phosphorylation (Greengard, 1976, 1978). We were also influenced by the observation that illuminated Rho was phosphorylated by an endogenous opsin kinase (Kühn and Dreyer, 1972). In spite of our strong experimental prejudice the search for the key phosphorylated intermediate was continuously discouraging and fruitless.

The phosphorylation odyssey began with Rho. There were, however, a number of good reasons for eliminating this substance from consideration. Perhaps the earliest evidence against its phosphorylation as an essential factor in light-mediated PDE activation was the finding that one could extensively inhibit Rho phosphorylation by including 20 m*M* adenosine in the reaction mixture. Although this achieved better than 95% inhibition of Rho, it had little if any inhibitory effect on the light activation of PDE. Subsequent evidence (see below) confirmed the fact that Rho was not the solution to the "phosphoprotein" problem.

We then began to analyze ROS proteins that had been incubated with [γ-^{32}P]ATP in SDS–polyacrylamide gel electrophoresis. Since these preparations supported light-mediated activation of PDE, we hoped to observe the specific phosphorylation of a unique protein to which we could assign responsibility for the activation step. The principal conceptual obstacle in these experiments was the selection of [γ-^{32}P]ATP at concentrations in the millimolar range. With this approach we succeeded in incorporating radioactive P_i into every segment of the analytical gels. Far from being in a position to select a particular phosphorylated intermediate and assign to it responsibility in the activation sequence, we were confronted by a morass of indistinguishable phosphoproteins. In the midst of such abundance it was not practicable to compare the rates or conditions of phosphorylation with the rates or conditions of PDE activation. Attempts to reduce background phosphorylation with adenosine were of little help in "clearing" the gels. Thus we continued to experience frustration and negative results in this line of experimentation.

B. GTPase

1. The Selective Requirement of Micromolar GTP for the Light-Dependent Activation of Rod Disk Phosphodiesterase

Both our theoretical perspective and the analytical gels were substantially improved by the decision to reexamine the nucleoside triphosphate requirement for light activation of PDE. On this occasion, however, it was decided to examine steadfastly a broad range of concentrations for each of the nucleoside triphosphates. It was soon realized that GTP was more efficient, by three orders of magnitude, than ATP or the other nucleoside triphosphates we had originally examined. GTP served as an effective cofactor in the light activation of disk membrane PDE at concentrations below 1 μM (Wheeler *et al.*, 1977). Moreover, further study suggested that the reason other nucleoside triphosphates appeared to support the light-mediated activation of PDE in earlier experiments probably resulted from contamination with 1 part per 1000 parts GTP. At about this time, Kimura and Nagata (1976) published findings demonstrating such GTP contamination of all ordinary commercial preparations of ATP.

The simple observation that GTP was the specific nucleoside triphosphate required as a cofactor in the light activation of rod disk PDE carried with it a variety of splendid dividends. It was possible to demonstrate a quite striking activation of PDE with as little as 1 μM GTP. It was also possible to reevaluate the hypothesis that there existed a phosphorylated intermediate. With this approach it was expected that it might be possible to examine the phosphorylation reaction under conditions where nonspecific phosphorylation was greatly reduced. With these experimental advantages it seemed, at last, possible either to identify the putative phosphoprotein or to lay the phosphorylation hypothesis to rest. Accordingly, analytical denaturing gels were prepared from rod outer segment materials in which PDE had been activated by light in the presence of 1 μM [γ-^{32}P]GTP. Gel slices were systematically counted in order to detect and identify all phosphorylated proteins. Clearly this was a legitimate experimental approach, since a 40-fold light activation of PDE was supported under the same conditions (Wheeler *et al.*, 1977).

These experiments clearly showed that neither PDE nor Rho was phosphorylated during the activation of PDE. Moreover, there was no observable phosphorylation of any ROS protein during the very rapid activation of PDE, using 0.2% Rho bleaches in the presence of 1 μM GTP. These findings led to the conclusion that a phosphorylated intermediate was not involved in light activation of ROS PDE. Additional and convincing evidence for this conclusion was provided by the simple expedient of demonstrating that light activation of disk PDE was effectively supported by nonhydrolyzable analogues of GTP such as

Gpp(NH)p. This analogue, when used at submicromolar concentrations, was even more effective than GTP as cofactor for the light-dependent activation of PDE. The long, fruitless search for the phosphorylated intermediate was finally abandoned.

2. A Light-Activated GTPase in Vertebrate Rods

A second dividend that followed from the correct assignment of GTP as cofactor was the observation that light-dependent hydrolysis of GTP also took place in suspensions of ROS disk membranes. This finding emerged as a result of our efforts to ascertain the fate of GTP which was added to support the activation of PDE. When GTP (at concentrations of 1 μM or below) was employed as the cofactor in support of PDE activation, one could observe a time-dependent decay of PDE activity, which was reversed by the addition of fresh increments of GTP. Moreover, by using [γ-^{32}P]GTP one could inventory the fate of the released inorganic phosphate during the course of PDE activation and the decay of activation. It became clear that GTP was being hydrolyzed and inorganic phosphate released at rates that were perfectly correlated with the rates of *inactivation of PDE*. When higher concentrations of GTP were used (for example, in the range of 50 μM), there was simply no observed decay in PDE activation for periods greater than ½ hr. Finally, when the nonhydrolyzable analogue Gpp(NH)p was used as cofactor, even at concentrations below 1 μM, the resulting PDE activation was persistent. These findings demonstrated that the ROS disk membrane suspensions contained a GTPase activity that terminated activation of PDE by light and GTP (Wheeler and Bitensky, 1977; also see Fig. 1).

It was reasonable to inquire whether, by analogy with PDE, the activity of GTPase was enhanced by the illumination of ROS disk membrane suspensions. This question led to the finding that disk membrane suspensions contained two GTPase activities. The GTPase with a K_m of about 1 μM appeared to be closely linked to PDE activation, since its K_m perfectly matched the K_m for PDE activation. This low-K_m GTPase showed a striking light-dependent activation and an action spectrum identical to that of PDE [i.e., Rho was the photopigment responsible for light activation of GTPase (Wheeler *et al.*, 1977)]. In addition to the low-K_m GTPase, which exhibited profound substrate specificity and light sensitivity, we found another nucleoside triphosphatase that operated at a K_m for GTP in the 90 μM range. The high-K_m enzyme did not exhibit nucleotide specificity (e.g., it appeared to hydrolyze ATP as well) nor did it show any dependence on light for its activity. Because of the perfect correspondence between the K_m for the light-dependent GTPase activity and the K_m for the activation of PDE by GTP, we postulated that the activation site was identical with or closely related to the site at which GTP was hydrolyzed (Wheeler *et al.*, 1977).

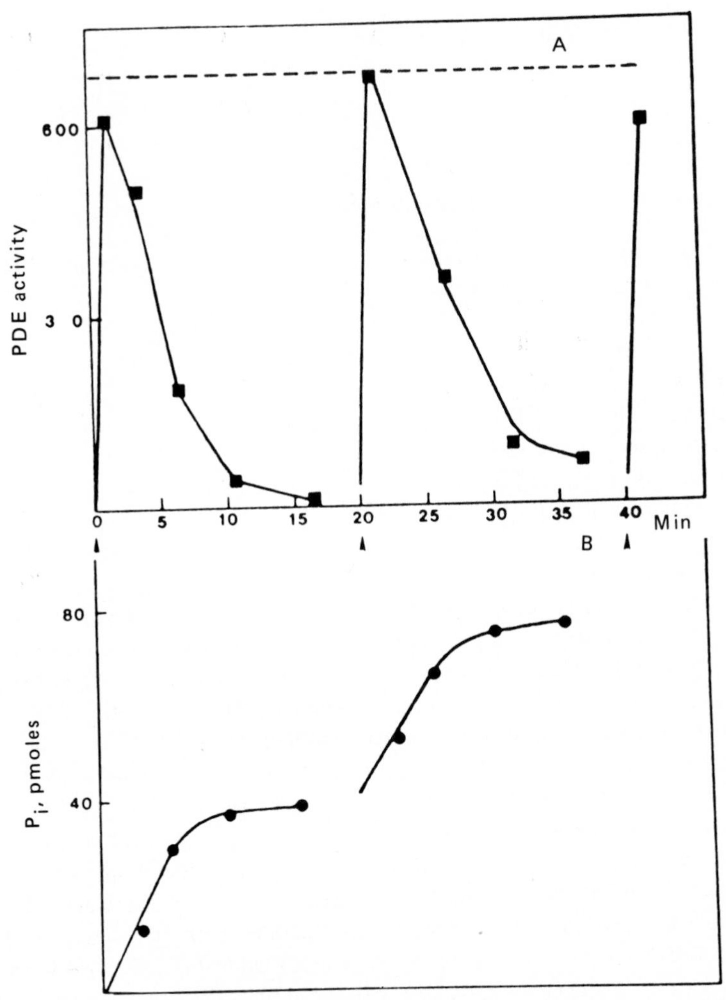

FIG. 1. A comparison of the rates of PDE activation–deactivation with the rates of P_i liberation by light-activated GTPase. (A) PDE activity, shown as a function of time, is expressed as the percentage activation above basal. (Basal activity was 0.55 μmoles of cAMP hydrolyzed per minute per milligram of protein.) GTP was added at 0, 20, and 40 min to a concentration of 0.4 μM. Assays were done at 30°C, and 0.05% bleached membranes were used for the experiments described in both (A) and (B). Dashed line indicates persistent PDE activity in the presence of Gpp(NH)p (2 μM). (B) Time course of the concomitant release of P_i during PDE activation–deactivation. Reactions took place in 100 μl volume, and 40 pmoles of [γ-^{32}P]GTP (specific activity 220 Ci/mmole) was added at 0, 20, and 40 min. The liberated P_i is expressed as picomoles of P_i accumulated at each of the times shown.

3. A Mechanism for Light- and GTP-Dependent Activation of Phosphodiesterase

In fact, these findings had provided the rudiments of a model explaining the light- and GTP-dependent activation of PDE. The GTP-binding locus (G protein or G) that participated in the GTP- and light-dependent activation was contiguous or identical with a GTPase locus that participated in the termination of PDE activation.[7] Furthermore, it appeared that binding of GTP to this locus was fully dependent on the bleaching of Rho, since the activation of PDE showed both a light sensitivity and an action spectrum that corresponded perfectly to the light sensitivity and action spectrum for the GTPase system (Keirns *et al.*, 1975). Following photon capture by Rho, a GTP molecule was admitted to the activation site on the G protein. This G protein–GTP complex formed an active aggregate with PDE. PDE activation persisted until the GTP was hydrolyzed. If more bleached Rho and GTP had been available, activation would have continued. If GTP had been limiting, activation would have decayed with the kinetics of GTP hydrolysis and P_i release (Wheeler and Bitensky, 1977) (Fig. 1).

4. Reconstitution of Light-Activated GTPase and Phosphodiesterase Activities

We then began to purify the light-activated GTPase associated with ROS membrane suspensions. GTPase behaves as a peripheral protein which is readily eluted from disk membranes with isotonic buffers and GTP (Kühn, 1980). The partially purified material has the capacity to hydrolyze GTP when added back to membranes containing bleached Rho (Shinozawa *et al.*, 1980). Furthermore, when we add partially *purified* PDE (entirely free of GTPase activity) to illuminated, heated bovine or amphibian disk membranes (also free of GTPase activity), we were unable to demonstrate PDE activity following the addition of light and GTP. However, when we added to such disk membrane–PDE combinations the partially purified GTPase fraction, we were successful in reconstituting light- and GTP-dependent PDE activity (Shinozawa *et al.*, 1980) as shown in Table I. These reconstitution experiments definitively implicate the G protein as an indispensable component of the light- and GTP-dependent PDE cascade. The data also demonstrate that the dependence of rod PDE on bleached Rho is more properly the dependence of the light-activated GTPase on bleached Rho. That is, the data support a role for the G unit as an intermediary between Rho and PDE. The participation of bleached Rho appears to facilitate interaction between GTP and the G protein. In both the turkey erythrocyte AC and light-sensitive PDE

[7] The G protein (or G) is the polypeptide that binds GTP. These terms are to be considered distinct from the term "G unit" which refers to the complex of G unit, helper (*vide infra*), and associated or as yet unidentified cofactors necessary for GTPase activity or other G-unit functions.

TABLE I
G-FRACTION REQUIREMENT FOR LIGHT- AND GTP-DEPENDENT PDE ACTIVITIES

Addition(s)	PDE activity[a] With 5.0 μM GTP	No GTP
Rho[b]	0	0
Rho + H[c]	0	0
Rho + G[d]	9.0	0
Rho + H + G	12.8	0

[a] Values are nanomoles of cyclic GMP hydrolyzed during a 7-min incubation at 30°C and are averages of two or more determinations that agreed within 5%. (The data are in full agreement with those obtained in four repetitions of this experiment.) Basal PDE activity [i.e., obtained with the PDE fraction alone (6.6 pmoles)] was subtracted in each case. PDE (0.2 μg) partially purified by sucrose density gradient centrifugation was added to each 40-μl assay mixture.

[b] Purified, reconstituted (bleached) rhodopsin (8.75 μg) in phosphatidylcholine (Hong and Hubbell, 1972) was added.

[c] H fraction (0.2 μg) from DEAE-Sephadex was added.

[d] G fraction (0.08 μg) from Blue Sepharose CL-6B was added.

systems, hormone–receptor interaction or light–Rho interaction appears necessary to facilitate exchange or binding of GTP at the G unit activator site. Moreover, formation of the active complex (which eventually will include the catalytic "readout") is initiated by binding of GTP to the G protein (Wheeler and Bitensky, 1977; Cassel and Selinger, 1976).

5. LUMI-LAMBAN: AN EXPLANATION OF PHOSPHODIESTERASE ACTIVATION BY ADMIXTURE OF BLEACHED RHODOPSIN

Recently, the activation mechanism for PDE was reexamined specifically with the aim of understanding the effects of bleached Rho on the GTP binding protein (G). Our working hypothesis predicted that G interaction with bleached Rho permitted binding of GTP to G. Furthermore, we inquired whether the complex of guanine nucleotide and G, when purified away from Rho, could produce activation of PDE. This experimental approach has provided a solution to the question

TABLE II
LUMI-LAMBAN ACTIVATION OF PDE IN DARK ROS[a]

Components preincubated			Activation of PDE assayed in dark (%)
G + Rho + Gpp(NH)p	RHO membranes removed	Soluble proteins added to dark disks	85
G + heated Rho + Gpp(NH)p			85
heated G + Rho + Gpp(NH)p			22
G + Rho			4
Rho + Gpp(NH)p			4
G + Gpp(NH)p			12

[a] G was purified from an EDTA extract of dark frog ROS by sequential chromatography using DEAE Sephadex A-50 and Blue Sepharose CL-6B columns. Rho equals ROS membranes washed prior to bleaching with 6 *M* urea, containing 2 m*M* EDTA and heated to 65° for 3 min. Various mixtures of 70 μg Rho, 75 μg G, Gpp(NH)p (2.5 μ*M*) in 10 m*M* Tris–HCl buffer (pH 2.5) containing 5 m*M* $MgSO_4$, 1 m*M* DTT, and 50% glycerol were incubated for 30 min at 30°. Then, Rho and unbound Gpp(NH)p were completely removed by centrifugation and Sephadex G-25 column chromatography. The soluble protein (15 μl; 0.6 μg protein) was combined with a 40 μl suspension of whole dark disk (5 μg protein) containing 1.3 m*M* [^{3}H]cGMP (12.5 Ci/mole). PDE assays were carried out for 3 min at 30° in total darkness.

PDE activities are given as percentage of the maximum activity observed (100%) when saturating amounts of light and Gpp(NH)p were added to a similar aliquot of dark ROS membranes. One hundred percent equals 2 μmoles of cGMP hydrolyzed per minute per milligram of total protein.

of how PDE is activated by an admixture of bleached Rho. Our working hypothesis explaining the mixing phenomenon was based on the expectation that the material in the dark reservoir was donated G protein which could interact with the bleached minority Rho population. Under the influence of bleached Rho, the donated G population could acquire GTP and activate PDE associated with the unbleached Rho majority. Using purified bovine Rho, a partially purified G protein, and Gpp(NH)p, we were able to show that a G·Gpp(NH)p complex was formed and could be purified free of Rho on Sephadex G-25. This G-Gpp(NH)p complex fully activated PDE associated with unilluminated Rho. We named this photoreceptor-derived activator complex ''lumi-lamban''[8] [from the Latin root *lumen* (light) and the Greek root *lambano* (to take)] because of its unique capacity to transmit the influence of light from Rho to other components of the PDE cascade (Uchida *et al.*, 1980) (Table II).

[8]While the G protein [complexed with Gpp(NH)p] of the ROS PDE system is here uniquely named, we wish to reiterate its close functional relationship with G components from other systems (see below).

C. Helper

1. Demonstration of an Additional Component Required for the Light-Activated GTPase of Rod Outer Segments

Purification of the light-sensitive GTPase activity began uneventfully with two column chromatographic separations. We then loaded the partially purified GTPase onto a DEAE-Sephadex column and eluted the enzyme activity with a continuous KCl gradient. At this point, there was an unexpected and unexplained loss of GTPase activity, i.e., failure to recover activity during elution with a continuous KCl gradient. We considered the possibility that a large fraction of the GTPase activity was being retained on or denatured within the DEAE-Sephadex column. We also considered a more attractive possibility; i.e., during the course of adsorption onto, and elution from, the DEAE-Sephadex column, we had separated the GTPase activity into multiple components, all essential for the expression of enzyme activity. Such a postulated separation would have occurred if different components of the GTPase exhibited different affinities for the ion-exchange matrix and differential elution with a KCl gradient. We then found that, by combining earlier and later eluting column fractions, we could indeed restore most of the GTPase activity that we had "lost" in the DEAE-Sephadex column. That is, the later eluting and very diminished GTPase activity peak was fully resuscitated by an "early" fraction which itself was entirely devoid of GTPase activity (Shinozawa *et al.*, 1980). This was convincing proof that the enzyme was neither retained nor denatured on the DEAE-Sephadex column.

The earlier eluting fraction, which restored activity to the devitalized GTPase fraction, was called helper (H). We were able to distinguish H from G by a variety of procedures in addition to DEAE-Sephadex column chromatography. The H fraction showed greater stability to trypsin digestion than the G fraction. It exhibited no intrinsic GTPase activity and, in contrast to the G fraction, had no capacity to bind GTP in the Millipore assay. The H fraction was entirely free of PDE activity. Nevertheless, the G protein showed the capacity to bind GTP (in the presence of purified bleached Rho) even in the absence of H component in a Millipore filter, binding assay (Shinozawa *et al.*, 1980). In recent studies, we have found that the G protein (in the presence of purified bleached Rho) can be selectively labeled by [8,5′-^{3}H]p^{3}-(4-azidoanilido)-p'-5′-GTP. The purified photoaffinity-labeled G protein had a molecular weight of about 42,000 (Uchida *et al.*, 1980).

2. The Indispensability of Helper Fraction in the Reconstitution of GTPase Activity

Using highly purified frog PDE and highly purified bovine Rho, reconstituted in phosphatidylcholine vesicles and partially purified G (free of Rho, PDE, and

H activities) and H (free of Rho, PDE, and G activities) fractions, we undertook a series of interesting reconstitution experiments. These experiments demonstrated that the reconstruction of light- and GTP-activated PDE showed an absolute dependence on the presence of the G and Rho fractions. The H fraction was not required for the binding of GTP to the G fraction or light activation of PDE (Shinozawa *et al.*, 1980). Although we could reconstitute PDE activity with Rho, PDE, and the G fraction, these components were not sufficient for the reconstitution of GTPase activity. Reconstitution of light-activated low K_m GTPase was absolutely dependent on the Rho, G, and H components (Fig. 2). These findings suggest a variety of new possibilities for understanding the mechanism of interaction between cholera enterotoxin and G units, since they allow study of the G unit as a complex of H and G fractions. It is therefore possible to inquire which of the two components (G or H) can be ADP-ribosylated by the A_1 subunit of cholera enterotoxin. Furthermore, it is possible to ask whether the *conformational* influence of H on the G unit can effect interaction of the G unit with cholera enterotoxin. Clearly the G fraction (without H) acted

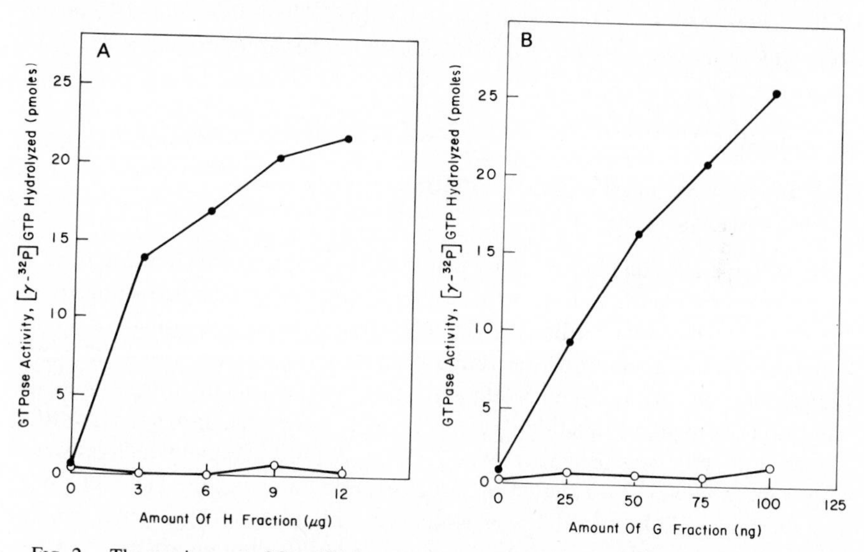

FIG. 2. The requirement of G and H fractions for the expression of GTPase activity. (A) Varying amounts of H fraction from the DEAE-Sephadex column fractions were incubated with 25 μg of LiCl–urea-washed disk membrane with (solid circles) and without (open circles) 75 ng of G-fraction protein extensively purified by DEAE-Sephadex, AH-Sepharose 6B, and Blue Sepharose CL-6B column chromatography. The amount of [γ-^{32}P]GTP hydrolyzed by LiCL–urea-washed membranes alone was zero. (B) Varying amounts of G fraction, extensively purified by DEAE-Sephadex, AH-Sepharose 6B, and Blue Sepharose CL-6B column chromatography were incubated with 25 μg of LiCl–urea-washed disk membranes with (solid circles) and without (open circles) 8.6 μg of H fraction from DEAE-Sephadex.

like a cholera-treated, AC-derived GTPase activity in the sense that it could promote activation of the catalytic readout but was unable to support signal-specific GTPase activity (Shinozawa *et al.*, 1980).

D. Opsin Kinase

Kühn and Dreyer (1972) were the first to describe an opsin kinase that phosphorylated illuminated Rho. A relationship of this protein to light-activated PDE was first suggested by Sitaramaya *et al.* (1977). They found that phosphorylation of Rho markedly diminished its capacity to activate PDE. In so doing they provided yet another clue to the modulation of light-activated PDE. Inasmuch as bleached Rho can potentially activate many PDE molecules (via many lumi-lambans) the phosphorylation of Rho provides a stop mechanism which prevents Rho from activating additional PDE molecules via lumi-lamban formation. Thus a full depiction of the turn-off mechanism for PDE must include, at a minimum, the hydrolysis of GTP bound to G protein and the phosphorylation of opsin. Should cyclic-GMP hydrolysis turn out to mediate transduction, the inactivation of ROS PDE both by Rho phosphorylation and GTP hydrolysis could be important steps in the recovery of resting membrane potential by excited rods.

E. Inhibitor

1. The Heat-Stable Inhibitor of Rod Outer Segment Phosphodiesterase

In earlier studies on the light-activated PDE system, we noted that a number of variables were important in the observation of light-dependent activation. First, the extent of the light-mediated PDE activation was clearly a function of the extent of dark adaptation of the retina prior to the preparation of ROS membranes. Second, the light-dependent activation was dependent on the systematic exclusion (by using infrared light) of visible light during the dark phases of ROS preparation and assay (Bitensky *et al.*, 1971). A third variable which appeared important, and which was noted in earlier experiments, was the fact that extensive dilution of ROS material was associated with a loss of light activation. At the time, we did not appreciate the mechanism of this dilution effect. Dummler and Etingof (1976) found, in association with the cyclic-GMP PDE of bovine photoreceptors, a heat-stable macromolecule which appeared capable of reversibly inhibiting PDE and whose effects were diminished by dilution. Although this material was not fully characterized, its apparent molecular weight was estimated to be in the range of 38,000. Our recent studies have suggested that this protein is not merely a passive inhibitor of ROS PDE. Rather, it appears to participate actively in the light-dependent cascade that regulates ROS PDE. The potency of the inhibitor appears substantially reduced or abolished in the presence of light and GTP [Yamazaki *et al.*, 1980b); also see below].

2. High-Affinity, Cyclic-GMP-Specific Binding Sites on Rod Outer Segment Phosphodiesterase

As an important sequel to our studies on cyclic-GMP metabolism in vertebrate rods, we were interested in identifying possible mechanisms by which light-mediated fluctuations in cyclic GMP could influence rod biochemistry and physiology. This interest inspired a systematic examination of the possibility that proteins associated with ROS materials could display saturable, high-affinity, cyclic-GMP-specific binding sites. A useful approach to this study was to use a standard Millipore filter assay and photoaffinity labeling techniques as a means of detecting putative cyclic-GMP-binding sites (Yamazaki *et al.*, 1980a).

The initial screening for ^{3}H-labeled cyclic-GMP-binding sites was carried out with two ROS fractions: the integral and the soluble-plus-peripheral proteins. Integral proteins are those that remain in the disk membrane of the ROS following EDTA and urea washing. Soluble-plus-peripheral proteins are those that are liberated by EDTA washing of the disk membranes. In this initial survey it became clear that high-affinity cyclic-GMP-binding sites were present, and that they partitioned with the soluble-plus-peripheral proteins (Yamazaki *et al.*, 1980a). The fact that cyclic-GMP-binding protein(s) was released from the disk membranes with EDTA and could be readsorbed onto the disk membranes with magnesium kindled the suspicion that the cyclic-GMP-binding sites might in some way be related to PDE.

3. Phosphodiesterase is the Locus of the Saturable, High-Affinity Cyclic-GMP-Binding Sites

We then attempted extensive purification of the protein responsible for this cyclic-GMP-binding activity. With each additional purification step, however, it became more apparent that the protein(s) we were seeking was identical with the light- and GTP-activated ROS PDE. The high-affinity cyclic-GMP-specific sites copurified with the photoreceptor PDE through a variety of steps. These purification steps included sucrose density gradient centrifugation, adsorption onto disk membranes with Mg^{2+}, purification with BioGel A 0.5 *M* column chromatography, and isoelectric focusing. In addition, the protein associated with the cyclic-GMP-binding sites could be covalently labeled with [^{32}P]8-azido-cyclic-IMP which effectively labeled ROS PDE. This labeling (as well as ^{3}H-labeled cyclic-GMP binding) was prevented by unlabeled cyclic GMP (but not cyclic AMP) and by limited trypsin proteolysis of the PDE (Yamazaki *et al.*, 1980a).

The possibility was also considered that the high-affinity cyclic-GMP-binding sites could be related to, or identical with, the catalytic site of PDE. Several types of evidence provided convincing proof that the cyclic-GMP-binding sites were independent of the PDE catalytic site. First, limited trypsin proteolysis rapidly abolished 80–95% of the high-affinity cyclic-GMP binding and at the same time, enhanced PDE activity. Second, 3-isobutyl-l-methylxanthine, which enhanced cyclic-GMP binding to high-affinity sites, at the same time significantly reduced PDE activity. Third, the high-affinity cyclic-GMP-binding sites on PDE and the PDE catalytic sites showed differential stability on storage. The PDE catalytic sites were stable when the enzyme was stored at pH 6 in phosphate buffer at ice temperature for more than 7 days. In contrast, the high-affinity cyclic-GMP-specific binding sites rapidly deteriorated under these conditions. Although hydrolyzed less rapidly than cyclic GMP, cyclic AMP was a functional substrate for the photoreceptor PDE. In contrast, cyclic AMP was not able to displace ^{3}H-labeled cyclic-GMP binding at high-affinity sites, even when cyclic AMP was in a 1000-fold excess. Finally, while protamine caused a striking activation of PDE, it had no effect on the binding of ^{3}H-labeled cyclic GMP to the high-affinity sites. This binding, in contrast to the catalytic activity was independent of Mg^{2+} (Yamazaki *et al.,* 1980a).

4. A Relationship between the Heat-Stable Inhibitor and the High-Affinity Cyclic-GMP-Binding Sites

In the course of our studies on the cyclic-GMP-binding site, we observed that binding appeared to be enhanced by soluble factors present in the "EDTA supernatant." The affinity constant for cyclic-GMP binding to crude PDE fractions was significantly lower than the affinity constant for cyclic GMP with purified PDE (Yamazaki *et al.,* 1980a). The factor that enhanced cyclic-GMP binding behaved as a macromolecule on sucrose density gradients and in Sephadex column chromatography. It was heat-stable, trypsin-labile, and not only stimulated cyclic-GMP binding to PDE but also displayed (or comigrated in sucrose gradients, Blue Sepharose CL-6B, and DEAE-Sephacel with a substance that displayed) the capacity to inhibit ROS PDE (Fig. 3). Thus a macromolecular heat-stable PDE inhibitor in frog ROS was also able to enhance cyclic-GMP-specific binding to PDE and appeared to be related to the heat-stable bovine ROS PDE inhibitor described by Dummler and Etingof (1976).

We next examined whether the effects of the amphibian PDE inhibitor were influenced by the known modulators of PDE, i.e., light and GTP. We found that the inhibitor effect was strikingly attenuated by light and GTP (Yamazaki *et al.,* 1980b). These findings suggested that inhibitor was a component functioning within the paradigm for the light regulation of PDE. Moreover, at the same time that the inhibitor diminished the "basal" activity of PDE, it stimulated cyclic-GMP binding to the specific, saturable, high-affinity binding sites on PDE.

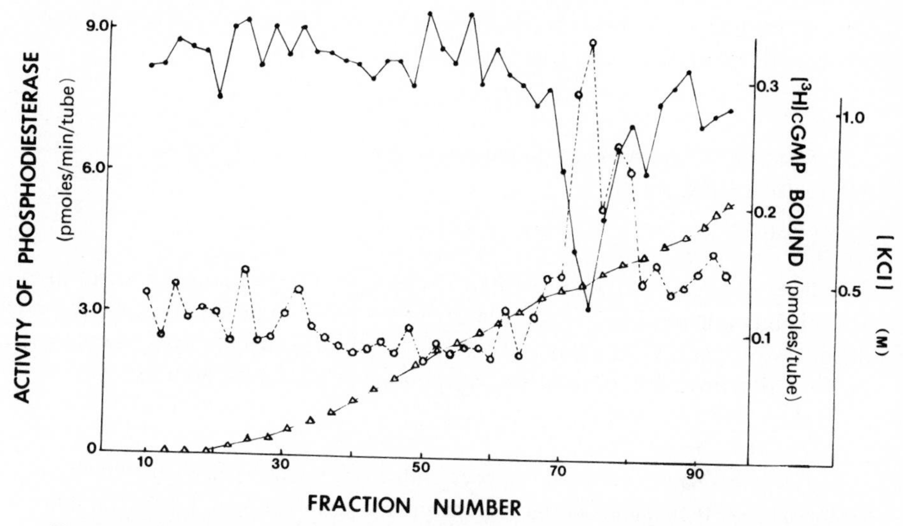

FIG. 3. Relationship of cGMP binding and PDE activity to inhibitor. ROS peripheral and soluble proteins were adsorbed with bleached frog disk membranes (in 200 m*M* Tris, pH 7.5, 20 m*M* Mg^{2+}, and 1 m*M* DTT). The membranes were removed by centrifugation, and the supernatant concentrated by ultrafiltration and loaded onto a Blue Sepharose CL-6B column. The column was washed with the above buffer and then washed with 10 m*M* Tris (pH 7.5), 1 m*M* DTT, and 2 m*M* EDTA. The column was then eluted with a 0–1.2 *M* KCl gradient (triangles). The experiment showed that inhibitor (or a uniquely comigrating factor) not only decreased PDE activity (solid circles) but enhanced (noncatalytic) cGMP binding (open circles) as well.

Furthermore, under conditions where inhibitor failed to inhibit PDE, i.e., in the presence of GTP and bleached Rho, inhibitor also failed to stimulate the binding of cyclic GMP to the saturable, high-affinity sites. These observations have given additional credence to the concept that the inhibitor protein and cyclic-GMP binding to the high-affinity cyclic-GMP sites are in some way interrelated (Yamazaki *et al.*, 1980b). It is, however, not yet possible to conclude that the high-affinity cyclic-GMP-binding sites function in the regulation of PDE activity. An equally important possibility would assign to such sites a function related to the modulation of sodium channels, calcium-binding sites, or both.

III. ASSEMBLING THE PIECES: A HYPOTHETICAL SEQUENCE FOR COMMUNICATION OF THE LIGHT SIGNAL ALONG THE CASCADE

The following steps represent a hypothetical scheme for the activation and reversal of activation of ROS PDE.

1. A photon is captured by Rho (R) initiating the series of light-induced conformational changes characteristic of this molecule.

$$h\nu \rightarrow R \rightarrow R^*$$

2. Rho* (R*), an illuminated conformer of Rho, interacts with G, forming a transient complex:

$$R^* + G \rightarrow R^* \cdot G$$

3. In the presence of illuminated Rho, G can bind GTP (forming lumi-lamban). This binding releases Rho* from its interaction with G. Thus Rho* is now free to interact with other G proteins, continuing to produce many active complexes until the illuminated Rho* is altered by phosphorylation (see Step 6).

$$R^* \cdot G + GTP \rightarrow R^* + G \cdot GTP$$

4. G (with bound GTP) can interact with PDE which is itself bound to the inhibitor (In). PDE contains the high-affinity cyclic-GMP (cG) sites. Inhibitor and cyclic GMP are released.

$$G \cdot GTP + cG \cdot PDE \cdot In \rightarrow G \cdot GTP \cdot PDE^{act} + cG + In$$

5. The complex of G, GTP, and PDE is active and begins to catalyze the hydrolysis of cyclic GMP. It then encounters H which forms a complex with G and thus facilitates hydrolysis of GTP to GDP and P_i. This hydrolytic step releases PDE which may still be active (see Step 7 for the subsequent turn-off).

$$G \cdot GTP \cdot PDE^{act} + H \rightarrow H \cdot G \cdot GDP + P_i + PDE^{act}$$

6. Rho*, which has been interacting with many sequential G proteins, allowing them to bind GTP, now interacts with an opsin kinase (OpK) and ATP, producing Rho* phosphate. This phosphorylated Rho* is no longer capable of interacting with G. It will now be out of the active cycle until it is both regenerated and dephosphorylated.

$$R^* + ATP \overset{OpK}{\rightarrow} R^*PO_4 + ADP$$

7. The PDE that has been activated as a result of its interaction with G protein now once again interacts with inhibitor protein and cyclic GMP, forming a complex. This PDE is now inactive and will remain inactive until it encounters another G protein which has formed a complex with GTP.

$$PDE^{act} + In + cG \rightarrow cG \cdot PDE \cdot In$$

A model summarizing these hypothetical steps is shown in Fig. 4.

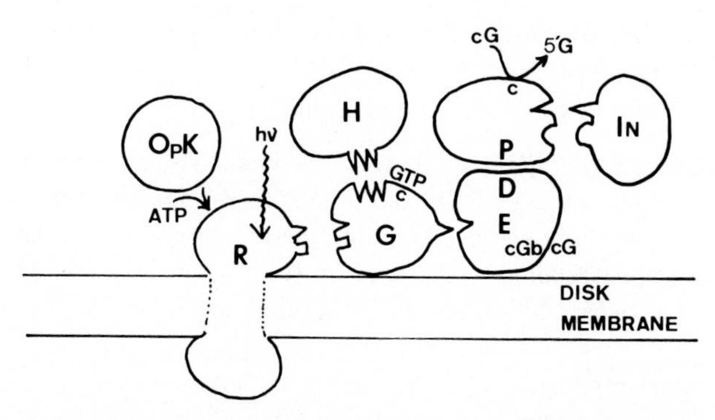

FIG. 4. The components of the photoreceptor PDE cascade. OpK, Opsin kinase; R, rhodopsin; H, helper; G, GTP-binding protein, GTP, catalytic site for GTPase; PDE, phosphodiesterase; c, catalytic site of PDE and GTPase; cGb/cG, cyclic GMP-binding site. Articulating facets indicate protein–protein interactions. See Section III for an explanation of component interactions.

IV. SOLVING OTHER PUZZLES

A. Analogy between Hormone-Sensitive Adenylate Cyclase and Light-Activated Phosphodiesterase

While these exciting findings were emerging from studies on rod biochemistry, equally interesting findings were being reported from studies on hormone-sensitive AC. It was known from the work of Schramm and Rodbell (1975), Spiegel *et al.* (1980) and Cassel and Selinger (1976) (and their co-workers) that activation of hormone-sensitive AC by a hormone involved the binding of GTP to a G unit. Furthermore, it was known that the GTP could be hydrolyzed (thus terminating activation) and that a nonhydrolyzable analogue of GTP such as Gpp(NH)p could be substituted for GTP to provide *stable* activation (Schramm and Rodbell, 1975). Moreover, in association with hormone-activated AC there is a hormone-activated GTPase in striking analogy with light-activated GTPase (Shinozawa *et al.*, 1979). The common features of the activation mechanism are so similar for both hormone-activated AC and light-activated PDE that we launched a search for additional points of similarity in the regulation of the two systems. We were impressed that the photoreceptor PDE components of light and Rho could be considered analogous to the components of hormone and hormone receptor. The G unit also appeared to be functionally identical in both systems.[9]

[9]Recent data (Abramowitz *et al.*, 1979) indicate that, for some AC systems, GDP may also function as an effective activator. However, in the most extensively studied case, i.e., the turkey erythrocyte, only the triphosphate form of the guanine nucleotide supports activation (Downs *et al.*, 1980).

Finally, for the photoreceptor PDE system, we could compare the catalytic readout (PDE) with the catalytic moiety of the AC system. The most recent compilation of this series of similarities includes:

1. Both light-activated PDE and hormone-activated AC require two signals for activation. For light-activated PDE, the signals are bleaching of the photoreceptor protein (Rho) by light and GTP binding, and for hormone-activated AC, the signals are hormone binding to the specific receptor protein and GTP binding (Aurbach *et al.*, 1975; Miki *et al.*, 1975; Rodbell *et al.*, 1975; Wheeler *et al.*, 1977).

2. The sequence of the signals is critical in both systems. For both light-activated PDE and hormone-activated AC, GTP binding is only effective following the receptor specific signal. Transient presentation of GTP (and its removal) followed by the receptor signal in the absence of GTP will not activate in either case (Aurbach *et al.*, 1975; Wheeler and Bitensky, 1977).

3. Both light-activated PDE and hormone-activated cyclases show a requirement for guanine nucleotides. Adenine nucleotide analogues free of guanine contaminants cannot be substituted (Aurbach *et al.*, 1975; Wheeler and Bitensky, 1977).

4. In both systems, GTP serves as an activator at concentrations below 1 μM (Aurbach *et al.*, 1975; Rodbell *et al.*, 1975; Wheeler and Bitensky, 1977).

5. In both systems, the nonhydrolyzable GTP analogue Gpp(NH)p can substitute for GTP as an activator (Aurbach *et al.*, 1975; Rodbell *et al.*, 1975; Schramm, 1975; Wheeler and Bitensky, 1977).

6. In both systems, the nonhydrolyzable GTP analogue Gpp(NH)p produces activation of the enzyme, which does not decay with time (Schramm and Rodbell, 1975; Wheeler and Bitensky, 1977).

7. In both systems, the primary system-specific signal (photobleaching of Rho or hormone binding to specific receptors) leads to an apparent increase in GTPase activity (Anderson *et al.*, 1978; Wheeler and Bitensky, 1977; Wheeler *et al.*, 1977).

8. In both systems, the signal-activated GTPase has a K_m for GTP below 1 μM and can be inhibited by 1 μM Gpp(NH)p. Interestingly, this is the same range of GTP concentration required for PDE or AC activation (Cassel and Selinger, 1976; Wheeler and Bitensky, 1977; Wheeler *et al.*, 1977).

9. In both systems, the hydrolysis of GTP is associated with inactivation of a signal-stimulated enzyme (PDE or cyclase). Activity may be restored by replenishing the GTP (Schramm and Rodbell, 1975; Wheeler and Bitensky, 1977).

10. In both systems, trypsin can activate the signal-specific enzyme (PDE or cyclase) in the absence of a signal, suggesting a release from conformational constraint on the catalytic moiety (Anderson *et al.*, 1978; Miki *et al.*, 1975).

11. In both systems, fluoride ion can activate the signal-specific enzyme (Perkins, 1974; Sitaramaya *et al.*, 1977).

12. In both systems, the activator signal may originate on a membrane fraction separate from, or independent of, the membrane fraction containing the signal-activated enzyme (Bitensky *et al.*, 1975; Miki *et al.*, 1975; Sahyoun *et al.*, 1977).

13. In both systems, the signal receptor complex can activate more than one catalytic complex (i.e., there can be significant amplification) (Tolkovsky and Levitzki, 1978; Yee and Liebman, 1978).

14. In both systems, there is evidence for receptor excess (i.e., the ratio of receptor to G units is a large number) (Miki *et al.*, 1975; Tolkovsky and Levitzky, 1978; Yee and Liebman, 1978).

15. In both systems, there is demonstrable interchangeability of receptor elements (e.g., the turkey erythrocyte catecholamine receptor can function to activate the Friend leukemia cyclase, and bovine Rho can function to activate frog disk phosphodiesterase) (Bitensky *et al.*, 1975; Shinozawa *et al.*, 1980).

16. In both systems, GDP is bound to the G unit when the system is inactive and is replaced by GTP with activation (Fong and Stryer, 1980; Rodbell, 1980).

17. In both systems, there is evidence that the G unit consists of more than one polypeptide chain (Kaslow *et al.*, 1980; Kuhn, 1980; also see below).

18. In both systems, introduction of the A-1 subunit of cholera enterotoxin (with cytoplasmic cofactors) and NAD results in the ADP-ribosylation of one or more components of the G unit (D. M. Gill and M. W. Bitensky, unpublished observations, 1980; Enomoto and Gill, 1980).

This remarkable collection of similarities between the two systems encouraged us to expect that data emerging from studies on photoreceptor PDE would have additional predictive value for the study and further understanding of the AC system (Shinozawa *et al.*, 1979). Furthermore, we were interested in the possibility of exchanging parts between the two systems in order to create hybrid complexes containing functionally interacting components of different origin (see below).

B. Interchange of Components between Light-Activated Photoreceptor Phosphodiesterase and Hormone-Sensitive Adenylate Cyclase

The striking analogy between the AC and PDE systems is most conveniently discussed when each of the systems is viewed as a cluster of three segments. The receptor segment includes either Rho or the hormone receptor, their postulated kinases and phosphatases, and specific signals. The second segment (G unit) includes the G and H components and GTP. The third segment includes the catalytic readout which may in both cases be associated with the high-affinity cyclic-GMP-binding sites (noncatalytic), the catalytic sites, and an inhibitor protein that can influence both cyclic-GMP binding and the catalytic moiety. The analogy is both exciting and useful, since predictions can be made about the AC system from studies on the photoreceptor system, and vice versa. For example, we anticipated that the cholera enterotoxin could ADP-ribosylate some aspect of the G unit. We have preliminary data that the ROS component that is ADP-ribosylated is in the 42,000-dalton range (D. M. Gill and M. W. Bitensky, unpublished observations, 1980). It is not yet possible to be certain whether it is the G or the H units (or both) that interact with the enterotoxin. Another exciting prediction is that there may be agonist binding-stimulated phosphorylation of the hormone receptor. Such a phosphorylation could have functional implications for the dynamics of hormone receptor interaction, especially as regards agonist-specific desensitization (Hoffman *et al.,* 1979). Some facets of this analogy may be experimentally tested by interchanging appropriate components of each system. Thus Rho* may substitute for the hormone receptor of the AC system, producing light-activated AC. Furthermore, the inhibitor of ROS PDE might inhibit AC. There is also the possibility that G units could be interchanged between the two systems.

Recent experiments in our laboratory have met with limited success in interchanging components between the AC and PDE systems. We attempted to take the G component from the photoreceptor system and donate it to the turkey erythrocyte, but this combination did not produce functional Gpp(NH)p enhancement of the turkey AC system.

We have successfully utilized illuminated amphibian and bovine Rho to produce Gpp(NH)p-dependent activation of particulate AC prepared from rat synaptic membranes or frog red cell ghosts (Table III). Unilluminated Rho *fails* to activate the particulate brain AC (M. M. Rasenick, M. Wheeler, and M. W. Bitensky, unpublished observations, 1979). In these recombination experiments, detergents are not required to facilitate interaction of the membrane components. Simple preincubation of Rho reconstituted in phosphatidylcholine vesicles with the particulate brain AC suffices to permit observation of Gpp(NH)p and light-dependent cyclase activation. The same experiment is successful with illuminated frog disk membranes from which all peripheral proteins have been eluted with 6 *M* urea. [The Rho–cyclase cross did not succeed when we used frog disks or reconstituted bovine Rho in 8 phosphatidylcholine vesicles and turkey erythrocyte ghosts as the source of AC (A. Spiegel, A. Yamazaki, and M. W. Bitensky, unpublished observations, 1980).] We have also been successful in showing that inhibitor protein from the photoreceptor PDE system can inhibit AC prepared from rat brain hypothalamus (M. Wheeler, A. Yamazaki, and M. W. Bitensky, unpublished observations, 1980). These observations imply that there are specific domains in the receptor proteins that can functionally interact with the adjacent G components (by analogy with the articulating facets of joints). It appears that the structure of these domains has been conserved (in an evolutionary sense) to a remarkable degree. One might also anticipate that amino acid

TABLE III
ACTIVATION OF PARTICULATE ADENYLATE CYCLASE BY BLEACHED RHODOPSIN

AC preparation	Rhodopsin preparation	AC activity[a]	
		No addition	Gpp(NH)p[b]
Rat cerebral cortex	None	32	96
	Unbleached reconstituted bovine[c]	34	90
	Bleached reconstituted bovine	34	186
	Bleached frog disks	30	280
Rat hypothalamus	None	22	72
	Bleached reconstituted bovine	62	285
Frog erythrocyte	None	1.9	1.6
	Bleached frog disks	1.3	4.2

[a] AC activity is expressed as picomoles cAMP formed per milligram protein per minute. Brain synaptic membrane particles and frog erythrocytes were incubated at 30°C.
[b] Gpp(NH)p was present at 5 μM for the brain synaptosome and at 2 μM for frog erythrocytes.
[c] The reconstituted rhodopsin preparation of Hong and Hubbell was used (1972).

sequences within these domains will reflect this conservation. It is exciting to contemplate the possibility that such sequence similarities may emerge from future studies on the interchangeable components.

C. The Ubiquitous G Unit

Recent developments in disparate systems emphasize the remarkable ubiquity of the G unit in regulatory mechanisms. There are at least five examples in which the binding of GTP to a G unit is critical as an activator of a physiological process. Further, in each case, hydrolysis of GTP terminates the activation or promotes oscillation of, and within, the activation cycle. These examples include hormone-sensitive AC (Rodbell, 1980), light-activated PDE (Bitensky *et al.*, 1977), the initiation step in mRNA translation (Weissbach, 1979), the elongation step in mRNA translation (Weissbach, 1979), and polymerization of tubulin (Jacobs, 1975). Furthermore, it is both fascinating and remarkable that at least four unrelated bacteria have in one way or another recognized and exploited this essential function. Thus a necrotizing *Pseudomonas* exotoxin, diphtheria toxin, cholera enterotoxin, and *Escherichia coli* enterotoxin all utilize, as their mechanism of action, ADP-ribosylation of one of these ubiquitous regulatory G units (Pollack, 1980). Moreover, the pathogenetic implications of the G unit appear to extend to inborn errors of metabolism. The findings of Levine *et al.* (1980) indicate that a congenital disorder of calcium metabolism, pseudohypoparathyroidism (which is associated with failure to respond to endogenous parathyroid hormone), can be attributed to a defect in the AC G unit.

Whether or not the pathogenesis of inherited retinal dystrophies (Schmidt and Lolly, 1973; Aguire *et al.*, 1978; Bitensky *et al.*, 1980; also see below) involves an abnormality in the G unit remains an open question.

V. AN OVERALL ASSESSMENT

A. Physiological Ponderings

We are not yet in a position to conclude that the rapid light-induced decline of cyclic GMP is actually a component of transduction or sensitivity control in rods. Even at this late date, and after much laborious and imaginative experimentation, the question still remains unanswered. Clearly the photoreceptor and Rho are profoundly committed to light regulation of cyclic-nucleotide metabolism. Two recent studies emphasize the rapidity with which cyclic nucleotides can fall after illumination of ROS (Yee and Liebman, 1978; Woodruff and Bounds, 1979). However, speed in itself does not necessarily distinguish between transduction and sensitivity control. The most rapid component of light adaptation that occurs on introduction of background light is expressed within a time frame comparable to that required for the production of a receptor potential by a light flash.

An appealing possibility is the concept that both cyclic nucleotides and calcium collaborate to produce excitation and sensitivity control. Studies on disk membrane Na^+ fluxes by Caretta *et al.* (1979) show that cyclic GMP can profoundly influence the sodium permeability of disk membranes. These authors suggest that cyclic GMP has the same effect on both disk and plasma membranes. Sorbi (this volume, Chapter 18) also reports that cyclic GMP releases calcium from disk membranes. Of course, the question remains whether disk membranes can be taken as a model for the plasma membrane, but this appears to be a reasonable possibility. Recent experiments by R. Sorbi, P. Stein, and M. W. Bitensky (unpublished observations) in this laboratory confirm that cyclic GMP can decrease the affinity of calcium for *binding sites* in disk membranes. These observations suggest that elevated levels of cyclic GMP facilitate increases in sodium conductance by reducing the affinity of calcium for sodium channels. Since the number of disk membrane binding sites far exceeds the number of sodium channels predicted (or measured) in the plasma membrane (Cone, 1973; Bader *et al.*, 1979; Oakley, 1980), it is possible that they represent a reservoir of calcium-buffering sites similarly influenced by cyclic GMP. Therefore, when cyclic GMP falls with illumination, the affinity of both the sodium channels and the buffering sites for calcium are increased, causing a decrease in the sodium conductance and a fall in free cytoplasmic $[Ca^{2+}]$. This might explain the need for larger photon fluxes to produce equivalent voltage responses as the background illumination increases. That is, there is less Ca^{2+} available to interact with the sodium channels as free [Ca^{2+}] is reduced. An associated and interesting possibility is that reducing cytoplasmic [Ca^{2+}] also stimulates guanylate cyclase. Thus hydrolysis of cyclic GMP (a force that reduces sodium conductance) is opposed by stimulated cyclic-GMP synthesis (a force that increases sodium conductance) and decreased cytoplasmic [Ca^{2+}] (also a force that increases sodium conductance).

B. Critical Questions That Remain Unanswered

The emerging physiological data obtained by measurements of membrane voltage and associated with intracellular iontophoresis of cyclic GMP appear to justify the view that cyclic GMP has the ability to depolarize the ROS membrane (Miller and Nicol, 1979; Waloga and Brown, 1979). Biochemical data appear to justify the view that the rod has the capability to hydrolyze rapidly large quantities of cyclic GMP as a consequence of illumination (Yee and Liebman, 1978). Recent data from studies with extracellular calcium electrodes also suggest that light somehow increases free [Ca^{2+}] in the medium bathing the rod (Yoshikami and Hagins, 1980; Gold and Korenbrot, 1980), but the exact consequences of such a change on the cytoplasmic free [Ca^{2+}] are not yet fully understood. Many previous studies support the notion that manipulation of calcium ion concentrations can profoundly influence the membrane sodium conductance in the rod (Hagins, 1972; Brown and Pinto, 1974; Lipton *et al.*, 1977). These studies are also accompanied by observations indicating that manipulation of extracellular calcium can greatly influence the levels of cyclic GMP, especially in the unilluminated rod (Cohen *et al.*, 1978).

Many critical questions need to be answered en route to a more confident understanding of the linkage between photoreceptor biochemistry and electrophysiology. For example, what does light do to the cytoplasmic free [Ca^{2+}] in the rod. If the effects on calcium are those suggested in the preceding discussion, i.e., that cyclic GMP can modulate calcium interactions with both sodium channels and calcium-buffering sites, then one would expect that light would produce a fall in the free calcium pool and an increase in calcium buffering. However, both Gold and Korenbrot (1980) and Yoshikami and Hagins (1980) have found that calcium concentrations increase in the external milieu of the rod with illumination. These may not be contradictory observations, providing that the light-mediated appearance of calcium in the extracellular fluid (as monitored with a calcium-sensitive electrode) can reflect changes in a Ca^{2+} pool other than rod outer segment cytoplasmic free [Ca^{2+}]. There is also an additional question: What are the effects of cyclic GMP on cytoplasmic free [Ca^{2+}]? Ultimately, we will wish to understand which of the putative modulators has the primary responsibility for regulating rod sodium conductance *in vivo*. If cyclic GMP has this direct role, which appears to be the case in studies on sodium permeability in disk membrane populations (Caretta *et al.*, 1979), to what extent does cyclic GMP act directly on sodium conductance channels? Or to what extent does this cyclic-GMP-mediated event interact with or depend on calcium?

Much additional information will be needed in order to understand the mechanisms by which guanylate cyclase is regulated. Although there is a biochemical suggestion that calcium can influence guanylate cyclase activity (Pannbacker, 1973; Krishnan *et al.*, 1978), the data in support of this idea are limited. Additional information is also needed concerning the localization of guanylate cyclase within the outer segment and axoneme. There is conflicting evidence as to whether the guanylate cyclase is diffusely distributed throughout the outer segment (de Azeredo and Passinneau, 1980) or more heavily concentrated in the axoneme (Fleischman and Denisevich, 1979).

A role for phosphorylation must still be considered a possible component in the cascade regulated by cyclic GMP. Are phosphorylation mechanisms involved in the regulation of sodium channel permeability and/or the regulation of calcium transport or calcium buffering? To what extent are phosphorylation mechanisms implicated in the regulation of transduction or sensitivity control, exclusive of the now well-documented phosphorylation of Rho which appears to be involved in the turn-off mechanism of PDE (Sitaramaya *et al.*, 1977)?

More data are needed to resolve these questions and finally to assign a detailed molecular mechanism for transduction and sensitivity control. A reasonable view of the existing data holds that both cyclic GMP and calcium collaborate in an intimate way in the generation of the receptor potential, as well as in the regulation of sensitivity in the vertebrate rod.

C. Inherited Retinal Dystrophies

A series of studies have recently suggested that cyclic-GMP metabolism in rods is important in explaining congenital retinal dystrophies in mice and dogs. Possible relationships to retinitis pigmentosa are also being sought (for a review, see Bitensky *et al.*, 1980). In these disorders the rods deteriorate early in life, resulting in total blindness with selective sparing of other retinal components. In the C3H mouse, the collie, and the Irish setter, retinal cyclic-GMP levels are known to rise prior to rod cytolysis and to fall dramatically after rod disappearance (Schmidt and Lolly, 1973; D. E. Wolf, personal communication, 1980; Aguire *et al.*, 1978). Some component is clearly malfunctioning in the diseased rods, which ultimately interferes with the hydrolysis of cyclic GMP. It is not known whether elevated cyclic-GMP levels cause the death of the rods, nor is it possible to assign an enzymatic defect to explain the lesion. Since a cascade is involved in the hydrolysis of cyclic GMP, it is conceivable that a lesion at any location (including the G unit) could result in functional impairment of the system.

D. Concluding Comments

We have attempted to describe some of the surprises encountered in the course of efforts to unravel and correlate the biochemistry and physiology of vertebrate photoreceptors. The story is still unfolding. This is clearly an area where biochemistry and physiology embrace and collaborate. We anticipate that continuing studies on the photoreceptor cyclic-nucleotide system will provide additional understanding of both photoreceptor biochemistry and physiology, and perhaps even of some pathophysiological mechanisms.

Clearly it was the G unit that first called our attention to the dramatic similarities between hormone-activated AC and light-activated PDE (Wheeler and Bitensky, 1977; Shinozawa *et al.*, 1979). Moreover, the success in interchanging components between the two systems (e.g., substituting bovine or amphibian Rho for the hormone receptor of rat brain AC) argues forcefully for evolutionary conservation of the G unit and consequently conservation of the articulating facets through which it interfaces with contiguous elements in the cascade.

It is quite reasonable to expect that other, equally fundamental, features of the regulatory cascade mechanism will also be conserved. The light-dependent PDE system is a veritable showcase of regulatory devices. Each of the major subsections of this three-segment molecular switch exemplifies an oscillating regulatory principle which may very well have wide application in biological systems. The phosphorylation–dephosphorylation mechanism that appears to modulate the participation of Rho in the PDE cascade (Sitaramaya *et al.*, 1977) has been implicated in a score of cyclic-nucleotide-directed responses (Greengard, 1978). Regulation of the G unit by nucleotide binding and hydrolysis has the potential for widespread application even beyond the switching mechanisms considered here. The same may be said for the reversible inhibition of the catalytic readout by a macromolecular heat-stable inhibitor. The combination of all three regulatory mechanisms within a single paradigm provides incredible possibilities for amplification and/or rate control. With this in mind, it is our expectation that many of the as yet unstudied G-unit-dependent regulatory mechanisms will also be found to utilize phosphorylation–dephosphorylation cycles, guanine nucleotide binding and hydrolysis, and reversible inhibition of the catalytic readout. It seems likely that, when primal evolutionary forces devise a uniquely useful and effective *combination* of gene products, such a combination is frequently exploited. Furthermore, it seems prudent to acknowledge that our current appreciation and understanding of the complexity, versatility, and potential applications for such molecular switching mechanisms is only in its earliest stages.

ACKNOWLEDGMENT

This research was largely supported by PHS-NIH AM EY 20179.

REFERENCES

Abramowitz, J., Iyengar, R., and Birnbaumer, L. (1979). Guanyl nucleotide regulation of hormonally responsive adenylyl cyclases. *Mol. Cell. Endocrinol.* **16,** 129–146.

Aguire, G., Farber, D., Lolly, R., Fletcher, R. T., and Chader, G. D. (1978). Rod-cone dysplasia in Irish setters: A defect in cyclic GMP metabolism in visual cells. *Science* **201** 1133–1134.

Anderson, W. B., Jaworski, C. J., and Vlahkis, G. (1978). Proteolytic activation of adenylate cyclase from cultured fibroblasts. *J. Biol. Chem.* **253,** 2921–2926.

Aurbach, G. D., Spiegel, A. M., and Garner, J. (1975). Beta-adrenergic receptors, cyclic AMP, and ion transport in the avian erythrocyte. *Adv. Cyclic Nucleotide Res.* **5,** 117–132.

Bader, C. R., MacLeish, P. R., and Schwartz, E. A. (1979). A voltage-clamp study of the light response in solitary rods of the tiger salamander. *J. Physiol. (London)* **296,** 1–26.

Bitensky, M. W., Gorman, R. E., and Miller, W. H. (1971). Adenyl cyclase as a link between photon capture and changes in membrane permeability of frog photoreceptors. *Proc. Natl. Acad. Sci. U.S.A.* **68,** 561–562.

Bitensky, M. W., Miller, W. H., Gorman, R. E., Neufeld, A. H., and Robinson, R. (1972). The role of cyclic AMP in visual excitation. *Adv. Cyclic Nucleotide Res.* **1,** 317–335.

Bitensky, M. W., Miki, N., Marcus, F. R., and Keirns, J. J. (1973). The role of cyclic nucleotides in visual excitation. *Life Sci.* **13,** 1451–1472.

Bitensky, M. W., Miki, N., Keirns, J. J., Baraban, J. A., Wheeler, M. A., Lacy, J., and Marcus, F. R. (1975). Activation of photoreceptor disc membrane phosphodiesterase by light and ATP: Cyclic GMP as a modulator of sense receptor function. *Adv. Cyclic Nucleotide Res.* **5,** 213–240.

Bitensky, M. W., Wheeler, G. L., Aloni, B., Vetury, S., and Matuo, Y. (1977). Light and GTP activated photoreceptor phosphodiesterase: Regulation by a light-activated GTPase and identification of rhodopsin as the PDE binding site. *Adv. Cyclic Nucleotide Res.* **9,** 553–572.

Bitensky, M. W., Rasenick, M. M., Shinozawa, T., Uchida, S., and Yamazaki, A. (1980). Altered cyclic nucleotide metabolism and the pathogenesis of hereditary dystrophies. *Adv. Cyclic Nucleotide Res.* **12,** 227–238.

Brown, J. E., and Pinto, L. H. (1974). Ionic mechanism for the photoreceptor potential of the retina of *Bufo marinus*. *J. Physiol. (London)* **236,** 575–591.

Caretta, A., Cavaggioni, A., and Sorbi, R. T. (1979). Cyclic GMP and the permeability of the disks of the frog photoreceptors. *J. Physiol. (London)* **295,** 171–178.

Cassel, D., and Selinger, Z. (1976). Catecholamine-stimulated GTPase activity in turkey erythrocyte membranes. *Biochim. Biophys. Acta* **452,** 538–551.

Cohen, A. I., Hall, I. A., and Ferrendelli, J. A. (1978). Calcium and cyclic nucleotide regulation in washed mouse retinas. *J. Gen. Physiol.* **71,** 595–612.

Cone, R. A. (1973). The internal transmitter model for visual excitation: Some quantitative implications. *In* "Visual Pigments: Biochemistry and Physiology" (H. Langer, ed.), pp. 275–282. Springer-Verlag, Berlin and New York.

de Azeredo, F. A. M., and Passinneau, J. (1980). Effects of *in vivo* light adaptation on guanylate cyclase within frog retinal layers. *Invest. Ophthalmol. Visual Sci.* **19,**S22.

Downs, R. W., Jr., Spiegel, A. M., Singer, M., Reen, S., and Aurbach, G. D. (1980). Fluoride stimulation of adenylate cyclase is dependent on the guanine nucleotide regulatory protein. *J. Biol. Chem.* **225,** 949–954.

Dummler, I. L., and Etingof, R. N. (1976). Protein inhibitor of cyclic adenosine 3′,5′-monophosphate phosphodiesterase in retina. *Biochim. Biophys. Acta* **429,** 474–484.

Enomoto, K., and Gill, D. M. (1980). Cholera toxin activation of adenylate cyclase. *J. Biol. Chem.* **255,** 1252–1258.

Fleischman, D., and Denisevich, M. (1979). Guanylate cyclase of isolated bovine retinal rod axonemes. *Biochemistry* **18,** 5060–5066.

Fong, B. K., and Stryer, L. (1980). Photolyzed rhodopsin catalyzes the exchange of GTP for bound GDP in retinal rod outer segments. *Proc. Natl. Acad. Sci. U.S.A.* **77,** 2500–2504.

Gold, G. H., and Korenbrot, J. I. (1980). Light-induced Ca efflux from intact rod cells in living retinas. *Fed. Proc., Fed. Am. Soc. Exp. Biol.* **39,** 1814.

Greengard, P. G. (1976). Possible role for cyclic nucleotides and phosphorylated membrane proteins in post synaptic actions of neurotransmitters. *Nature (London)* **260,** 101–108.

Greengard, P. G. (1978). "Cyclic Nucleotides, Phosphorylated Proteins, and Neuronal Function." Raven, New York.

Hagins, W. A. (1972). The visual process: Excitatory mechanisms in the primary receptor cells. *Annu. Rev. Biophys. Bio-eng.* **1,** 131–158.

Hagins, W. A., Penn, R. D., and Yoshikomi, S. (1970). Dark current and photocurrent in retinal rods. *Biophys. J.* **10,** 380–412.

Hoffman, B. B., Kilpatrick, D., and Lefkowitz, R. J. (1979). Desensitization of beta-adrenergic stimulated adenylate cyclase in turkey erythrocytes. *J. Cyclic Nucleotide Res.* **5,** 355–366.

Hong, K., and Hubbell, W. L. (1972). Preparation and properties of phospholipid bilayers containing rhodopsin. *Proc. Natl. Acad. Sci. U.S.A.* **69,** 2617–2621.

Jacobs, M. (1975). Tubulin nucleotide reactions and their role in microtubule assembly and dissociation. *Ann. N.Y. Acad. Sci.* **253,** 562–572.

Kakiuchi, S., Yamazaki, R., Teshima, Y., and Venishi, K. (1973). Regulation of nucleoside cyclic 3′,5′-monophosphate phosphodiesterase activity from rat brain by a modulator and Ca^{2+}. *Proc. Natl. Acad. Sci. U.S.A.* **70,** 3526–3530.

Kaslow, H. R., Johnson, G. L., Brothers, V. M., and Bourne, H. R. (1980). A regulatory component of adenylate cyclase from human erythrocyte membranes. *J. Biol. Chem.* **255,** 3736–3741.

Keirns, J. J., Miki, N., Bitensky, M. W., and Keirns, M. (1975). A link between rhodopsin and disc membrane cyclic nucleotide phosphodiesterase action spectrum and sensitivity to illumination. *Biochemistry* **14,** 2760–2765.

Kimura, N., and Negata, N. (1976). The requirement of guanine nucleotides for glucagon stimulation of adenylate cyclase in rat liver plasma membranes. *J. Biol. Chem.* **252,** 3829–3835.

Krishnan, N., Fletcher, R. T., Chader, G. J., and Krisha, G. (1978). Characterization of guanylate cyclase of rod outer segments of the bovine retina. *Biochim. Biophys. Acta.* **523,** 506–515.

Kühn, H. (1980). Light- and GTP-regulated interaction of GTPase and other proteins with bovine photoreceptor membranes. *Nature (London)* **283,** 587–589.

Kühn, H., and Dreyer, W. J. (1972). Light dependent phosphorylation of rhodopsin by ATP. *FEBS Lett.* **20,** 1–6.

Kühne, W. (1878). "On the Photochemistry of the Retina and Visual purple. Ewalda a Kuhne, W. Untersuchungen über den Sehpurpur," Vols. I–IV. Unters. Physiol. Inst., Heidelburg.

Levine, M. A., Downs, R. W., Jr., Singer, M., Marx, S. J., Aurbach, G. D., and Spiegel, A. M. (1980). Deficient activity of guanine nucleotide regulatory protein in erythrocytes from patients with pseudohypoparathyroidism. *Biochem. Biophys. Res. Commun.* (in press).

Lipton, S. A., Ostroy, S. E., and Dowling, J. E. (1977). Electrical and adaptive properties of rod photoreceptors in *Bufo marinus*. I. Effects of altered extracellular Ca^{+2} levels. *J. Gen. Physiol.* **70,** 747–770.

Miki, N., Keirns, J. J., Marcus, F. R., Freeman, J., and Bitensky, M. W. (1973). Regulation of cyclic nucleotide concentrations in photoreceptors: An ATP-dependent stimulation of cyclic nucleotide phosphodiesterase by light. *Proc. Natl. Acad. Sci. U.S.A.* **76,** 3820–3824.

Miki, N., Keirns, J. J., Marcus, F. R., and Bitensky, M. W. (1974). Light regulation of adenosine 3′,5′-cyclic monophosphate levels in vertebrate photoreceptors. *Exp. Eye Res.* **18,** 281–297.

Miki, N., Baraban, J. M., Keirns, J. J., Boyce, J. J., and Bitensky, M. W. (1975). Purification and properties of light-activated cyclic nucleotide phosphodiesterase of rod outer segments. *J. Biol. Chem.* **250,** 6320–6327.

Miller, W. H., and Nicol, G. D. (1979). Evidence that cyclic GMP regulates membrane potential in rod photoreceptors. *Nature (London)* **280,** 64–66.

Miller, W. H., Gorman, R. E., and Bitensky, M. W. (1971). Cyclic adenosine monophosphate: Function in photoreceptors. *Science* **174,** 295–297.

Oakley, B., and Pinto, L. H. (1980). $[Ca^{+2}]_i$ Modulates membrane sodium conductance in rod outer segments. *Invest. Ophthalmol. Visual Sci.* **19,** S-102.

Pannbacker, R. C. (1973). Control of guanylate cyclase activity in the rod outer segment. *Science* **182,** 1138–1140.

Perkins, J. P. (1974). Adenyl cyclase. *Adv. Cyclic Nucleotide Res.* **3,** 1–64.

Pollack, M. (1980). *Pseudomonas aeruginosa* exotoxin A. *N. Engl. J. Med.* **302,** 1360–1366.

Rodbell, M. (1980). The role of hormone receptors and GTP regulatory proteins in membrane transduction. *Nature (London)* **284,** 17–22.

Rodbell, M., Krans, M., Pohl, S., and Birnbaumer, L. (1971), The glucagon sensitive adenyl cyclase system in plasma membrane of rat liver. *J. Biol. Chem.* **246,** 1861–1871.

Rodbell, M., Lin, M. C., Salomon, Y., Londos, C., Harwood, J. P., Martin, B. R., Rendell, M., and Berman, M. (1975). Role of adenine and guanine nucleotides in the activity and response of adenylate cyclase systems to hormones: Evidence for multisite transition states. *Adv. Cyclic Nucleotide Res.* **5,** 3–29.

Sahyoun, N., Hollenberg, M. D., Bennett, V., and Cuatrescasas, P. (1977). Topographic separation of adenylate cyclase and hormone receptors in the plasma membrane of toad erythrocyte ghosts. *Proc. Natl. Acad. Sci. U.S.A.* **74,** 3795–3810.

Schmidt, S. Y., and Lolly, R. N. (1973). An early defect in inherited retinal degeneration of C3H mice. *J. Cell Biol.* **57,** 117–123.

Schramm, M. (1975). The catecholamine-responsive adenylate cyclase system and its modification by 5′-guanylylimidophosphate. *Adv. Cyclic Nucleotide Res.* **5,** 105–115.

Schramm, M., and Rodbell, M. (1975). A persistent active state of the adenylate cyclase system produced by the combination of isoproterenol and guanylylimidophosphate in frog erythrocyte membranes. *J. Biol. Chem.* **250,** 2232–2237.

Shinozawa, T., Sen, I., Wheeler, G. L., and Bitensky, M. W. (1979). Predictive value of the analogy between hormone-sensitive adenylate cyclase and light-sensitive photoreceptor cyclic GMP phosphodiesterase: a specific role for light-sensitive GTPase as a component in the activation sequence. *J. Supramol. Struct.* **10,** 185–190.

Shinozawa, T., Uchida, S., Martin, E., Cafisco, D., Hubbell, W., and Bitensky, M. W. (1980). Additional component required for activity and reconstitution of light-activated vertebrate photoreceptor GTPase. *Proc. Natl. Acad. Sci. U.S.A.* **77,** 1408–1411.

Sitaramaya, A., Virmaux, N., and Mandel, P. (1977). On a soluble system for studying light activation of rod outer segment cyclic GMP phosphodiesterase. *Neurochem. Res.* **2,** 1–10.

Spiegel, A. M., Downs, R. W., and Aurbach, G. D. (1980). Separation of a guanine nucleotide regulatory unit from the adenylate cyclase complex with GTP affinity chromatography. *J. Cyclic Nucleotide Res.* **5,** 3–17.

Tolkovsky, A. M., and Levitzki, A. (1978). Mode of coupling between the β-adrenergic receptor and adenylate cyclase in turkey erythrocytes. *Biochemistry* **17,** 3795–3810.

Uchida, S., Wheeler, G. L., Yamazaki, A., and Bitensky, M. W. (1980). Lumi-lamban: A GTP-protein activator of PDE which forms in response to bleached rhodopsin (in press).

Waloga, G., and Brown, J. E. (1979). Effects of cyclic nucleotides and calcium ions on *Bufo* rods. *Invest. Ophthalmol.* and ARVO Abstracts No. 3, pp. 5–6 (in press).

Weissbach, H. (1979). Soluble factors in protein synthesis. *In* "The Ribosomes" (G. Chambliss, G. R. Cramer, J. Davies, K. Davis, L. Kahan, and M. Nomura, eds.), pp. 377–411. Univ. Park Press, Baltimore, Maryland.

Wheeler, G. L., and Bitensky, M. W. (1977). A light-activated GTPase in vertebrate photoreceptors:

Regulation of light-activated cyclic GMP phosphodiesterase. *Proc. Natl. Acad. Sci. U.S.A.* **74,** 4238–4242.

Wheeler, G. L., Matuo, Y., and Bitensky, M. W. (1977). Light-activated GTPase in vertebrate photoreceptors. *Nature (London)* **269,** 822–824.

Woodruff, M. L., and Bownds, M. D. (1979). Amplitude, kinetics, and reversibility of a light-induced decrease in guanosine 3′,5′-cyclic monophosphate in frog photoreceptor membranes. *J. Gen. Physiol.* **73,** 629–653.

Yamazaki, A., Sen, I., and Bitensky, M. W. (1980a). Cyclic GMP specific, high affinity, non catalytic, binding sites on light-activated phosphodiesterase. *J. Biol. Chem.* **255,** 11619–11624.

Yamazaki, A., Ting, A., and Bitensky, M. W. (1980b). A single protein which can regulate both cyclic GMP binding to and catalytic activity of photoreceptor phosphodiesterase. (In preparation.)

Yee, R., and Liebman, P. A. (1978). Light-activated phosphodiesterase of the rod outer segment. *J. Biol. Chem.* **253,** 8902–8909.

Yoshikami, S., and Hagins, W. A. (1980). Kinetics of control of the dark current of retinal rods by Ca^{++} and light. *Fed. Proc., Fed. Am. Soc. Exp. Biol.* **39,** 1814.

Chapter 15

Guanosine Nucleotide Metabolism in the Bovine Rod Outer Segment: Distribution of Enzymes and a Role of GTP

HITOSHI SHICHI

Laboratory of Vision Research
National Eye Institute
National Institutes of Health
U.S. Department of Health and Human Services
Bethesda, Maryland

I. INTRODUCTION

As exemplified by the electron transport systems of mitochondria and chloroplasts and hormone-activated adenylate cyclase systems, membrane functions are usually catalyzed by multiple protein components which interact with each other in an orderly sequence. The phospholipid bilayer of biomembranes also plays an important role in various reactions catalyzed by protein components. The concept of dynamic membranes (Singer and Nicolson, 1972) has introduced a new dimension to our understanding of the function and regulation of membrane phenomena.

Vertebrate rod photoreceptor membranes were long believed to be biochemically inert, being composed of phospholipids, the visual pigment rhodopsin, and

ISBN 0-12-153315-8

trace amounts of enzymes for vitamin A metabolism. In fact, thoroughly washed rod outer segments show little or no enzymatic activity, and rhodopsin accounts for over 90% of the total protein. However, the discovery during the last decade of the highly fluid nature of the membrane and a number of membrane-associated enzymes involved in nucleotide metabolism aroused renewed interest in this membrane, especially in the role of rhodopsin in the regulation of these membrane enzymes. The reported enzymes include adenylate and guanylate cyclases (Krishnan *et al.*, 1978; Fleischman and Denisevich, 1979), cyclic-nucleotide phosphodiesterase (Baehr *et al.*, 1979; Bitensky *et al.*, 1978), guanosinetriphosphatase (GTPase) (Robinson and Hagins, 1979; Wheeler and Bitensky, 1977), rhodopsin kinase (Kühn *et al.*, 1973; Shichi and Somers, 1978), and 5′-nucleotidase (Shichi and Somers, 1981). Both cyclic-nucleotide phosphodiesterase and GTPase require photobleached rhodopsin for activation (Bitensky *et al.*, 1978; Robinson and Hagins, 1979; Wheeler and Bitensky, 1977). Rhodopsin kinase does not phosphorylate rhodopsin in the dark (Kühn *et al.*, 1973; Shichi and Somers, 1978). The enzyme seems to require an intermediate of the rhodopsin bleaching process as a substrate. Virtually nothing is known at present concerning how bleached rhodopsin or opsin activates cyclic-nucleotide phosphodiesterase and GTPase, or what changes light causes in the opsin protein so that it can be phosphorylated. Whatever the mechanism of activation, there is little doubt that these enzymes and rhodopsin maintain close interactions in rod outer segment membranes.

In order to understand the mode of interaction as well as regulation of the reactions, we first attempted to determine the location of enzymes in the membrane. A role in the visual transduction process has been suggested for cyclic-nucleotide phosphodiesterase (Yee and Liebman, 1978). Whatever enzyme it may be, the enzyme that plays a cardinal role in the visual transduction process exerts its effect through a putative transmitter. It therefore follows that the site of action and movement of a putative transmitter substance across the membrane must be consistent with the location of the enzymes involved.

Guanosine 5′-triphosphate (GTP) is known to be a cofactor for well-characterized enzymatic reactions such as those catalyzed by succinyl-CoA synthetase and phosphoenolpyruvate carboxykinase. In addition, GTP is a requirement in various reactions involving interactions of macromolecules. Typical examples are peptide elongation (Kaziro, 1978), tubulin assembly (Snyder and McIntosh, 1976), and hormone activation of adenylate cyclase (Abramowitz *et al.*, 1979; Rodbell, 1980). Light activation of cyclic-nucleotide phosphodiesterase in the presence of GTP is one of the latest additions (Bitensky *et al.*, 1978). Compared to other systems, the photoreceptor system seems to be advantageous in that GTP hydrolysis (the GTPase system) and the regulatory enzyme (cyclic-nucleotide phosphodiesterase) can be separately characterized. GTPase proteins associated with hormone receptor sites are separated only with detergents (Kaslow *et al.*, 1980), and adenylate cyclase has not been isolated in a homogeneous state. We have recently purified from the photoreceptor membrane a soluble protein having both GTP-binding and GTPase activities (Somers and Shichi, 1980). In this chapter we will describe purification of the GTP-binding protein (GTPase) and discuss a possible function in the rod membranes. The properties of the protein will also be compared with those of proteins found in other reactions in the hope of unraveling a common mechanism of GTP action in various systems in which macromolecular interactions occur.

II. LOCALIZATION IN ROD MEMBRANES OF RHODOPSIN AND ENZYMES INVOLVED IN NUCLEOTIDE METABOLISM

We have developed a technique for the preparation of intact disks and inverted disks (Adams *et al.*, 1978). The method involves (1) isolation of closed, osmotically active rod outer segments in a continuous gradient of metrizamide [2-(3-acetamido-5-*N*-methylacetamide-2,4,6-triiodobenzamide)-2-deoxy-D-glucose], (2) release of disks by rupturing the outer segment plasma membrane in 5% Ficoll (Smith *et al.*, 1975), (3) chromatography of disks on a concanavalin A-Sepharose column, and (4) inversion of disks by freezing and thawing, followed by chromatography on a concanavalin A column. Metrizamide is a low-osmolality, low-viscosity gradient material useful for maintaining the intactness of the outer segment membrane. The separation of disks is based on the fact that intact disks have no affinity for concanavalin A, while inverted disks bind the lectin (Fig. 1).

Inverted disks bind wheat germ agglutinin also but not *Ricinus* lectin. The affinity of inverted disks for these lectins indicates that sugar moieties such as mannose (concanavalin A binding) and *N*-acetylglucosamine (wheat germ lectin binding) are exposed on the surface of inverted disks. The chemical structure of the oligosaccharide moiety of rhodopsin has been found to be (Liang *et al.*, 1979; Fukuda *et al.*, 1979)

$$
\begin{array}{l}
\text{Man} \xrightarrow{\alpha 1\to 6} \\
\qquad\qquad\qquad \text{Man} \xrightarrow{\beta 1\to 4} \text{GlcNac} \xrightarrow{\beta 1\to 4} \text{GlucNac} \longrightarrow \text{Asn} \\
\text{GlcNac} \xrightarrow{\beta 1\to 2} \text{Man} \xrightarrow{\alpha 1\to 3}
\end{array}
$$

which is compatible with the lectin-binding properties of the disk membrane. The sugar moiety of rhodopsin is therefore exposed on the internal surface of the intact disk and is absent on the external surface. When intact disks are incubated with purified rhodopsin kinase and ATP in the light, the opsin protein is phosphorylated (Adams *et al.*, 1979). Little phosphorylation occurs when inverted

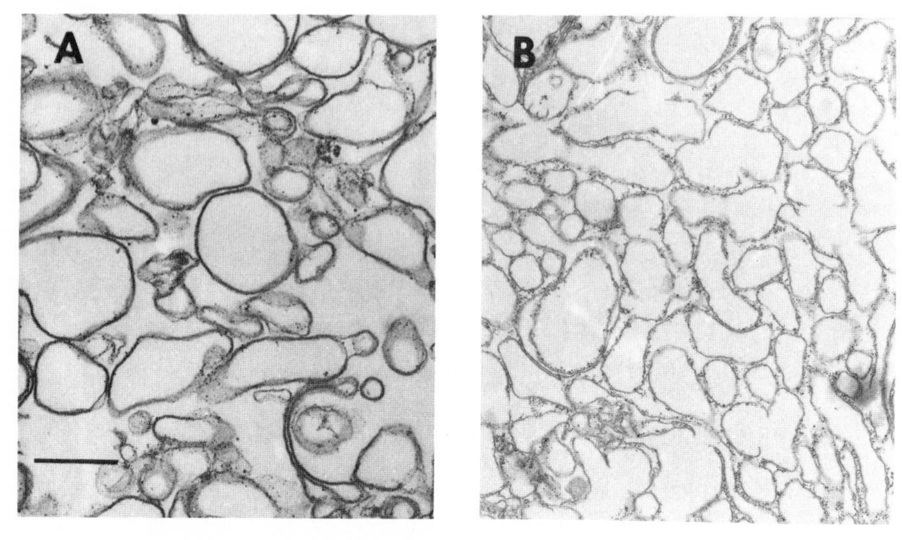

FIG. 1. Binding of concanavalin A to disks. Intact disks (A) and inverted disks (B) after labeling with ferritin–concanavalin A. The bar designates 0.50 μm.

disks are treated under similar conditions. Since neither adenosine 5′-triphosphate (ATP) nor rhodopsin kinase can permeate the disk membrane, the phosphorylation sites of rhodopsin must be exposed on the external surface of disks. These results lead us to conclude that rhodopsin spans the disk membrane (Adams *et al.,* 1979). The transmembrane nature of rhodopsin has also been suggested by other investigators from the results of chemical labeling of disks (Mas *et al.,* 1980; Nemes *et al.,* 1980).

Similar studies were carried out with intact and inverted disks to determine the localization of enzymes. The methods of enzyme assays have been described elsewhere (Shichi and Somers, 1981). During the isolation of disks by centrifugation according to the method of Smith *et al.* (1975), membranous pellets are obtained which contain membrane vesicles (fragments of the plasma membrane?) and the connecting cilium. The enzyme activities of the pelleted fraction were also determined. The results are summarized in Table I. It is noted that activities of peripheral enzymes such as rhodopsin kinase, cyclic-nucleotide phosphodiesterase, and GTPase (GTP-binding protein) are not extracted from the membrane in the presence of 2 m*M* Mg^{2+}. At 0.1 m*M* or lower concentrations of Mg^{2+} (i.e., in the presence of EDTA), the enzymes dissociate from the membrane. The magnesium dependence of enzyme association is in agreement with the previous observation that cyclic-nucleotide phosphodiesterase is solubilized from the membranes by depleting Mg^{2+} (Baehr *et al.,* 1979) and that rhodopsin kinase is extracted with EDTA from the membrane (Kühn *et al.,* 1973). Thus Mg^{2+} acts

TABLE I

DISTRIBUTION OF ENZYMES IN ROD MEMBRANES[a]

Preparation	Rhodopsin kinase (pmoles phosphate incorporated in 5 min)	Guanyl cyclase (nmoles cGMP formed per mg rhodopsin)	Cyclic-nucleotide phosphodiesterase (nmoles cGMP hydrolyzed per mg rhodopsin)	5′-Nucleotidase (nmoles GMP hydrolyzed per mg rhodopsin)	GTP-binding protein (nmoles GppNp bound per mole rhodopsin)
Whole rod outer segment	48.2 ± 5.3[b]	9.0 ± 1.0	390 ± 28	—	40 ± 3 (2 m*M* Mg^{2+})[d]
					3.2 ± 0.2 (0.1 m*M* Mg^{2+})[e]
Intact disks	83.8 ± 7.1[c,d]	0.19 ± 0.06	90 ± 8 (2 m*M* Mg^{2+})[d]	15.0 ± 1.3	130 ± 2.1 (2 m*M* Mg^{2+})[d]
	32.5 ± 0.3[b,d]		20 ± 1 (0.1 m*M* Mg^{2+})[e]		1.9 ± 0.1 (0.1 m*M* Mg^{2+})[e]
	0.9 ± 0.6[b,e]				
Inverted disks	18.9 ± 3.2[c]	0.06 ± 0.04	23 ± 2 (0.1 m*M* Mg^{2+})[e]	58.0 ± 4.1	1.9 ± 0.1 (0.1 m*M* Mg^{2+})[e]
	1.2 ± 0.5[b]				
Membranous pellet[d]	—	6.00 ± 0.40	530 ± 43 (2 m*M* Mg^{2+})[d]	500.0 ± 4.6	92.0 ± 8 (2 m*M* Mg^{2+})[d]
			74 ± 7 (0.1 m*M* Mg^{2+})[e]		8.4 ± 5 (0.1 m*M* Mg^{2+})[e]

[a] Activities are given in terms of the mean values plus or minus deviations for four assays.

[b] Purified kinase was not added.

[c] Purified kinase was added.

[d] Membranes were prepared in the presence of 2 m*M* Mg^{2+}.

[e] Membranes were prepared in the presence of 0.1 m*M* Mg^{2+}.

as an *in vivo* regulator to determine the state of enzymes, i.e., whether the enzymes exist in a membrane-associated or -dissociated state. Since these enzymes dissociate from intact (not inverted) disks at low Mg^{2+} (Table I), it is concluded that, at physiological concentrations of magnesium (about 4 m*M*; Long, 1961), the peripheral enzymes are associated on the external surface of the disk membrane. The intrinsic membrane protein 5′-nucleotidase does not show much activity with intact disks. The activity increases significantly on membrane inversion, supporting the location of the enzyme (or the catalytic site of the enzyme) on the internal surface of the disk. Of interest is the finding of high levels of enzyme activity in the membranous pellet. The activity is probably associated with the plasma membranes expected to be contained in the pellet, as the enzyme is generally considered a marker enzyme for the plasma membrane (DePierre and Karnovsky, 1973). 5′-Nucleotidase is reported to be an ectoenzyme (Newby *et al.,* 1975), but it remains to be determined whether it spans the membrane. In contrast to the enzymes described above, guanylate cyclase does not seem to be associated with the disk. Although the enzyme is tightly bound to membranes, neither intact nor inverted disks show appreciable guanylate cyclase activity. The activity is almost exclusively associated with the membranous pellet containing the connecting cilium. This enzyme has been recently reported to be associated with the axoneme (cilium) isolated from bovine rod outer segments (Fleischman and Denisevich, 1979). It should be noted that most of the enzymes studied here, whether peripheral enzymes or integral enzymes, are highly concentrated in the cilium-containing pellet. In fact, the specific activities of both cyclic-nucleotide phosphodiesterase and GTPase of the pellet are significantly higher than those of the whole outer segment. This suggests that metabolically active membranes are concentrated in the pellet. However, the pellet is heterogeneous in composition, and it remains to be seen with which membrane component the individual enzymes are associated. From this work and published data by others (Fleischman and Denisevich, 1979; Kühn, 1980), the location of enzymes is concluded to be as follows: rhodopsin kinase, cyclic-nucleotide phosphodiesterase, and GTP-binding protein are located on the external surface of the disk, while the catalytic site of 5′-nucleotidase is present on the internal surface. Guanylate cyclase is associated with the connecting cilium. If the disk is formed by infolding of the plasma membrane (Young, 1976), the external surface of the disk corresponds to the cytoplasmic surface of the plasma membrane. It is therefore possible to infer the appropriate location of the enzymes in the plasma membrane.

The location of the enzymes allows us to draw several interesting conclusions: (1) Since cyclic guanosine 3′,5′-monophosphate (cyclic GMP) is synthesized by the cyclase associated with the connection cilium extending from the base of the outer segment less than halfway toward the apex, there will be a gradient of cyclic GMP in the cytoplasm of the outer segment, with the highest concentration in the basal region. (2) If cyclic GMP is involved, be it directly or indirectly, in keeping the "Na^+ channel" of the plasma membrane open in the dark, it may regulate the channels in the basal region more effectively than those in the distal region. Biochemical heterogeneities of the rod have been suggested by our previous finding that the level of rhodopsin phosphorylation was higher in the basal region of the outer segment and decreased gradually as the point of measurement moved toward the apex (Shichi and Williams, 1979). The heterogeneity of the rod membrane system has been indicated also by intrinsic birefringence changes (Kaplan *et al.*, 1978) and electron microscopic observations of the disks (Andrews and Cohen, 1979). These results, taken together, suggest that newly assembled disks in the basal region may be functionally more important than older disks. This possibility seems to be supported also by electrophysiological evidence (Baylor *et al.*, 1979). (3) The location of the enzymes and the disposition of rhodopsin in the disk membrane indicate that these components are able to interact readily with each other on the external surface of the disk (or by inference on the cytoplasmic surface of the plasma membrane). According to a model derived from neutron diffraction patterns of the disk membrane (Saibil *et al.*, 1976), a considerable mass (about 30%) of rhodopsin in the carboxyl-terminal region is exposed on the external surface of the disk. The exposed portion of the molecule must be large enough to form a functional complex with cyclic-nucleotide phosphodiesterase and GTPase.

III. LIGHT-ACTIVATED GTP BINDING PROTEIN (GTPASE)

ATP and GTP have different functions in the rod outer segments. ATP, through the action of rhodopsin kinase (Shichi and Somers, 1978), quenches the ability of activated rhodopsin to activate cyclic-nucleotide phosphodiesterase (Liebman and Pugh, 1979). On the other hand, GTP is involved in activation of the phosphodiesterase. We have previously observed that the binding of GTP [as studied by the binding of the nonhydrolyzable GTP analogue guanosine 5′-(β,γ-imido)triphosphate (GppNp)] to bovine rod membranes ($K_d = 0.3\ \mu M$) is markedly stimulated by light (Somers and Shichi, 1979). The amount of GppNp bound is about 3 mmoles/mole rhodopsin. As described above, dissociation and association of the GTP binding protein is regulated by magnesium concentrations. At 2 mM Mg^{2+}, the GTP binding protein can be extracted with aqueous EDTA solution. The increased binding of GTP to membranes is due to hydrolysis of GTP by light-activated GTPase and exchange of guanosine 5′-diphosphate (GDP) for GTP. Early experiments have indicated that GTP-binding and GTPase activities may be attributed to the same protein. In fact, the extraction of GTP bp and GTPase occurs in an approximately parallel fashion under different conditions. Purification of the extract on a Sephacryl S-300 column results in the separation of at least three activity peaks which are eluted close to each other (Fig. 2). The first peak shows cyclic-nucleotide phosphodiesterase activity, and the second peak both GTP-binding and GTPase activities. The third peak contains rhodopsin kinase. When the first and second peaks are combined and purified further on a Blue Sepharose column (Fig. 3), four protein peaks are separated which are designated peaks I, II, III, and IV, respectively, in the order of elution. Peak III contains cyclic-nucleotide phosphodiesterase which is detected as double bands with similar mobility in sodium dodecyl sulfate (SDS)–polyacrylamide gel electrophoresis and has a holoenzyme molecular weight of about 180,000, whereas peak IV shows both GTPase and GTP-binding activities. Elution of the latter two activities coincides well, supporting the association of these activities with a single protein. Table II summarizes the purification of GTP binding protein (GTPase) from aqueous extracts of bovine rod outer segments. In terms of specific activities, the degree of purification was less than three fold. After a repetition of the Blue Sepharose chromatography, the purified protein was found to be homogeneous and to migrate as a single protein with a molecular weight of 39,000 in SDS–polyacrylamide gel electrophoresis (Fig. 4).

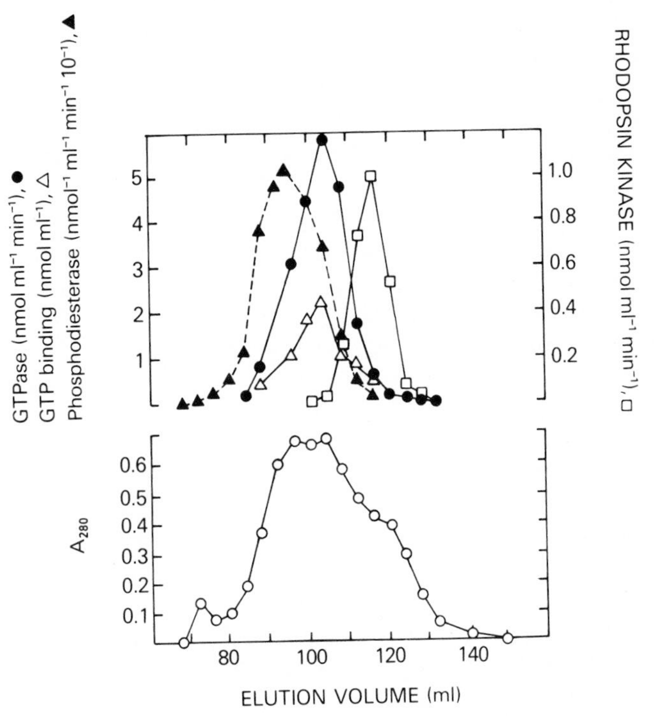

FIG. 2. Separation of three enzymatic activities on a Sephacryl S-300 column.

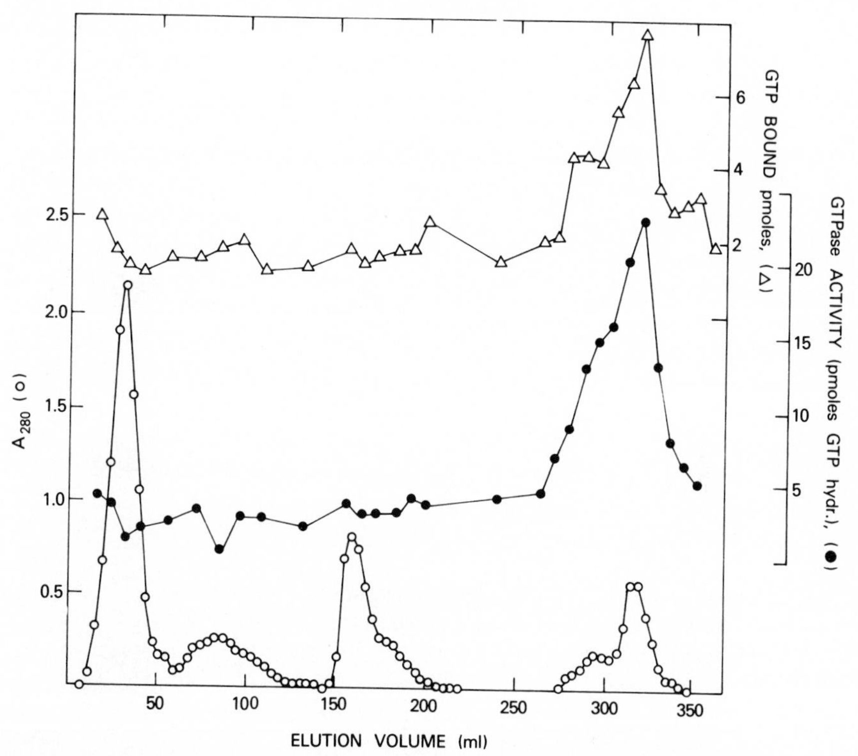

FIG. 3. Purification of GTP binding protein (GTPase) on a Blue Sepharose column. Elution with a linear gradient of NaCl (0–1.5 *M*) started from elution volume 140 ml and ended at elution volume 380 ml. Peak I (10–50 ml) and peak II (60–120 ml) show no enzymatic activity. Peak III (150–200 ml) and peak IV (280–350 ml) demonstrate cyclic-nucleotide phosphodiesterase and GTPase activities, respectively.

TABLE II
PURIFICATION OF GTPASE AND GTP BINDING ACTIVITIES

Purification steps	GTPase: Specific activity (nmoles GTP hydrolyzed per mg per 10 min)	GTPase: Total activity (nmoles GTP hydrolyzed per 10 min)	GTPase: Recovery (%)	GTP binding: Specific activity (nmoles GTP bound per mg protein)	GTP binding: Total activity (nmoles GTP bound)	GTP binding: Recovery (%)
Extract	40	900	100	0.70	64	100
Sephadex S-200	82	870	97	1.40	60	94
Blue Sepharose	110	290	32	6.50	30	47

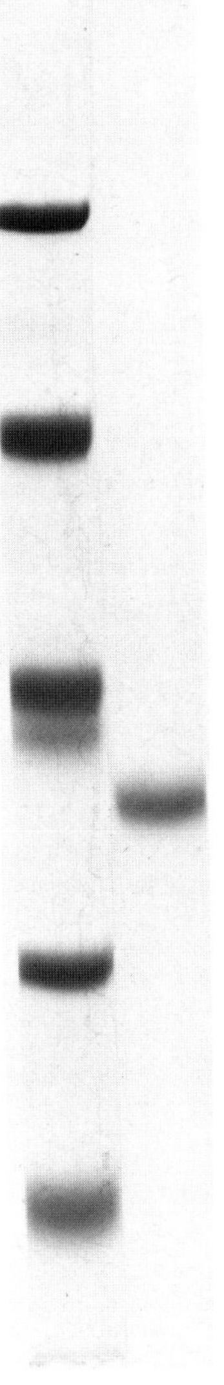

FIG. 4. SDS-polyacrylamide gel electrophoresis of purified GTP binding protein (GTPase). (A) Molecular-weight markers (94,000, 67,000, 43,000, 30,000, and 20,100, from top to bottom). (B) GTP binding protein (GTPase).

Biochemical activities of peak I (MW 37,000) and peak II (MW 37,000) were not identified. Either one of the peaks may correspond to the cyclic-nucleotide phosphodiesterase inhibitor (MW 38,000) reported by Dumber and Etingoff (Dumler and Etingof, 1976). In the frog GTPase system, a component designated helper protein has been reported to be an absolute requirement for light activation of the enzyme (Shinozawa *et al.*, 1980). This is not the case with the bovine GTPase system: GTPase purified to homogeneity is fully activated by bleached rhodopsin in the absence of an additional component. Bovine retinal GTP binding protein has been reported to be composed of two subunits (MW 37,000 and 39,000) (Kühn, 1980; Godchaux and Zimmerman, 1979). If the subunit having a lower molecular weight corresponds to peak I or peak II protein, our study indicates that the higher-molecular-weight subunit (probably identical with peak IV) alone is capable of binding opsin and hydrolyzing GTP.

It is interesting to note that GTPase from Sephacryl S-300 has a molecular weight of about 250,000. A molecular weight of 39,000 for the monomeric form suggests that the enzyme may be present *in vivo* as an oligomer (hexamer?). Furthermore, GTPase and cyclic-nucleotide phosphodiesterase may exist as a complex within the cell. The *in vivo* concentrations of cyclic-nucleotide phosphodiesterase [about 1 enzyme molecule per 500 rhodopsin molecules (Miki *et al.*, 1975)] and GTPase [about 1 enzyme molecule per 300 rhodopsin molecules (Liebman and Pugh, 1979)] support the possibility of a 1:1 molar complex formation. (These considerations are incorporated in our proposal presented in Fig. 8.)

Light-activated GTP-binding and GTPase activities of purified enzyme in the presence of purified rhodopsin are shown in Fig. 5. Rhodopsin used for this experiment was purified to homogeneity by the method described previously (Shichi *et al.*, 1969). Neither GTPase nor GTP-binding activity is observed when purified enzyme and purified rhodopsin are incubated in the dark. Both activities become measurable only in the light. The level of GTPase activity determined with purified rhodopsin (in dilute detergent) is about $^1/_{10}$ of that determined with the disk membrane (data not shown). This is because 0.01% Emulphogene, included as a solvent for rhodopsin in the reaction mixture, is highly toxic to GTPase and also because the opsin protein undergoes conformational denaturation on bleaching in the detergent (Shichi, 1971). Of significance is the fact that membranes of the phospholipid bilayer are not required for the manifestation of these activities. More recent experiments show that similar light activation of GTP binding and GTPase occurs when the purified protein and rhodopsin-containing phospholipid vesicles are incubated.

The bovine retinal GTPase system shows several distinct properties: (1) A purified protein (peak IV from a Blue Sepharose column) with a molecular weight of 39,000, as such, shows neither GTP-binding nor GTPase activity. (2) When the protein is incubated with disks or purified rhodopsin in the light, both

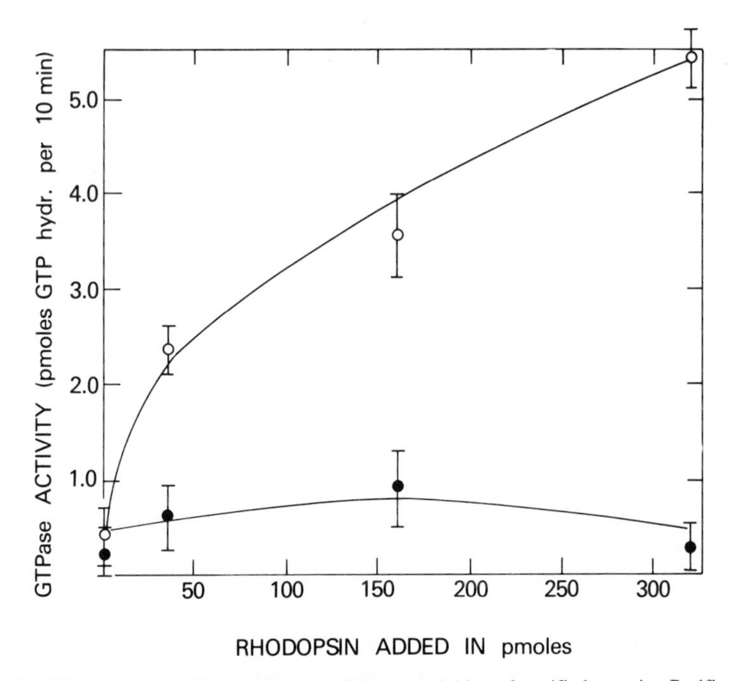

FIG. 5. Light-activated GTP binding and GTPase activities of purified protein. Purified GTP binding protein (GTPase) and rhodopsin were incubated in the dark (solid circles) or in the light (open circles). (A) GTP binding to opsin. (B) GTPase activation by rhodopsin in the light.

GTP-binding and GTPase activities are manifested. (3) Prebleached disk membranes are also capable of activating GTP-binding and GTPase activities. (4) Magnesium is required for binding of GTPase to disk membranes. (5) GTP binding probably occurs as a result of the exchange of bound GDP for GTP. There characteristics are incorporated in the scheme in Fig. 6.

Guanine nucleotide-binding proteins extracted from bovine outer segments have been suggested to contain bound GDP which is exchanged with GTP when the proteins associate with the disk membrane (Shinozawa *et al.*, 1980). The absence of bound nucleotide in purified GTPase (as judged from ultraviolet absorption spectra) may be due to dissociation of bound nucleotide from the protein during purification procedures. The extent of GTP binding becomes greater with an increase in the amount of rhodopsin bleached (Fig. 5). The ratio of GTP bound to rhodopsin bleached is low, and the binding reaction does not seem to be an amplification process. GTPase activity has a turnover in the range of 1 to 5 and is too slow to be involved in the transduction process.

Although information on the retinal GTPase system is still limited, it is helpful to compare properties of the retinal system with those of other systems in which GTP is involved. I will first summarize below what is known about the role of

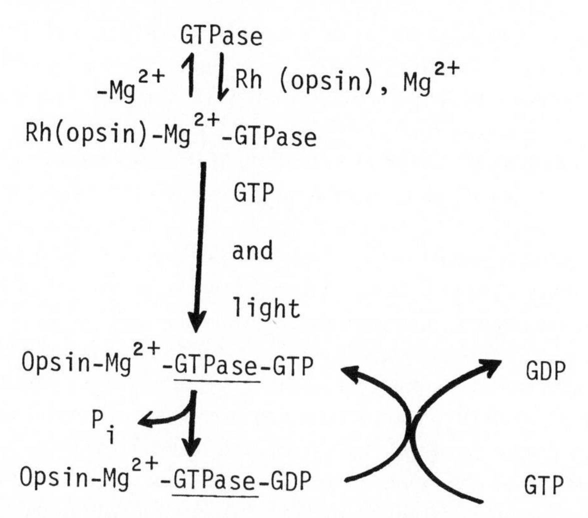

FIG. 6. Schematic presentation of light-activated GTP binding and GTPase activities of rod membrane protein.

GTP in different systems and then discuss the significance of GTP binding and hydrolysis in the retinal system.

Peptide elongation in prokaryotic cells begins with the formation of an aminoacyl-tRNA–Tu factor–GTP complex (Kaziro, 1978). The complex binds to the acceptor site of mRNA on the ribosome and receives a peptide chain from the preexisting peptidyl-tRNA at the donor site (Fig. 7). GTP hydrolysis is essential for fast release of Tu factor from the ribosome. Another GTP molecule is hydrolyzed while the elongated peptidyl-tRNA translocates to the donor site. Peptide elongation in eukaryotic cells also proceeds in a similar manner. It should be noted that the elongation factor alone shows no GTPase activity; the presence of

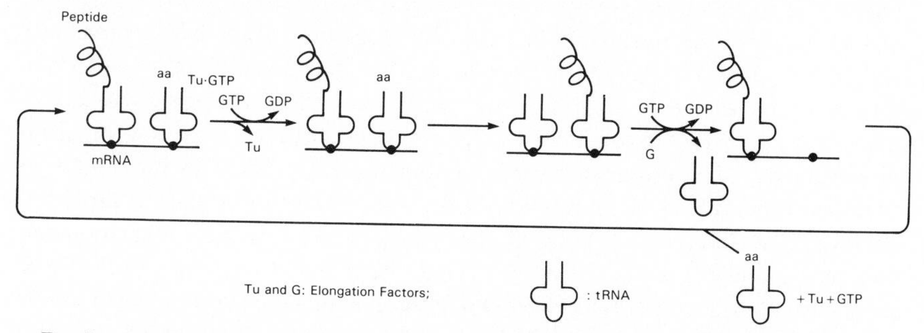

FIG. 7. Requirements for GTP in peptide elongation. GTP is required for the formation of (1) the tRNA·Tu·mRNA complex and (2) release of tRNA from the ribosome.

the ribosome is essential for the manifestation of activity. In regard to the role of guanine nucleotides, GTP and GDP have been suggested to act as conformational effectors of the elongation factor so that the GTP complex may have a higher affinity for the ribosomes than the GDP complex (Kaziro, 1978).

In tubulin assembly (Snyder and McIntosh, 1976), the binding of two guanine nucleotides per tubulin dimer (α,β) at nonexchangeable and exchangeable sites is essential. Polymerization of tubulins occurs as an elongation of preexisting polymers and is promoted by GTP but not by GDP (Jameson and Caplow, 1980). Under physiological conditions GTP hydrolysis occurs concurrently with polymerization. Isolated microtubules contain only GDP at the exchangeable site. GppNp is also known to cause tubulin assembly (Arai and Kaziro, 1976). However, microtubules assembled with the GTP analogue are stable to the concentration of calcium at which microtubules formed with GTP depolymerize completely (Weisenberg *et al.*, 1976), and are therefore different from the microtubules assembled under physiological conditions. GTP hydrolysis may be necessary to maintain (or produce) a microtubule conformation readily capable of deploymerization in the presence of Ca^{2+}.

An feature apparently common to both peptide elongation and tubulin assembly is that the affinity of nucleotide-binding proteins (elongation factors and tubulin proteins) for interacting macromolecules (ribosomes and preformed tubulin polymers) is modulated by binding of either GDP or GTP.

In hormone-activated adenylate cyclase systems, GTP exerts its effects by (1) accelerating the rate of hormone binding to and release from its receptor, and (2) promoting coupling of receptors to cyclase. The effects have been described by the following set of reactions (Rodbell, 1980).

$$H + R \cdot N \rightarrow H \cdot R \cdot N$$

$$H \cdot R \cdot N + GTP \rightarrow H \cdot R \cdot N \cdot GTP \rightarrow H + R \cdot N \cdot GTP$$

$$H \cdot R \cdot N \cdot GTP + C \rightarrow H \cdot R \cdot N \cdot C + GTP$$

where N is the nucleotide-binding protein, H is the hormone, R is the receptor, and C is adenylate cyclase. The regulatory complex R·N has a molecular weight of several hundred thousands and is considerably larger than the estimated size of the R·N·C unit, i.e., R·N associated with the enzyme. On this basis, Rodbell (1980) has recently proposed that the role of GTP is to dissociate the N·R oligomer to the monomeric form which in turn binds to cyclase. The GTP N·R monomeric form thus formed may take a preferred conformation. In other words, as has been suggested for peptide elongation and tubulin assembly (Kaziro, 1978), GTP may be required to induce in the N·R regulatory unit a conformation essential for binding to adenylate cyclase. It should be pointed out that Rodbell's model considers GTP a type of allosteric effector for the R·N unit and does not require GTP hydrolysis for the activation of adenylate cyclase.

IV. MODEL FOR THE ROLE OF GTP BINDING PROTEIN (GTPase) IN THE PHOTORECEPTOR SYSTEM

How can this model be useful in understanding the role of GTP in the retinal photoreceptor system? Since the action of light is to isomerize and release 11-*cis*-retinal from the receptor protein (opsin), it is not unreasonable to assume that 11-*cis*-retinal is an inhibitory hormone. The signal of isomerization (i.e., release of the inhibitory hormone) is somehow transmitted to the N unit (GTPase). Since the N unit in crude extracts is severalfold larger in size than purified N, the N unit *in vivo* probably exists as an oligomer composed of identical subunits each of which shows both GTP binding and GTPase. Although clear evidence is not available, GTP hydrolysis is probably not required for the activation of cyclic-nucleotide phosphodiesterase. With these assumptions I propose the following scheme (Fig. 8) to explain a role of GTP in the retinal system. As in the case of the hormonal system, GTP binding causes dissociation of the N unit and sub-

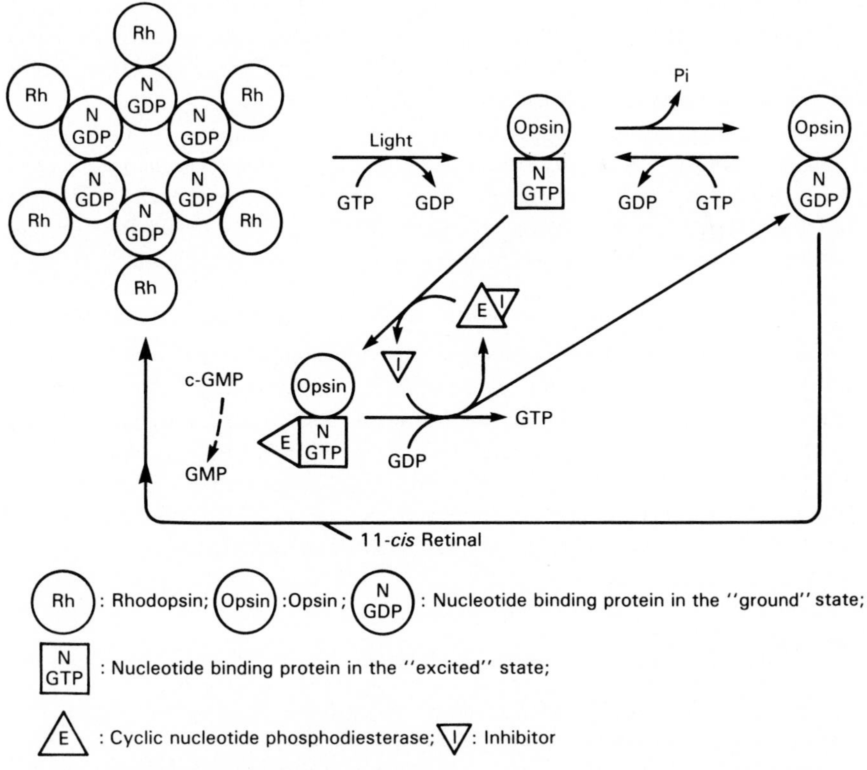

FIG. 8. A proposed scheme for light-initiated reactions involving GDP–GTP exchange, GTPase, and cyclic-nucleotide phosphodiesterase activation.

sequent conformational change in such a manner that the affinity of the N unit for cyclic-nucleotide phosphodiesterase is increased. This is somewhat similar to the excitation of an electron by a photon from the ground state to an excited state. In this analogy, GTP and the N unit correspond to the photon and the electron, respectively. The free enzyme (phosphodiesterase) is probably associated with an inhibitor (Dumler and Etingof, 1976; Hurley, 1980) and is enzymatically inactive. The basic feature of the model does not change whether the activation process involves association of an activator or a GTPase-induced conformational change in phosphodiesterase rather than dissociation of an inhibitor. The opsin-N-GTP (in excited state) follows two paths simultaneously, either decaying to opsin-N-GDP (ground state) or activating the enzyme by complexing with it. According to this hypothesis, continuous hydrolysis of GTP is not required for maintaining phosophodiesterase in active form. Since the decay process hydrolyzes GTP, the GDP level in the environment increases and an exchange of GTP for GDP in opsin-N-phosphodiesterase is facilitated. This results in the dissociation and inactivation of the the enzyme. It should be noted that the hypothesis assumes the photic bleaching of rhodopsin to be required solely for the activation of GTP-binding protein which subsequently activates phosophodiesterase. The GTP binding protein (N) therefore serves as a *coupler* between the signal receptor (rhodopsin) and the intracellular regulator (cyclic-nucleotide phosphodiesterase).

A number of questions are raised concerning the feasibility of the proposed model. Does the oligomeric form of GTP-binding protein (N) bind GDP? Does GTP dissociate N-GDP into the monomeric form? How fast does the GDP–GTP exchange occur? How does opsin induce GTPase activity in N? Does N take different conformations between the ground state and excited state? Is phosphodiesterase in dark-adapted rods inactive because of an associated inhibitor? Is an inhibitor released when the enzyme is activated? Does the opsin–N–GTP complex induce a conformational change in phosphodiesterase during activation? Is there an amplification of the photon signal? In other words, does one bleached pigment activate more than one molecule of GTP-binding protein and phosphodiesterase? Does the regeneration of rhodopsin reverse the ability of the membrane to activate N? Is N protein (coupler) specific for the rod enzyme? Or, does it also couple to intracellular regulator enzymes in other systems? It is hoped that answers to these questions will be found in the near future.

REFERENCES

Abramowitz, J., Iyengar, R., and Birnbaumer, L. (1979). *Mol. Cell. Endocrinol.* **16,** 129–146.
Adams, A. J., Tanaka, M., and Shichi, H. (1978). *Exp. Eye Res.* **27,** 595–605.
Adams, A. J., Somers, R. L., and Shichi, H. (1979). *Photochem. Photobiol.* **29,** 687–692.
Andrews, L. D., and Cohen, A. I. (1979). *J. Cell Biol.* **81,** 215–228.
Arai, T., and Kaziro, Y. (1976). *Biochem. Biophys. Res. Commun.* **69,** 369–376.

Baehr, W., Devlin, M. J., and Applebury, M. L., (1979). *J. Biol. Chem.* **254,** 11669-11677.
Baylor, D. A., Lamb, T. D., and Yau, K.-W. (1979). *J. Physiol. (London)* **288,** 589-611.
Bitensky, M. W., Wheeler, G. L., Aloni, B., Vetury, S., and Matuo, Y. (1978). *Adv. Cyclic Nucleotide Res.* **9,** 553-572.
DePierre, J. W., and Karnovsky, M. L. (1973). *J. Cell Biol.* **56,** 275-303.
Dumler, I. L., and Etingof, R. N. (1976). *Biochim. Biophys. Acta* **429,** 474-484.
Fleischman, D. E., and Denisevich, M. (1979). *Biochemistry* **18,** 5060-5066.
Fukuda, M., Papermaster, D. S., and Hargrave, P. A. (1979). *J. Biol. Chem.* **254,** 8201-8207.
Godchaux, W., and Zimmerman, W. F. (1979). *J. Biol. Chem.* **254,** 7874-7884.
Hurley, J. B. (1980). *Biochem. Biophys. Res. Commun.* **92,** 505-510.
Jameson, L., and Caplow, M. (1980). *J. Biol. Chem.* **255,** 2284-2292.
Kaplan, M. W., Deffebach, M. E., and Liebman, P. A. (1978). *Biophys. J.* **23,** 59-70.
Kaslow, H. R., Johnson, G. L., Brothers, V. M., and Bourne, H. R. (1980). *J. Biol. Chem.* **255,** 3736-3741.
Kaziro, Y. (1978). *Biochim. Biophys. Acta* **505,** 95-127.
Krishnan, N., Fletcher, R. T., Chader, G. J., and Kirshna, G. (1978) *Biochim. Biophys. Acta* **523,** 506-515.
Kühn, H. (1980). *Nature (London)* **283,** 587-589.
Kühn, H., Cook, J. H., and Dreyer, W. (1973). *Biochemistry* **12,** 2495-2502.
Liang, C.-J., Yamashita, K., Muellenberg, C. G., Shichi, H., and Kobata, A. (1979). *J. Biol. Chem.* **254,** 6414-6418.
Liebman, P. A., and Pugh, E. N. (1979). *Vision Res.* **19,** 375-380.
Long, C. (1961). "Biochemists' Handbook," pp. 706-714. Van Nostrand-Reinhold, New Jersey.
Mas, M. T., Wang, J. K., and Hargrave, P. A. (1980). *Biochemistry* **19,** 684-692.
Miki, N., Baraban, J. M., Keirns, J. J., Boyce, J. J., and Bitensky, M. W. (1975). *J. Biol. Chem.* **250,** 6320-6327.
Nemes, P. P., Miljanich, G. P., White, D. L., and Dratz, E. A. (1980). *Biochemistry* **19,** 2067-2074.
Newby, A. C., Luzio, J. P., and Hales, C. N. (1975). *Biochem. J.* **146,** 625-633.
Robinson, W. E., and Hagins, W. A. (1979). *Photochem. Photobiol.* **29,** 693.
Rodbell, M. (1980). *Nature (London)* **284,** 17-22.
Saibil, H., Chabre, M., and Worcester, D. (1976). *Nature (London)* **262,** 266-270.
Shichi, H. (1971). *J. Biol. Chem.* **246,** 6178-6182.
Shichi, H., and Somers, R. L. (1978). *J. Biol. Chem.* **253,** 7040-7046.
Shichi, H., and Somers, R. L. (1980). *Photochem. Photobiol.* **32,** 491-495.
Shichi, H., and Williams, T. C. (1979). *J. Supramol. Struct.* **12,** 419-424.
Shichi, H., Lewis, M. S., Irreverre, F., and Stone, A. L. (1969). *J. Biol. Chem.* **244,** 529-536.
Shinozawa, T., Uchida, S., Martin, E., Cafiso, D., Hubbell, W., and Bitensky, M. (1980). *Proc. Natl. Acad. Sci. U.S.A.* **77,** 1408-1411.
Singer, S. J., and Nicolson, G. L. (1972). *Science* **175,** 720-731.
Smith, H. G., Stubbs, G. W., and Litman, B. J. (1975). *Exp. Eye Res.* **20,** 211-217.
Snyder, J. A., and McIntosh, J. R. (1976). *Annu. Rev. Biochem.* **45,** 699-720.
Somers, R. L., and Shichi, H. (1979). *Biochem. Biophys. Res. Commun.* **89,** 479-485.
Somers, R. L., and Shichi, H. (1980). *Fed. Proc., Fed. Am. Soc. Exp. Biol.* **39,** 2070.
Weisenberg, R. C., Deery, W. J., and Dickinson, P. J. (1976). *Biochemistry* **15,** 4248-4254.
Wheeler, G., and Bitensky, M. W. (1977). *Proc. Natl. Acad. Sci. U.S.A.* **74,** 4238-4242.
Yee, R., and Liebman, P. A. (1978). *J. Biol. Chem.* **253,** 8902-8909.
Young, R. W. (1976). *Invest. Ophthalmol.* **15,** 700-725.

CURRENT TOPICS IN MEMBRANES AND TRANSPORT, VOLUME 15

Chapter 16

Calcium Tracer Exchange in the Rods of Excised Retinas

ETE Z. SZUTS

Laboratory of Sensory Physiology
Marine Biological Laboratory
Woods Hole, Massachusetts

I. INTRODUCTION

The actual role of calcium in phototransduction remains as elusive today as it was a decade ago when Yoshikami and Hagins (1971) proposed that calcium was the required intracellular transmitter that mediated receptor excitation. According to their hypothesis (see also review by Hagins, 1972), two properties were predicted for photoreceptor disks: a light-induced calcium release and an energy-dependent calcium uptake. The latter uptake process is a necessary requirement if repeated light releases are to occur from individual disks. In spite of an initial spate of published reports, it is now generally accepted that the predicted release has not yet been verified. Whether measured within intact recep-

ISBN 0-12-153315-8

tors (Szuts and Cone, 1977) or in cell fragments (Smith *et al.*, 1977; Szuts and Cone, 1977; Liebman, 1978), calcium releases are either undetectable or many orders of magnitude too small. Similarly, compelling evidence for active calcium uptake is also absent. Unfortunately, this lack of experimental verification is equivocal on the role of the predicted processes in transduction, since the cited experiments were performed either under nonphysiological conditions or with insufficiently sensitive techniques. However, recent measurements with tracer exchange indicate that even within intact receptors calcium uptake by disks is too slow to be compatible with the "calcium hypothesis." Some of these results have been recently published elsewhere (Szuts, 1980).

II. EXPERIMENTAL DESIGN AND EXCHANGE THEORY

Excised bullfrog retinas, freed of their pigment epithelium, were incubated with physiological saline solutions containing trace amounts of ^{45}Ca. It is now common knowledge that, in such a preparation, photoreceptors retain their electrophysiological and biochemical properties (see review by Sickel, 1973). Indeed, when the spectral sensitivity of the excised bullfrog retinas was routinely measured in this study by a criterion electroretinogram response, rods were found to be light-sensitive for at least several hours under the above incubation conditions.

Receptors accumulate radioactive calcium during the above incubations. In rod outer segments, radioactive calcium first enters the cytoplasmic and then the intradiskal space, its rate of entry into either of these compartments being determined by the exchange rate across the respective membranes. As commonly used, "exchange" here refers to the general process by which ions of separate "pools" intermingle with each other. Such pools are formed whenever membranes or binding sites, respectively, separate calcium into distinct cellular compartments or into populations of free and bound ions. The diverse processes responsible for exchange across membranes (e.g., calcium–anion cotransport or calcium–cation countertransport) may or may not be energy-dependent. Irrespective of the identity of the processes, calcium flux across any membrane can be separated into two opposing unidirectional fluxes: influx and efflux. At steady state, these fluxes are balanced, leading to zero net flux, hence to no net gain or loss of calcium. Thus for cells in the steady state, the calcium content is constant for any subcellular compartment. Under physiological conditions, dark-adapted photoreceptors exist in such a steady state. Light flashes, which may transiently imbalance the fluxes, need not substantially alter the steady state over the long run or the time-averaged calcium content of a compartment. However, with increased intensity or repetitiveness, light stimuli could create a new steady state. The rate of exchange for the cytoplasmic and intradiskal compartments is essentially determined by the magnitude of the unidirectional fluxes across the respective membranes. If the unidirectional flux across the plasma membrane of bullfrog rod outer segments is about 1 pmole cm^{-2} sec^{-1}, as reported for cattle (Schnetkamp, 1979), frog outer segments exchange or turn over their cytoplasmic calcium content with a half-time of less than 5 min (Section VII). Thus, if excised retinas are incubated for a long period, say an hour, disks will be exposed for most of this period to a cytoplasmic environment whose specific radioactivity (essentially counts per minute per mole of total calcium) is nearly the same as that of the external medium. Measurement of the intradiskal ^{45}Ca content (in counts per minute) then permits calculation of the amount of calcium accumulated by disks.

Disks have to be isolated before their ^{45}Ca content can be measured. In these experiments, this was accomplished by vortexing the retinas in ice-cold solutions that contained no added ^{45}Ca. Vortexing not only breaks off the outer segment from the rest of the cell but also fragments the outer segments so that their plasma membranes but not their disk membranes become leaky. The evidence for the nonleakiness of disks comes from cation measurements on fragmented and hypotonically shocked outer segments. After such a manipulation, disks still retain a high content of the common cations; total (both bound and free) concentrations were measured to be about 40 m*M* sodium, 80 m*M* potassium, 5 m*M* calcium, and 8 m*M* magnesium (Szuts, 1975; Szuts and Cone, 1977). Thus, with vortexing, calcium within the cytoplasm but not necessarily within the intradiskal space is lost into the suspending solution. The radioactive content of disks can then be measured provided that they are adequately separated from other cellular contaminants.

III. SIZE OF THE EXCHANGEABLE POOL WITHIN DISKS

The need to separate disks from cellular contaminants is shown in the experiment described in Fig. 1, where a suspension of fragmented outer segments, freshly obtained from a ^{45}Ca-incubated retina, was centrifuged on a density gradient. Calcium that is either within the disks or strongly bound to the cytoplasmic disk surface should comigrate with rhodopsin. Note the very poor match between ^{45}Ca and rhodopsin distribution throughout the gradient even though rhodopsin, the membrane protein of disks, constitutes most of the protein in the sample. The large ^{45}Ca levels within contaminating organelles nearly mask the ^{45}Ca associated with the disks. Indeed, in other experiments background contamination was found to be even greater. The amount of ^{45}Ca comigrating with the disks in Fig. 1 is so small that, after the specific activity of the incubating solutions is taken into account, the radioactive content is equivalent to an exchange of only 0.004 mole calcium/mole of rhodopsin (or simply 0.004 Ca/Rho). Since the disks contain a total of 0.13 Ca/Rho (Szuts and Cone, 1977), the

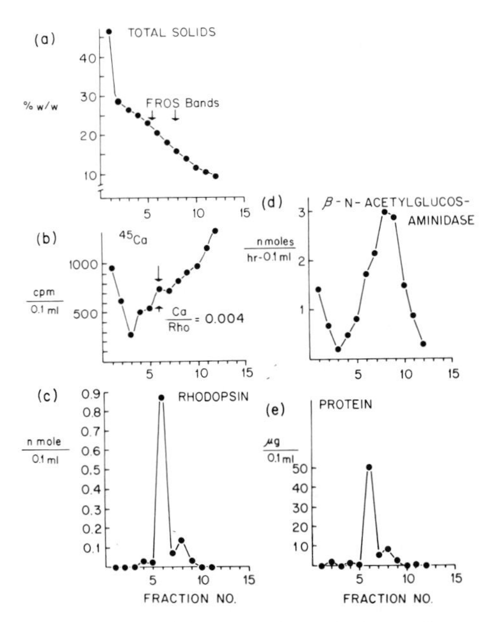

FIG. 1. Separation of disks from cellular contaminants by isopycnic centrifugation for the measurement of intradiskal ^{45}Ca content in freshly isolated fragmented outer segments. The distribution of metrizamide (the density material, expressed as percentage of total solids) (a), ^{45}Ca (b), rhodopsin (c), β-*N*-acetylglucosaminidase (d), and protein (e) is shown throughout the gradient. Two retinas were initially incubated for 1 hr in a solution (90 m*M* NaCl, 25 m*M* $NaHCO_3$, 2.5 m*M* KCl, 1.5 m*M* $MgSO_4$, 3 m*M* $CaCl_2$, 0.1 m*M* Ca-EGTA, 10 m*M* glucose, gassed with 95% O_2–5% CO_2, pH 7.4) containing ^{45}Ca with a specific activity of 27.7×10^{12}cpm/mole calcium. The retinas were transferred to 2 ml of chilled stock solution (115 m*M* NaCl, 2.5 m*M* KCl, 1.5 m*M* $MgSO_4$, 5 mM HEPES, pH 7.4) and vortexed to isolate and fragment the outer segments. The final calcium concentration of the solution was about 100 μM because of incubating solution adhering to the retinas. The suspension was filtered through a nylon mesh to remove the tissue and then placed atop a metrizamide density gradient and centrifuged for 30 min at 160,000 *g* (2°C). The tube bottom was pierced, and 0.7-ml fractions were collected and assayed for the listed constituents. Thus fractions 1 and 15, respectively, represent the bottom and top of the gradient. As previously reported (Szuts, 1980), disks are separated into two subpopulations of slightly different density. Their relative proportion changes with time following disk isolation. (Reprinted by permission from the *Journal of General Physiology* **76,** pp. 253–286. Copyright 1980, The Rockefeller University Press.)

measured exchange represents an exchangeable pool that is less than 10% of the entire content. Essentially the same results were obtained when the ionic composition of the isolation solutions was altered and when the retinas were exposed to light stimulation during incubation.

The time resolution for ^{45}Ca measurements was about 40 min in Fig. 1. This refers to the interval between cell rupture by vortexing and disk arrival at the final equilibrium position on the gradient. Once centrifugation began, about 5 min was required for the disks to reach their equilibrium position. Any ^{45}Ca losses during this 40-min time interval would have artificially lowered the radioactivity for the disk fraction. Losses during subsequent centrifugation and fraction collection could not have significantly affected the results, as such losses were limited by aqueous diffusion within the centrifuge tube. Since the axial distance per fraction was 5 mm within the centrifuge tube, several hours would have been required for released ^{45}Ca to diffuse out of the disk fraction (calculation based on the diffusion rate being 1 μm/msec and the diffusion time being proportional to the square of the distance). Of course, the time resolution for ^{45}Ca measurement could have been reduced to 5–10 min by quickly pelleting freshly isolated fragments in a table-top centrifuge. Although larger ^{45}Ca contents were measured in such pellets, the increased amounts could be solely attributed to the increased level of contamination.

IV. ABSENCE OF SIGNIFICANT INTRADISKAL ^{45}Ca LOSSES FOLLOWING CELL RUPTURE

A. Losses by Net Efflux

Since the time resolution for ^{45}Ca measurements was as much as 40 min, the occurrence of possible ^{45}Ca loss prior to centrifugation must be considered. Radioactive calcium can be lost in only two ways—net efflux and exchange. The major difference between these processes is that, with the former, the calcium content within disks should continuously drop with time. However, disks are relatively resistant to net losses. Their initial total content is 0.1–0.2 Ca/Rho within intact isolated outer segments and remains at this level even after cell rupture in the presence of extremely low calcium activity (Szuts and Cone, 1977). The inability of disks to lose all their calcium has been verified by several investigators (Table I). Since the experimental conditions shown in Fig. 1 would have favored calcium retention (calcium activity in the isolation solutions was about 100 μM, much greater than the physiological activity within the cell; the ionic composition of the solution created a membrane potential that made the inside of the disks more negative), it is unlikely that greater net losses occurred in the experiments Fig. 1 than in those cited in Table I. Thus total calcium within the disks of Fig. 1 most likely remained at about 0.1 Ca/Rho, even though the exchanged pool was less than 0.004 Ca/Rho. Such considerations make it unlikely that massive net releases were a source of ^{45}Ca losses.

B. Losses by Exchange

Calcium exchange across disk membranes was measured two different ways. The first method, employing the same procedure as in Fig. 1, was based on the rationale that disks could just as easily gain as lose calcium if exchange existed.

TABLE I
TOTAL CALCIUM CONTENT OF ROD DISKS IN LOW-CALCIUM SOLUTIONS

Animal	Preparative procedure	Content (Ca/Rho)	Reference
Frog	Purification with sucrose gradients; 1 mM EGTA, free calcium < 10^{-9} M	0.20 ± 0.02[a]	Liebman, 1974
Bullfrog	Minimal purification, free calcium < 3×10^{-6} M	0.07–0.27[b]	Hess, 1975
Bullfrog	Minimal purification with hypotonic shock; 3 mM EDTA, free calcium < 10^{-10} M	0.13 ± 0.08	Szuts and Cone, 1977
Cattle	Purification with sucrose gradients; 1 mM EGTA	0.43 ± 0.20	Schnetkamp *et al.*, 1977
Cattle	Purification with sucrose gradients; calcium-free	0.1–0.2	Noll *et al.*, 1979

[a] Published data were converted using a molecular weight of 40,000 for frog rhodopsin.

[b] Published data were converted using the color equivalency between 15 nmoles rhodopsin and 1 mg bovine serum albumin by the Lowry assay.

Therefore vortexing nonradioactive retinas in a solution that contained trace amounts of ^{45}Ca should lead to ^{45}Ca accumulation. The accumulated amount then could be measured with a time resolution of about 5 min using density gradient centrifugation. Figure 2 presents the results of such an experiment. The ^{45}Ca associated with the disks in this experiment was more completely swamped by the radioactivity of the contaminants. Even though exchange may have been favored by the higher experimental temperatures, the disks apparently accumulated less than 0.007 Ca/Rho in 30 min.

Since the time resolution was about 5 min with the above procedure, a second method was employed to determine exchange on a faster time scale. Disks were purified and passively loaded with ^{45}Ca. The loading procedure and the necessary controls for showing that ^{45}Ca was inside the disks rather than bound to the external surface are described in detail elsewhere (Szuts, 1980). Efflux from such ^{45}Ca-loaded disks was studied as a function of intradiskal calcium concentration. Irrespective of the conditions, the exchange rate across the disk membranes was immeasurably small. A representative experiment is shown in Fig. 3. At time zero, ^{45}Ca-loaded disks were resuspended in a solution whose ionic composition was the same as in Figs. 1 and 2. At various time intervals, identical aliquots were filtered and the radioactive content of the disks remaining on the filter was measured. The total calcium content of these disks was 0.4 Ca/Rho and, by the procedures and criteria described elsewhere (Szuts, 1980), intradiskal free calcium was estimated to be 0.7 mM. As observed, ^{45}Ca losses

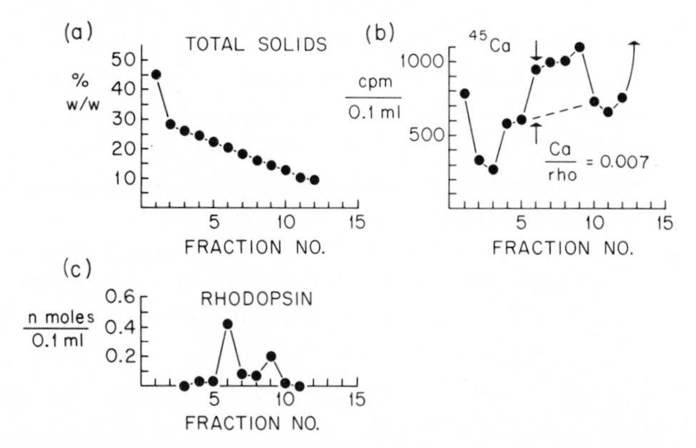

FIG. 2. Separation of disks from cellular contaminants by isopycnic centrifugation for the measurement of ^{45}Ca accumulation by exchange within freshly isolated fragmented outer segments. The distribution of metrizamide (a), ^{45}Ca (b), and rhodopsin (c) is shown throughout the gradient. Outer segments were isolated at room temperature by vortexing retinas in 2 ml of the stock solution in Fig. 1. The solution also contained 100 μM $CaCl_2$ and ^{45}Ca with a specific activity of 1.05×10^{14} cpm/mole calcium. The suspension was filtered through a nylon mesh and after a 30-min wait at room temperature was placed atop a chilled metrizamide density gradient and centrifuged at 2°C as before. Fractions were collected as described for Fig. 1. Notice that the time resolution for ^{45}Ca measurement in this experiment was only about 5 min (the time needed for the disks to reach their equilibrium position), since the disks were exposed to the same medium from the time of isolation until the onset of centrifugation.

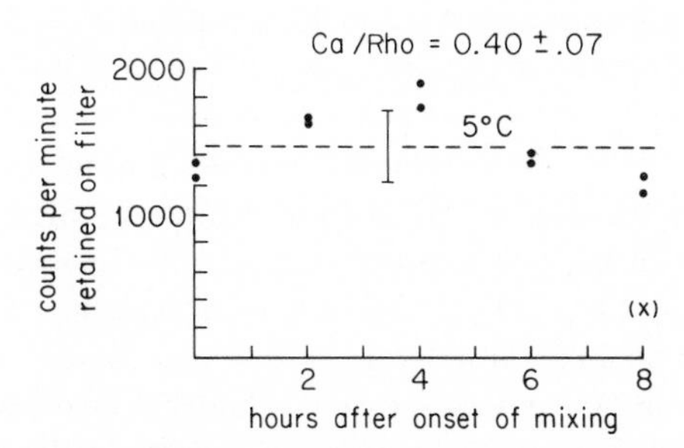

FIG. 3. ^{45}Ca efflux from preloaded disks at 5°C. Purified disks were initially loaded with ^{45}Ca using the ionophore X537A (specific activity during incubation, 1.90×10^{12}cpm/mole calcium). After the removal of extracellular ^{45}Ca and X537A from the membranes, the disks were resuspended in the stock solution in Fig. 1. The solution also contained 100 μM $CaCl_2$. The suspension of disks was divided into 0.2-ml aliquots each containing an average of 1.93 nmoles rhodopsin. At various time intervals, the aliquots were drawn by suction through a 0.45-μm Millipore filter. Disks were retained by the filters so that intradiskal ^{45}Ca could be directly measured. After 8 hr had elapsed, one of the aliquots was warmed to room temperature and mixed with 2.0 osmolal sucrose solution. The suspension of hypertonically shocked disks was filtered 10 min later. The radioactive content of these disks, marked as (x), was used to estimate intradiskal free calcium concentration.

(by either exchange or net loss) were insignificant over 8 hr. Given the measurement errors, ^{45}Ca efflux and hence calcium exchange is calculated to be less than 0.01 fmole cm^{-2} sec^{-1}. The results are consistent with the data in Fig. 2, where the observed exchange rate is less than 0.02 fmole/cm^{-2} sec^{-1}. Therefore it appears that calcium exchange is comparable in both freshly isolated and purified disks. The purification procedure does not appear to alter this property of disks.

With an exchange flux of 0.01 fmole cm^{-2} sec^{-1}, more than 6 hr is required before disks release half of their initial ^{45}Ca. Thus, if the disks in Fig. 1 lost any ^{45}Ca by an exchange process, their loss was less than 6% of their initial value. In conclusion, it appears unlikely that either net loss or exchange could have released significant amounts of ^{45}Ca from the disks of ^{45}Ca-incubated retinas prior to measurement.

V. EFFECT OF INTRADISKAL BINDING ON THE EXCHANGEABLE POOL

Disk lumens may contain calcium-binding sites that in principle could affect both the rate and magnitude of intradiskal exchange. Binding sites separate intradiskal calcium into pools of bound and free ions which are physically distinct from the exchangeable pool measured with tracers. If the turnover rate at the binding sites were much faster than the exchange rate across the disk membrane, tracer calcium would be proportionately well represented among the bound and free ions and the entire intradiskal calcium content would constitute the exchangeable pool. If, however, adsorption–desorption from the sites were rate-limiting, the magnitude of the exchangeable pool could be restricted to the free pool size. For adsorption–desorption to be rate-limiting in experiments with excised retinas, the half-life of the calcium complexes must be comparable to or longer than the 1-hr incubation period. Such a long half-life would be three to four orders of magnitude greater than for other calcium-chelating biological molecules (e.g., troponin; Johnson *et al.*, 1979) and would reflect an extraordinary affinity for calcium. For example, if the half-life of the complex were 1 hr, its affinity would be $\geq 10^{13}$ M^{-1} (calculation based on a diffusion-limited association rate of 2.5×10^{9} M^{-1} sec^{-1}; Burgen, 1966). This predicted affinity is much higher than the measured 10^{4}–10^{5} M^{-1} for the strongly binding sites of bovine disks (Hendriks *et al.*, 1977). More importantly, previous observations with ^{45}Ca-loaded disks do not support the existence of long-lived complexes (Szuts, 1980). When the ionophore X537A was added to a suspension of loaded disks, all intradiskal ^{45}Ca was rapidly released, indicating that the half-life of the complexes were less than a few minutes. Thus calcium binding does not seem to limit the magnitude of the observed exchangeable pool.

VI. LIGHT AND CALCIUM EFFECTS ON INTRADISKAL EXCHANGE

According to the calcium hypothesis, light-induced calcium release and subsequent uptake should increase the turnover or the exchange rate of intradiskal calcium. This prediction was tested in a series of experiments. Control and test retinas were dissected and handled under infrared illumination. While both sets of tissues were incubated in the dark, test retinas were also exposed to repetitive light flashes every 10 sec throughout the 1-hr-long period. The retinas were chilled, and their fragmented outer segments isolated while still under infrared illumination. To increase detectability, contaminating organelles were removed with a 2- to 3-hr-long purification procedure. Based on previous measurements on calcium exchange, disks probably retained more than 95% of their initial content during this lengthy process. As shown in Table II, light stimuli had no significant effect on intradiskal exchange. This was always the case, irrespective of light intensity. Note that, according to the hypothesis, the disks of light-exposed retinas should show an *increased* ^{45}Ca content. Surprisingly, incubating retinas in 10 m*M* calcium, which has been reported to increase cytoplasmic calcium activity about 100-fold (Yoshikami and Hagins, 1973; Wormington and Cone, 1978), also failed to increase the exchange rate. Thus calcium accumulation seems to be insensitive to cytoplasmic calcium activity. For all the experiments listed in Table II, the average exchange was about 0.008 Ca/Rho per hour

TABLE II
EFFECT OF ILLUMINATION AND EXTRACELLULAR CALCIUM ($[Ca^{2+}]_o$) ON ^{45}Ca EXCHANGE IN THE DISKS OF INCUBATED RETINAS[a]

Type of experiment	$[Ca^{2+}]_o$ during incubation (m*M*)	Calcium exchanged per Rho per hour		Light − dark (mean ± SD)
		Dark	Light	
	3	0.0072	0.0062	
Control experi-	3	0.0134	0.0076	
ments with mock	3	0.0050	0.0080	−0.0011 ± 0.0031
illumination	3	0.0119	0.0108	
	3[b]	0.0031	0.0027	
Test experiments	3	0.0059	0.0114	
with flashes that	10	0.0099	0.0091	
photoactivated	3[b]	0.0040	0.0080	−0.0012 ± 0.0082
4×10^4 Rho* per outer segment per flash	3[b]	0.0030	0.0033	
(7×10^3 Rho* per disk per hour)	10[b]	0.0226	0.0075	
Overall mean ± SD		0.008 ± 0.005		

[a] Reprinted by permission of the *Journal of General Physiology* **76,** pp. 253–286. Copyright 1980, The Rockefeller University Press.

[b] Purification performed with metrizamide gradients.

of incubation—a result very similar to that observed in Fig. 1. This agreement strengthens the previous conclusion that only insignificant amounts of intradiskal ^{45}Ca could have been lost as a result of the 2- to 3-hour-long procedure.

VII. CALCULATION OF THE UNIDIRECTIONAL FLUX ACROSS DISK MEMBRANES

The specific structure of the outer segment and the distribution of its total calcium content must be considered before the unidirectional flux across disk membranes can be calculated from the tracer exchange data. For these studies, the organelle can be considered equivalent to two compartments (intradiskal and cytoplasmic) in series with an extracellular reservoir. In accordance with this model, any ^{45}Ca accumulated by the disks must initially pass through the plasma membrane. Ion movement through the ciliary bridge and mitochondrial regions is assumed to be negligible compared to the total amount crossing the outer segment plasma membrane. The rate of exchange into the two subcellular compartments can be mathematically expressed by two simultaneous differential equations requiring six parameters. These are the membrane area, total calcium content, and unidirectional flux of the two compartments. For a bullfrog outer segment with dimensions of 7×70 μm, the total surface area of the disks and of the plasma membrane is respectively, 1.8×10^{-3} and 1.6×10^{-5} cm^2, while the total intradiskal calcium is 0.75 fmole (equivalent to a content of 0.1 Ca/Rho). The total cytoplasmic calcium has not yet been directly measured. For calculation purposes, an upper limit was sufficient and was obtained by subtracting disk content (0.1 Ca/Rho; Szuts and Cone, 1977) from the total content of the outer segment layer of an excised retina (0.7 Ca/Rho; Yoshikami and Hagins, 1976), yielding a value of 0.6 Ca/Rho or 4.5 fmole per outer segment. As shown later, the exact value for the cytoplasmic calcium is actually of little significance in the calculation of disk membrane flux. Given the above numerical values, the solution of the differential equations for various specified fluxes is presented in Fig. 4 in graphical form. Note that the solution for intradiskal exchange reduces to a single (intradiskal) compartment in contact with an external reservoir if unidirectional flux across the plasma membrane exceeds that of the disk membrane by more than three orders of magnitude.

If the unidirectional flux across the plasma membrane were about 1 pmole cm^{-2} sec^{-1}, as in cattle (Schnetkamp, 1979), analysis indicates that about 10% of the intradiskal calcium would be exchanged in 1 hr with a unidirectional disk flux of 0.01 fmole cm^{-2} sec^{-1}. Since the observed exchangeable pool was less than 10%, the flux across disk membranes is commensurately lower. With such a low flux across the disks, the outer segment behaves as a single compartment with

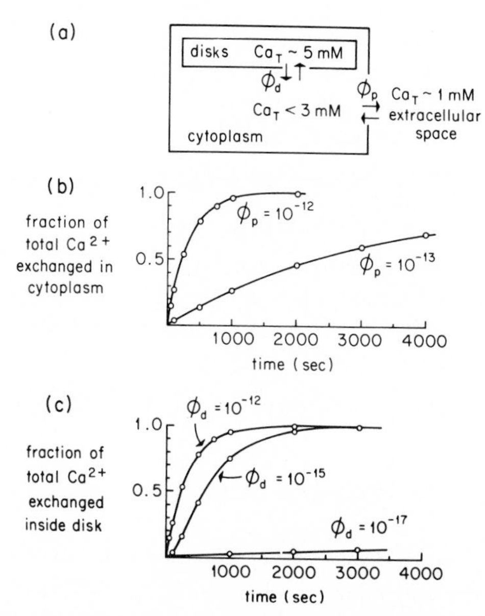

FIG. 4. The equivalent model for tracer exchange in rod outer segments—two compartments in series with a reservoir. (a) Schematic representation of the compartments with their approximate *total* calcium concentrations. (b) Fraction of exchanged cytoplasmic calcium as a function of time and unidirectional flux across the plasma membrane ($\phi_p = 10^{-12}$ or 10^{-13} mole cm^{-2} sec^{-1}). In both cases, unidirectional flux across the disk membranes is set at 10^{-12} mole cm^{-2} sec.$^{-1}$. (c) Fraction of exchanged intradiskal calcium as a function of time and unidirectional flux across the disk membranes ($\phi_d = 10^{-12}$, 10^{-15}, or 10^{-17} mole cm^{-2} sec^{-1}). For all three cases, unidirectional flux across the plasma membrane is set at 10^{-12} mole cm^{-2} sec^{-1}.

The more than 2000 disks of a frog outer segment, presumed to be equivalent in function, can be represented by a single compartment whose surface area and calcium content is proportionately increased. Intradiskal and cytoplasmic exchange rates are interdependent on six parameters: the surface area, total calcium content, and unidirectional membrane flux for each compartment. For a bullfrog outer segment with dimensions of 7×10 μm, the total surface area of disks is 1.8×10^{-3} cm^2, the surface area of the plasma membrane is 1.6×10^{-5} cm^2, the total intradiskal calcium is 0.75 fmole (equivalent to a content of 0.1 Ca/Rho; Szuts and Cone, 1977), and the total cytoplasmic calcium is 4.5 fmoles (this represents an upper limit; for its derivation see text). The aqueous volumes of the cytoplasm and of the cumulative intradiskal spaces are, respectively, 50 and 7% of the total outer segment volume (Chabre and Cavaggioni, 1975). (Reprinted by permission of the *Journal of General Physiology*, **76**, pp. 253-286. Copyright 1980, The Rockefeller University Press.)

disk exchange solely limited by flux across its own membrane. Two consequences of this are that intradiskal exchange would remain relatively unchanged even if (1) total cytoplasmic calcium were lower than the stated upper limit, and (2) flux across the plasma membrane were to decrease by one or two orders of magnitude.

A flux of 0.01 fmole cm^{-2} sec^{-1} across disks corresponds to 5 ions per disk per second or to 0.07 ions μm^{-2} sec^{-1}. The insensitivity of this flux to light sets an upper limit for light-induced calcium release. For the experiments in Table II, the exchange corresponds to a release of less than 3 calcium ions per photoactivated rhodopsin molecule. It is two to three orders of magnitude lower than that predicted by the hypothesis if the stoichiometry of release were 10^2–10^3 ions per activated molecule (Cone, 1973; Yoshikami and Hagins, 1973). As previously stated, exchange was also insensitive to cytoplasmic calcium. Thus, if each disk contained a minimum of one pump, the observed maximum velocity of this putative enzyme would be less than 5 ions/sec, an unusually low rate.

VIII. CURRENT STATUS OF THE CALCIUM HYPOTHESIS

A. Predicted Role of Disks Still Unverified

Neither the light-induced calcium release nor the active calcium accumulation predicted by the calcium hypothesis has yet been verified. The absence of the predicted light-induced release has been thoroughly reviewed elsewhere (Hubbell and Bownds, 1979). What is less well known is that reproducible evidence is equally lacking for active accumulation of calcium—and this property is nearly as crucial a requirement for the calcium hypothesis as light-induced release. Two previously published reports purported to show adenosine 5′-triphosphate (ATP)-supported calcium uptake (Bownds *et al.*, 1971; Schnetkamp *et al.*, 1977). The first observation, however, has since been attributed to subcellular organelles other than disks (Szuts, 1980). The reliability or the interpretation of the second report has been found to be similarly less certain (Schnetkamp, 1979). The study described in this chapter also failed to find any evidence for the presence of the predicted sequestering reaction under physiological conditions. Specifically, the uptake rate was less than 5 ions per disk per second and was insensitive to light stimuli. Such a low rate of accumulation is inconsistent with the calcium hypothesis and implies that light-induced releases from disks are rather small. To be sure, the hypothesis could still be defended against these results by postulating that this study underestimated the exchangeable pool because the cell disruption procedure itself destroyed either a rapid disk exchange or a rapid but limited net loss, which was then undetectable in the fragmented outer segments. Though this could be the case, the more properties disks are postulated to lose on cell rupture (light-induced efflux, active uptake, and large exchange), the less verifiable and credible the predicted role of disks becomes.

In addition to ion flux studies, nucleotide triphosphatase measurements have also failed to supply evidence for a calcium pump. Because phosphate splitting need not reflect the activity of an ion-translocating process, such enzymatic measurements can only form an indirect test of the hypothesis. Frog outer segments appear to possess some type of guanosine 5′-triphosphate (GTP)-dependent phosphatase, since light preferentially hydrolyzes GTP over other nucleotides (Biernbaum and Bownds, 1979; Robinson and Hagins, 1979). It is not yet known what fraction of the overall activity observed in the organelle is actually restricted to the disks. Though Robinson and Hagins interpret the effect of light on GTP levels as evidence for a calcium-sequestering reaction, their light effect could be partially, if not completely, attributed to the guanosine 5′-diphosphate (GDP)–GTP exchange protein that seems to regulate phosphodiesterase activity (Fung and Stryer, 1980; Stryer *et al.*, this volume, Chapter 5).

B. Cytoplasm as Possible Site of Light-Sensitive Pool

For calcium to remain the internal transmitter of phototransduction, I propose that the required light-sensitive pool resides within the cytoplasm in bound form. Although the cytoplasm is known to contain substances that may bind calcium (such as nucleotide- and protein-bound phosphates, along with most of the phosphatidylserines within the outer segments; Dratz *et al.*, 1979), the actual magnitude of the bound calcium is not known.

The proposed modification of the original hypothesis is consistent with most data previously published on calcium fluxes in disks and on extracellular calcium effects in cones (Bertrand *et al.*, 1978; Arden and Low, 1978), with the possible exception of data reported by Yoshikami and Hagins (1978) and by Schnetkamp (1980). The proposal is also compatible with recent observations on light-induced calcium efflux from receptors (Gold and Korenbrot, 1980, and this volume, Chapter 17; Yoshikami *et al.*, 1980). In these experiments the efflux was monitored extracellularly, so that the subcellular source of the ion could not be identified. If the cytoplasm were the ultimate origin of the reported efflux, the size of the light-sensitive bound pool could be estimated from the observed maximum release. Such a calculation would yield a pool size of 10^7 ions per cell (Gold and Korenbrot, 1980) or 0.005 Ca/Rho. This pool would then be somewhat larger than the pool of free cytoplasmic calcium, which in frog is of the order of 10^{-6} M or 0.001 Ca/Rho (Hagins and Yoshikami, 1974; Wormington and Cone, 1978), and would still be only a small fraction of the overall calcium content of the outer segment. The presence of the proposed light-sensitive bound pool remains to be tested. The observation by Wormington and Cone (1978) that calcium ionophores do not abolish light responses even in the presence of low extracellular free calcium suggests that the proposal may have some merit, since calcium-binding sites (in contrast to free calcium concentrations within disks) are expected to be relatively unaffected by ionophores.

In conclusion, the results of this study do not support the calcium hypothesis and specifically suggest that the rate of calcium uptake by the disks of intact receptors is too slow for them to reaccumulate calcium previously released by light. The study suggests that, if calcium is the intracellular transmitter of transduction, the light-sensitive storage site is bound calcium within the cytoplasm rather than free (or bound) calcium within the disks.

REFERENCES

Arden, G. B., and Low, J. C. (1978). Changes in pigeon cone photocurrent caused by reduction in extracellular calcium activity. *J. Physiol. (London)* **280,** 55–76.

Bertrand, D., Fuortes, M. G. F., and Pochobradsky, J. (1978). Actions of EGTA and high calcium on the cones in the turtle retina. *J. Physiol. (London)* **275,** 419–437.

Biernbaum, M. S., and Bownds, M. D. (1979). Influence of light and calcium on guanosine-5′-triphosphate in isolated frog rod outer segments. *J. Gen. Physiol.* **74,** 649–669.

Bownds, D., Gordon-Walker, A., Gaide-Huguenin, A., and Robinson, W. (1971). Characterization and analysis of frog photoreceptor membranes. *J. Gen. Physiol.* **58,** 225–237.

Burgen, A. S. V. (1966). The drug-receptor complex. *J. Pharm. Pharmacol.,* **18,** 137–149.

Chabre, M., and Cavaggioni, A. (1975). X-ray diffraction studies of retinal rods. II. Light effect on the osmotic properties. *Biochim. Biophys. Acta* **382,** 336–343.

Cone, R. A. (1973). The internal transmitter model for visual excitation: Some quantitative implications. *In* "Biochemistry and Physiology of Visual Pigments" (H. Langer, ed.), pp. 275–282. Springer-Verlag, Berlin and New York.

Dratz, E. A., Miljanich, G. P., Nemes, P. P., Gaw, J. E., and Schwartz, S. (1979). The structure of rhodopsin and its disposition in the rod outer segment disk membrane. *Photochem. Photobiol.* **29,** 661–670.

Fung, B. K.-K., and Stryer, L. (1980). Photolyzed rhodopsin catalyzes the exchange of GTP for bound GDP in retinal rod outer segments. *Proc. Natl. Acad. Sci. U.S.A.* **77,** 2500–2504.

Gold, G. H., and Korenbrot, J. I. (1980). Light-induced calcium release by intact retinal rods. *Proc. Natl. Acad. Sci. U.S.A.* **77,** 5557–5561.

Hagins, W. A. (1972). The visual process: Excitatory mechanisms in the primary receptor cells. *Annu. Rev. Biophys. Bioeng.* **1,** 131–158.

Hagins, W. A., and Yoshikami, S. (1974). A role for Ca^{++} in excitation of retinal rods and cones. *Exp. Eye Res.* **18,** 299–305.

Hendriks, T., Van Haard, P. M. M., Daemen, F. J. M., and Bonting, S. L. (1977). Biochemical aspects of the visual process. XXXV. Calcium binding by cattle rod outer segment membranes studied by means of equilibrium dialysis. *Biochim. Biophys. Acta* **467,** 175–184.

Hess, H. (1975). The high calcium content of retinal pigmented epithelium. *Exp. Eye Res.* **21,** 471–479.

Hubbell, W. L., and Bownds, M. D. (1979). Visual transduction in vertebrate photoreceptors. *Annu. Rev. Neurosci.* **2,** 17–34.

Johnson, J. D., Charlton, S. C., and Potter, J. C. (1979). A fluorescence stopped flow analysis of Ca^{2+} exchange with troponin C. *J. Biol. Chem.* **254,** 3497–3502.

Liebman, P. A. (1974). Light-dependent Ca^{++} content of rod outer segment disc membranes. *Invest. Ophthalmol.* **13,** 700–701.

Liebman, P. A. (1978). Rod disk calcium movement and transduction: A poorly illuminated story. *Ann. N.Y. Acad. Sci.* **307,** 642–644.

Noll, G., Stieve, H., and Winterhager, J. (1979). Interaction of bovine rhodopsin with calcium ions.

II. Calcium release in bovine rod outer segments upon bleaching. *Biophys. Struct. Mech.* **5,** 43–53.

Robinson, W. E., and Hagins, W. A. (1979). GTP hydrolysis in intact rod outer segments and the transmitter cycle in visual excitation. *Nature (London)* **280,** 398–400.

Schnetkamp, P. P. M. (1979). Calcium translocation and storage of isolated intact cattle rod outer segments in darkness. *Biochim. Biophys. Acta* **554,** 441–459.

Schnetkamp, P. P. M. (1980). Ion selectivity of the cation transport system of isolated intact cattle rod outer segments: Evidence for a direct communication between the rod plasma membrane and the rod disk membranes. *Biochim. Biophys. Acta* **598,** 66–90.

Schnetkamp, P. P. M., Daemen, F. J. M., and Bonting, S. L. (1977). Biochemical aspects of the visual process. XXXVI. Calcium accumulation in cattle rod outer segments: Evidence for a calcium-sodium exchange carrier in the rod sac membrane. *Biochim. Biophys. Acta* **468,** 259–270.

Sickel, W. (1973). Energy in vertebrate photoreceptor function. *In* "Biochemistry and Physiology of Visual Pigments" (H. Langer, ed.), pp. 195–202. Springer-Verlag, Berlin and New York.

Smith, H. G., Fager, R. S., and Litman, B. J. (1977). Light-activated calcium release from sonicated bovine retinal rod outer segment disks. *Biochemistry* **16,** 1399–1405.

Szuts, E. Z. (1975). Calcium content of the frog rod outer segments: Effects of light and hypotonic lysis. Ph.D. Thesis, Johns Hopkins University, Baltimore, Maryland.

Szuts, E. Z. (1980). Calcium flux across disk membranes: Studies with intact rod photoreceptors and purified discs. *J. Gen. Physiol.* **76,** 253–286.

Szuts, E. Z., and Cone, R. A. (1977). Calcium content of frog rod outer segments and discs. *Biochim. Biophys. Acta* **468,** 194–208.

Wormington, C. M., and Cone, R. A. (1978). Ionic blockage of the light-regulated sodium channels in isolated rod outer segments. *J. Gen. Physiol.* **71,** 657–681.

Yoshikami, S., and Hagins, W. A. (1971). Light, calcium and the photocurrent of rods and cones. *Biophys. J.* **11,** 47a (abstr.).

Yoshikami, S., and Hagins, W. A. (1973). Control of the dark current in vertebrate rods and cones. *In* Biochemistry and Physiology of Visual Pigments" (H. Langer, ed.), pp. 245–255. Springer-Verlag, Berlin and New York.

Yoshikami, S., and Hagins, W. A. (1976). Ionic composition of vertebrate photoreceptors by electron probe analysis. *Biophys. J.* **16,**35a (abstr.).

Yoshikami, S., and Hagins, W. A. (1978). Calcium in excitation of vertebrate rods and cones: Retinal efflux of calcium studied with dichlorophosphonazo III. *Ann. N.Y. Acad. Sci.* **307,** 545–561.

Yoshikami, S., George, J., and Hagins, W. A. (1980). Light-induced calcium fluxes from outer segment layer of vertebrate retinas. *Nature (London)* **286,** 395–398.

Chapter 17

The Regulation of Calcium in the Intact Retinal Rod: A Study of Light-Induced Calcium Release by the Outer Segment

GEOFFREY H. GOLD[1]

Department of Physiology
University of California School of Medicine
San Francisco, California

AND

JUAN I. KORENBROT

Departments of Physiology and Biochemistry
University of California School of Medicine
San Francisco, California

I. INTRODUCTION

The coordination and regulation of many of the processes occurring within a single cell are performed by intracellular messengers. These messengers are generally of two classes: ions, such as Ca^{2+}, H^{+}, and Na^{+}, and small organic molecules, such as 3′,5′-cyclic adenosine monophosphate (cAMP) and cyclic 3′,5′-guanosine monophosphate (cGMP). Where the intricacies of intercellular communication have been investigated, for example, in the sychronization

[1]Present address: Department of Physiology, Yale University School of Medicine, New Haven, Connecticut 06510.

ISBN 0-12-153315-8

of events following fertilization (Epel, 1980; Jaffe, 1980) or cell proliferation (Whitfield *et al.*, 1980), it has been learned that cells can utilize both classes of messengers simultaneously, and also that cells may use more than one member of a single class.

The vertebrate rod photoreceptor represents a class of cells in which the need for an internal messenger is well established. In the rod outer segment, illumination produces an electrical response which is due primarily to a decrease in the Na permeability of the outer segment plasma membrane (Hagins, 1972; Korenbrot and Cone, 1972; Brown and Pinto, 1974). Yet the visual pigment that absorbs light is bound to internal membrane disks that are structurally (Cohen, 1970), osmotically (Korenbrot *et al.*, 1973), and electrically (Hagins and Ruppel, 1971) isolated from the plasma membrane. Because the Na permeability of the rod plasma membrane is reduced by externally applied Ca ions, Yoshikami and Hagins (1971) have proposed that Ca is the internal messenger, or transmitter, responsible for visual transduction. Specifically, they have proposed that Ca released by the disks communicates light absorption in the disk membrane to the Na channels in the plasma membrane. This hypothesis has since been supported by the observation that increased intracellular Ca decreases plasma membrane Na permeability (Hagins and Yoshikami, 1973; Wormington and Cone, 1978) and also that the introduction of Ca chelators into the rod cytoplasm attenuates responses to light (Hagins and Yoshikami, 1977; Brown *et al.*, 1977). While these data suggest that a rise in intracellular free Ca could mediate the rod's light response, evidence is lacking as to whether or not a sufficiently large rise in free cytoplasmic Ca does occur *in vivo* [at least 100 Ca^{2+} per Rh* per second (Rh* is used here to indicate bleached rhodopsin molecules); Cone, 1973; Hagins and Yoshikami, 1975]. Attempts to infer how intracellular Ca is regulated from studies on isolated or reconstituted disk membranes have so far met with little success (see below, and reviews by Hubbell and Bownds, 1979, and Kaupp *et al.*, 1979).

The observation of light-dependent changes in cAMP levels in isolated rod outer segments (Bitensky *et al.*, 1971) introduced the possibility that cyclic nucleotides also act as internal messengers in the rod outer segment. Subsequent studies have shown that cGMP, not cAMP, is the nucleotide that is regulated *in vivo,* and that this regulation is mediated by a light-activated phosphodiesterase (PDE) (Miki *et al.*, 1973; Goridis and Virmaux, 1974; Fletcher and Chader, 1976; Orr *et al.*, 1976; Yee and Liebman, 1978). In addition, elevated intracellular cGMP depolarizes the plasma membrane (Miller and Nicol, 1979, and this volume, Chapter 24), so that activation of PDE, like illumination, should produce membrane hyperpolarization. Therefore cGMP, in addition to Ca, is also a possible candidate for the internal transmitter role in rod phototransduction. This proposition was most quantitatively stated by Liebman and Pugh (1979), who showed that light activation of PDE was both large enough and fast enough to mediate rod transduction. Also, Woodruff and Bownds (1979) have reported large, rapid light-activated changes in the cGMP content of isolated rod outer segments (10^5 cGMP molecules per Rh* in less than 1 sec). However, such large, rapid changes are not observed with measurements of the cGMP content of whole retinas (Kilbride and Ebrey, 1979; Goridis *et al.*, 1977). This variability may reflect different levels of molecular components and cofactors known to regulate PDE (review in Pober and Bitensky, 1979). For example, Liebman and Pugh (1980) have shown that, whereas PDE requires low levels of guanosine 5′-triphosphate (GTP) for activation, adenosine 5′-triphosphate (ATP) acts as an inhibitory factor. Furthermore, at physiological levels of ATP, about 1 m*M* (Biernbaum and Bownds, 1979; Robinson and Hagins, 1979), the extent of PDE activation in isolated disk membranes is reduced by more than 15 times. These more recent observations question the ability of cGMP to mediate rod transduction as proposed in the model of Liebman and Pugh (1979). Therefore the role of cGMP in mediating rod transduction is still uncertain, because the amplitude and time course of cGMP changes *in vivo* are not known with certainty.

It is evident that measurements of the Ca and cGMP levels in intact cells are needed to resolve the roles of these substances in mediating rod phototransduction. We have addressed this problem, with respect to Ca, by measuring the extracellular Ca concentration in the photoreceptor layer of a living retina with an ion-selective electrode. We reasoned that active transport mechanisms in the outer segment plasma membranes might communicate changes in intracellular free Ca to the extracellular space. In fact, we have discovered a large, rapid light-induced efflux of Ca from the rod outer segments in the isolated retina of the toad *Bufo marinus*. These measurements do not directly reveal the changes occurring inside the cell but, from our data, we can infer that a large store of intracellular Ca must rapidly be made available to the cytoplasm in response to light. This release of Ca into the cytoplasm is a required step in the Yoshikami and Hagins hypothesis. However, some of our findings may be inconsistent with their hypothesis. The resolution of these issues will require futher experimentation. Preliminary reports of these findings have appeared elsewhere (Gold and Korenbrot, 1980a,b,c).

II. METHOD OF MEASURING CHANGES IN EXTRACELLULAR CALCIUM

The measurement of the expected small changes in the extracellular Ca concentration required a method of high sensitivity. We found that a new type of Ca-selective electrode (Ruzicka *et al.*, 1973) was very well suited for solving this problem. Our experimental method is described in detail elsewhere (Gold and Korenbrot, 1980c). Briefly, the measurement was made by placing the photoreceptor surface of an isolated toad retina in contact with a planar Ca-selective membrane electrode, the design of which is shown in Fig. 1. The electrode consisted of a Ca-selective poly(vinyl chloride) (PVC) membrane (Ruzicka *et al.*, 1973) glued onto a solid PVC support. The Ca concentration in the space between the retina and the electrode membrane was determined by measuring the potential across the PVC membrane. The PVC membrane potential was measured between two Ag–AgCl electrodes: one, in a Ca standard solution, was in contact with the lower surface of the membrane, and the other, via an agar salt bridge, was in contact with the upper surface of the membrane. The latter served as a reference electrode and was also used in conjunction with a third electrode, positioned above the retina, to record the transretinal potential, or electroretinogram (ERG). The upper electrode, in addition, was used to superfuse the retina with oxygenated Ringer's solution. The experiments were performed by transferring an isolated toad retina, receptor side down, onto the upper surface of the Ca electrode. Since the receptor surface of the toad retina consists almost entirely of rod outer segments, this method is ideally suited for detecting changes in the extracellular Ca concentration produced by the rods. In addition, the volume of solution between the rod outer segments and the electrode surface is very small (29 ± 2 μm thick as measured microscopically), so that the release or uptake of Ca by the rod outer segments may be detected with minimal dilution. Another feature of our experimental design is the low intrinsic noise level (4 μV

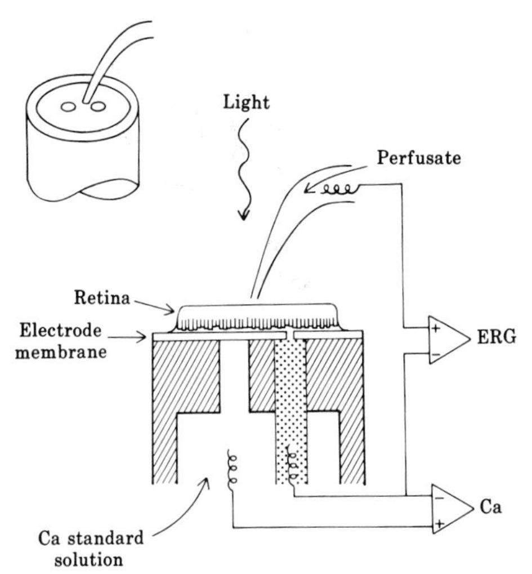

FIG. 1. Diagram of the Ca electrode (cross section) and electrical recording configuration used for measuring changes in extracellular Ca concentration and the receptor potential from the isolated retina (approximately to scale). The PVC electrode support is indicated by hatched shading and the agar–salt bridge for the reference elecetrode is indicated by stippled shading. (Inset) Perspective view of the Ca electrode assembly.

peak to peak, 5 Hz bandwidth) which results from using a macroscopic (1 mm^2 active area), and therefore low-impedance (4–5 $M\Omega$) electrode. The intrinsic noise level of the membrane, in conjunction with the electrode's nearly Nernstian behavior, predicts the minimal detectable concentration change, without signal averaging, to be 2×10^{-7} *M* on a 10^{-3} *M* Ca background. The response time constant of the electrode to step changes in Ca concentration is about 100 msec. Therefore this experimental method has allowed us to measure extracellular Ca changes in the space surrounding the rod outer segments with excellent sensitivity and time resolution.

The control Ringer's solution used in our experiments was modified from conventional toad Ringer's solution (Brown and Pinto, 1974) and consisted of 27.5 m*M* NaCl, 25 m*M* KCl, 63 m*M* choline Cl, 1 m*M* $CaCl_2$, 1 m*M* $MgCl_2$, 5 m*M* Na L-aspartate, 0.13 m*M* $NaHCO_3$, 5 m*M* glucose, and 3 m*M* HEPES, pH 7.4, which was bubbled with 100% oxygen. The aspartate was added to block chemical synaptic transmission between the receptors and second-order neurons (Sillman *et al.*, 1969). This eliminated ERG components arising from second-order neurons and simplified the interpretation of the resulting ERG waveforms. Under these conditions, the ERG consisted only of the isolated receptor potential, which will be described in detail below. The Na and K concentrations were modified from their normal values of 110 and 2.5 m*M*, respectively, in order to minimize ERG contamination of the Ca electrode signal. The existence of this contamination was inferred from the observation that the Ca electrode signal in normal Ringer's solution had the same short latency as the ERG; this is not consistent with the minimum diffusion time needed for Ca released from the retinal surface to reach the electrode membrane. The purpose of our Ringer's solution modification was to reduce the amplitude of the receptor potential, and consequently the electrical contamination of the Ca signal. In the modified Ringer's solution, the receptor potential amplitude was decreased by a factor of 30, and the latency of the Ca signal was then consistent with the time needed for Ca to diffuse to the electrode surface. Therefore, with the modified Ringer's solution, we concluded that ERG contamination of the Ca signal was reduced below detectability.

An important consequence of the decreased external Na concentration in the modified Ringer's solution was that the Ca concentration in the receptor layer, as indicated by the electrode, was less than that in the perfusate. In normal Ringer's solution, containing 110 m*M* Na, the Ca concentrations in the perfusate and in the receptor layer were identical (small differences may have occurred, but would not have been detectable against electrode drift). This apparent Ca uptake by the retina which occurs in decreased extracellular Na may reflect the reduced ability of the retinal neurons to expel Ca via Na–Ca exchange. In the data presented below, the actual Ca concentration in the receptor layer for each experiment is indicated in the figure legends. Our use of a modified Ringer's solution may cause some features of the Ca changes to differ from those in conventional Ringer's solution. However, the similarity between the isolated receptor potential in our modified Ringer's solution and the rod photocurrent in normal Ringer's solution (Baylor *et al.*, 1979) argues that the physiology of the rod was not drastically affected by our modified solution.

III. CHARACTERISTICS OF A LIGHT-INDUCED EXTRACELLULAR CALCIUM INCREASE

The Ca and ERG responses to brief flashes of light are shown in Fig. 2a. The ERG (bottom traces) is typical of the aspartate isolated receptor potential. At dim light levels, the receptor potential reached a peak in about 2 sec and increased in amplitude linearly with light intensity. At higher light levels, this fast component (fast PIII) saturated, revealing the presence of a slower component (slow PIII) which peaked at 5 and 10 sec at the two highest intensities. The slow PIII is thought to arise from the polarization of glial cells by changes in extracellular potassium (Witkovsky *et al.*, 1975; Lurie and Marmor, 1980). In normal Ringer's solution, the receptor potential is dominated by the slow PIII. However, in the modified Ringer's solution, the slow PIII was reduced in amplitude, providing a more accurate record of the photoreceptor currents (e.g., Baylor *et al.*, 1979).

The Ca electrode signal (top traces in Fig. 2a) differed from the receptor potential both in its kinetics and intensity dependence. At dim intensities, the electrode signal indicated a transient increase in extracellular Ca, which peaked in about 5 sec and decayed to the dark level in 15–20 sec. At dim intensities, the peak amplitude was proportional to the stimulus intensity. At higher intensities, the Ca electrode signal began to saturate, but in a way quite different from the receptor potential. For example, the initial rate of Ca efflux saturated at much dimmer intensities than the peak amplitude of the Ca concentration change (Fig. 8). By comparison, the receptor potential exhibited the opposite behavior; i.e., the peak amplitude saturated at much dimmer intensities than the initial rate of rise. In these respects the Ca signal and photocurrent were quite different. The Ca and slow PIII signals exhibited some similarity in time course, but we know of no causal relationship between them, since the slow PIII can be selectively abolished with high K^+, or the addition of Ba^{2+} (Bolnick *et al.*, 1979), with little effect on the Ca signal.

Because the Ca electrode was not indifferent to the electrical activity of the retina (see above), and also because the electrode could respond to ions other than Ca, we endeavored to demonstate that the Ca electrode signals did in fact reflect changes in the extracellular concentration of Ca. This was done in three ways. First, we confirmed that the amplitude of the Ca electrode signal was

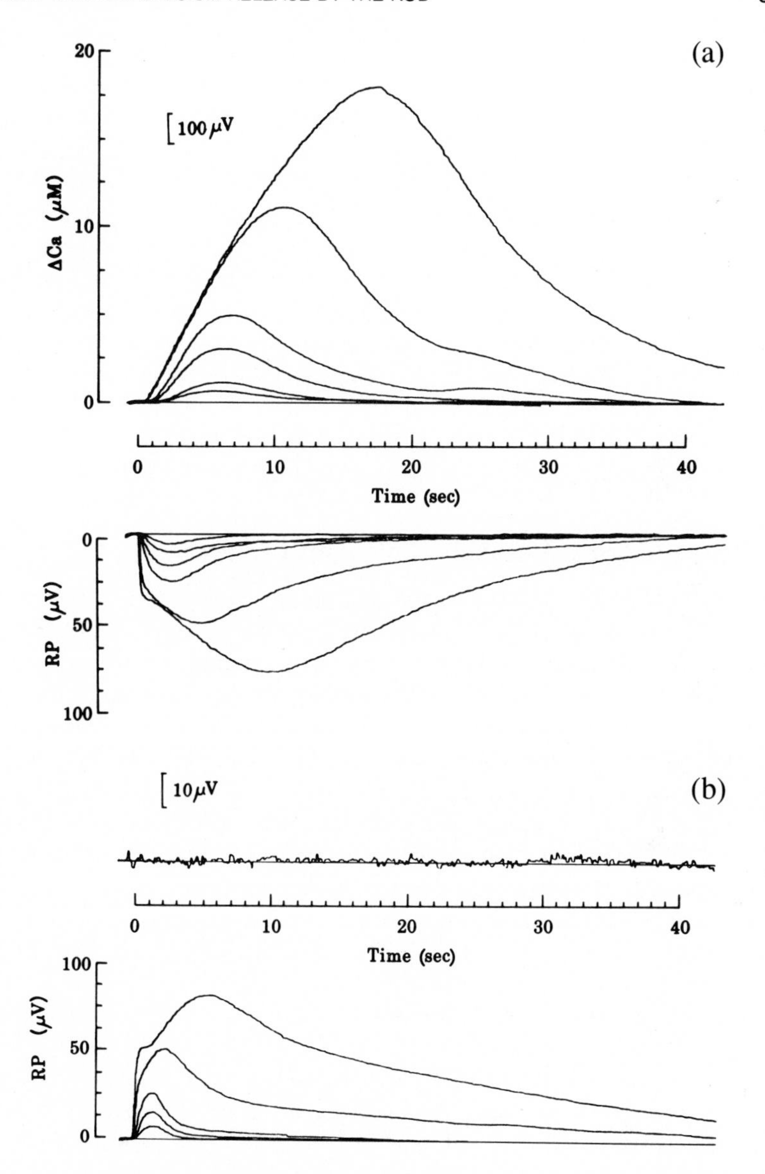

FIG. 2. Responses to 20-msec flashes of 500-nm light in a 0.2 m*M* Ca background. The upper trace shows the change in extracellular Ca concentration, and the lower trace shows changes in the receptor potential (RP). Shown in both sets of traces are responses to 42, 113, 216, 544, 7×10^3, and 94×10^3 absorbed photons per rod. (a) Retina receptor side down; (b) retina receptor side up (inverted). Note that the kinetics of the RP were markedly faster when the retina was inverted. This probably reflects the better metabolic state of the receptors when they are in direct contact with the oxygenated Ringer's solution.

attenuated by the Ca buffer nitrilotriacetic acid (NTA) to the extent quantitatively predicted by the measured Ca-buffering capacity NTA (see Gold and Korenbrot, 1980c, Fig. 3). We further established that the Ca electrode signalled changes in Ca (and not Mg) by exploiting the logarithmic behavior of the ion-selective electrode; i.e., the sensitivity of the electrode to a given ion is inversely proportional to the background concentration of the ion. Complete omission of Mg from the Ringer's solution did not affect the electrode response or the receptor potential. Thus the electrode did not respond to Mg, a possibility made unlikely by the high selectivity of the electrode ligand for Ca over other divalent cations (Ruzicka *et al.*, 1973). Moreover, when the background concentration of Ca was varied between 1 and 0.1 m*M*, the electrode signal (peak voltage amplitude) was inversely proportional to background Ca concentration. Thus the electrode indeed responded to changes in extracellular Ca, and illumination of the retina resulted in a transient increase in the extracellular Ca concentration.

A. Source of the Extracellular Calcium Changes

The records in Fig. 2a demonstrate that illumination leads to a transient increase in the extracellular Ca concentration along the distal (receptor) surface of the retina. We demonstrated that these changes were unique to the receptor side of the retina by transferring a retina onto the electrode with the receptor side facing up rather than down. In this case, the polarity of the receptor potential was inverted and the Ca electrode did not generate a detectable response (Fig 2b). Therefore the extracellular Ca changes must arise from structures along the distal surface of the retina. However, this surface consists of both the inner and outer segments of the red rods, green rods, and cones, as well as terminations of the retinal glial cells (see Gold, this volume, Chapter 4). The exact source of the Ca changes was investigated as follows. The action spectrum for the Ca increase was measured to determine which photoreceptor types initiated the changes. In addition, we analyzed the time required for Ca to diffuse from its source to the electrode surface. Both methods showed that the major source of the Ca increase was the red rod outer segments.

The three receptor types in the toad retina may be readily distinguished by their different visual pigments and corresponding action spectra. The maximal absorbance and sensitivity for these cell types occur at the following wavelengths: red rods at 502 nm, green rods at 433 nm, and cones 575 nm (Harosi, 1975).[2] A typical action spectrum for the peak of the Ca concentration change is shown in Fig. 3. The data exhibit a peak at about 500 nm and are much closer to fitting the absorbance curve of the red rods than to fitting the absorbance curves of either

[2]The accessory member of double cones may have a pigment with λ_{max} = 502 nm (Liebman and Entine, 1968), but cone contributions are ruled out by the latency analysis described below.

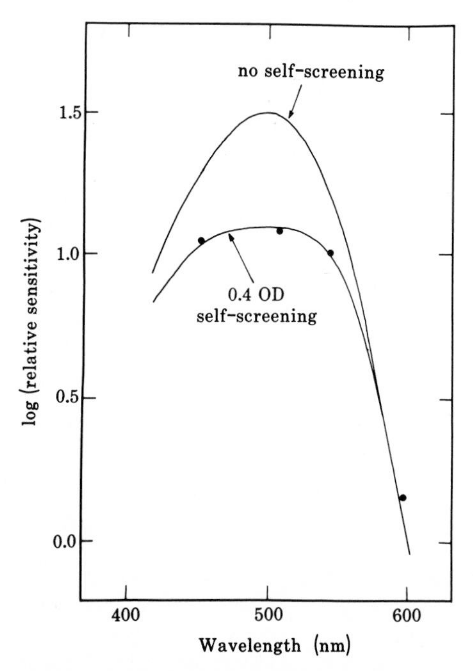

FIG. 3. Action spectrum of the Ca increase. Experimental data (solid circles) were obtained by determining the amplitude of the Ca increase in response to a fixed number of incident photons as a function of wavelength (all measurements were made within the proportional response range). The predicted action spectrum in the absence of self-screening was derived from a Dartnall nomogram with maximal absorbance at 502 nm. To correct for the 0.4 OD of pigment self-screening, the Dartnall function was multiplied by $10^{-0.4A(\lambda)}$, where $A(\lambda)$ is the Dartnall function normalized to 1.0 at 502 nm.

the green rod or cone photopigments. However, the data do not fit the absorbance curve of the red rod photopigment as it is measured in dilute solution (curve denoted "no self-screening"). Rather, the data fit a curve obtained by correcting for 0.4 OD of self-screening.[3] Self-screening occurs in the outer segment because it contains a high concentration of visual pigment (about 3 mM), which at 500 nm results in a total axial absorbance of 1.2 OD (Harosi, 1975). Since light reaching the distal end of the outer segment must pass through this pigment, its spectral content will be distorted by absorption in the proximal portion of the outer segment. The Ca action spectrum fits a Dartnall nomogram corrected for 0.4 OD of self-screening, i.e., a nomogram multiplied by the attenuation due to 0.4 OD of visual pigment. Therefore the Ca that contributes to the peak of the

[3]Self-screening is here defined as the effect of light absorption in one part of the rod outer segment on the spectral content of light in other parts of the rod outer segment.

response must originate in the distal two-thirds of the outer segment. In contrast, self-screening has not been observed in the electrical responses of the rods (Fain, 1976). This occurs primarily because all parts of the outer segment contribute equally to the electrical response, whereas the Ca electrode, due to diffusion effects, should be most sensitive to Ca changes produced at the distal end. We expect that, during axial illumination, local photocurrents generated in the distal part of the outer segment exhibit self-screening similar to that described here. There is little doubt, in comparing the action spectrum data with the absorbance curves of the green rod and cone photopigments, that the Ca signals are dominated by the red rods. However, these data do not exclude the possibility that other receptors also release Ca. For example, since there are six times fewer green rods than red rods, an equal Ca release per cell from the green rods would not significantly alter the observed action spectrum.

The action spectrum data plus our interpretation of the self-screening effect, argue that the Ca changes originate primarily in the red rod outer segments. We have confirmed this by spatially localizing the source of the Ca concentration changes. This was done by measuring the latency of the Ca signal and comparing it with the estimated time required for Ca to diffuse from the red rod outer segments to the electrode surface. The two are in excellent agreement. Figure 4

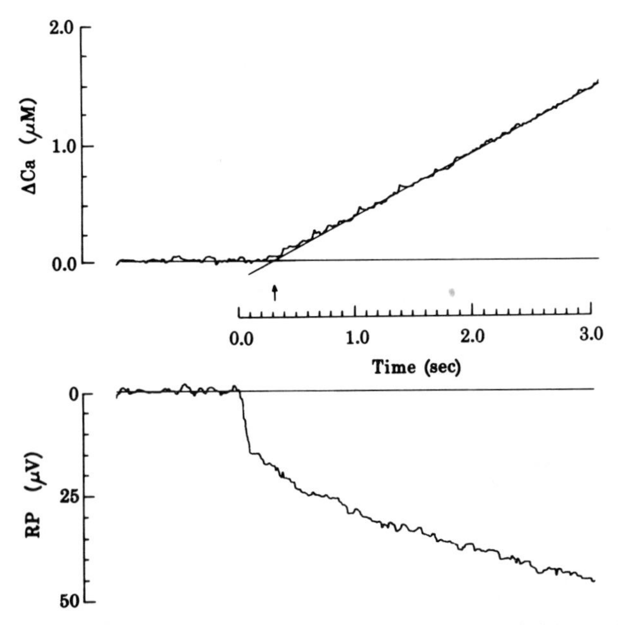

FIG. 4. The Ca signal and receptor potential (RP) shown on an expanded time scale to illustrate their different latencies. The 20-msec stimulus flash of 500-nm light delivered 10^5 absorbed photons per rod. The arrow indicates the latency of the Ca signal as defined in the text. The background Ca concentration was 0.2 m*M*.

shows that, at nearly saturating intensities, the receptor potential exhibited a latency of about 25 msec, whereas the Ca electrode signal began to rise linearly after a delay of about 320 msec. The linear rise in the Ca electrode signal persisted for as long as 15 sec (Fig. 2a), so that the Ca concentration within the receptor layer must have been increasing linearly over approximately the same time interval. With these facts, the diffusion time necessary for Ca to reach the electrode surface may be estimated in the following way. It can be shown that, if the Ca concentration begins to rise linearly at time $t = 0$ along a planar source and diffuses to a planar barrier (the electrode surface) a distance x away, then the Ca concentration at the barrier will begin to rise linearly with a latency t given by $t = x^2/2D$ (Williamson and Adams, 1919). The latency is defined by the intercept between the linearly rising portion of the signal and the initial baseline (arrow in Fig. 4). With a diffusion coefficient D for Ca of 10^{-5} cm^2/sec (Robinson and Stokes, 1959), the experimentally observed mean latency of 0.52 ± 0.22 sec indicated that the source was 36 ± 9 μm away from the electrode surface.

The actual separation between the red rod outer segments and the electrode surface was measured by transferring an isolated retina onto a separate piece of the PVC membrane, just as in the electrophysiological experiments, and viewing it from below with an inverted microscope. This procedure also confirmed that the majority of the rod outer segments had remained intact and well oriented following the transfer. By focusing between the tips of the red rod outer segments and the membrane surface, a separation of 29 ± 2 μm was determined. The small variability of this separation probably reflects the fact that, although representing only about one-sixth of the rods, the green rods in the toad retina project 20 ± 5 μm beyond the tips of the red rods and must help to support the retina. This measured distance between the red rod outer segments and the electrode surface is within experimental error of the distance predicted from the Ca signal latency. Therefore the red rod outer segments must be the source of the Ca increase. Because the diffusion delay increases with the square of separation, the observed latency also rules out contributions to the initial wave of Ca from deeper parts of the retina. For example, if the Ca source were in the rod inner segments, an additional 50 μm away from the electrode, the signal would have a latency of at least 3 sec.

The Ca latency may also be used to estimate the delay between the light flash and activation of the Ca efflux. If the source of the Ca is assumed to coincide with the tips of the outer segments, then the time for Ca to reach the electrode surface would be 0.41 ± 0.03 sec. Since the observed latency is 0.52 ± 0.22 sec, this determines the latency for initiation of the Ca increase to be 0.11 ± 0.18 sec, i.e., between 0 and 0.22 sec ($p > 0.95$). Therefore, in the dark-adapted toad rod, initiation of the Ca increase follows closely and may even precede the changes in membrane potential. Recent measurements of the Ca efflux by Yoshikami *et al.* (1980) also indicate that its latency is similar to that of the receptor potential.

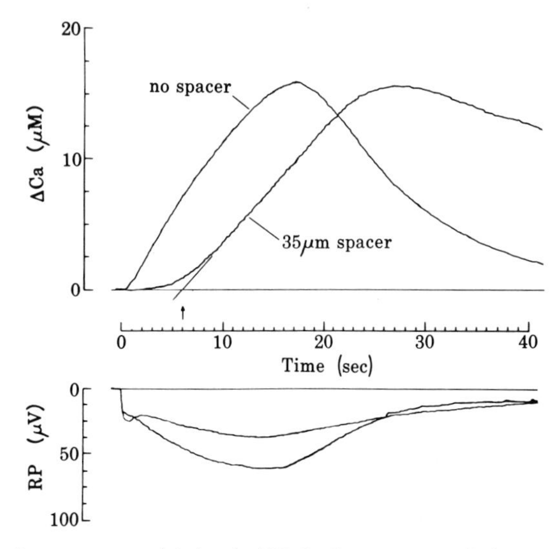

FIG. 5. Ca and receptor potential signals (RP) in the presence and absence of a 35-μm-thick cellulose acetate spacer. The spacer was placed between the retina and the electrode to demonstrate that the latency of the Ca signal was diffusion-limited. The Ca signal obtained in the presence of the spacer has been enlarged to match the peak amplitude of the signal in the absence of the spacer.

The above analysis rests on the assumption that Ca diffuses passively toward the electrode surface. This assumption was shown to be valid by placing an additional diffusion barrier between the retinal surface and the electrode. This barrier consisted of a 35-μm-thick piece of hydrated cellulose acetate in which we measured the Ca diffusion coefficient to be 1.2×10^{-6} cm^2/sec (determined by measuring the electrode response time to a step change in Ca concentration on top of the cellulose membrane (Carslaw and Jaeger, 1948). In the presence of this barrier, the kinetics of the receptor potential were unaffected (Fig. 5); in contrast, the latency of the Ca electrode response increased to 6 sec. This is in excellent agreement with the 5.6-sec latency predicted from the latency equation described above. Therefore the Ca changes we observed were due to an increase in Ca concentration, which occurred in the layer of the red rod outer segments and diffused passively toward the Ca electrode.

B. Mechanism of the Extracellular Calcium Changes

The mechanism of the light-dependent Ca increase has not yet been determined. Nonetheless, our data suggest that the increase reflects the activation of a Ca efflux mediated by active outward transport. The light-activated appearance of Ca in the extracellular space could be due either to the initiation of a Ca efflux or to the cessation of a Ca influx existing in parallel with a light-independent efflux. If the latter alternative were true, Ca would most likely enter the outer segments through the light-dependent Na channels which are open in the dark and closed in the light. To investigate this possibility, we blocked the Na channels (and the light response) by increasing the extracellular Ca concentration (Yoshikami and Hagins, 1973; Brown and Pinto, 1974). In modified Ringer's solution, when the extracellular Ca concentration was increased from 0.2 to 0.88 m*M*, the amplitude of the receptor potential was reduced to 12% of the control amplitude (Fig. 6). In contrast, the amplitude of the Ca increase was 65% of the control amplitude. In addition, high Ca prolonged the duration of the Ca increase but had little effect on the duration of the receptor potential (see also Fig. 10 for Ca dependence of the time to peak). Taken together, these results are not consistent with the Na channels being the sole cause of the Ca increase, although they may contribute to the total increase observed. Therefore the increase in extracellular Ca must be due largely to light activation of a Ca efflux.

The low intracellular free Ca concentration in the rod outer segments, about 1 μM (Hagins and Yoshikami, 1973; Wormington and Cone, 1978), requires that the Ca efflux be mediated by active outward transport. As in other cells of the nervous system (Blaustein, 1974), active outward Ca transport is accomplished primarily through a Na–Ca exchange process, and to a lesser extent through a

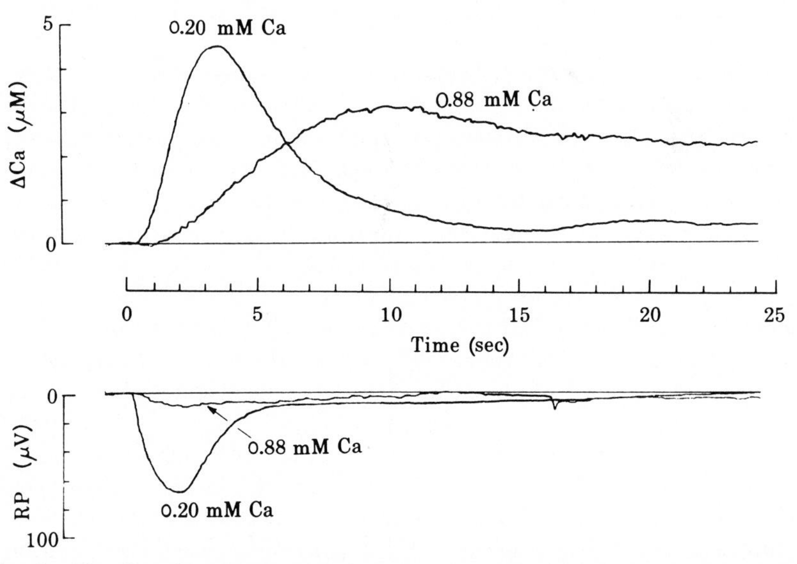

FIG. 6. The effect of varying the extracellular Ca concentration on the responses of the isolated retina. The Ca concentration in the perfusate was increased from 1 to 6 m*M* in order to increase the extracellular Ca concentration in the receptor layer from 0.2 to 0.88 m*M*. Shown are signals recorded in response to flashes that delivered 600 absorbed photons per rod. The apparent lack of return to the dark Ca level in 0.88 m*M* Ca may be due to electrode drift.

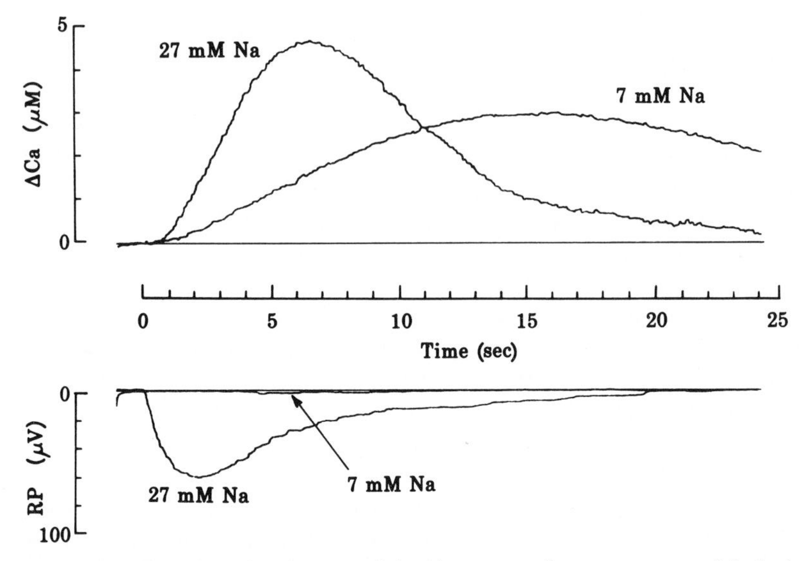

FIG. 7. The effect of varying the extracellular Na concentration on responses of the isolated retina. Shown are signals recorded in the presence of 27 or 7 m*M* NaCl and 0.6 m*M* extracellular Ca. The stimulus intensity was 600 absorbed photons per rod.

Na-independent mechanism. The rate of Ca efflux is reduced considerably when external Na is lowered from 27 to 7 m*M* (Fig. 7). This is indicative of a Na–Ca exchange (Blaustein, 1974). However, in the absence of added external Na, Ca release can still be detected. Thus a Na-independent mechanism for active Ca transport may exist in parallel with the Na–Ca exchange.

Schnetkamp (1980) has presented evidence for the existence of a Na–Ca exchange in the plasma membranes of isolated bovine rod outer segments. Biernbaum and Bownds (1979) have reported a light-activated guanosinetriphosphatase (GTPase)[4] in isolated frog outer segments, which reduces the GTP content of isolated outer segments by up to 70% within about 20 sec following illumination; additional evidence has led them to suggest that this GTPase is a Ca pump in the rod plasma membrane. Robinson and Hagins (1979) reported a similar light-activated GTPase in isolated frog rod outer segments and argued that the changes in GTP levels they observed were due to a Ca transport process that reloaded Ca into the disks. It is of course possible that part of the GTP hydrolysis reported by Robinson and Hagins was involved in the active extrusion we observed; this phenomenon was unknown at the time of their suggestion.

[4]Note that this GTPase is distinct from the GTPase associated with the activation of PDE (Pober and Bitensky, 1979).

Thus biochemical evidence supports the proposition that parallel Na-dependent and Na-independent mechanisms for active Ca transport may exist in the plasma membrane. Unfortunately, Schnetkamp's data were not expressed as absolute flux rates nor was their light sensitivity investigated, so we cannot estimate the contribution his Na–Ca exchange would make to the efflux reported here. On the other hand, both reports of GTPase activity reveal GTP hydrolysis rates of 100–200 GTP per Rh* per second, which could only account for about 0.5% of the transport we observed. Therefore these data suggest that Na–Ca exchange accounts for most of the transport we observed.

The mechanisms by which the extracellular Ca concentration returns to the initial dark level can be suggested based on our data. Two extreme cases are that the released Ca was either taken back up by the rod outer segments or that it simply diffused away via the extracellular space. If Ca diffused away, the time course of the Ca changes would depend solely on the geometry of the extracellular space and not on physiological parameters. The fact that the Ca kinetics depend on extracellular Na and Ca (Figs. 6 and 7) argues that much if not all of the Ca is taken back up by the retina. However, if all the released Ca were taken up on the same time scale as the Ca signal, the net flux of Ca toward the electrode would be small relative to the plasma membrane Ca fluxes. In this case, most of the Ca reaching the electrode membrane by necessity would originate from the rod outer segment tips, and the effective self-screening would greatly exceed the 0.4 OD we observed. In reality, both diffusion and reuptake must contribute to the observed time course of the Ca signal. It may seem surprising that the extracellular Ca returns so precisely to its dark level. Actually, this is not unreasonable if the return of the retina to its fully dark-adapted state necessitates the return of intracellular Ca pools to their initial levels. Therefore we suggest that the decaying phase of the Ca signal may be related to dark adaptation of the rods. This interpretation of course implies that the released Ca is being taken up by the rod outer segments.

IV. EXTRACELLULAR CALCIUM CHANGES AND PHOTOTRANSDUCTION

The fact that the Ca efflux is both large and rapidly initiated by light is not sufficient to establish its role in generating the rod's electrical response. Nevertheless, several of its features provide important insights. For example, the Ca efflux persists even in the absence of the receptor potential [in low Na (Fig. 7) or in high Ca (not shown)]. Therefore the Ca efflux is not simply a consequence of the electrical response and may reflect an earlier stage in the transduction process. This conclusion is also supported by the intensity dependence of the Ca signal and receptor potential as shown in Fig. 8. In this figure the peak Ca

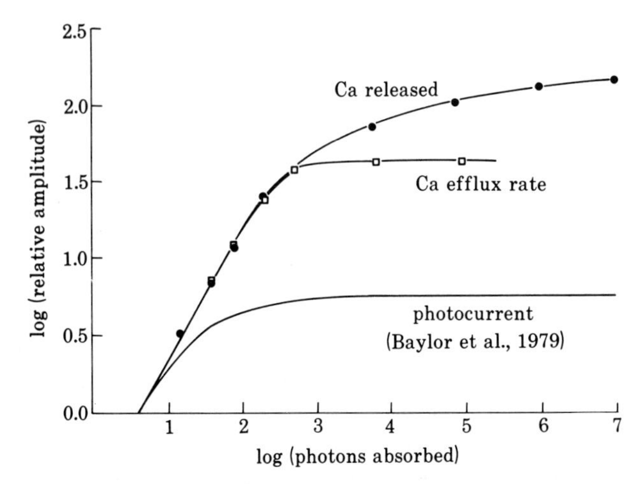

FIG. 8. The intensity dependence of Ca release and the Ca efflux rate. At log(relative ampiitude) = 1.0, the Ca release was 5.6×10^5 Ca^{2+} per rod and the efflux rate was 3×10^5 Ca^{2+} per rod per second. For comparison, the data of Baylor *et al.* (1979) on the relationship between the normalized peak amplitude of the photocurrent and light intensity is also shown. The vertical position of this curve is arbitrary.

concentration change and maximal Ca efflux rate are plotted as a function of stimulus intensity and are compared with the peak rod photocurrent. For the photocurrent, we have used the current measurements of Baylor *et al.* (1979) rather than our receptor potential measurements, because the latter may be contaminated by slow PIII or cone contributions. At the dimmest light intensities all three signals are proportional to light intensity, as indicated by a slope of 1 on these log–log coordinates. However, the peak Ca concentration change and the maximum efflux rate continue to increase linearly with intensity, well above intensities that begin to saturate the photocurrent. This suggests that the Ca efflux reflects a process occurring prior to control of Na permeability. Therefore saturation of the photocurrent may simply reflect saturation of sites that regulate the Na permeability of the plasma membrane, which was suggested several years ago by Baylor and Fuortes (1970).

A key feature of the internal transmitter hypothesis is that it provides high gain to account for the high sensistivity of the rod photoreceptor. In the range of proportional Ca responses, we found that the peak concentration change was $2.1 \pm 0.8 \times 10^{-8}$ *M* Ca per photon per rod. Assuming that the concentration increase occurred uniformly in the layer of solution between the rod outer segments and the membrane electrode, which was about 30 μm thick, and taking the density of red rods to be $1.7 \times 10^6/cm^2$ (Fain, 1976), the measured change corresponded to $2.1 \pm 0.8 \times 10^4$ Ca^{2+} per Rh* per rod. This amount of Ca is over 10 times greater than previous estimates of the minimum number of transmitter molecules necessary for the transduction process. However, this number is even more surprising because it must underestimate the Ca concentration changes occurring inside the cell. Although the Ca changes we observed may appear unreasonably large, there is no difficulty in accounting for the maximal Ca release from a single flash (about 10^7 Ca^{2+} per rod) in terms of the estimated Ca content of frog rods (about 10^9 Ca^{2+} per rod: Hagins and Yoshikami, 1975; Szuts and Cone, 1977).

The amplitude and kinetics of the extracellular Ca changes, as well as their independence of the receptor potential, suggest that these changes are intimately involved in the transduction process. The magnitude of the Ca release we observed indicates that a pool of sequestered Ca must enter the cytoplasm in the light. The intracellular free Ca concentration in the outer segment cytoplasm is not known precisely, but Hagins and Yoshikami (1973) in the rat, and Wormington and Cone (1978) in the frog, have estimated it to be about 1 μM from titration of the plasma membrane Na permeability by intracellular Ca. In a toad rod outer segment [6 μm diameter and 50 μm long, and assuming about half of this volume to be taken up by the disks (Korenbrot *et al.*, 1973)], this corresponds to about 25×10^4 Ca ions. Given the magnitude of the single-photon response, a flash delivering about 10 photons would exhaust the cytoplasmic pool of free Ca. Yet we have observed that the Ca release increases linearly up to several hundred absorbed photons and nonlinearly above that. Thus a pool of sequestered Ca must be made available in the cytoplasmic space in response to illumination. However, it could be argued that the source of this intracellular Ca is a Ca-binding ligand and that this ligand keeps the free Ca concentration nearly constant by virtue of its buffering capacity. This possibility cannot be ruled out but seems unlikely in light of the pronounced effects of artificially introduced Ca buffers (Brown *et al.*, 1977; Hagins and Yoshikami, 1977). The results of the buffer experiments in fact argue that a change in intracellular free Ca is a necessary step in the transduction process.

If it is assumed that the Ca efflux reflects an increase in intracellular free Ca, the origin of this Ca remains to be established. Yoshikami and Hagins (1971) originally proposed that Ca was sequestered in the disk interior and released by a light-induced increase in the disk membrane permeability to Ca. The inability to demonstrate light-induced Ca fluxes across isolated disk membranes of sufficient magnitude and appropriate kinetics remains one of the major experimental failures of the hypothesis. A multitude of studies (reviews in Hubbell and Bownds, 1979; Kaupp *et al.*, 1979; Nöll *et al.*, 1979) have attempted to measure light-dependent release of Ca trapped within isolated disk membrane vesicles. Various membrane preparations, ranging from freshly dissected and unpurified to frozen and extensively purified disk membranes, combined with various Ca detection techniques have yielded results indicating either no release or a release that is too small ($^1/_{70}-1$ Ca^{2+}/Rh^*) or too slow to be of relevance in visual excitation.

Under the premise that 85–90% of the disk membrane protein is rhodopsin, and pursuing the hypothesis that rhodopsin may be a light-activated ionic channel, reconstituted membranes consisting of purified rhodopsin and phospholipids have also been investigated. Initial studies reported that such membranes exhibited large light-dependent Ca permeability changes (Hubbell *et al.*, 1977; Darszon *et al.*, 1977; O'Brien, 1979). These data were taken to support the proposition that rhodopsin molecules act as light-activated ionic channels. However, these initial studies did not investigate the kinetics of the permeability changes. Therefore we measured the time course of the light-dependent Ca fluxes across reconsitituted rhodopsin–phospholipid membranes in order to assess the physiological significance of this phenomenon (Gold and Korenbrot, 1979). Some of our data are shown in Fig. 9 (experimental methods are described in the figure legend). While we agree with others that illumination results in the release of Ca from these membranes, we find the properties of this release to be inconsistent with a role in transduction, i.e., the release has a latency of several seconds and a half-time at the highest intensity of 3-5 min. Analysis of the data in Fig. 9

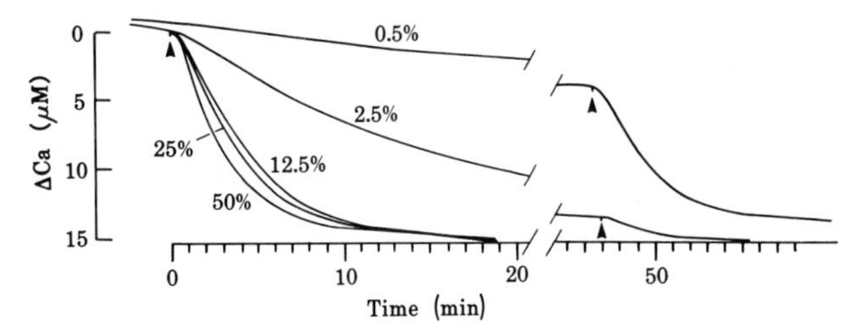

FIG. 9. The time course and intensity dependence of the light-induced Ca efflux from membrane vesicles containing purified cattle rhodopsin. The sample was illuminated at the times indicated by the arrows with 25-msec flashes of yellow light. The flash intensity is indicated in the figure by the percentage of rhodopsin bleached. At the two dimmest intensities, not all of the Ca was released after 50 min. The fraction of Ca remaining within the vesicles was consistent with the number of vesicles in which no rhodopsin was bleached, as predicted by the Poisson distribution (each vesicle contained about 100 rhodopsin molecules). Therefore it appears that bleaching a single rhodopsin molecule produces a permanent increase in membrane permeability. The rhodopsin-phosopholipid recombinants were prepared by the method of Hong and Hubbell, as described by Korenbrot and Pramik (1977). Single-walled vesicles were then prepared in a 10 mM $CaCl_2$ salt solution by the cholate dialysis technique of Brunner *et al.* (1976). Ca was then removed from the extravesicular space by exclusion chromatography (Sephadex G-50). The release of Ca into the extravesicular space was monitored by adding the Ca indicator dye arsenazo III to the vesicle suspension [arsenazo III was purified chromatographically by the method of Kendrick (1976), followed by acid precipitation (Savvin, 1961); acid precipitation was necessary to remove all of the chromatographic solvents]. The kinetics of the Ca release were determined by monitoring the sample absorbance at 650 nm with a kinetic flash photometer. Absorbance changes were converted into Ca concentration changes using our value of the Ca–arsenazo III extinction coefficient, $e_{650} = 1.4 \times 10^{-4}\ M^{-1}\ cm^{-1}$, measured in an excess of dye.

reveals a transport rate of 0.1 Ca^{2+} per Rh* per second. These charcteristics indicate that the permeability change mediated by rhodopsin in reconstituted membranes is not relevant to the transduction process. Furthermore, this conclusion does not necessarily reflect deleterious effects of the reconstitution procedure, since similar results were also obtained from fresh, sonicated disk membranes. This rate of transport is 10^4–10^5 times smaller than that observed in other Ca channels (Krishtal *et al.*, 1980) and therefore is inconsistent with rhodopsin functioning as a light-activated channel in these preparations.

Our observations on the intact retina suggest either that the isolation and manipulation of disk membranes result in the loss of components critical to the release of Ca or that the light-regulated Ca pool is not sequestered within the disks (see Szuts, this volume, Chapter 16). The former conclusion would not be surprising in light of recent evidence concerning the structure of the disk membrane surface. Although the a majority of the disk membrane protein is rhodopsin, the remaining 10–15% includes a number of loosely bound enzymes known to be regulated by rhodopsin. Since most membrane isolation procedures result in the loss of some or all of these other proteins, it is not surprising that these isolated membranes fail to exhibit all the physiological properties they do *in vivo* (Godchaux and Zimmerman, 1979; Kuhn, 1980). Since the disks are much more complex structures than originally supposed, much greater care will need to be taken before their function can be studied *in vitro*.

It is evident in all the traces shown that extracellular Ca changes occur on roughly the same time scale as the receptor potential; this suggests a possible relationship between them, and this point will be further refined. According to the original hypothesis of Yoshikami and Hagins (1971; Hagins, 1972), the recovery of the receptor potential after a brief flash of light reflects the pumping of released Ca back into the disks (and, as we now know, into the extracellular space as well). Therefore, according to their hypothesis, the time course of the Ca efflux and the receptor potential must be causally related. In particular, we expect that the efflux rate, i.e., the slope of the Ca concentration signal, will reach a maximum at about the same time that the receptor potential reaches its peak amplitude. Similarly, it might be expected that the receptor potential will return to its dark level at about the same time that all of the Ca has been pumped out, i.e., when the Ca signal reaches it maximum amplitude. These predictions are in fact supported by the traces in Fig. 6, obtained in 0.2 mM Ca. This observation also agrees with the findings of Yoshikami *et al.* (1981), who used a Ca-selective microelectrode to detect extracellular Ca changes in the rat retina on Ca backgrounds of 100 μM or less. However, in 0.88 mM Ca, the kinetic relationship between the Ca efflux and receptor potential is quite different. In fact, it appears that the Ca efflux rate is nearly maximal after the receptor potential (and presumably intracellular Ca) has returned to its dark level. The fact

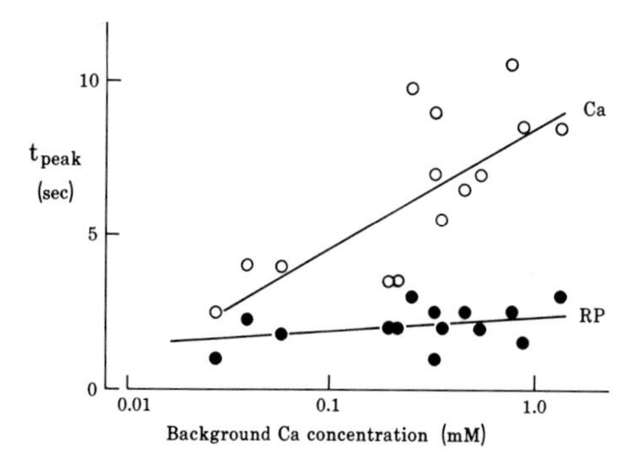

FIG. 10. The time to peak of the Ca signal [open circles] and receptor potential [solid circles] as a function of the extracellular Ca concentration. The solid lines were fit by eye. The stimulus intensity was 600 absorbed photons per rod.

that the kinetics of the Ca efflux and receptor potential exhibit different dependences on extracellular Ca is further illustrated in Fig. 10, which shows the time to peak of the receptor potential and the Ca signal as a function of extracellular Ca concentration. There is no significant change in receptor potential kinetics over the range of 1–0.03 m*M* Ca. The weak Ca dependence reported elsewhere (see Miller and Nicol, this volume, Chapter 24) is within the variability of our data (each point in Fig. 10 represents a measurement from a single retina; the variability reflects variability among retinas). However, there is a significant trend in the time to peak of the Ca signal, which reflects the dependence on extracellular Ca seen in Fig. 6. The relative independence of the Ca and receptor potential kinetics have two possible explanations: Either Ca in fact controls the membrane permeability but its efflux rate is unrelated to the kinetics of intracellular Ca changes, or alternatively, the efflux rate reflects the intracellular Ca concentration changes and factors other than Ca also control the membrane permeability. New experimental data are needed to resolve this issue.

In summary, our extracellular Ca measurements lead to the almost inescapable conclusion that light regulates the intracellular Ca concentration in the vertebrate rod. In particular, illumination leads to an efflux of Ca from the rod outer segment that is sufficiently large and rapid to play a role in phototransduction. Future work must elucidate the specific relationship between the Ca efflux and changes in intracellular free Ca. Intracellular Ca can regulate the Na permeability, hence the membrane potential of rods. However, the fact that cGMP can also produce this effect, and the relative independence of extracellular Ca and receptor potential kinetics, raises doubt as to how simple a model may be needed to explain phototransduction.

ACKNOWLEDGMENTS

This work was supported by National Institutes of Health grants EY-01586, EY-00050, and EY-05286. G.H.G. received a postdoctoral research fellowship from Fight-for-Sight, Inc., New York City. J.I.K. is a Sloan Research Fellow.

REFERENCES

Baylor, D. A., and Fuortes, M. G. F. (1970). Electrical responses of single cones in the retina of the turtle. *J. Physiol.* (*London*) **207,** 77–92.

Baylor, D. A., Lamb, T. D., and Yau, K. W. (1979). The membrane current of single rod outer segments. *J. Physiol.* (*London*) **288,** 589–611.

Biernbaum, M. S., and Bownds, M. D. (1979). Influence of light and calcium on guanosine 5′-triphosphate in isolated frog rod outer segments. *J. Gen. Physiol.* **74,** 649–669.

Bitensky, M. W., Gorman, R. E., and Miller, W. H. (1971). Adenyl cyclase as a link between photon capture and changes in membrane permeability of frog photoreceptors. *Proc. Natl. Acad. Sci. U.S.A.* **68,** 561–562.

Blaustein, M. P. (1974). The interrelationship between sodium and calcium fluxes across cell membranes. *Rev. Physiol., Biochem. Pharmacol.* **70,** 33–74.

Bolnick, D. A., Walter, A. E., and Sillman, A. J. (1979). Barium suppresses slow PIII in perfused bullfrog retina. *Vision Res.* **19,** 1117–1119.

Brown, J. E., and Pinto, L. H. (1974). Ionic mechanism for the photoreceptor potential of the retina of *Bufo marinus*. *J. Physiol.* (*London*) **236,** 575–591.

Brown, J. E., Coles, J. A., and Pinto, L. (1977). Effects of injection of calcium and EGTA into the outer segments of retinal rods of *Bufo marinus*. *J. Physiol.* (*London*) **269,** 707–722.

Brunner, J., Skrabaal, P., and Hauser, H. (1976). Single bilayer vesicles prepared without sonication: Physico-chemical properties. *Biochim. Biophys. Acta* **455,** 322–331.

Carslaw, H. S., and Jaegar, J. C. (1948). "Conduction of Heat in Solids." pp. 83–84. Oxford Univ. Press (Clarendon), London and New York.

Cohen, A. I. (1970). Further studies on the question of the patency of saccules in outer segments of vertebrate photoreceptors. *Vision Res.* **10,** 445–453.

Cone, R. A. (1973). The internal transmitter model for visual excitation: Some quantitative implications. *In* "Physiology and Biochemistry of Visual Pigments" (H. Langer, ed.), pp. 275–282. Springer-Verlag, Berlin and New York.

Darszon, A., Montal, M., and Zarco, J. (1977). Light increases the ion and non-electrolyte permeability of rhodopsin-phospholipid vesicles. *Biochem. Biophys. Res. Commun.* **76,** 820–827.

Epel, D. (1980). Ionic triggers in the fertilization of sea urchin eggs. *Ann. N.Y. Acad. Sci.* **339,** 75–85.

Fain, G. L. (1976). Sensitivity of toad rods: Dependence on wavelength and background illumination. *J. Physiol.* (*London*) **261,** 71–101.

Fletcher, R. T., and Chader, G. J. (1976). Cyclic GMP: Control of concentration by light in retinal photoreceptors. *Biochem. Biophys. Res. Commun.* **70,** 1297–1302.

Godchaux, W., and Zimmerman, W. F. (1979). Membrane dependent guanine nucleotide binding and GTPase activities of soluble proteins from bovine rod cell outer segments. *J. Biol. Chem.* **254,** 7874.

Gold, G. H., and Korenbrot, J. I. (1979). Light-evoked calcium fluxes across native and reconstituted rod disc membranes. *Invest. Ophthalmol. Visual Sci.* **18,** 1, Suppl., 4.

Gold, G. H., and Korenbrot, J. I. (1980a). Light-induced Ca efflux from rod cells in living retinas. *Invest. Ophthalmol. Visual Sci.* **19,** 1, Suppl., 281.

Gold, G. H., and Korenbrot, J. I. (1980b). Light-induced Ca efflux from intact rod cells in living retinas. *Fed. Proc., Fed. Am. Soc. Exp. Biol.* **39,** 1814.

Gold, G. H., and Korenbrot, J. I. (1980c). Light-induced Ca release by intact retinal rods. *Proc. Natl. Acad. Sci. U.S.A.* **77,** 5557-5561.

Goridis, C., and Virmaux, N. (1974). Light-regulated guanosine 3', 5'-monophosphate phosphodiesterase of bovine retina. *Nature (London)* **248,** 57-58.

Goridis, C., Urban, P. F., and Mandel, P. (1977). The effect of flash illumination on the endogenous cyclic GMP content of isolated frog retinae. *Exp. Eye Res.* **24,** 171-177.

Hagins, W. A. (1972). The visual process: Excitatory mechanisms in the primary receptor cells. *Annu. Rev. Biophys. Bioeng.* **1,** 131-158.

Hagins, W. A., and Ruppel, H. (1971). Fast photoelectric effects and the properties of vertebrate photoreceptors as electric cables. *Fed. Proc., Fed. Am. Soc. Exp. Biol.* **30,** 64-68.

Hagins, W. A., and Yoshikami, S. (1973). A role for Ca^{2+} in excitation of retinal rods and cones. *Exp. Eye Res.* **18,** 299-306.

Hagins, W. A., and Yoshikami, S. (1975). Ionic mechanisms in excitation of photoreceptors. *Ann. N.Y. Acad. Sci.* **264,** 314-325.

Hagins, W. A., and Yoshikami, S. (1977). Intracellular transmission of visual excitation in vertebrate photoreceptors: Electrical effects of chelating agents introduced into rods by vesicle fusion. *In* "Vertebrate Photoreception" (H. B. Barlow and P. Fatt, eds.), pp. 97-139. Academic Press, New York.

Harosi, F. I. (1975). Absorption spectra and linear dichroism of some amphibian photoreceptors. *J. Gen. Physiol.* **66,** 357-382.

Hubbell, W. L., and Bownds, D. M. (1979). Visual transduction in vertebrate photoreceptors. *Annu. Rev. Neurosci.* **2,** 17-34.

Hubbell, W. L., Fung, K.-K., Hong, K., and Chen, Y.-S. (1977). Molecular anatomy and light-dependent processes in photoreceptor membranes. *In* "Vertebrate Photoreception" (H. B. Barlow and P. Fatt, eds.), pp. 41-59. Academic Press, New York.

Jaffe, L. F. (1980). Calcium explosions as triggers of development. *Ann. N.Y. Acad. Sci.* **339,** 86-101.

Kaupp, U. B., Schnetkamp, P. P. M., and Junge, W. (1979). Flash-spectrophotometry with arsenazo III in vertebrate photoreceptor cells. *In* "Detection and Measurement of Free Ca in Cells" (C. C. Ashley and A. K. Campbell, eds.), pp. 287-308. Elsevier/North Holland, Amsterdam.

Kendrick, N. C. (1976). Purification of arsenazo III, a Ca sensitive dye. *Anal. Biochem.* **76,** 487-501.

Kilbride, P., and Ebrey, T. G. (1979). Light-initiated changes of cyclic guanosine monophosphate levels in the frog retina measured with quick freezing techniques. *J. Gen. Physiol.* **74,** 415-426.

Korenbrot, J. I., and Cone, R. A. (1972). Dark ionic flux and the effects of light in isolated rod outer segments. *J. Gen. Physiol.* **60,** 20-45.

Korenbrot, J. I., and Pramik, M. J. (1977). Formation, structure, and spectrophotometry of air-water interface films containing rhodopsin. *J. Membr. Biol.* **37,** 235-262.

Korenbrot, J. I., Brown, D. T., and Cone, R. A. (1973). Membrane characteristics and osmotic behavior of isolated rod outer segments. *J. Cell Biol.* **56,** 389-398.

Krishtal, O. A., Pidoplichko, V. I., and Shakhovalov, Y. A. (1980). Properties of single calcium channels in the neuronal membrane. *Bioelectrochem. Bioenerg.* **7,** 195-213.

Kuhn, H. (1980). Light- and GTP-regulated interaction of GTPase and other proteins with bovine photoreceptor membranes. *Nature (London)* **283,** 587–589.

Liebman, P. A., and Entine, G. (1968). Visual pigments of frog and tadpole. *Vision Res.* **8,** 761–775.

Liebman, P. A., and Pugh, E. N. (1979). The control of phosphodiesterase in rod disc membranes: Kinetics, possible mechanisms and significance for vision. *Vision Res.* **19,** 375–380.

Liebman, P. A., and Pugh, E. N. (1980). ATP mediates rapid reversal of cyclic GMP phosphodiesterase activation in visual receptor membranes. *Nature (London)* **287,** 734–736.

Lurie, M., and Marmor, M. F. (1980). Similarities between the c-wave and slow PIII in the rabbit eye. *Invest. Ophthalmol. Visual Sci.* **19,** 1113–1117.

Miki, N., Keirns, J. J., Marcus, F. R., Freeman, J., and Bitensky, M. W. (1973). Regulation of cyclic nucleotide concentrations in photoreceptors: An ATP-dependent stimulation of cyclic nucleotide phosphodiesterase by light. *Proc. Natl. Acad. Sci. U.S.A.* **70,** 3820–3824.

Miller, W. H., and Nicol, G. D. (1979). Evidence that cyclic GMP regulates membrane potential in rod photoreceptors. *Nature (London)* **280,** 64–66.

Nöll, G., Stieve, H., and Winterhager, J. (1979). Interaction of bovine rhodopsin with calcium ions. *Biophys. Struct. Mech.* **5,** 43–53.

O'Brien, D. F. (1979). Light-regulated permeability of rhodopsin-phospholipid membrane vesicles. *Photochem. Photobiol.* **29,** 679–685.

Orr, H. T., Lowry, O. H., Cohen, A. I., and Ferrendelli, J. A. (1976). Distribution of 3′,5′-cAMP and 3′,5′-cGMP in rabbit retina *in vivo:* Selective effects of dark and light adaptation and ischemia. *Proc. Natl. Acad. Sci. U.S.A.* **73,** 442–445.

Pober, J. S., and Bitensky, M. W. (1979). Light-regulated enzymes of vertebrate retinal rods. *Adv. Cyclic Nucleotide Res.* **11,** 266–295.

Robinson, R. A., and Stokes, R. H. (1959). "Electrolyte Solutions." Butterworth, London.

Robinson, W. E., and Hagins, W. A. (1979). GTP hydrolysis in intact rod outer segments and the transmitter cycle in visual excitation. *Nature (London)* **280,** 398–400.

Ruzicka, J., Hansen, E. H., and Tjell, J. C. (1973). Selectrode—The universal ion-selective electrode. Part III. The calcium selectrode. *Anal. Chim. Acta* **67,** 155–178.

Savvin, S. B. (1961). Analytical use of arsenazo III. *Talanta* **8,** 673–685.

Schnetkamp, P. P. M. (1980). Ion selectivity of the cation transport system of isolated intact cattle rod outer segments. *Biochim. Biophys. Acta* **598,** 66–90.

Sillman, A. J., Ito, H., and Tomita, T. (1969). Studies on the mass receptor potential of the isolated frog retina. *Vision Res.* **9,** 1435–1442.

Szutz, E. Z., and Cone, R. A. (1977). Calcium content of frog rod outer segments and discs. *Biochim. Biophys. Acta* **468,** 194–208.

Whitfield, J. F., Boynton, A. L., McMannus, J. P., Rixon, R. H., Sikorska, M., Tsang, B., and Walker, P. R. (1980). The roles of calcium and cyclic AMP in cell proliferation. *Ann. N.Y. Acad. Sci.* **339,** 216–240.

Williamson, E. D., and Adams, L. H. (1919). Temperature distribution in solids during heating or cooling. *Phys. Rev.* **14,** 99–114.

Witkovsky, P., Dudek, F. E., and Ripps, H. (1975). Slow PIII component of the carp electroretinogram. *J. Gen. Physiol.* **65,** 119–134.

Woodruff, M. L., and Bownds, M. D. (1979). Amplitude, kinetics, and reversibility of a light-induced decrease in guanosine 3′,5′-monophosphate in frog photoreceptor membranes. *J. Gen. Physiol.* **73,** 629–653.

Wormington, C. M., and Cone, R. A. (1978). Ionic blockage of the light-regulated sodium channels in isolated rod outer segments. *J. Gen. Physiol.* **71,** 657–681.

Yee, R., and Liebman, P. A. (1978). Light-activated phosphodiesterase of the rod outer segment. *J. Biol. Chem.* **253,** 8902–8909.

Yoshikami, S., and Hagins, W. A. (1971). Light, calcium and the photocurrent of rods and cones. *Abstr. Biophys. Soc. Meet.* p. 47a.

Yoshikami, S., and Hagins, W. A. (1973). Control of the dark current in vertebrate rods and cones. *In* "Biochemistry and Physiology of Visual Pigments" (H. Langer, ed.), pp. 325–336. Springer-Verlag, Berlin and New York.

Yoshikami, S., George, J. S., and Hagins, W. A. (1980). Light-induced Ca fluxes from outer segment layer of vertebrate retinas. *Nature (London)* **286,** 395–398.

Chapter 18

Modulation of Sodium Conductance in Photoreceptor Membranes by Calcium Ions and cGMP

ROBERT T. SORBI[1]

Istituto di Fisiologia Umana
University of Parma
Parma, Italy

It has been suggested (Hagins, 1972) that Ca^{2+} is the soluble transmitter needed to link the disk membranes with the plasma membrane of the rod outer segment (ROS) in the process of photoreceptor excitation (Baylor and Fuortes, 1970). According to this hypothesis, in the dark calcium ions are sequestered and maintained within the disk spaces by active Ca^{2+} transport and plasma membrane permeability to Na^+ is high, allowing a "Na^+ dark current" to flow into the cytoplasm. When light bleaches disk membrane rhodopsin, disk permeability to Ca^{2+} is enhanced, causing increased Ca^{2+} activity in the cytoplasm and closing the Na^+ channels. The data obtained by either increasing or lowering the cytoplasmic Ca^{2+} activity strongly support the idea that Ca^{2+} can modulate plasma membrane permeability to Na^+ (Hagins and Yoshikami, 1974, 1977; Brown *et al.*, 1977; Oakley and Pinto, 1980). However, many experiments have been attempted in search of light-dependent Ca^{2+} release from the disk (for a review, see Nöll *et al.*, 1979; Schnetkamp, 1979), and still no direct evidence indicates that *in vivo* physiological levels of illumination are able to promote the release of Ca^{2+} from the disks. Recently a light-induced increase of the calcium activity has been recorded in the extracellular fluid surrounding the outer segments in isolated retinas (Gold and Korenbrot, 1980; this volume, Chapter 17; Yoshikami *et al.*, 1980). This observation raises the possibility but does not give direct proof, that

[1]Present address: Department of Pathology, Yale University School of Medicine, New Haven, Connecticut.

ISBN 0-12-153315-8

the increase in the extracellular activity reflects a light-induced increase of the cytoplasmic activity of calcium ions in the outer segment of the rod.

In our laboratory, light-induced effects on the ionic fluxes from the ROS were detected after bleaching less than 1% of the rhodopsin, but no change was observed in calcium fluxes (Cavaggioni *et al.*, 1973). We could not observe the dim light effects when the ROS were broken (Fig. 1A), which suggests that broken membranes allow the leakage of some necessary endogenous factor (Sorbi and Cavaggioni, 1975). In the broken ROS preparation, however, a bright light bleaching more than 5% of the rhodopsin increased the effluxes of Na^+, K^+, and Rb^+, but still not those of Ca^{2+} (Fig. 1B). Since this effect of light was

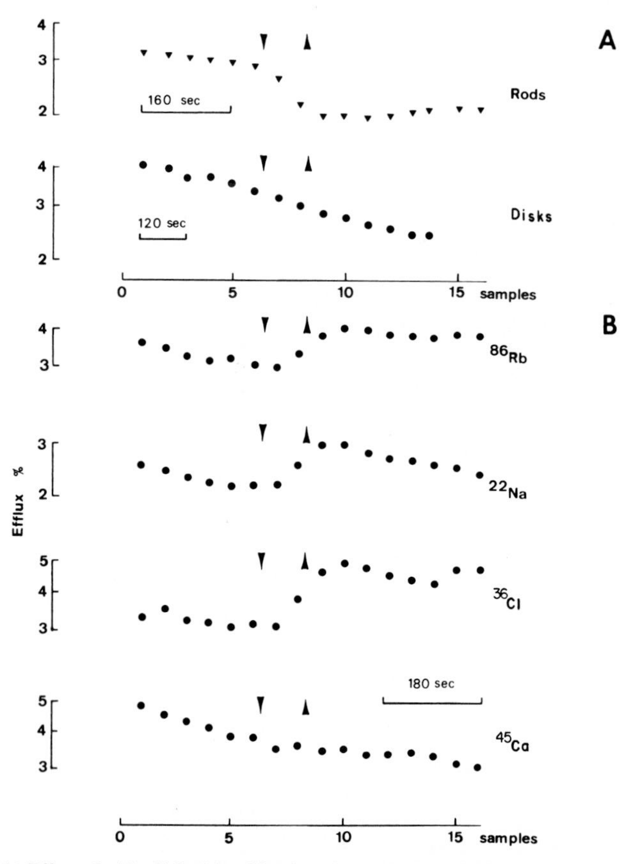

FIG. 1. (A) Effect of white light, bleaching less than 1% of the rhodopsin, on the rate of efflux of ^{86}Rb from isolated rods and isolated disks. (B) Effect of white light, bleaching more than 5% of the rhodopsin, on the rate of efflux of ^{86}Rb, ^{22}Na, ^{36}Cl, and ^{45}Ca from isolated disks. In both (A) and (B), the arrows indicate the onset and the end of illumination. (From Sorbi and Cavaggioni, 1975.)

proportional to the amount of bleached rhodopsin, it might be due to nonspecific changes in disk membrane properties induced by the modification of such a large proportion of the proteins within the membrane and by the release of a large number of retinal molecules.

In 1971, a possible role for cyclic-nucleotide metabolism in visual excitation was proposed for the first time (Bitensky *et al.*, 1971). Since then much work has been done in this area (Bitensky *et al.*, this volume, Chapter 14), and currently cyclic guanosine 3′,5′-monophosphate (cGMP) seems to be a viable candidate for the internal transmitter role (for a review, see Hubbell and Bownds, 1979; Pober and Bitensky, 1979; Bitensky *et al.*, this volume, Chapter 14). The hypothesis involving cGMP in visual excitation suggests that in the dark plasma membrane permeability is high because of cGMP-dependent phosphorylation of some protein(s) related to membrane permeability to Na^+ (Polans *et al.*, 1979). In the presence of GTP, cGMP-phosphodiesterase (cGMP-PDE) is activated by bleached rhodopsin (Wheeler and Bitensky, 1977; Bignetti *et al.*, 1978) with a rate constant that is consistent with the time required for excitation (Yee and Liebman, 1978; Caretta *et al.*, 1979a). Consequently the internal cGMP concentration, $[cGMP]_i$, falls with a rate that is compatible with its involvement in excitation (Woodruff *et al.*, 1977; Kilbride, 1980) and causes dephosphorylation of the protein(s) related to membrane permeability (Polans *et al.*, 1979).

A direct relationship between plasma membrane permeability to Na^+ and the endogenous cGMP content of isolated ROS has been shown (Woodruff *et al.*, 1977), and we have demonstrated that cGMP, in the micromolar range, can modulate the Na^+ permeability of the disk membrane (Fig. 2) (Caretta *et al.*, 1979b). Thus the endogenous factor we lost in 1975 was probably cGMP. Moreover, iontophoretically injecting cGMP into the ROS causes depolarization of the plasma membrane and increases the latency of the photoresponse (Miller and Nicol, 1979). This evidence suggests that cGMP may affect the plasma membrane in the same way as the disk membranes; i.e., it increases the Na^+ conductance. Since the effect of cGMP on Na^+ permeability is still present on thoroughly washed disk membranes, where proteins dephosphorylated by light in a cGMP-dependent way are presumed to be absent (Polans *et al.*, 1979), it is worthwhile to propose that a phosphorylation–dephosphorylation process may not be involved.

One of the major criticisms that can be raised against a role of cGMP in visual excitation is the relative inefficiency of a negative transmitter. In fact, it seems to be less efficient for the plasma membrane to sense a decrease in $[cGMP]_i$ which in the dark is approximately 40–90 μM in the ROS (Woodruff *et al.*, 1977; Robinson and Hagins, 1980), than to detect the presence of a few hundred calcium ions above the background which is postulated to be zero. However, the plasma membrane should be able to detect fast and pronounced decreases of [cGMP] in the microdomain of excitation if the recharge of the transmitter from

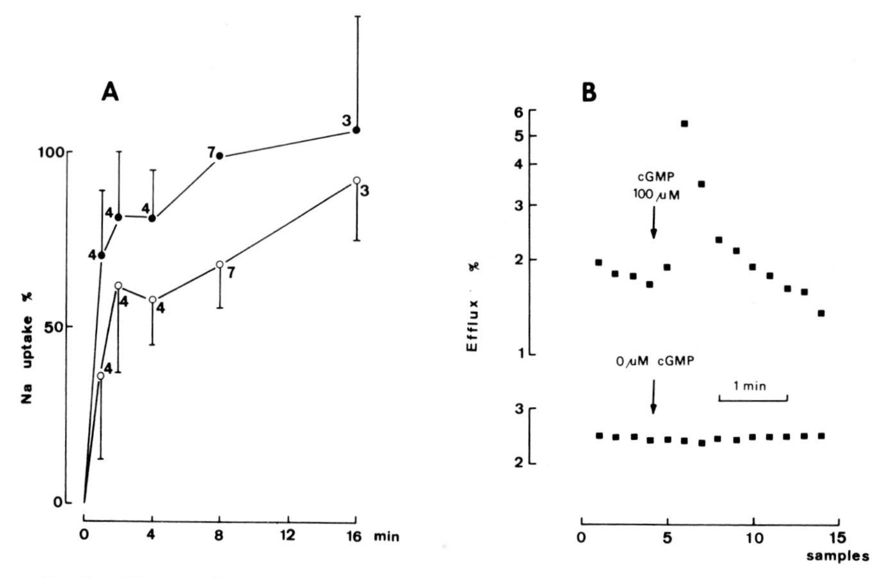

FIG. 2. Effect of cGMP on the uptake (A) and on the release (B) of ^{22}Na from disk vesicles. (A) Loading of ^{22}Na by the disk vesicles. Solid circles, in the presence of 1 m*M* cGMP; open circles, in the absence of cGMP. The values are normalized in order to have the point at 8 min (with cGMP) equal to 100, to make a comparison between different experiments possible. The values are the mean of the number of experiments indicated at each point, and the bars indicate the standard deviation. (B) Effect of 100 μM cGMP on the rate of efflux of ^{22}Na from disk vesicles.

the surrounding nonexcited spaces is slow compared to the PDE activity. Measurements of the space constant of excitation in the rod do not rule out such a hypothesis (Yau *et al.*, 1980; this volume, Chapter 2).

The question now arises, How is it possible to combine the data that favor Ca^{2+} as the internal transmitter with those that favor cGMP. The finding that Ca^{2+} influences the cGMP content in the ROS (Cohen *et al.*, 1978), presumably by acting on a guanylate cyclase (Pannbacker, 1973; Krishnan *et al.*, 1978; de Vries and Ferrendelli, 1980), could lead to the conclusion that calcium modulates the sodium conductance via cGMP. However, Yoshikami *et al.* (1980) have shown that Ca^{2+} affects sodium conductance directly and without the time lag needed to enhance $[cGMP]_i$. Recently, we have reported that cGMP influences Ca^{2+} movement to and from the disk membranes (Cavaggioni and Sorbi, 1980). At physiological concentrations, cGMP stimulates the release and inhibits the uptake of calcium ions by the disk vesicles (Fig. 3). Increasing the cGMP concentration in the perfusing solution (in the micromolar range) causes the effect to become larger. Both the Ca^{2+} uptake and the cGMP effect on the Ca^{2+} uptake seem completely ATP-independent (Fig. 3). Thus cGMP may affect the binding sites for Ca^{2+} on the disk membranes, as has already been suggested for

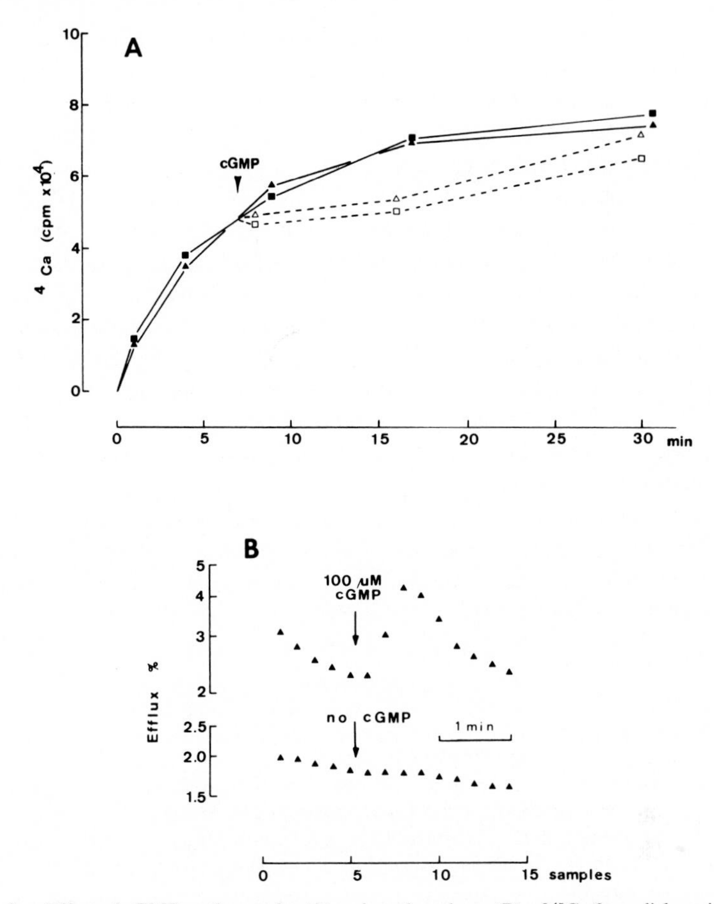

FIG. 3. Effect of cGMP on the uptake (A) and on the release (B) of ^{45}Ca from disk vesicles. (A) Squares, in the presence of 1 m*M* ATP; triangles, in the absence of ATP. Open symbols, in the presence of 400 μM cGMP; solid symbols, in the absence of cGMP. The disk vesicles were from the same stock. (B) Effect of 100 μM cGMP on the rate of ^{45}Ca release from disk vesicles.

other systems (Weler and Laing, 1979). Since this is the case, it is not surprising that a light-induced increase in the internal Ca^{2+} activity has been so difficult to detect. These new data suggest, on the contrary, a light-induced decrease in the Ca^{2+} activity inside the ROS as a function of cGMP modulation.

As a consequence, a model may now be proposed in which both Ca^{2+} and cGMP are coprimary and complementary factors in the process of visual excitation, if we suppose that the Ca^{2+}-binding sites affected by cGMP are functional gates for the Na^+ channels. Stieve and Bruns (1978) have suggested that in invertebrate photoreceptors the sodium channels are transmembrane proteins with binding sites for which Ca^{2+} and Na^+ may compete antagonistically. When Na^+ is bound to these sites, the channels are open for Na^+ conductance. When Ca^{2+} is bound to these sites, the channels are closed for Na^+ conductance. Similar arguments have been presented by Schnetkamp (1980) for the Ca^{2+}-translocating system in the plasma membrane of vertebrate photoreceptors. When the above-mentioned hypothesis is considered in conjunction with the cGMP effect on Ca^{2+}-binding sites, I feel it likely that, when cGMP decreases the binding site affinity for Ca^{2+}, it may favor the binding of Na^+ at the same sites. Thus, in the dark, when the cGMP concentration is elevated, Na^+ binds to and opens the channels. In the light, the cGMP concentration in the microdomain falls because of the light-activated cGMP-PDE, and Ca^{2+} binding to the channels is favored. Such calcium binding switches the channel to the closed conformation. Moreover, since the guanylate cyclase seems to be Ca^{2+}-modulated, when the Ca^{2+} activity is diminished, this enzyme is stimulated in order to restore the normal cGMP concentration.

It is possible to develop this model in order to deal with sensitivity regulation, taking into account the idea that the Ca^{2+}-binding sites on the disk membrane are probably ineffective in modulating the Na^+ dark current (Bitensky *et al.*, this volume, Chapter 14).

REFERENCES

Baylor, D., and Fuortes, M. G. F. (1970). Electrical responses of single cones in the retina of the turtle. *J. Physiol.* (*London*) **207,** 77–92.

Bignetti, E., Cavaggioni, A., and Sorbi, R. T. (1978). Light-activated hydrolysis of GTP and cyclic GMP in the rod outer segments. *J. Physiol.* (*London*) **279,** 55–69.

Bitensky, M. W., Gorman, R. E., and Miller, W. H. (1971). Adenyl cyclase: A link between photon capture and changes in membrane permeability of frog photoreceptors. *Proc. Natl. Acad. Sci. U.S.A.* **63,** 561-562.

Brown, J. E., Coles, J. A., and Pinto, L. H. (1977). Effect of injections of calcium and EGTA into the outer segments of retinal rods of *Bufo marinus*. *J. Gen. Physiol.* **269,** 707–722.

Caretta, A., Cavaggioni, A., and Sorbi, R. T. (1979a). Phosphodiesterase and GTPase in rod outer segments: Kinetics *in vitro*. *Biochim. Bophys. Acta* **583,** 1–13.

Caretta, A., Cavaggioni, A., and Sorbi, R. T. (1979b). Cyclic GMP and permeability of the disc membrane of the photoreceptors. *J. Physiol.* (*London*) **295,** 171–178.

Cavaggioni, A., and Sorbi, R. T. (1980). Cyclic GMP affects calcium movements in the disc membrane of photoreceptors. *Proc. Natl. Acad. Sci. U.S.A.* **78,** 3964-3968.

Cavaggioni, A., Sorbi, R. T., and Turini, S. (1973). Efflux of potassium from isolated rod outer segments: A photic effect. *J. Physiol.* (*London*) **232,** 609–620.

Cohen, A. I., Hall, I. A., and Ferrendelli, J. A. (1978). Calcium and cyclic nucleotide regulation in incubated mouse retinas. *J. Gen. Physiol.* **71,** 595–612.

de Vries, G. W., and Ferrendelli, J. A. (1980). Evidence for multiple guanylate cyclase in a vertebrate photoreceptor. ARVO Abstr. in *Invest. Ophthalmol. Visual Sci.* **19,** 21.

Gold, G. H., and Korenbrot, J. I. (1980). Light-induced calcium release by intact retinal rods. *Proc. Natl. Acad. Sci. U.S.A.* **77,** 5557–5561.

Hagins, W. A. (1972). The visual process: Excitatory mechanism in the primary receptor cells. *Annu. Rev. Biophys. Bioeng.* **1,** 131–159.

Hagins, W. A., and Yoshikami, S. (1974). A role for Ca^{2+} in excitation of retinal rods and cones. *Exp. Eye Res.* **18,** 299–305.

Hagins, W. A., and Yoshikami, S. (1977). Intracellular transmission of visual excitation in photoreceptors: Electrical effects of chelating agents introduced into rods by vesicles fusion. *In* "Vetebrate photoreception" (H. D. Barlow and P. Fatt, eds.), pp. 97–138. Academic Press, New York.

Hubbell, W. L., and Bownds, M. D. (1979). Visual transduction in vertebrate photoreceptors. *Annu. Rev. Neurosci.* **2,** 17–34.

Kilbride, P. (1980). Calcium effects on frog retinal cGMP levels and their light-initiated rate of decay. *J. Gen. Physiol.* **75,** 457–465.

Krishnan, N., Fletcher, R. T., Chader, G. J., and Krishna, G. (1978). Characterization of guanylate cyclase of rod outer segments of the bovine retina. *Biochim. Biophys. Acta* **523,** 506–515.

Miller, W. H., and Nicol. G. D. (1979). Evidence that cyclic GMP regulates membrane potential in rod photoreceptors. *Nature (London)* **280,** 64–66.

Nöll, G., Stieve, H., and Winterhager, J. (1979). Interaction of bovine rhodopsin with calcium. II. Calcium release in bovine rod outer segments upon bleaching. *Biophys. Struct. Mech.* **5,** 43–53.

Oakley, B., II, and Pinto, L. H. (1980). $[Ca^{2+}]_i$ modulates membrane sodium conductance in rod outer segments. ARVO Abstr. in *Invest. Ophthalmol. Visual Sci.* **19,** 102.

Pannbacker, R. G. (1973). Control of guanylate cyclase activity in the rod outer segment. *Science* **182,** 1138–1140.

Pober, J. S., and Bitensky, M. W. (1979). Light-regulated enzymes of vertebrate retinal rods. *Adv. Cyclic Nucleotide Res.* **11,** 265–301.

Polans, A. S., Hermolin, J., and Bownds, M. D. (1979). Light-induced dephosphorylation of two proteins in frog rod outer segments. *J. Gen. Physiol.* **74,** 595–613.

Robinson, W. E., and Hagins, W. A. (1980). Bound and free nucleotides in rod outer segments. Biophys. Soc. Abstr. in *Fed. Proc., Fed. Am. Soc. Exp. Biol.* **39,** 2067.

Schnetkamp, P. P. M. (1979). Calcium translocation and storage of isolated intact cattle outer segments in darkness. *Biochim. Biophys. Acta* **554,** 441–459.

Schnetkamp, P. P. M. (1980). Ion selectivity of the cation transport system of isolated intact cattle rod outer segments: Evidence for a direct communication between the rod plasma membrane and the rod disk membranes. *Biochim. Biophys. Acta* **598,** 66–90.

Sorbi, R. T., and Cavaggioni, A. (1975). Effect of strong illumination on the ion efflux from isolated discs of frog photoreceptors. *Biochim. Biophys. Acta* **394,** 577–585.

Stieve, H., and Bruns, M. (1978). Extracellular calcium, magnesium and sodium ion competition in the conductance control of the photosensory membrane of *Limulus* ventral nerve photoreceptor. *Z. Naturforsch.: Biosci.* **33C,** 574–579.

Weller, M., and Laing, W. (1979). The effect of cyclic nucleotides and protein phosphorylation on calcium permeability and binding in the sarcoplasmic reticulum. *Biochim. Biophys. Acta* **551,** 406–419.

Wheeler, G. L., and Bitensky, M. W. (1977). A light-activated GTPase in vertebrate photoreceptors: Regulation of light-activated cyclic GMP phosphodiesterase. *Proc. Natl. Acad. Sci. U.S.A.* **74,** 4238–4242.

Woodruff, M. L., Bownds, M. D., Green, S. H., Morrissey, J. L., and Shedlovsky, A. (1977). Guanosine 3′,5′-cyclic monophosphate and *in vitro* physiology of frog photoreceptor membranes. *J. Gen. Physiol.* **69,** 667–679.

Yau, K. W., Baylor, D., Matthews, G., Lamb, T., and McNaughton, P. (1980). Background adaptation in toad rods. ARVO Abstr. in *Invest. Ophthalmol. Visual Sci.* **19,** 192.

Yee, R., and Liebman, P. A. (1978). Light-activated phosphodiesterase of the rod outer segment: Kinetics and parameters of activation and deactivation. *J. Biol. Chem.* **253,** 8902–8909.

Yoshikami, S., George, J. S., and Hagins, W. A. (1980). Light-induced calcium fluxes from outer segment layer of vertebrate retinas. *Nature (London)* **286,** 395–398.

Part III
Calcium, Cyclic Nucleotides, and the Membrane Potential

To what extent do calcium and/or cyclic nucleotides affect the rod outer segment sodium permeability as light and darkness?

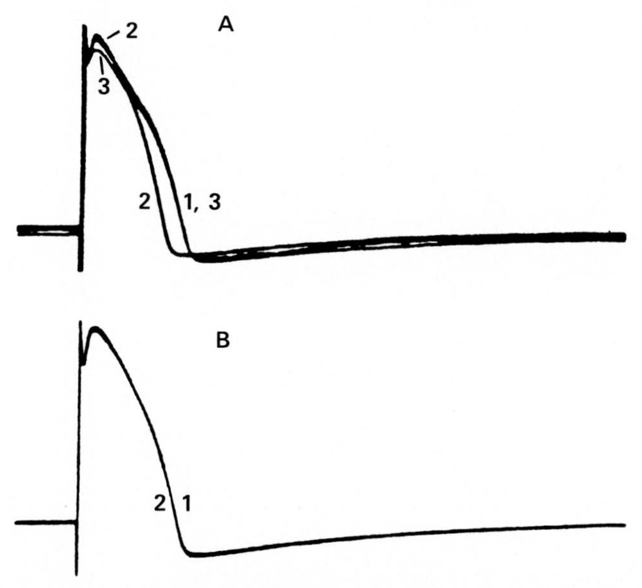

First intracellular injection of a cyclic nucleotide. Transmembrane potentials recorded from calf Purkinje fibers. In record A, trace 1 is before the iontophoretic injection of cyclic AMP, 2 is 4 min after, and 3 is 23 min after the injection. In record B, trace 1 is before injection of 5′-AMP and 2 is 3 min after the injection. The injected cyclic AMP causes a marked decrease in action potential duration, a rise of the plateau level, and a slight steepening of the pacemaker depolarization. [From R. W. Tsien, Adrenaline-like effects of intracellular iontophoresis of cyclic AMP in cardiac Purkinje fibres. *Nature* (*London*), *New Biol*. **245**, 120–121, 1973. Reprinted by permission. Copyright 1973 Macmillan Journals Ltd.]

Chapter 19

Calcium and the Mechanism of Light Adaptation in Rods

BRUCE L. BASTIAN AND GORDON L. FAIN

Department of Ophthalmology
Jules Stein Eye Institute
UCLA School of Medicine
Los Angeles, California

I. INTRODUCTION

Since most of the rhodopsin in a rod is in the membranes of the disks, which are not electrically coupled to the plasma membrane, the absorption of a photon must somehow alter the concentration of a diffusible messenger in order to produce the resultant change in the conductance and voltage of the rod. The majority of chapters in this volume describe experiments that favor or disfavor one or another of the substances that have been proposed as the internal messenger (or transmitter) for excitation. Unfortunately we are still unable to identify this substance with certainty, even though, as this book demonstrates, we have learned quite a lot about the effects of calcium, cyclic nucleotides, and other

ISBN 0-12-153315-8

substances on photoreceptors, giving us a fair beginning in obtaining an understanding of the light-dependent biochemistry of rods.

Although there is now nearly universal agreement that an internal messenger is necessary for excitation, it is perhaps less widely known that a diffusible substance also regulates the sensitivity of the receptor during light adaptation. This substance may or may not be the same as the messenger for excitation. In Section II of this chapter, we shall describe the evidence that sensitivity is regulated by a diffusible messenger. In Section III, we shall present experiments which we believe show that sensitivity is not normally controlled by increases in Ca^{2+}. In Section IV, we shall describe the effects of exposing rods to low-Ca^{2+} solutions. In the final section, we shall discuss some of the implications of our experiments, which we believe exclude for toad rods some of the models for light adaptation in receptors that have been previously proposed (e.g., Kleinschmidt and Dowling, 1975; Baylor *et al.,* 1974) and suggest experiments combining biochemical and physiological approaches to unravel the processes responsible for the regulation of sensitivity.

II. REGULATION OF SENSITIVITY BY A DIFFUSIBLE SUBSTANCE

When the retina is exposed to continuous background illumination, the sensitivity of the receptors to brief flashes decreases (Kleinschmidt, 1973; Kleinschmidt and Dowling, 1975; Baylor and Hodgkin, 1974; Fain, 1976). We shall define the flash sensitivity (S_F) as the peak amplitude of the voltage change per unit light intensity for small-amplitude responses. Experiments on toad rods with both the eyecup (Fain, 1976) and isolated retina (Bastian and Fain, 1979) show that their sensitivity during light adaptation can be described by a modified Weber–Fechner law:

$$1/S_F = 1/S_F^D + kI_B/\Delta V_T \tag{1}$$

In this equation, S_F^D is the flash sensitivity in the absence of a background, k is the behavioral Weber fraction in seconds per flash, ΔV_T is the voltage response of the receptor at the behavioral threshold (Fain *et al.,* 1977), and I_B is the background intensity. According to this equation, inverse sensitivity is proportional to background intensity, and this has been found to be the case for toad rods over a range for I_B of 4–5 log units (Fain, 1976). At an I_B of approximately 3×10^{11} quanta cm^{-2} sec^{-1}, the rods of toads and most other animals show increment saturation ($S_F \rightarrow 0$), and rod responses to flashes no longer can be recorded (Fain, 1976).

One of the most remarkable things about the rod is its extraordinary sensitivity to light. As Hecht *et al.* (1942) first showed, excitation can be triggered in a receptor by the absorption of only a single photon. The process that regulates receptor sensitivity during light adaptation is also extremely sensitive. To estimate its sensitivity quantitatively, we shall define a quantity I_0 as the intensity of the background that reduces S_F to half its value in the dark, that is, to ½ S_F^D. I_0 is the receptor equivalent of the psychophysical "dark light" (Barlow, 1972).

According to Eq. (1), $S_F = ½ S_F^D$ when $I_B = I_0 = \Delta V_T / k S_F^D$. Hence we can estimate the value of I_0 by plotting the sensitivity of a rod as a function of the background intensity, as in Fig. 1A. The data points in this figure give the mean

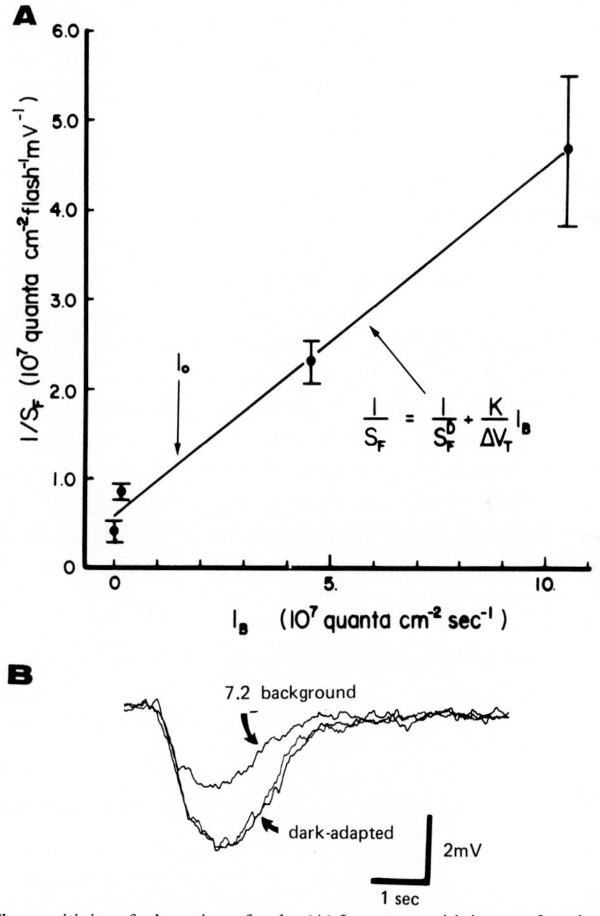

FIG. 1. The sensitivity of adaptation of rods. (A) Inverse sensitivity as a function of background intensity for five cells. Error bars give 1 SE. Means fitted with Eq. (1) by linear regression. (B) Small-amplitude responses before, in the presence of, and following a background I_0 of intensity (7.2 log quanta cm^{-2} sec^{-1}) necessary to halve the dark sensitivity as determined from (A). Each trace represents an average of eight responses to flashes of 8.0 log incident quanta cm^{-2} flash^{-1}. All flashes were 100 msec of 501-nm light. Reprinted with permission from Bastian and Fain (1979). Copyright by The Physiological Society.

flash sensitivity for five toad rods in the dark and in three dim backgrounds. The best-fitting line for these points has a slope $k/\Delta V_T = 3.9 \times 10^{-4}$ sec per flash per microvolt and an intercept at $S_F = S_F^D$ of 1.8×10^{-4} μV per incident quantum per square centimeter per flash. I_0 is therefore 1.4×10^7 quanta cm^{-2} sec^{-1}, or 4.1 rhodopsin molecules bleached per rod per second (Bastian and Fain, 1979).

The effect of a background of this intensity on the responses of a rod can be seen from the data in Fig. 1B. There are three traces in this figure, each the average of eight responses all at the same flash intensity. The two larger traces were recorded in the dark before and after exposure to the background. The smaller trace is the averaged response in the presence of a continuous background of intensity equal to the value for I_0 calculated in the previous paragraph. The background reduces the response amplitude from 5 to 2.5 mV, that is, by half. Since for these responses the peak amplitude was nearly proportional to the flash intensity, the halving of the amplitude in the presence of the background represents an approximately twofold decrease in sensitivity.

When a background of intensity I_0 is presented to the retina, the sensitivity of the rod reaches a steady state quickly, certainly in less than 10 sec (B. L. Bastian and G. L. Fain, unpublished observations). Therefore, by the time the sensitivity has stabilized, a background at I_0 has bleached no more than $4.1 \times 10 = 41$ rhodopsin molecules per rod. Since a toad rod contains well over 2000 disks, at most only 2% of these will contain bleached rhodopsin molecules, even though the sensitivity of the rod is reduced by a factor of 2.

These experiments show that the disks that absorb photons must somehow communicate this event and alter the transduction process initiated when photons are absorbed by the other disks. Since disks are physically isolated from one another and from the plasma membrane, the absorption of a photon must alter the concentration of some messenger substance that diffuses to the disks or the plasma membrane and alters the sensitivity of the rod. Changes in sensitivity are not restricted to the disks that catch the quanta but spread for some distance down the length of the outer segment.

III. EVIDENCE THAT SENSITIVITY IS NOT REGULATED BY INCREASES IN Ca^{2+}

A. Increases in $[Ca^{2+}]_o$

An obvious and attractive candidate for the substance that regulates rod sensitivity is Ca^{2+}. Calcium ions were first proposed by Yoshikami and Hagins (1971) to be released by disks during excitation and to diffuse to the plasma membrane and decrease the light-dependent conductance. Although the evidence that calcium actually plays this role has so far been discouraging, there can be no question that Ca^{2+} has profound effects on the physiological properties of vertebrate receptors. Yoshikami and Hagins (1973) have shown that increases in the external Ca^{2+} concentration ($[Ca^{2+}]_o$) reduce both the dark current and photocurrent of rat rods. Furthermore, Brown and Pinto (1974) have demonstrated that increases in $[Ca^{2+}]_o$ hyperpolarize the membrane potential and reduce the amplitude of rod responses, just as background light does.

The similarity of the effects of Ca^{2+} and background light can be seen in Fig. 2. The presentation of a background and an increase in $[Ca^{2+}]_o$ both cause the rod to hyperpolarize and the responses to become smaller. However, as we shall see in more detail below, this resemblance is superficial. The changes in sensitivity produced by background light can be orders of magnitude larger than those produced by increases in $[Ca^{2+}]_o$. Furthermore, backgrounds produce a pronounced quickening of the decay of the responses, which can be clearly seen in Fig. 2A for the flashes at 8.4 log incident quanta cm^{-2} sec^{-1} but not in Fig. 2B where Ca^{2+} levels were increased.

The effects of Ca^{2+} and background light on receptor sensitivity are shown in more detail in Fig. 3. Here we compare for a single cell the changes in the intensity–response function produced by background light and elevated Ca^{2+} levels. The solid circles in Fig. 3A show the dark-adapted intensity–response curve in normal Ringer's solution. After making these measurements, we presented the retina with a dim background. The intensity–response curve shows

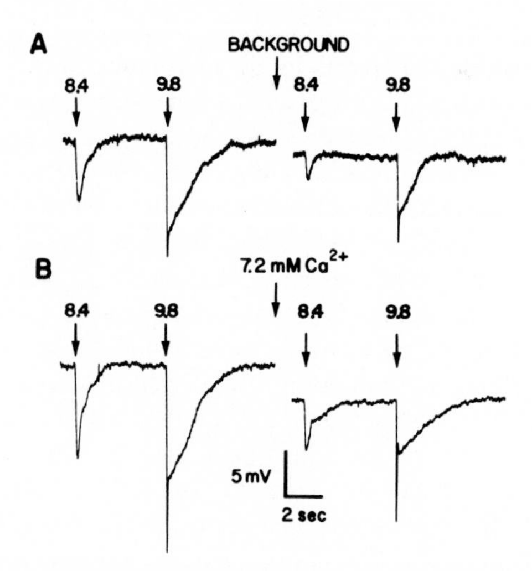

FIG. 2. Rod responses in the presence of background light and high $[Ca^{2+}]_o$. (A) Responses to flashes of 8.4 and 9.8 log incident quanta/cm^2 per flash in the dark (left) and in the presence of 8.0 log quanta cm^{-2} sec^{-1} background (right). (B) Dark-adapted responses in normal (left) and 7.2 m*M* Ca^{2+} Ringer's solution (right) in a different cell. Reprinted with permission from Bastian and Fain (1979). Copyright by The Physiological Society.

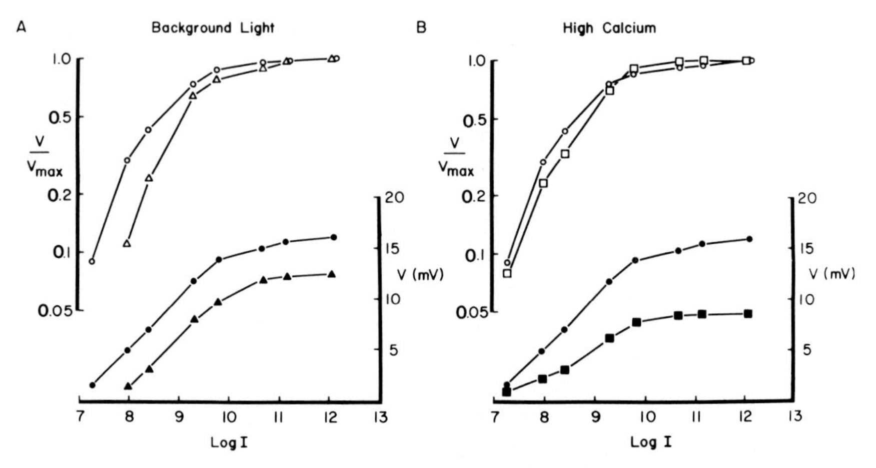

FIG. 3. Intensity–response relations in light and in high external Ca^{2+}. Peak response amplitudes in millivolts (solid symbols) and normalized responses (V/V_{max}, open symbols) are plotted against log intensity (quanta per square centimeter per flash). In (A), an adapting light (8.0 log quanta cm^{-2} sec^{-1}) shifts the response curves (triangles) rightward along the intensity axis with respect to the dark-adapted curves (circles). In (B), for the same cell as in (A), 7.2 m*M* $[Ca^{2+}]_o$ (squares) compresses the response function more than the background (see unnormalized scale) but causes only a slight shift rightward (see normalized scale). Reprinted with permission from Bastian and Fain (1979). Copyright by The Physiological Society.

that the background decreased the saturating response amplitude (V_{max}) by about 4 mV and decreased the rod sensitivity by approximately 0.7 log units, that is, about fivefold. This change in sensitivity is accounted for in part by the reduction in V_{max}, but mostly by a shifting of the intensity–response curve along the intensity axis (Kleinschmidt and Dowling, 1975; Baylor and Hodgkin, 1974; Fain, 1976). The shifting of the intensity–response curves can be more clearly seen in the upper curves in Fig. 3A, where the dark-adapted and light-adapted responses have been normalized to their respective V_{max} values.

After turning off the background and allowing the cell to dark-adapt to its previous sensitivity, the retina was superfused with 7.2 m*M* (four times normal) Ca^{2+} Ringer's solution. A comparison of the intensity–response curves in high Ca^{2+} and in normal Ringer's solution is given in Fig. 3B. Although the high-Ca^{2+} solution produced an even greater reduction in V_{max} than the background light, it caused a much smaller change in sensitivity. The principal effect of Ca^{2+} was to reduce the scale of the responses, as can be seen more clearly in the normalized curves in Fig. 3B. Increased $[Ca^{2+}]_o$ produced small but consistently measured shifts toward higher intensities in the normalized intensity–response curves. At 7.2 m*M* $[Ca^{2+}]_o$, the mean shift was 0.18 log unit, (n = 18), and shifts were observed in 17 of the 18 cells, which is significant at the 0.0001 level. However, these shifts were much smaller in high-Ca^{2+} solutions than in background light for comparable reductions in V_{max}.

The effects of background light and high Ca^{2+} solutions on rod sensitivity and response amplitude are compared quantitatively in Fig. 4. The data in this figure were collected from experiments on 63 cells in the presence of a variety of background intensities and Ca^{2+} concentrations. For each cell, the ratio of the flash sensitivity under the experimental conditions (S_F) to that in normal Ringer's solution in the dark (S_F^D) is plotted against the ratio of maximum response amplitude under the experimental and control conditions ($\Delta V_{max}/V_{max}$). For the smallest values of $\Delta V_{max}/V_{max}$, the sensitivities with background light are nearly two orders of magnitude lower than those with high Ca^{2+}.

Background light and elevated Ca^{2+} concentrations also have different effects on the response waveform. This can be seen in Fig. 5. The two traces in Fig. 5A (left) are responses to the same test intensity under dark-adapted and light-adapted conditions. The light-adapted responses are smaller and faster. The

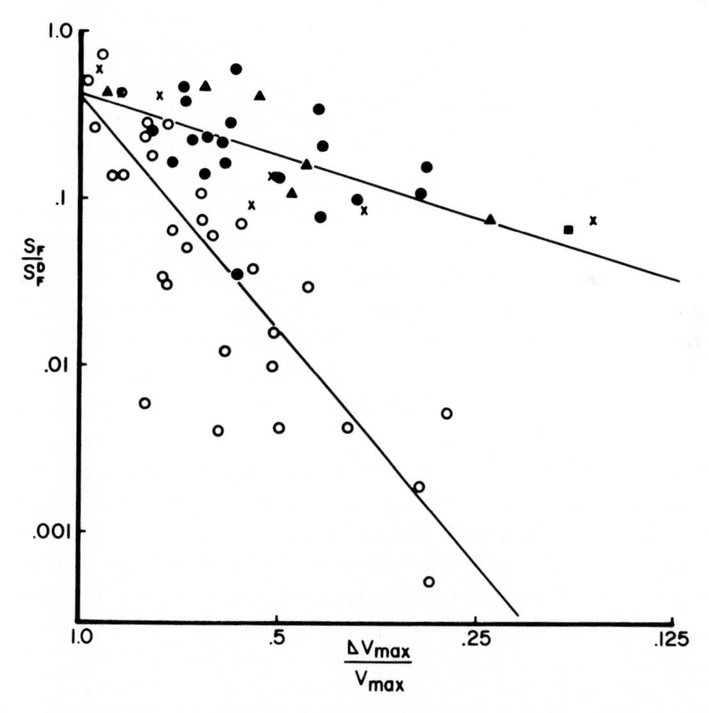

FIG. 4. Light and Ca^{2+} effects on the relation between sensitivity and maximum amplitude. Sensitivity and maximum amplitudes in normal Ringer's solution, S_F^D and V_{max}, are used to normalize the corresponding values, S_F and ΔV_{max}, measured in the presence of background light (open symbols), high $[Ca^{2+}]_o$ (solid symbols), or ionophore-induced high $[Ca^{2+}]_i$ (x's). Values of $[Ca^{2+}]_o$ represented are 3.6 m*M* (●), 7.2 m*M* (▲), and 14.4 m*M* (■). Light and high $[Ca^{2+}]_o$ data, plotted on log scales, are fit with regression lines constrained to pass through the point (1,1). Reprinted with permission from Bastian and Fain (1979). Copyright by The Physiological Society.

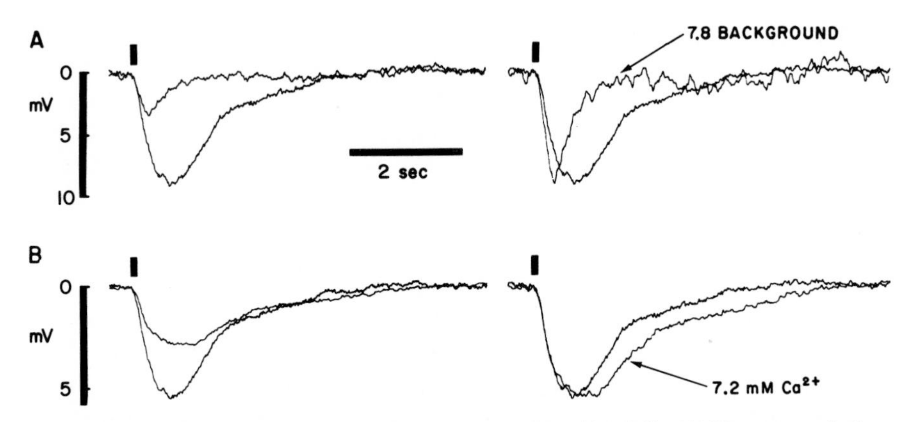

FIG. 5. Small-amplitude responses compared in light and in high Ca^{2+}. (A) The response in the presence of background (8.0 log quanta cm^{-2} sec^{-1}) is smaller than that in the dark (left). The same responses normalized to equal amplitude (right) illustrate a quickening of the kinetics for the light-adapted response. (B) 7.2 m*M* $[Ca^{2+}]_o$ reduces the response amplitude (left), but normalization of the same responses (right) shows that high Ca^{2+} slows the kinetics. All responses were to test flashes of 8.0 log quanta cm^{-2} $flash^{-1}$; the light-adapted response is a single trace, all others are averages of four each. Reprinted with permission from Bastian and Fain. Copyright by The Physiological Society.

difference in waveform can be seen more clearly in the traces on the right, which have been normalized to the same peak amplitude. Both the time to peak and decay time were much smaller in the presence of background light. These changes in response kinetics cannot be mimicked by elevated Ca^{2+} levels. Figure 5B shows, for the same cell, the effect of superfusing the retina with 7.2 m*M* Ca^{2+} Ringer's solution. The responses in high Ca^{2+} are somewhat slower than those in normal Ringer's solution, although the extent of this effect varied from cell to cell. In no instance did high-Ca^{2+} levels speed up the kinetics of the responses, as in Fig. 5A.

In summary, the changes in rod sensitivity and waveform that occur during light adaptation cannot be matched by changing extracellular calcium concentrations. Similar experiments have been done by Bertrand *et al.* (1978) on turtle cones with nearly the same results. Since changing extracellular calcium levels should also alter intracellular free calcium concentrations, our inability to produce large changes in the sensitivity and waveform by increasing $[Ca^{2+}]_o$ suggests that calcium cannot be the messenger substance for adaptation.

B. Increases in $[Ca^{2+}]_i$: Experiments with X537A

Unfortunately, there is one difficulty with the experiments we have described so far. Although increasing $[Ca^{2+}]_o$ should increase $[Ca^{2+}]_i$, we do not know whether the effects shown in Figs. 2–5 are caused by calcium increases inside or outside the rod. For example, it is possible that increasing $[Ca^{2+}]_o$ produces hyperpolarization of the rod membrane potential and a decrease in the amplitude of receptor responses by some effect on the external surface of the plasma membrane. Sensitivity might be affected only by changes in $[Ca^{2+}]_i$, and the increases in $[Ca^{2+}]_o$ in our experiments might have been too small to produce significant changes in the intracellular concentration.

To avoid this problem, we repeated the experiments in Figs. 2–5 using the ionophore X537A, in an attempt to change intracellular calcium levels directly. X537A is a relatively nonspecific ionophore which has been shown in several systems to facilitate the permeability of biological membranes to Ca^{2+} (see McLaughlin and Eisenberg, 1975). For some reason we do not understand, the simple addition of X537A to the Ringer's solution superfusing the retina produces no effect (Hagins and Yoshikami, 1974; Bastian and Fain, 1979). However, when X537A is added in the presence of low extracellular Ca^{2+} (10^{-7} *M*) and the retina then returned to normal Ringer's solution, rod responses can be completely suppressed.

The protocol for such experiments is shown in Fig. 6. In both parts of this figure, the ordinate gives the absolute membrane potential and the heights of the horizontal axes indicate the potential of the rod membrane at a steady state in each of the various Ringer's solutions to which the rod was exposed. In each Ringer's solution responses were recorded at several light intensities. The intensity is plotted along the horizontal axes in units of log incident quanta per square centimeter per flash, and the vertical bars descending from these axes give the peak response amplitude for each flash. Thus the dotted lines trace out intensity–response functions, inverted from their usual format.

Figure 6A represents a control experiment showing the effect on a typical cell of 10^{-7} *M* $[Ca^{2+}]_o$ in the absence of X537A. The introduction of 10^{-7} *M* Ca^{2+} Ringer's solution produced a large depolarization of the rod and an increase in the maximum peak response amplitude (V_{max}). The return of the retina from 10^{-7} *M* Ca^{2+} to normal Ringer's solution caused the rod membrane potential to hyperpolarize to a point negative to the original resting potential. Figure 6 shows that, after 30 sec in normal Ringer's solution, the membrane potential can be almost 20 mV below its normal value, and that this hyperpolarization is accompanied by marked suppression of the light responses. With time this effect disappears, and within 3–4 min the rod membrane potential and light responses recover nearly to their original amplitudes.

Figure 6B shows for a different cell the effects of X537A. In Ringer's solution containing the ionophore and 10^{-7} *M* Ca^{2+}, the rod depolarized and the peak response amplitude greatly increased, much as in Fig. 6A. When, after perfusion with ionophore in 10^{-7} *M* Ca^{2+} Ringer's solution, the retina was returned to normal Ringer's solution, the rod membrane hyperpolarized below its original dark value, but in this case no light responses could be recorded from the rod,

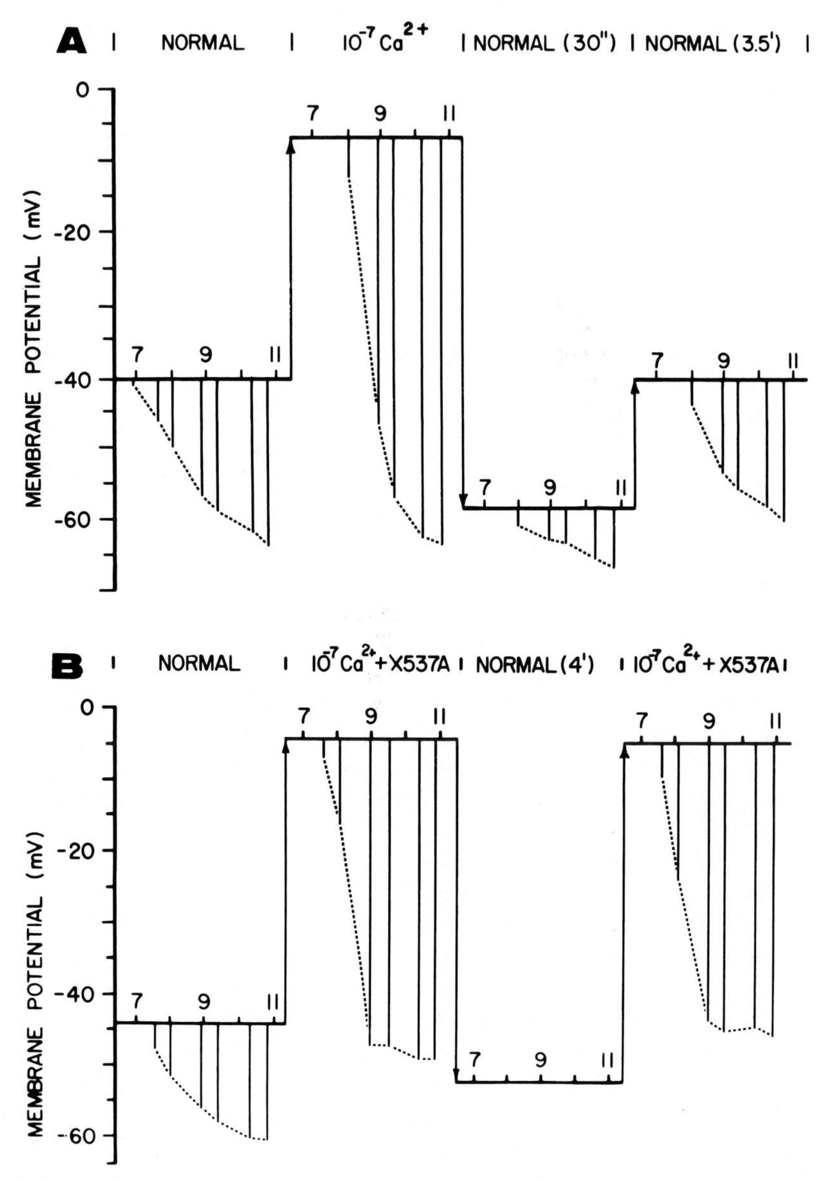

FIG. 6. Effects of the ionophore X537A on rod responses. The absolute membrane potential is plotted against the test flash intensity (log quanta per square centimeter) for the listed experimental conditions. Downward vertical lines give the maximum amplitude of responses measured from the resting potential to the indicated flash intensities. In a control experiment (A) intensity-response functions (dotted lines) are shown for normal, 10^{-7} *M* Ca^{2+}, and normal Ringer's solution 30 sec and 3.5 min after the solution change. (B) When 100 μM X537A was added to 10^{-7} *M* Ca^{2+} Ringer's solution, no responses were seen 4 min after a return to normal Ringer's solution, although full recovery was seen after a return to low-Ca^{2+} Ringer's solution. Reprinted with permission from Bastian and Fain (1979). Copyright by The Physiological Society.

regardless of the intensity of the stimulus. Notice that this was true even 4 min after the return to normal Ringer's solution, sufficient time for the cell in Fig. 6A to have almost fully recovered. The suppression was not the result of injury to the rod or deterioration of the preparation, since, when the retina was returned to low-Ca^{2+} Ringer's solution, the membrane potential and photoresponses returned nearly to their previous amplitudes.

The experiment in Fig. 6 demonstrates the feasibility of using X537A to alter intracellular Ca^{2+} levels, but it does not permit us to compare increased $[Ca^{2+}]_i$ with background light. The effect produced by exposing the retina to 100 μM X537A and then to normal Ca^{2+} levels is to suppress rod responses *completely*, much like exposing the retina to extremely bright background light (Fain, 1976) or to extracellular Ca^{2+} concentrations of 15–20 m*M* (G. L. Fain and H. M. Gerschenfeld, unpublished observations). To compare the effects of ionophore perfusion with those of background light, we must find a means of producing only partial suppression. We have discovered two ways of doing this: first, exposing the retina to the same ionophore concentration as in Fig. 6 but to an intermediate Ca^{2+} level, say 10^{-4} *M*; and second, reducing the ionophore concentration. Both procedures produce similar effects. We shall recount here only

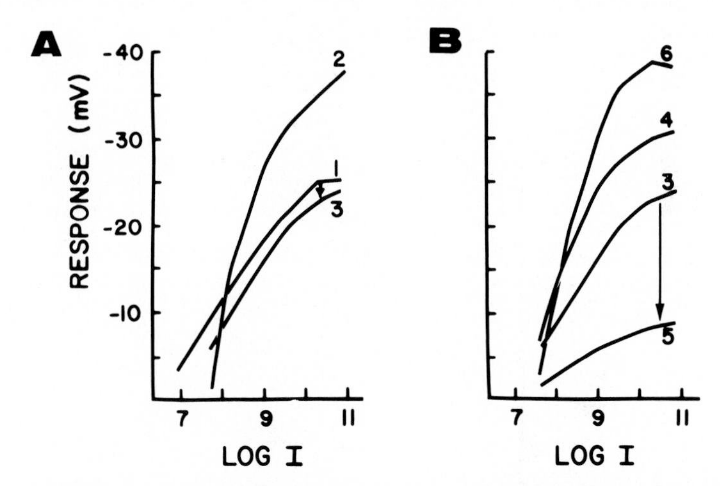

FIG. 7. Partial suppression of rod responses in normal Ringer's solution after exposure to X537A. (A) Intensity–response functions of one cell in the presence of the following solutions, in the order in which they were given: normal Ringer's solution (1), 10^{-7} *M* Ca^{2+} Ringer's solution (2), normal Ringer's solution (3), 10^{-7} *M* Ca^{2+} Ringer's solution with 10 μM X537A (4), normal Ringer's solution (5), and 10^{-7} *M* Ca^{2+} Ringer's solution without ionophore (6). Arrows indicate the difference between maximum responses in normal Ringer's solution before and after exposure to 10^{-7} *M* Ca^{2+}. The 10^{-7} *M* Ca^{2+} without ionophore (left) caused less suppression than the 10^{-7} *M* Ca^{2+} plus X537A (right). Reprinted with permission by Bastian and Fain (1979). Copyright by The Physiological Society.

the experiments in which we reduced the ionophore concentration, since they are somewhat simpler to describe.

In Fig. 7 we show an experiment identical to that in Fig. 6, except that the ionophore concentration was 10 μM. Since the changes in membrane potential were similar to those in Fig. 6, we have omitted the absolute potential indications and plotted only the intensity-response curves, numbered in the order of the solution changes. All the data in Fig. 7 are from the same cell. Figure 7A is a control experiment identical to the one in Fig. 6A. Prior exposure to 10^{-7} M Ca^{2+} has, by itself, little effect on the intensity-response relation in normal Ringer's solution (compare curves 1 and 3). In Fig. 7B, we show the effect of the ionophore. The arrow from curve 3 to curve 5 indicates the extent of suppression produced by prior exposure to 10 μM X537A. The effect of the ionophore is similar to that shown in Fig. 3 for increases in $[Ca^{2+}]_o$: In both cases there is a reduction in the scale of the responses but little shifting of the curves along the abscissa. This similarity is even more evident when the intensity-response curves in Fig. 7 are plotted on a normalized scale (Fig. 8). In Fig. 4, we have plotted as the x's the relative change in sensitivity (S_F/S_F^D) as a function of the reduction in maximum response amplitude ($\Delta V_{max}/V_{max}$) for all the ionophore experiments. The changes in sensitivity produced by treating the retina with X537A closely resemble those produced by increasing $[Ca^{2+}]_o$ and do not appear to resemble those that occur during the presentation of background light.

The effects of the ionophore on the response waveform also resemble those

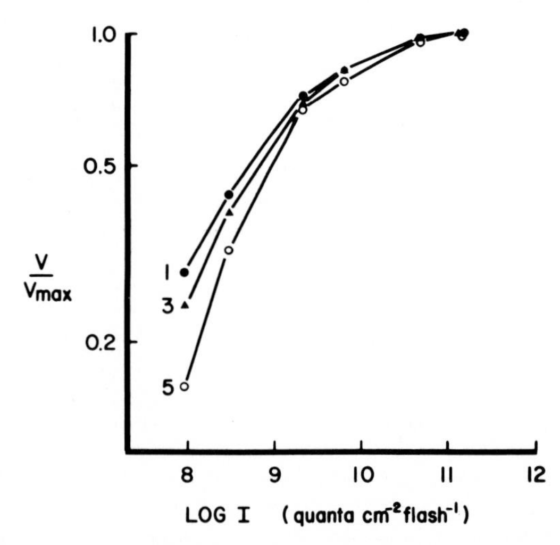

FIG. 8. Intensity-response curves of the rod in Fig. 7 on a normalized voltage scale. The solutions (labeled 1, 3, and 5) in which the curves were taken are described in the legend for Fig. 7. Reprinted with permission from Bastian and Fain (1979). Copyright by The Physiological Society.

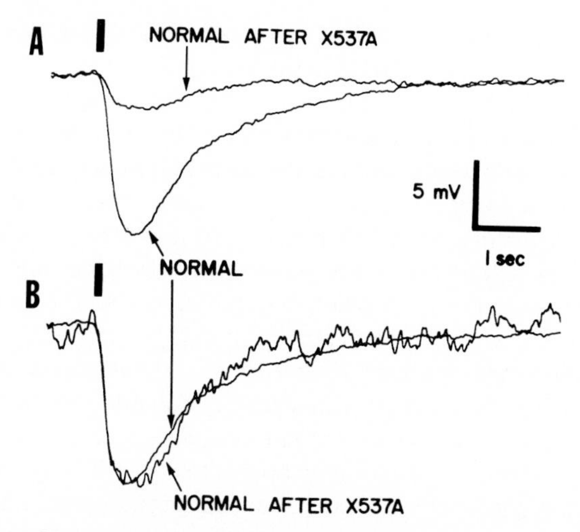

FIG. 9. Effects of ionophore-induced suppression on the small-amplitude response waveform of the cell in Figs. 7 and 8. (A) Traces compare responses to a flash of 8.0 log incident quanta cm^{-2} $flash^{-1}$ during conditions 1 (normal) and 5 (normal after X537A), described in the legend for Fig. 7. (B) Traces compare the same waveforms as in (A) but normalized at their peak amplitudes. Reprinted with permission from Bastian and Fain (1979). Copyright by The Physiological Society.

produced by high external Ca^{2+} levels. This can be seen in Fig. 9, which shows responses of the cell in Figs. 7 and 8 in normal Ringer's solution before and after treatment with X537A. The difference in the waveform is best seen in Fig. 9B, where the responses have been normalized to the same peak amplitude. For this cell there was little effect of the ionophore on the time to peak and the decay time, although in other cells responses after exposure to X537A showed a slowing of the response kinetics much like that in Fig. 5B.

In conclusion, we have attempted to alter the intracellular free calcium concentration, both by increasing extracellular calcium levels and by facilitating the passage of calcium through the plasma membrane with an ionophore. Both these procedures have the same effect. They produce hyperpolarization of the membrane potential and a decrease in sensitivity which superficially resemble the effects produced by background light. However, closer examination shows that the sensitivity changes produced by increasing Ca^{2+} levels are much too small to resemble light adaptation. Furthermore, we have been unable to mimic the effect of backgrounds on the receptor waveform by increasing intracellular calcium levels. These results are consistent with the hypothesis of Yoshikami and Hagins (1971) that Ca^{2+} is the internal messenger for excitation, since we have shown that increases in internal Ca^{2+} hyperpolarize the rod much as light does. However, they provide rather strong evidence against the possibility that increases in Ca^{2+} regulate receptor sensitivity during light adaptation.

IV. THE EFFECT OF DECREASES IN EXTERNAL Ca^{2+} ON SENSITIVITY

Although sensitivity is little altered by increases in Ca^{2+}, there is evidence that it *can* be affected by lowering Ca^{2+} levels. Yoshikami and Hagins (1973) have shown that exposure of rat retina to Ringer's solution in which the Ca^{2+} concentration has been lowered to 10^{-7} *M* produces a 10-fold decrease in the sensitivity of the extracellular light-induced voltage response, which they measure with two electrodes at the tips and bases of the rod outer segments. They claim that even larger desensitizations can be produced by prolonged treatment with solutions containing calcium concentrations below 10^{-8} *M*. Lipton *et al.* (1977) examined the effects of Ringer's solutions containing nominally 10^{-9} *M* Ca^{2+} on the intracellular responses of toad rods. They observed little effect on sensitivity during the first 4–5 min of exposure but a gradually increasing desensitization during the subsequent 5–10 min, which reduced sensitivity by as much as three or four orders of magnitude. The changes in sensitivity were accompanied by a progressive decrease in the maximum amplitude of the receptor response.

These observations, though suggestive, leave many unanswered questions. For example, the measurements of Yoshikami and Hagins in low-calcium solutions were not accompanied by a source–sink analysis. There is no evidence that the sources and sinks of the voltages Hagins and his colleagues recorded extracellularly were the same in normal and low Ca^{2+}, and it is therefore unclear whether the sensitivities they measured under the two circumstances were comparable. The measurements of Lipton *et al.* do not suffer from this disadvantage, since they were recorded intracellularly. However, there are other difficulties with their experiments. In the first place, the turnover time of their chamber (time for the exchange of one chamber volume) was approximately 1 min. It seems possible that the slow changes in sensitivity they observed were the result of the slow rate of perfusion in their experiments. In the second place, they took no precautions to prevent the rods from swelling. Rods in low-Ca^{2+} solutions show high Na^{+} permeability (Woodruff and Fain, 1980). If the receptors are also permeable to Cl^{-}, there will be a net influx of NaCl into the rods, which will lead to swelling. An effect of this kind has been observed in x-ray diffraction measurements (Chabre and Cavaggioni, 1973). Swelling can be reduced by replacing Cl^{-} by an impermeant anion, as first suggested by Hagins and Yoshikami (1977). In our experience this greatly improves the stability of intracellular recordings in low-Ca^{2+} solutions. Since Lipton *et al.* did not remove the extracellular Cl^{-}, their results in low Ca^{2+} are probably vitiated by a progressive deterioration of their preparations. They argue against this possibility by showing that the receptor membrane potential remained constant during their recordings. Since, however, the membrane potential of rods in low-Ca^{2+} solutions is near 0 mV (Fig. 6), the constancy of the membrane potential in their experiments is not strong evidence against cell damage.

A. Reexamination of Sensitivity Changes in Low Ca^{2+}

Because of these difficulties, we reexamined the effects of low-Ca^{2+} solutions on receptor sensitivity. Like Lipton *et al.*, we made intracellular recordings from the isolated, superfused retina of the toad *Bufo marinus*. However, the solutions in our chamber had a turnover time of less than 5 sec (see Bastian and Fain, 1979). In our experiments, we first measured the intensity–response relation of the rod in normal Ringer's solution. We then switched to a solution identical to the normal one except that most of the Cl^- was replaced by methane sulfonate. Exposure to this low-Cl^- solution, even for long periods of time, produced little change in sensitivity. We waited 2–3 min in the low-Cl^- Ringer's solution, to permit the cell to reach a steady state, and then changed to a low-Cl^- solution in which the Ca^{2+} was buffered with EGTA. A more complete description of the solutions, perfusion system, and procedures will be given elsewhere (Bastian and Fain, 1982).

When we exposed rods to Ca^{2+} concentrations below 10^{-3} *M*, the membrane potential depolarized and the maximum amplitude of the light responses increased. These two effects occurred roughly in parallel. We have described them for a 10^{-7} *M* Ca^{2+} solution in Fig. 6A. Similar changes occur for a variety of Ca^{2+} concentrations, from 0.4 m*M* to as low as 10^{-9} *M* (Bastian and Fain, 1979; Bastian and Fain, 1982), but we shall make no attempt to characterize them in detail in this chapter. Instead, we shall concentrate on the changes in sensitivity that occur in low-Ca^{2+} solutions.

As before, we shall define sensitivity as the peak amplitude of the voltage change per unit light intensity for small-amplitude responses. We find that, as the Ca^{2+} levels are reduced, sensitivity declines roughly in proportion to the decrease in Ca^{2+} concentration. This effect begins at about 10^{-6} *M*, where we can detect at most a 0.5–0.7-log unit reduction (three- to fivefold). As the Ca^{2+} concentration is reduced to even lower levels, the effects on the sensitivity can become quite dramatic. To illustrate these effects, we have chosen two typical experiments which we show in Fig. 10. In Fig. 10A, the solid circles give the intensity–response relation in normal Ringer's solution, and the other symbols the relations at different times after switching to a solution with a free Ca^{2+} concentration of 3×10^{-8} *M*. The first set of responses were measured beginning 2 min after changing solutions. By this time, sensitivity had decreased by almost 2 log units. During the succeeding 9 min, there were further decreases in sensitivity amounting to about 0.3 log unit. Although not shown for this cell, in another experiment we returned the retina to normal Ringer's solution after a 15-min

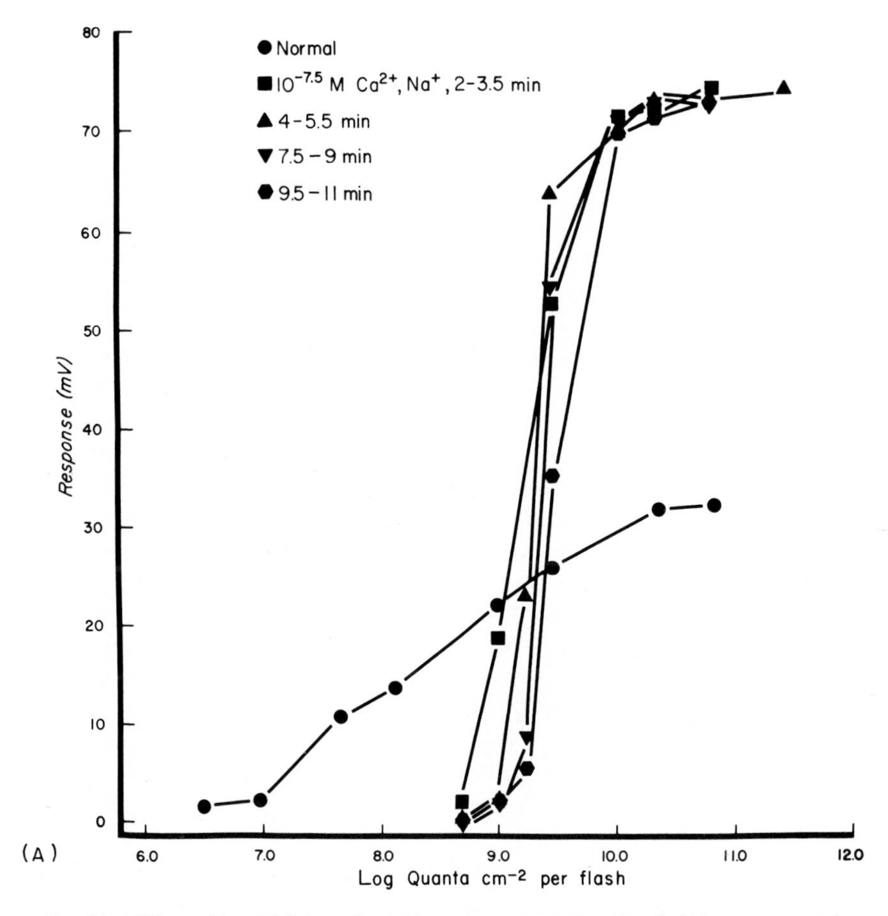

FIG. 10. Effects of low $[Ca^{2+}]_o$ on the rod intensity-response function. In (A) curves were taken in normal Ringer's solution and at various times after the introduction of Ringer's solution with 3×10^{-7} M Ca^{2+}. A similar display for Ringer's solution containing 10^{-9} M Ca^{2+} appears in (B).

exposure to a 3×10^{-8} M Ca^{2+} solution and observed a complete return of the sensitivity to its normal value. Furthermore, R. E. Greenblatt has demonstrated in our laboratory that the sensitivity of the extracellular photovoltage, measured between the tips and bases of the outer segments as in the experiments of Hagins and his colleagues, returns completely to normal in 3-5 min, even after long exposures to 3×10^{-8} M Ca^{2+} solutions (R. E. Greenblatt, in preparation). Thus the changes in sensitivity shown in Fig. 10A appear to be entirely reversible.

In Fig. 10B we show similar data for 10^{-9} M Ca^{2+}. The solid circles again show the intensity-response curve in normal Ringer's solution. After switching to the low-Ca^{2+} solution, we waited for the membrane potential to stabilize and

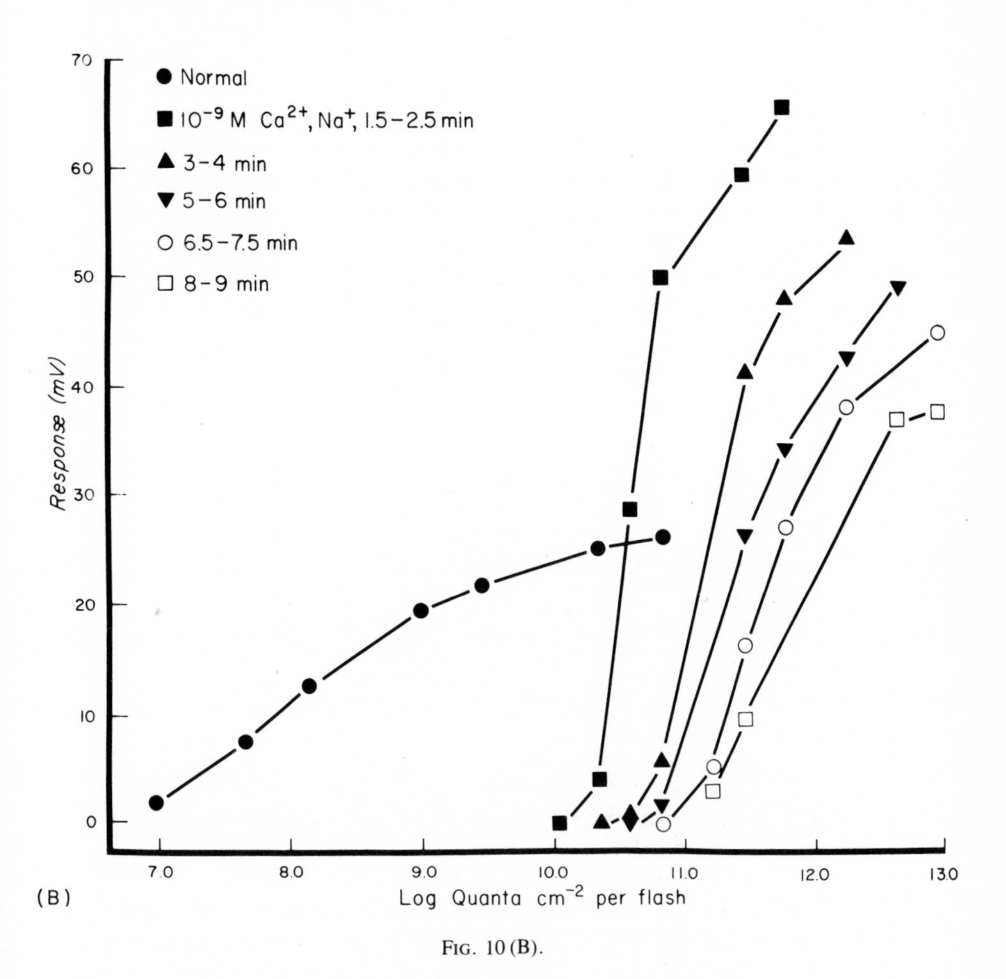

FIG. 10 (B).

then began immediately to record an intensity–response curve. These data, plotted as the solid squares, show that, as soon as we could measure stable responses in the cell, the sensitivity had decreased by about three orders of magnitude. During the next 1½–2 min, the sensitivity declined by about another ½ log unit; and in the following 5–6 min, there was a further ½-log unit decrease. Notice that the changes in sensitivity with time were larger in the 10^{-9} M solution than in 3×10^{-8} M Ca^{2+}. Since we changed the external instead of the internal Ca^{2+} concentration, it is possible that the slow changes in sensitivity reflect slow alterations in the intracellular Ca^{2+} levels, perhaps as the result of an alteration in the rate of Ca^{2+} transport across the plasma membrane.

As we have said, the change in receptor sensitivity produced by exposing the rod to 3×10^{-8} M Ca^{2+} can be completely reversed on returning the retina to normal Ringer's solution. Complete reversal also occurred in 10^{-8} M Ca^{2+} and in all the other low-Ca^{2+} solutions we used in our experiments up to the 10^{-9} M Ca^{2+} Ringer's solution used in Fig. 10B. In this case, the return to normal Ringer's solution caused sensitivity to recover only to within about 1 log unit of its normal value, even after extensive washing (R. E. Greenblatt, in preparation).

In Fig. 11 we have summarized all our experiments on the effects of low-Ca^{2+} solutions on sensitivity. On the ordinate, we have plotted the log of the ratio of the dark-adapted sensitivity in the low-Ca^{2+} solution (S_F) to the dark-adapted

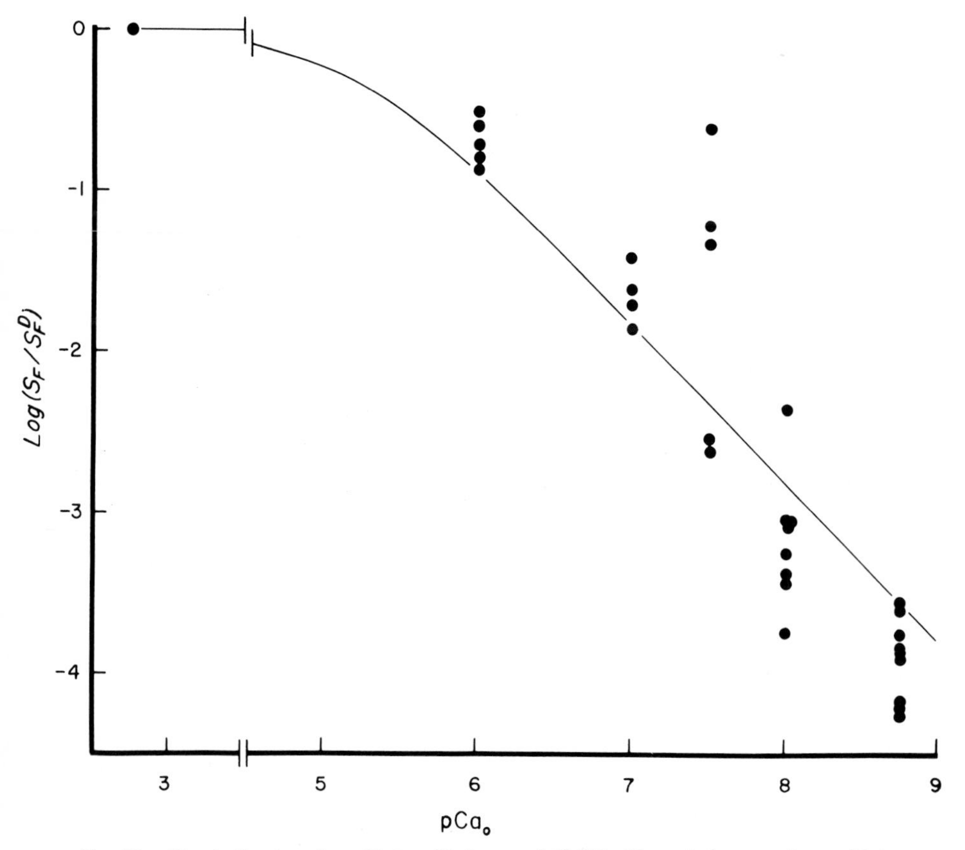

FIG. 11. The decline in rod sensitivity with decreased $[Ca^{2+}]_o$. The ratio between the sensitivity measured in low-Ca^{2+} Ringer's solution, S_F, and in normal Ringer's solution S_F^D, is plotted against pCa_o. Cells for which a time series was collected (as in Fig. 10) have sensitivity ratios plotted for both the most and the least sensitive curves measured. The number of separate cells represented for pCa_o values of 6, 7, 7.5, 8, and 9 were 3, 3, 3, 7, and 5, respectively. The smooth curve is given by the equation $S_F/S_F^D = [Ca]^n/([Ca]^n + K_D)$, with $n = 1$ and $K_D = 7 \times 10^{-6}$ M (see text).

sensitivity in normal Ringer's solution (S_F^D). The abscissa gives the negative logarithm of the Ca^{2+} concentration. Each point represents a single intensity–response curve. For some cells, several intensity–response curves were measured at different times after switching to the low-Ca^{2+} solution (as in Fig. 10), and these have been included in Fig. 11. The number of different cells represented at each concentration is given in the figure legend. Sensitivity was estimated from the intensity–response data by fitting them with a least-squares nonlinear regression algorithm and then using the best-fitting parameters to interpolate the sensitivity at a response amplitude of 1 mV (see Bastian and Fain, 1979). The curve we have fitted to the data in Fig. 11 is of the form

$$S_F/S_F^D = [Ca^{2+}]^n/([Ca^{2+}]^n + K_s) \qquad (2)$$

which is a modification of the Hill equation (Hill, 1910). A statistical analysis of the data in Fig. 11 (Bastian and Fain, 1982) gives a best-fitting value for n that is very close to 1. Thus over most of the range of calcium concentrations we measured, sensitivity declined nearly linearly with the extracellular Ca^{2+} concentration. The implications of this finding are at present problematic, since we do not know where the calcium acts or the nature of the relation between the extracellular and intracellular calcium concentrations.

B. Effects of Low Ca^{2+} on the Response Waveform

The experiments in Figs. 10 and 11 show that low-calcium solutions mimic background lights in their effect on receptor sensitivity. However, background lights, in addition to reducing sensitivity, also produce large changes in the response waveform (Fig. 5). In Fig. 12, we compare the effects of background light and low Ca^{2+} on rod responses. The superimposed waveforms in Fig. 12A were recorded in normal Ringer's solution in the dark-adapted retina (labeled "Normal, Dark") and in the presence of a 10.9 log quanta cm^{-2} sec^{-1} background (labeled "Bkgnd"). The intensity of the test flash was nearly 1000 times brighter in the presence of the background than in the dark, since the background decreased the sensitivity of the rod. The response to this brighter test flash, although comparable in amplitude to the one recorded in the dark, is shorter in latency, shorter in time to peak, and faster in its decay to the baseline. After recording these data, we turned off the background and waited for the cell to return to its dark-adapted sensitivity. We then switched to a low-Cl^- Ringer's solution and then to a solution containing low Cl^- and a free Ca^{2+} concentration of 10^{-8} M. Though this solution desensitized the rod by nearly the same amount as the background, its effects on the response waveform do not appear to be similar to those produced by light adaptation (Fig. 12B). When responses of nearly equal amplitude were compared, a low-Ca^{2+} solution produced a pro-

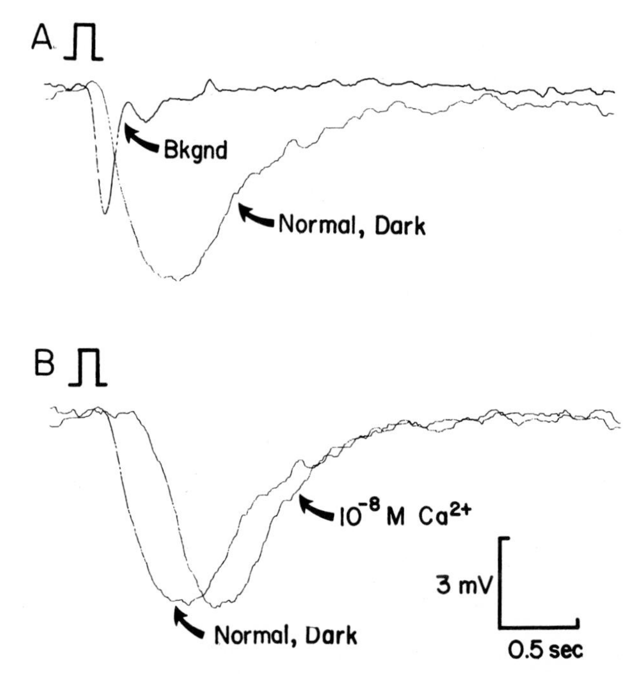

FIG. 12. The effects of background light and low-Ca^{2+} Ringer's solution on small-amplitude response kinetics. In (A) the dark-adapted response of a rod in normal Ringer's solution is compared to the response in the presence of a background (10.9 log quanta cm^{-2} sec^{-1}). The flash intensity for the dark-adapted response was 7.7 log quanta/cm^2 per flash, and for the light-adapted, 10.9 log quanta/cm^2 per flash. The onset time, time to peak and time to decay are all much shorter for the adapted response. In (B) for the same cell as in (A), the dark-adapted response in normal Ringer's solution [taken from (A)] is compared to a dark-adapted response (9.8 log quanta/cm^2 per flash) in 10^{-8} Ca^{2+} Ringer's solution.

nounced increase in the absolute latency, which was in a direction opposite that produced by background light (Fig. 12A). A latency increase might actually have been anticipated, since exposure to low-Ca^{2+} solutions has been shown to increase the cyclic guanosine 3′,5′-monophosphate (cGMP) concentration in rods (Woodruff and Bownds, 1979; Bownds, this volume, Chapter 11; Cohen, this volume, Chapter 6), and the injection of cGMP into rods has been shown to increase response latency (Miller and Nicol, this volume, Chapter 24). The low-Ca^{2+} solution also increased the time to peak, though most of this effect appears to be a result of the increase in absolute latency. Backgrounds, on the other hand, considerably decrease the time to peak (Fig. 12A). The low-Ca^{2+} Ringer's solution also somewhat shortened the time to decay of the response. This occurred in the same direction as that produced by background light, but the effect was very much smaller (Fig. 12A).

The data in Fig. 12 show that low-Ca^{2+} solutions do not mimic background light in their effect on the response waveform. Large changes in sensitivity can occur in low-Ca^{2+} solutions without the pronounced decreases in time to peak and time to decay that usually accompany desensitization during light adaptation. The converse appears also to be true; that is, large changes in time to peak and time to decay can occur with minimal effects on the sensitivity. This can be seen in the data in Figs. 13 and 14. Here we first exposed the rod to 10^{-8} Ca^{2+} Ringer's solution for 3 min to permit the greater part of the desensitization in this solution to occur. We then turned on background illumination at a series of intensities in the low-Ca^{2+} solution. In the dimmest background (7.3 log quanta cm^{-2} sec^{-1}), the peak response amplitude was unaffected but the sensitivity was

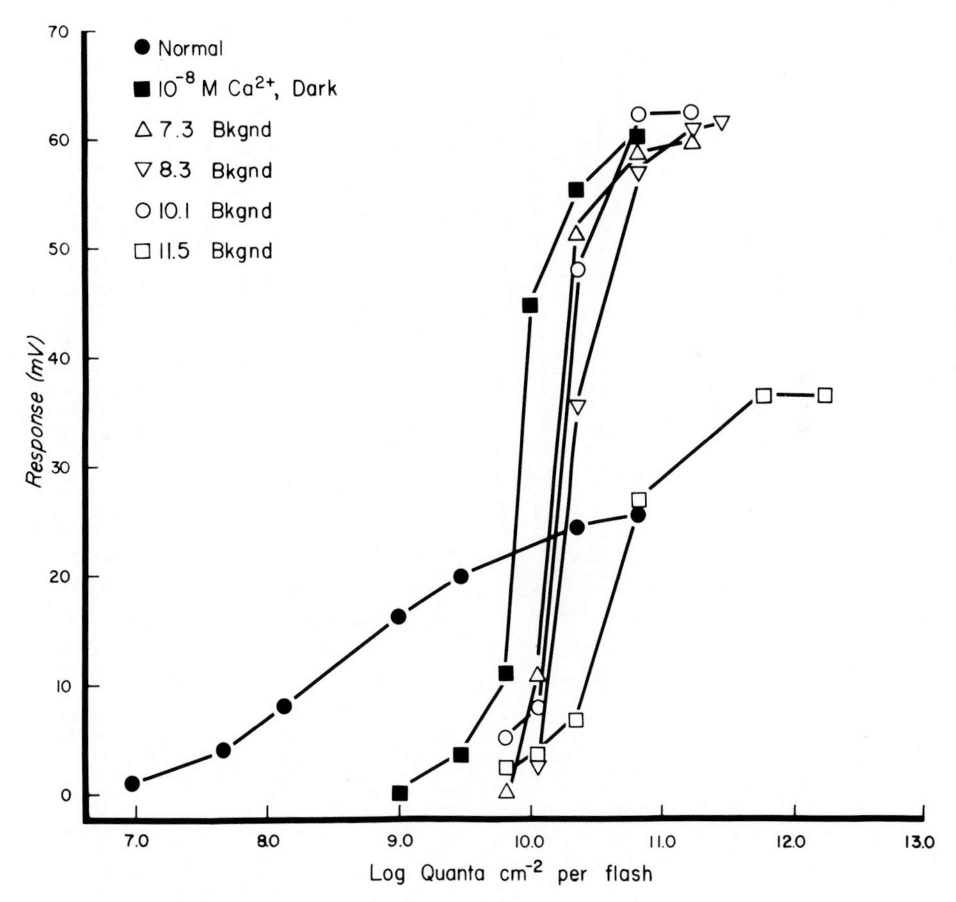

FIG. 13. Effect of background light on the intensity–response function in low-Ca^{2+} Ringer's solution. See text for details.

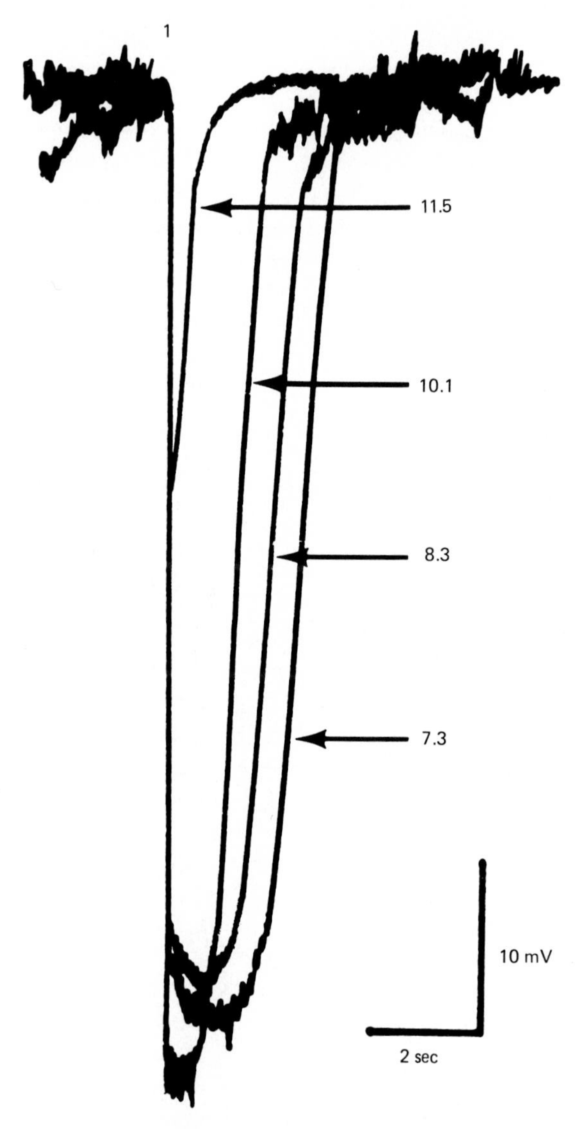

FIG. 14. The effects of background light on response kinetics in 10^{-8} M Ca^{2+} Ringer's solution for the same cell as in Fig. 13. The intensity for all the responses was 10.8 log quanta/cm^2 per flash. Arrows indicate the responses at four different background levels, and the numbers adjacent to the arrows are the intensities of the backgrounds in units of log quanta per square centimeter per second.

reduced by about ½ log unit. Because the cell had only been exposed to low Ca^{2+} for 3 min before the background was turned on, it is possible that some of this loss of sensitivity was due to the continued exposure to the 10^{-8} *M* solution (Fig. 10). As the backgrounds were made successively brighter, there was little if any effect on the response amplitude or sensitivity until the rod approached increment saturation. At a background of 11.5 log quanta cm^{-2} sec^{-1}, V_{max} was reduced by about one-half, although the sensitivity was still nearly the same as it was with the 7.3 log quanta cm^{-2} sec^{-1} background (see foot of curve). When the background light was increased from this level, even by only ½ log unit, the rod became completely unresponsive. This effect, called increment saturation (see Fain, 1976), appears to occur at the same background intensity in low-Ca^{2+} solutions as in normal Ringer's solution in spite of the large desensitization produced by exposure to reduced Ca^{2+} levels.

Figure 13 shows that, in a low-Ca^{2+} solution, background lights produce much smaller changes in sensitivity than in normal Ringer's solution. In particular, an increase in background levels of over 10,000-fold (from 7.3 to 11.5 log quanta cm^{-2} sec^{-1}) has almost no effect on the light intensity necessary to produce a just-detectable response. However, these backgrounds can have appreciable effects on the waveform. In Fig. 14 we show for the cell in Fig. 13 the responses to a 10.8 log quanta cm^{-2} $flash^{-1}$ test flash at four background intensities, from 7.3 to 11.5 log quanta cm^{-2} sec^{-1}. As the background was made brighter, a progressive decrease in time to peak and time to decay occurred, just as in normal Ringer's solution. Although not resolvable in this figure, brighter backgrounds also produced a progressive decrease in the latency of the response in low-Ca^{2+} solutions. It should be noted that these kinetic changes occurred even though all but the brightest of the backgrounds shown in Fig. 14 produced no perceptible change in the rod's resting potential. Since the peak amplitudes of the responses recorded in the presence of these backgrounds were nearly identical (Figs. 13 and 14), the large kinetic changes observed cannot be an effect mediated by voltage-dependent conductances.

In summary, decreases in external Ca^{2+} levels can produce large reductions in receptor sensitivity without the quickening of the response waveform characteristic of light adaptation. Conversely, when rods in low Ca^{2+} are exposed to background light, there may be large changes in the response waveform and little effect on sensitivity. The implications of these findings for the mechanism of sensitivity regulation will be discussed in the concluding section.

V. THE MECHANISM OF LIGHT ADAPTATION IN RECEPTORS

When the toad retina is exposed to continuous background illumination, the sensitivity of the rods decreases. It is now widely accepted that light adaptation is not the result of a simple reduction in the probability of photon absorption (Hecht, 1937), since even at increment saturation only a small fraction of the rhodopsin of a rod is bleached by background light (see Fain, 1976, for references). It is also unlikely that changes in input conductance make a significant contribution, since rod outer segment currents can be adapted by background light in much the same fashion as intracellular voltage responses (see Baylor and Matthew, this volume, Chapter 1). It therefore seems likely that adaptation occurs by some alteration in the mechanism of transduction itself.

Several theories have been proposed to explain light adaptation in vertebrate photoreceptors. A short discussion of these theories may help clarify the significance of our own experiments. To simplify our discussion, we shall assume that light causes the release of an internal messenger, call it X, which diffuses to the plasma membrane and blocks the light-dependent conductance. Note that X is the internal messenger for excitation. With this in mind, we can classify theories of light adaptation into two groups. One postulates that background lights produce a depletion of X from the disks. We shall call these "depletion theories." The other postulates that light alters the rate of reuptake or destruction of X; we shall call these "reuptake theories."

A. Depletion Theories

In their simplest form, depletion theories propose that a steady background light directly decreases the rate of release of the internal messenger. A theory of this kind has been proposed by Kleinschmidt and Dowling (1975), who refer to X as the blocking substance:

> On this view, the amount of blocking substance released per visual pigment molecule bleached is a constant fraction of the amount of substance contained in the saccules. Further, in steady light the blocking substance is depleted from the saccules and the extent of depletion is intensity dependent. Thus, more visual pigment must be activated during light adaptation to cause the release of a given (e.g., threshold) number of blocking molecules.

It is now easy to see that Kleinschmidt and Dowling's hypothesis cannot be true, at least in its simplest form. As we and Donner and Hemilä (1978) have shown, significant changes in receptor sensitivity can occur at backgrounds so dim that only a small fraction of the disks (or saccules) in a rod could have absorbed even one quantum. This does not exclude Kleinschmidt and Dowling's theory altogether, since one could imagine that the bleaching of a pigment molecule in one rod causes the release of a diffusible substance which reduces the concentration of X, even in disks that do not contain bleached rhodopsin molecules. However, there is a more powerful argument, first given for *Limulus* by Fuortes and Hodgkin (1964), which seems to exclude depletion theories of all types. When a rod or cone in normal Ringer's solution is stimulated with a flash of constant intensity, the initial waveform of the voltage change is, to a first approximation, the same in the presence of background light as in the dark. If the background light had depleted the concentration of X, one would expect the initial rise time of the response to be slower, since at any given flash intensity fewer molecules of X would be released. Furthermore, one would expect the initial rise times to be the same only when the amplitudes of the responses were the same, since only then would the rates of release of X be expected to be comparable. In the few cases where responses to constant-intensity flashes have been compared in dark-adapted and light-adapted receptors, there is little or no difference in the initial rise time (see Figs. 1B and 5A; Baylor and Hodgkin, 1974; Coles and Yamane, 1975). Constant-amplitude responses, on the other hand, are markedly faster in the presence of background light (Fig. 12A). We have observed that in low-Ca^{2+} Ringer's solution, changes in the initial kinetics of adapted responses to constant-intensity flashes can actually occur. However, backgrounds in this case apparently reduce latency without altering the rate of rise of the response, a result equally inconsistent with depletion theories. In our opinion the above observations make any form of depletion theory improbable, but it will clearly be necessary to investigate initial rise times in more detail before discarding these theories altogether.

B. Reuptake Theories

In contrast to depletion theories, reuptake theories postulate that there is no change in the rate of release of X at any given flash intensity when the retina is exposed to background light. This assumption is made to account for the constancy of the initial rise time of the response for constant-intensity flashes, described in the previous section. Changes in sensitivity are postulated to be produced by changes in the rate of reuptake or destruction of X. The most detailed of the reuptake theories is the one proposed by Baylor *et al.* (1974) for turtle cones. They postulate that X (which they call z_1) is inactivated by a reversible reaction to some substance z_2. This reaction could be an initial step in the resequestration of X back into the disks, an enzymatic conversion of X to an inactive form, or some other process. The central assumption of their theory is that the inactivation of X is autocatalytic; that is, the rate of removal of X increases as z_2 increases. In steady background light, the concentration of X and therefore of z_2 will increase. As a result, the rate of reuptake or inactivation of X will increase. In this way, Baylor *et al.* account for the changes in time to peak and decay time of responses in the presence of background illumination, and for desensitization.

The motivating concept in most reuptake theories, including that of Baylor *et al.*, is the notion that changes in the response waveform and in sensitivity are produced by the same mechanism, namely, the increase in the rate of transmitter inactivation. Our experiments show that there are circumstances for which this is not the case for toad rods. When rods in low-calcium solutions are exposed to background light (Figs. 13 and 14), it is possible to observe large changes in the receptor waveform with virtually no change in the sensitivity. Admittedly, we do not know how low Ca^{2+} desensitizes the rod. It is possible, for example, that exposure to low Ca^{2+} produces a decrease in the sensitivity by some process that is not normally part of the transduction machinery. Nevertheless, when the background is turned on, the decay time, especially for large-amplitude responses, shortens dramatically (Fig. 14). Background light, even in low Ca^{2+}, appears to change the rate of reuptake or destruction of the internal messenger. However, these changes in the waveform are not accompanied by significant changes in receptor sensitivity. Somehow, the rate of removal of X can be increased with very little change in the total number of X molecules released per flash.

While evidence already exists indicating that the Baylor–Hodgkin–Lamb model is unsuitable for rods (Coles and Yamane, 1975; Cervetto *et al.*, 1977), our experiments show for the first time that changes in the waveform can be dissociated from changes in the sensitivity. However, our results cannot be interpreted to exclude all forms of the reuptake hypothesis. Since our experiments were done under very nonphysiological conditions (10^{-8} *M* extracellular Ca^{2+}), they do not necessarily indicate that the sensitivity is unaffected by changes in reuptake in normal Ringer's solution. The most that can be said at present is that desensitization in toad rods is probably not entirely the result of changes in the rate of reuptake of the internal messenger.

C. Light Adaptation and the Molecular Mechanism of Transduction

Light initiates a series of biochemical changes in the rod outer segment, which are somehow responsible for visual transduction. As the other chapters in this book demonstrate, the absorption of a photon by a rhodopsin molecule leads to the binding and eventual hydrolysis of guanosine 5′-triphosphate (GTP); the stimulation of cyclic GMP-phosphodiesterase and a decrease in cyclic GMP; the phosphorylation of rhodopsin and certain other, low-molecular-weight proteins; an efflux of Ca^{2+}; and the closing of a light-dependent conductance channel. These processes, and others yet to be discovered, are likely to be involved not only in producing the change in receptor membrane potential but also in modulating the sensitivity of the rod during light and dark adaptation.

If we are to understand how adaptation occurs, we must discover how changes in the rates of the various light-dependent reactions in rod outer segments result in changes in the receptor waveform and sensitivity. Our approach to this problem has been pharmacological. That is, we attempted to alter the adaptation of rods by perfusing them with ions, nucleotides, and drugs of various kinds. Having characterized conditions that alter the sensitivity or the waveform of the receptor, we would next like to know how these conditions alter the biochemistry of the rods. If, as we show in Fig. 11, the sensitivity of the rod declines in proportion to the extracellular Ca^{2+} concentration, we would next hope to discover which of the light-dependent reactions are also affected by these conditions. It is not enough to show, as Bownds and his collaborators have done (this volume, Chapter 11) that several light-dependent reactions can be affected by Ca^{2+} in isolated rod outer segments. We must also show which *are* affected under the defined experimental conditions and whether these effects are sufficient to account for the observed desensitization. It is our hope that studies on the effects of Ca^{2+} and other agents on rod sensitivity can be combined with observations on their effects on the biochemistry of intact photoreceptors in order to provide a more complete understanding of the mechanism of light adaptation.

ACKNOWLEDGMENTS

We are grateful to R. E. Greenblatt for a careful reading of our manuscript. This work was supported in part by a grant from the National Eye Institute (EY 02728).

REFERENCES

Bader, C. R., MacLeish, P. R., and Schwartz, E. A. (1979). A voltage-clamp study of the light response in solitary rods of the tiger salamander. *J. Physiol. (London)* **296,** 1–26.

Barlow, H. B. (1972). Dark and light adaptation: Psychophysics. *In* "Handbook of Physiology" (D. Jameson and L. M. Hurvich, eds.), Vol. VII/4, pp. 1–28. Springer-Verlag, Berlin and New York.

Bastian, B. L., and Fain, G. L. (1979). Light adaptation in toad rods: Requirement for an internal messenger which is not calcium. *J. Physiol. (London)* **297,** 493–520.

Bastian, B. L., and Fain, G. L. (1982). Submitted.

Baylor, D. A., and Hodgkin, A. L. (1974). Changes in time scale and sensitivity in turtle photoreceptors. *J. Physiol. (London)* **242,** 729–758.

Baylor, D. A., Hodgkin, A. L., and Lamb, T. D. (1974). Reconstruction of the electrical responses of turtle cones to flashes and steps of light. *J. Physiol. (London)* **242,** 759–791.

Bertrand, D., Fuortes, M. G. F., and Pochobradsky, J. (1978). Actions of EGTA and high calcium on the cones in the turtle retina. *J. Physiol. (London)* **275,** 419–437.

Brown, J. E., and Pinto, L. H. (1974). Ionic mechanism for the photoreceptor potential of the retina of *Bufo marinus. J. Physiol. (London)* **236,** 575–591.

Cervetto, L., Pasino, E., and Torre, V. (1977). Electrical responses of rods in the retina of *Bufo marinus. J. Physiol. (London)* **267,** 17–51.

Chabre, M., and Cavaggioni, A. (1973). Light induced changes of ionic flux in the retinal rod. *Nature (London), New Biol.* **224,** 118–120.

Coles, J. A., and Yamane, S. (1975). Effects of adapting lights on the time course of the receptor potential of the anuran retinal rod. *J. Physiol. (London)* **247,** 189–207.

Donner, K. O., and Hemilä, S. (1978). Excitation and adaptation in the vertebrate rod photoreceptor. *Med. Biol.* **56,** 52–63.

Fain, G. L. (1976). Sensitivity of toad rods: Dependence on wave-length and background illumination. *J. Physiol. (London)* **261,** 71–101.

Fain, G. L., Granda, A. M., and Maxwell, J. H. (1977). Voltage signal of photoreceptors at visual threshold. *Nature (London)* **265,** 181–183.

Fuortes, M. G. F., and Hodgkin, A. L. (1964). Changes in time scale and sensitivity in the ommatidia of *Limulus. J. Physiol. (London)* **172,** 239–263.

Hagins, W. A., and Yoshikami, S. (1974). A role for Ca^{2+} in excitation of retinal rods and cones. *Exp. Eye Res.* **18,** 299–305.

Hagins, W. A., and Yoshikami, S. (1977). Intracellular transmission of visual excitation in photoreceptors: Electrical effects of chelating agents introduced into rods by vesicle fusion. *In* "Vertebrate Photoreception" (H. A. Barlow and P. Fatt, eds.), pp. 97–138. Academic Press, New York.

Hecht, S. (1937). Rods, cones and the chemical basis of vision. *Physiol. Rev.* **17,** 239–290.

Hecht, S., Schlaer, S., and Pirenne, M. H. (1942). Energy, quanta, and vision. *J. Gen. Physiol.* **25,** 819–840.

Hill, A. V. (1910). The possible effects of the aggregation of the molecule of haemoglobin on its dissociation curves. *J. Physiol. (London)* **40,** iv-vii.

Kleinschmidt, J. (1973). Adaptation properties of intracellularly recorded *Gekko* photoreceptor potentials. *In* "Biochemistry and Physiology of Visual Pigments" (H. Langer, ed.), pp. 219–224. Springer-Verlag, Berlin and New York.

Kleinschmidt, J., and Dowling, J. E. (1975). Intracellular recording from *Gekko* photoreceptors during light and dark adaptation. *J. Gen. Physiol.* **66,** 617–648.

Lipton, S. A., Ostroy, S. E., and Dowling, J. E. (1977). Electrical and adaptive properties of rod photoreceptors in *Bufo marinus*. I. Effects of altered extracellular Ca^{2+} levels. *J. Gen. Physiol.* **70,** 747–770.

McLaughlin, S., and Eisenberg, M. (1975). Antibiotics and membrane biology. *Annu. Rev. Biophys. Bioeng.* **4,** 335–366.

Woodruff, M. L., and Bownds, M. D. (1979). Amplitude, kinetics, and reversibility of a light-induced decrease in guanosine 3′,5′-cyclic monophosphate in frog photoreceptor membranes. *J. Gen. Physiol.* **73,** 629–653.

Woodruff, M. L., and Fain, G. L. (1980). Ion selectivity of the light-sensitive conductance of vertebrate photoreceptors. *Fed. Proc., Fed. Am. Soc. Exp. Biol.* **39,** 2071 (abstr.).

Yoshikami, S., and Hagins, W. A. (1971). Light, calcium, and the photocurrent of rods and cones. *Biophys. J.* **15,** 47a.

Yoshikami, S., and Hagins, W. A. (1973). Control of the dark current in vertebrate rods and cones. *In* "Biochemistry and Physiology of Visual Pigments" (H. Langer, ed.), pp. 245–255. Springer-Verlag, Berlin and New York.

Chapter 20

Effects of Cyclic Nucleotides and Calcium Ions on *Bufo* Rods

JOEL E. BROWN AND GERALDINE WALOGA[1]

Department of Physiology and Biophysics
State University of New York at Stony Brook
Stony Brook, New York

I. INTRODUCTION

In vertebrate rods, both calcium ions (Hagins, 1972) and cyclic nucleotides (Goridis *et al.*, 1974; Hubbell and Bownds, 1979; Liebman and Pugh, 1979; Pober and Bitensky, 1979) have been proposed to be the diffusible messenger that mediates excitation between the rhodopsin that absorbs a photon and the conductance channels of the outer segment plasma membrane. Consistent with these hypotheses, several studies have shown that alteration of the intracellular concentration of calcium ions, $[Ca^{2+}]_i$ (Hagins and Yoshikami, 1974; Hagins and Yoshikami, 1977; Brown *et al.*, 1977), or cyclic nucleotides (Lipton, *et al.*, 1977b; Miller and Nicol, 1979) can alter the electrical activity of rods. It is known in several other systems (Berridge, 1975; Rasmussen *et al.*, 1975) that the

[1]Present address: Department of Physiology, Boston University School of Medicine, Boston, Massachusetts.

ISBN 0-12-153315-8

mechanisms that control the intracellular concentration of calcium ions and of cyclic nucleotides interact. By analogy, it has been proposed that $[Ca^{2+}]_i$ and intracellular cyclic nucleotides also interact in rod outer segments so that a change in either produces similar changes in the electrical activity of the rods (Lipton *et al.*, 1977b).

To examine further the possibility that changes in the concentration of both Ca^{2+} and cyclic nucleotides produce the same changes in the electrical activity of rods, we have recorded intracellularly from rod outer segments in the isolated superfused retina of *Bufo marinus*. We have examined the sensitivity of the rods to light, the kinetics of the receptor potential, and the membrane voltage in the dark during procedures that ought to change the intracellular concentration of either Ca^{2+} or cyclic nucleotides.

II. METHODS

Pieces of retina from *B. marinus* were dissected free of pigment epithelium and mounted receptor side up on the bottom of a perfusion dish as described previously (Brown and Pinto, 1974; Brown, Coles, and Pinto, 1977). All dissections and subsequent manipulations were viewed under infrared illumination with the aid of image converter tubes. The retinas were continuously bathed in a solution saturated with oxygen and containing 108 m*M* NaCl, 2.5 m*M* KCl, 1.2 m*M* $MgCl_2$, 1.6 m*M* $CaCl_2$, 5.0 m*M* dextrose, and 3 m*M* *N*-2-hydroxyethylpiperazine *N'*-2-ethanesulfonic acid (HEPES) adjusted to pH 7.8 with NaOH. To lower the calcium concentration, $CaCl_2$ was omitted from the solution (denoted as the 0.Ca-added solution); the only calcium supplied was as impurities in the other salts. The volume of the perfusion dish was 0.3 ml, and the dish was perfused at a rate of at least 2 ml/min.

Rods were impaled with electrodes filled with 0.5 *M* potassium acetate unless otherwise noted. Methods for stimulating and recording have been described previously (Brown and Pinto, 1974; Brown *et al.*, 1977). The unattenuated intensity of the stimulus light falling on the retina was approximately 6×10^{-6} W/cm^2.

III. RESULTS

A. Reduction in Extracellular Calcium Ion Concentration

We constructed a graph of the amplitude of the receptor potential versus log stimulus light intensity both before and after the concentration of calcium ions in the bath, $[Ca^{2+}]_o$, was reduced. The graphs represent the following experiment. First, a rod was impaled, and both membrane voltage in the dark and the amplitude of receptor potentials elicited by dim stimuli were judged to be stable. Then a series of flashes was given, each 0.5 log unit brighter than its predecessor, until the amplitude of the receptor potential saturated. The time between flashes was adjusted to allow the membrane voltage to return to its steady level in the dark. Then the dish was perfused with O·Ca-added solution. After the membrane voltage reached a new steady level, the same series of flashes of progressively increasing intensity was given.

The results of such an experiment are shown in Fig. 1. When the retina was bathed in O·Ca-added solution, membrane voltage in the dark became more positive. The peak-to-peak amplitudes of the receptor potentials elicited by bright stimuli became larger, as previously described (Brown and Pinto, 1974; Lipton *et al.*, 1977a; Bastian and Fain, 1979). However, the peak-to-peak amplitudes of the receptor potentials elicited by dim stimuli were smaller than those in the control solution; for the responses elicited by dim stimuli, the slope of the curve measured in O·Ca-added solution was less than in the control solution. Thus the

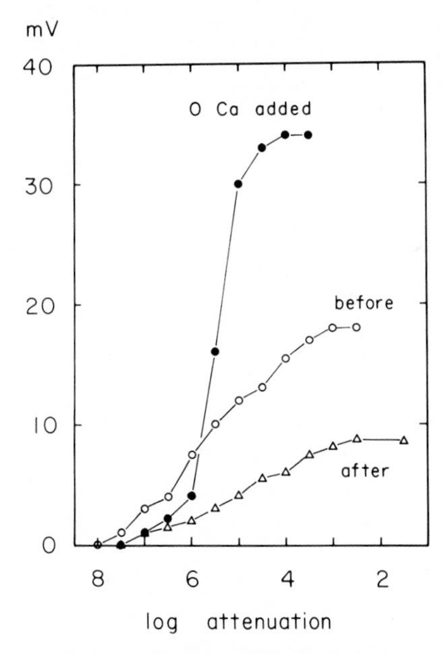

FIG. 1. The effects of reducing extracellular calcium on the receptor potential amplitude as a function of stimulus light intensity. To construct each graph, the cell was dark-adapted and then stimulated sequentially by flashes of increasing intensity. The cell was first bathed in 1.6 mM $[Ca^{2+}]_o$ (open circles), then bathed in O·Ca-added solution for 15 min (closed circles), and then returned to 1.6 mM $[Ca^{2+}]_o$ (open triangles). Note that the graph constructed while the retina was bathed in O·Ca-added solution crosses the control curve and that the intensity at half-saturation (arrows) was displaced to a dimmer intensity. Unattenuated intensity: 6.4 10^{-6} W/cm^2 on the retina. The half-saturation intensity for the cell when initially dark-adapted and bathed in 1.6 mM $[Ca^{2+}]_o$ was 40 photons per rod per 1-sec flash.

stimulus-response relation measured in O·Ca-added solution crossed that measured in 1.6 m*M* calcium. This finding was consistent in toad rods, even for cells that were very dark-adapted and had never been exposed to stimuli brighter than that necessary to reach the crossover point of the curves. When the $[Ca^{2+}]_o$ was returned to 1.6 m*M*, the membrane hyperpolarized in the dark, and the peak-to-peak amplitudes of the receptor potentials became smaller than those recorded before the retina was bathed in O·Ca-added solution. This attenuation of the response amplitudes persisted for many minutes after the low-$[Ca^{2+}]_o$ bath was ended. Despite this attenuation, the responses elicited by dim flashes in normal (1.6 m*M*) $[Ca^{2+}]_o$ were usually larger than those in O·Ca-added solution; that is, stimulus-response relations still crossed.

We attempted to replicate the recordings of Lipton *et al.* (1977a), reported to have been made in a solution containing 1 m*M* EGTA and with no added calcium salts. We were unable to record intracellularly from rods bathed in such a solution for more than 2–3 min without finding irreversible loss of membrane voltage and receptor potentials.

Bathing the retina in O·Ca-added solution changed not only the stimulus-response relation but also the kinetics of the receptor potentials. These changes are illustrated in Fig. 2. In the control solution containing 1.6 m*M* $[Ca^{2+}]_o$,

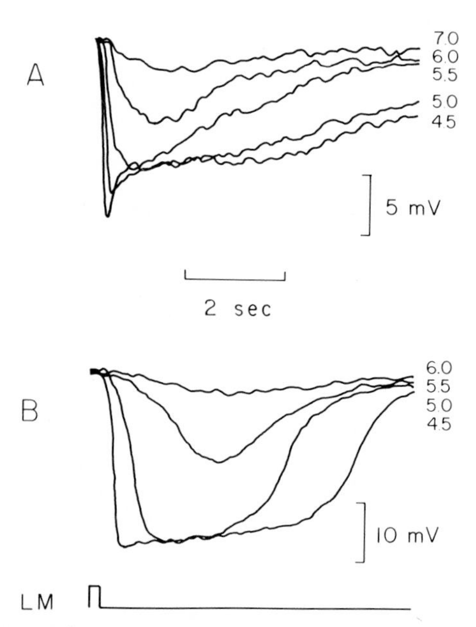

FIG. 2. Effects of low $[Ca^{2+}]_o$ on the time course of receptor potentials. (A) Retina bathed in 1.6 m*M* $[Ca^{2+}]_o$. (B) Retina bathed in O·Ca-added solution. Attenuation (density units) of stimulus light is given to the right of each trace.

responses elicited by progressively brighter flashes became progressively larger, and the times to peak of the responses became progressively shorter. Also, the responses elicited by brighter flashes had a pronounced early transient undershoot. In O·Ca-added solution, the time to peak of the response elicited at each flash intensity was longer, the rise times were slower, and the early transient undershoot was much less pronounced. However, for responses to the brighter stimuli, the return of the membrane voltage to its value in the dark occurred more suddenly and rapidly in O·Ca-added solution than in the control solution.

B. Addition of Phosphodiesterase Inhibitor to the Bath

We attempted to alter the cyclic-nucleotide metabolism of the rods by bathing the retina in solutions containing the phosphodiesterase inhibitor isobutylmethylxanthine (IBMX). When the isolated retina was bathed in a solution containing 0.05–0.1 m*M* IBMX, the membrane depolarized in the dark and the receptor potentials elicited by bright lights became larger, as previously reported by Lipton *et al.* (1977b). The graph of the amplitude of receptor potential versus log stimulus light intensity shows that the responses became larger at all light intensities, including very dim ones (Fig. 3). This finding is also illustrated for a very dark-adapted cell in Fig. 4. In this experiment, the retina had never been stimulated by a bright flash; the stimuli were alternated between two intensities

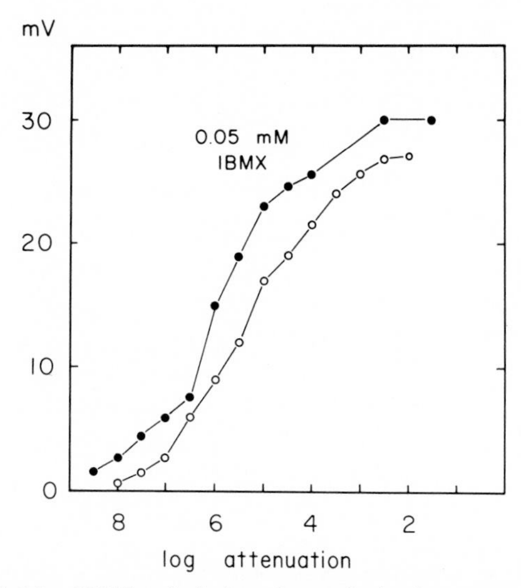

FIG. 3. Effects of adding IBMX to the bath on the amplitude of receptor potentials as a function of stimulus light intensity. To construct each graph, the cell was dark-adapted and then stimulated sequentially by flashes of increasing intensity. The graph (solid circles) was constructed 10 min after bathing in 0.05 m*M* IBMX.

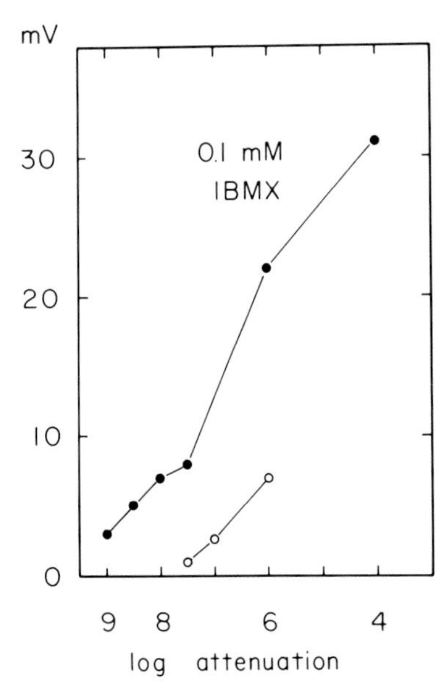

FIG. 4. Effects of adding IBMX on the amplitude of receptor potentials elicited by dim stimuli in a very dark-adapted rod. The graph (solid circles) was constructed (as in Fig. 3) 12 min after bathing the retina in 0.1 m*M* IBMX.

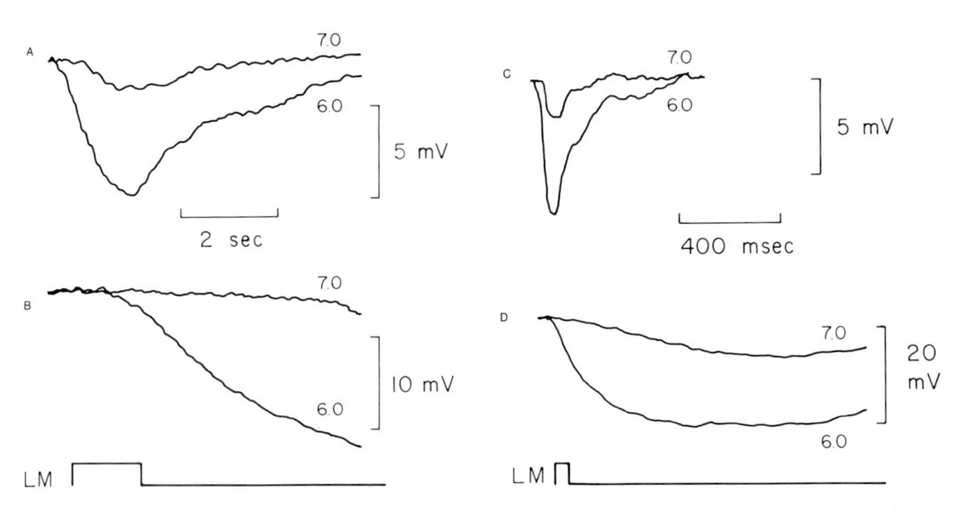

FIG. 5. Effects of IBMX on the time course of receptor potentials. Retina was bathed in control solution (A and C) and in the same solution to which 0.1 m*M* IBMX had been added (B and D). Attenuation (in density units) of the stimulus light is given to the right of each trace. The traces in (A) and (B) are identical to those shown in (C) and (D), respectively, except for the change in sweep time.

each dimmer than the half-saturation intensity for toad rods. The amplitudes of the receptor potentials elicited by these dim flashes became larger after the retina was bathed in 0.1 m*M* IBMX. While the cell was bathed in 0.1 m*M* IBMX, the entire graph of receptor potential amplitude versus log light intensity curve was constructed (Fig. 4). In Fig. 3, the half-saturation intensity shifted to a dimmer intensity while the cell was bathed in IBMX; the same finding is indicated by extrapolation from the curve for normal $[Ca^{2+}]_o$ in Fig. 4.

The kinetics of the receptor potential were also changed by the addition of IBMX to the bath. As shown in Fig. 5, the time scale of the responses to dim flashes became much slower when the retina was bathed in 0.1 m*M* IBMX. That is, although the amplitudes of the receptor potentials increased, the latency, time to peak, and decay time all became dramatically slower. These changes were also observed for responses elicited by brighter stimuli. After the retina was returned to the control solution, the responses recovered both their original amplitudes and kinetics.

C. Intracellular Injection of Cyclic Nucleotides

We attempted to alter the intracellular concentration of cyclic nucleotides directly by iontophoretic injection through the recording micropipet. The electrode was filled with a solution containing cyclic guanosine 3′,5′-monophosphate (cGMP) (either 100 m*M* cGMP, sodium salt, or 20–100 m*M* cGMP, sodium salt, plus 100 m*M* potassium acetate). Iontophoretic injection of cGMP into an impaled rod outer segment caused the membrane to depolarize. The extent of this depolarization varied from cell to cell but sometimes exceeded 30 mV. During the prolonged depolarization, receptor potentials became larger (Fig. 6B). These findings were similar to those reported previously (Nicol and Miller, 1978; Miller and Nicol, 1979). We occasionally observed receptor potentials as larger as 60 mV elicited by bright flashes. Also, the responses to bright flashes became prolonged (Fig. 6B and C). These prolonged receptor potentials consisted of an initial transient phase that lasted 10 or more seconds. This slow transient declined to a voltage maintained for 60 sec or more (the "intermediate" voltage), after which the voltage slowly declined to its original value in the dark. When the cell was stimulated repetitively during the intermediate phase of the decay of the light response, the membrane voltage of the cell in the dark could be maintained steadily. Moreover, the receptor potentials elicited while the cell remained at this intermediate voltage had kinetics similar to those of responses elicited by identical stimuli given before the injection of cGMP (Fig. 6C). When the iontophoretic injection of cGMP was given while stimulating the cell repetitively ($^1/_{10}$ sec), we often observed no marked changes in the light responses; however, the effects of the injection were readily observed when the stimuli were given infrequently.

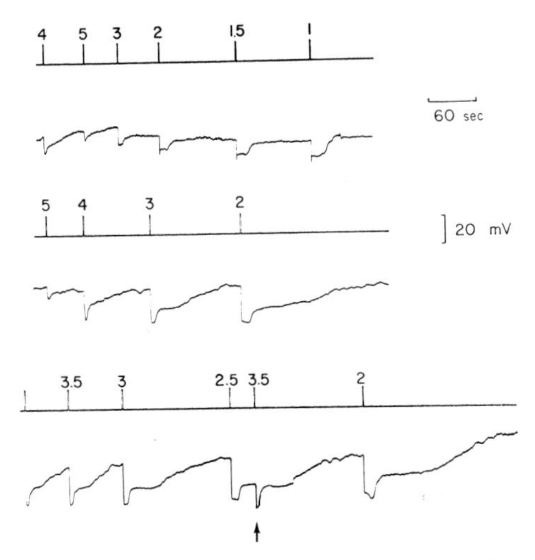

FIG. 6. Intracellular injection of cGMP into toad rod outer segments. Receptor potentials elicited before intracellular iontophoretic injection of cGMP (A) and after injection (B). Note the increased amplitude and prolonged time course of receptor potentials after injection. (C) In another cell after injection of cGMP, notice the prolonged transient phase and intermediate phase of the recovery after bright stimuli. A stimulus (arrow) given during the intermediate phase elicited a receptor potential similar to those recorded before injection of cGMP.

We attempted to inject cyclic adenosine 3′,5′-monophosphate (cAMP) or guanosine 5′-monophosphate (5′-GMP) in the same way as described for cGMP, but these procedures did not cause any of the changes in either receptor potential or steady voltage that were found after cGMP injections. Also, changes in the receptor responses such as those seen after cGMP injection were not observed after we attempted to inject acetate, chloride, or sulfate ions and were also very different from the effects of EGTA injection (Brown *et al.*, 1977).

In most respects our findings were similar to those reported previously (Nicol and Miller, 1978; Miller and Nichol, 1979), although there were some differences. Nicol and Miller (1978) reported that iontophoresis of 5′-GMP caused depolarization of the rod membrane in the dark, whereas we observed none. Also, they reported that the effects of iontophoretic injection of cGMP were reversed by bright light (i.e., both the membrane voltage in the dark and the receptor potentials appeared normal after a bright flash). In our experiments, after iontophoretic injection of cGMP both membrane voltage and receptor potentials appeared normal during the intermediate phase of the receptor response to a bright stimulus, but when a cell was allowed to dark-adapt long enough, the membrane voltage returned to its depolarized level. A large, slow receptor potential could then again be elicited without an additional iontophoretic injection of cGMP.

A possible explanation of this discrepancy is that in our experiments cGMP slowly but continuously diffused out of the intracellular micropipet after the initial iontophoretic injection. Therefore the level of membrane voltage would have been determined by the steady-state relationship of the diffusional influx of cGMP and its hydrolysis by phosphodiesterase in the dark. We have been unable to examine this explanation, because we have no independent way to determine the efflux of cGMP from our intracellular micropipet.

IV. DISCUSSION

Bathing intact rods either in a solution containing a low concentration of calcium ions (O · Ca-added) or an inhibitor of phosphodiesterase (IBMX) might be expected to cause an increase in the cytoplasmic levels of cGMP (for IBMX, Miki *et al.*, 1973; for low $[Ca^{2+}]_o$, Cohen *et al.*, 1978). Thus we might anticipate that both treatments would produce similar changes in membrane voltage and receptor potentials. Indeed, both treatments lead to depolarization of the cell membrane in the dark and large increases in the peak amplitudes of the responses elicited by bright stimuli. However, there are notable differences in the effects of these two treatments. Whereas O · Ca-added solution depresses the amplitudes of the responses elicited by dim stimuli, IBMX increases these amplitudes. The sensitivity of a photoreceptor cell can be defined as the change in membrane voltage produced per unit of stimulus intensity (e.g., Fuortes and Hodgkin, 1964) or, alternatively, as the slope of the graph of ΔV versus log stimulus intensity for dim stimuli. By either definition, the sensitivity of the cell is decreased in O · Ca-added solution and raised by IBMX.

In IBMX the response–intensity relation is shifted toward dimmer stimulus intensities relative to the control. These findings differ from those of Lipton *et al.* (1977b). Also, the response–intensity relation in low $[Ca^{2+}]_o$ crosses the control graph. Because the curves cross, there is neither a simple proportionality between the experimental and control curves nor can the curves be made congruent by a shift along the intensity axis. This finding is in agreement with those of Bastian and Fain (1979), in toad rods, and Bertrand *et al.* (1978), in turtle cones. It differs from the finding of Lipton *et al.* (1977a) in toad rods. Also, we have typically found that bathing a dark-adapted retina in O · Ca-added solution causes a shift in the intensity required to evoke a response of half-maximal amplitude, in agreement with Bastian and Fain (1979) and Flaming and Brown (1979). This finding also does not confirm that of Lipton *et al.* (1977a).

The kinetics of the light responses recorded in O · Ca-added solution differ markedly from those observed with IBMX. Although both treatments cause the receptor potentials to rise more slowly and have a much slower time to peak, after bright stimuli the recovery of the membrane voltage toward a dark "resting'' voltage is abrupt and rapid in O·Ca-added solution, whereas it is very slow in IBMX. Because the two treatments produce differences in the kinetics of the receptor potentials, in the sensitivity of the photoreceptor, and in the response–intensity relation, it seems unlikely that they act through an identical mechanism (although such suggestions have been made previously, cf. Lipton *et al.*, 1977b; Hubbell and Bownds, 1979).

Both morphological and physiological evidence suggests that excitation in vertebrate rods is mediated by a diffusible intracellular transmitter (Baylor and Fuortes, 1970; Hagins, 1972). Calcium ion has been proposed as the putative transmitter (Yoshikami and Hagins, 1971), and a variety of indirect experiments have produced findings consistent with this ''calcium hypothesis'' (Yoshikami and Hagins, 1973; Hagins and Yoshikami, 1974, 1977; Brown *et al.*, 1977; Gold and Korenbrot, 1980). On the other hand, changes in the intracellular concentration of a cyclic nucleotide have also been proposed to participate in photoreceptor function (Bitensky *et al.*, 1971; Miller *et al.*, 1971). It is known that the intracellular concentration of cyclic GMP decreases after illumination of rods (Goridis and Weller, 1976; Woodruff *et al.*, 1977; Woodruff and Bownds, 1978), largely because of light-activated phosphodiesterase (Keirns *et al.*, 1975; Miki *et al.*, 1973; Goridis and Weller, 1976; Sitaramayya *et al.*, 1977; Yee and Liebman, 1978). This phosphodiesterase requires a divalent cation (e.g., Mg^{2+}) and a nucleoside triphosphate (e.g., GTP) for expression of activation by light (Wheeler and Bitensky, 1977). Specific proposals that cGMP is the putative intracellular transmitter for excitation have been constructed on the accumulated knowledge of light-induced changes in rod cell biochemistry (for examples, see Hubbell and Bownds, 1979; Liebman and Pugh, 1979). In these hypotheses, the intracellular concentration of cGMP is high in the dark; the light-activated increase in phosphodiesterase activity causes a rapid decrease in the intracellular concentration of cGMP and ultimately causes a change in phosphorylation of some site appropriate to the control of ionic conductances of the plasma membrane. We have observed that the sensitivity of the rods is increased when the retina is bathed in IBMX. Miki *et al.* (1973) showed that IBMX partially inhibited (about 65% at 0.05 m*M*) light-activated phosphodiesterase in membrane vesicles prepared from frog rod outer segments. Assuming that IBMX has a similar action on phosphodiesterase in *Bufo* rods, any hypothesis incorporating a light-induced increase in phosphodiesterase activity as the mediator of excitation must account for the increase in photoreceptor sensitivity that accompanies partial inhibition of light-activated phosphodiesterase.

In summary, although (1) changing $[Ca^{2+}]_o$ may do more to the physiology of rods than can be attributed to a concomitant change in $[Ca^{2+}]_i$ and (2) IBMX may have effects other than inhibition of phosphodiesterase, it seems unlikely that changes in rod responses produced by lowering $[Ca^{2+}]_o$ or by adding IBMX are caused by a common mechanism. Comparisons of experiments such as those described in this chapter, as well as examination of the speculative hypotheses constructed on such experiments, will require both a more detailed understanding of the pharmacology of the inhibitors on rods and of the effects of changing $[Ca^{2+}]_o$ on the rod plasma membrane.

ACKNOWLEDGMENT

We wish to acknowledge the participation of Dr. L. H. Pinto in our first experiments involving the injection of cyclic nucleotides into rods and to thank him and Dr. Ann E. Stuart for helpful discussions. Supported by NIH grants EY01914 and EY01915.

REFERENCES

Bastian, B. L., and Fain, G. L. (1979). Light adaptation in toad rods: Requirements for an internal messenger which is not calcium. *J. Physiol.* (*London*) **297,** 493-520.

Baylor, D. A., and Fuortes, M. G. F. (1970). Electrical responses of single cones in the retina of the turtle. *J. Physiol.* (*London*) **207,** 77-92.

Berridge, M. J. (1975). The interaction of cyclic nucleotides and calcium in the control of cellular activity. *Adv. Cyclic Nucleotide Res.* **6,** 1-98.

Bertrand, D., Fuortes, M. G. F., and Pochobradsky, J. (1978). Actions of high calcium on the cones in the turtle retina. *J. Physiol.* (*London*) **275,** 419-437.

Bitensky, M. W., Gorman, R. E., and Miller, W. H. (1971). Adenyl cyclase as a link between photon capture and changes in membrane permeability of frog photoreceptors. *Proc. Natl. Acad. Sci. U.S.A.* **68,** 561-562.

Brown, J. E., and Pinto, L. H. (1974). Ionic mechanism for the photoreceptor potential of the retina of *Bufo marinus*. *J. Physiol.* (*London*) **236,** 575-591.

Brown, J. E., Coles, J. A., and Pinto, L. H. (1977). Effect of injection of calcium and EGTA into the outer segments of retinal rods of *Bufo marinus*. *J. Physiol.* (*London*) **269,** 707-722.

Cohen, A. I., Hall, I. A., and Ferendelli, J. A. (1978). Calcium and cyclic nucleotide regulation in incubated mouse retinas. *J. Gen. Physiol.* **71,** 595-612.

Flaming, D., and Brown, K. T. (1979). Effects of calcium on the intensity-response curve of toad rods. *Nature* (*London*) **278,** 852-853.

Fuortes, M. G. F., and Hodgkin, A. L. (1964). Changes in time scale and sensitivity in the ommatidia of *Limulus*. *J. Physiol.* (*London*) **172,** 239-263.

Gold, G. H., and Korenbrot, J. I. (1980). Ca release in the light by intact retinal rods. *Proc. Natl. Acad. Sci. U.S.A.* **77,** 5557-5561.

Goridis, C., and Weller, M. (1976). A role for cyclic nucleotides and protein kinase in vertebrate photoreception. *Adv. Biochem. Psychopharmacol.* **15,** 391-412.

Goridis, C., Virmaux, N., Cailla, H. L., and Delaage, M. A. (1974). Rapid, light-induced changes of retinal cyclic GMP levels. *FEBS Lett.* **49,** 167-169.

Hagins, W. A. (1972). The visual process: Excitatory mechanism in the primary receptor cells. *Annu. Rev. Biophys. Bioeng.* **1,** 131-158.

Hagins, W. A., and Yoshikami, S. (1974). A role for Ca^{++} in excitation of retinal rods and cones. *Exp. Eye Res.* **18,** 299-305.

Hagins, W. A., and Yoshikami, S. (1977). Intracellular transmission of visual excitation in photoreceptors: Electrical effects of chelating agents introduced into rods by vesicle fusion. *In* "Vertebrate Photoreception" (H. B. Barlow and P. Fatt, eds.), p. 97. Academic Press, New York.

Hubbell, W. L., and Bownds, M. D. (1979). Visual transduction in vertebrate photoreceptors. *Annu. Rev. Neurosci.* **2,** 17–34.

Keirns, J. J., Miki, N., Bitensky, M. W., and Keirns, M. (1975). A link between rhodopsin and disc membrane cyclic nucleotide phosphodiesterase, action spectrum and sensitivity to illumination. *Biochemistry* **14,** 2760–2766.

Liebman, P. A., and Pugh, E. N., Jr. (1979). The control of phosphodiesterase in rod disk membranes: Kinetics, possible mechanisms and significance for vision. *Vision Res.* **19,** 375–380.

Lipton, S. A., Ostroy, S. E., and Dowling, J. E. (1977a). Electrical and adaptive properties of rod photoreceptors in *Bufo marinus*. I. Effects of altered extracellular Ca^{++} levels. *J. Gen. Physiol.* **70,** 771–791.

Lipton, S. A., Rasmussen, H., and Dowling, J. E. (1977). Electrical and adaptive properties of rod photoreceptors in *Bufo marinus*. II. Effects of cyclic nucleotides and prostaglandins. *J. Gen. Physiol.* **70,** 771–791.

Miki, N., Keirns, J. J., Marcus, F. R., Freeman, J., and Bitensky, M. W. (1973). Regulation of cyclic nucleotide concentration in photoreceptors: An ATP-dependent stimulation of cyclic nucleotide phosphodiesterase by light. *Proc. Natl. Acad. Sci. U.S.A.* **76,** 3820–3824.

Miller, W. H., and Nicol, G. D. (1979). Evidence that cyclic GMP regulates membrane potential in rod photoreceptors. *Nature (London)* **280,** 64–66.

Miller, W. H., Gorman, R. E., and Bitensky, M. W. (1971). Cyclic adenosine monophosphate: Function in photoreceptors. *Science* **174,** 295–297.

Nicol, G. D., and Miller, W. H. (1978). Cyclic GMP injected into retinal rod outer segments increases latency and amplitude of response to illumination. *Proc. Natl. Acad. Sci. U.S.A.* **75,** 5217–5220.

Pober, J. S., and Bitensky, M. W. (1979). Light-regulated enzymes of vertebrate retinal rods. *Adv. Cyclic Nucleotide Res.* **11,** 265–301.

Rasmussen, H., Jensen, P., Lake, W., Friedman, N., and Goodman, D. B. P. (1975). Cyclic nucleotides and cellular calcium metabolism. *Adv. Cyclic Nucleotide Res.* **5,** 375–394.

Sitaramayya, A., Virmaux, N., and Mandel, P. (1977). On the mechanism of light activation of retinal rod outer segments cyclic GMP phosphodiesterase. (Light activation-influence of bleached rhodopsin and KF-inhibition.) *Exp. Eye Res.* **25,** 163–169.

Wheeler, G. L., and Bitensky, M. W. (1977). A light-activated GTPase in vertebrate photoreceptors: Regulation of light-activated cyclic GMP phosphodiesterase. *Proc. Natl. Acad. Sci. U.S.A.* **74,** 4238–4242.

Woodruff, M. L., and Bownds, D. (1978). Amplitude, kinetics and reversibility of a light-induced decrease in guanosine 3′,5′-cyclic monophosphate in isolated frog retinal rod outer segments. *J. Gen. Physiol.* **73,** 629–653.

Woodruff, M. L., Bownds, D., Green, S. H., Morrisey, J. L., and Shedlovsky, A. (1977). Guanosine 3′,5′-cyclic monophosphate and the *in vitro* physiology of frog photoreceptor membranes. *J. Gen. Physiol.* **69,** 667–679.

Yee, R., and Liebman, P. (1978). Light-activated phosphodiesterase of the rod outer segment. *J. Biol. Chem.* **253,** 8902–8909.

Yoshikami, S., and Hagins, W. A. (1971). Light, calcium, and the photocurrent of rods and cones. *Biophys. J.* **11,** 47a.

Yoshikami, S., and Hagins, W. A. (1973). Control of the dark current in vertebrate rods and cones. *In* "Biochemistry and Physiology of Visual Pigments" (H. Langer, ed.), p. 245–255. Springer-Verlag, New York.

Chapter 21

The Relation between Ca^{2+} and Cyclic GMP in Rod Photoreceptors

STUART A. LIPTON

Departments of Neurology
and Neurobiology
Harvard Medical School
Boston, Massachusetts

AND

JOHN E. DOWLING

Department of Biology
Harvard University
Cambridge, Massachusetts

I. INVERSE RELATION BETWEEN EFFECTS OF Ca^{2+} AND CYCLIC GMP

Four years ago, we published evidence that Ca^{2+} and cyclic guanosine 3′, 5′-monophosphate (cyclic GMP) may act as interrelated second messengers in the rod photoreceptors of the toad *Bufo marinus* (Lipton *et al.*, 1977a,b). We found that lowering extracellular Ca^{2+} or increasing intracellular levels of cyclic GMP by superfusion of the retina with the phosphodiesterase inhibitor isobutyl-methylxanthine (IBMX) caused a qualitatively similar sequence of effects on the intracellularly recorded responses of the rods to light. For example, when low Ca^{2+} (10^{-9} *M*) Ringer's solution or Ringer's solution containing 5 m*M* IBMX

ISBN 0-12-153315-8

was infused into the perfusion chamber, the receptors initially depolarized and showed a proportional increase in the response amplitude without a significant change in the response waveform. This is shown in Fig. 1 for responses recorded 2 min following introduction of the low calcium Ringer's solution or Ringer's solution containing IBMX.

After 4 min of infusion of these solutions, the rods depolarized further, the response amplitudes increased even more, and the response waveforms became altered (Fig. 1). The initial transient seen in the control and at 2 min of superfusion disappeared, the time to peak lengthened, and the duration of the response to the 200-msec flash was greatly prolonged. We also noted in the first 4–6 min of superfusion that, although the rods were depolarized and the response amplitudes were increased significantly, very little change in the intensity of light required to elicit a half-maximal response (σ) was observed (Figs. 2a and b). We found this to be the case in both dark-adapted and partially light-adapted cells. These results led us to conclude that, although lowering Ca^{2+} or raising cyclic GMP in the cell could initially cause a drastic alteration in both membrane potential and response

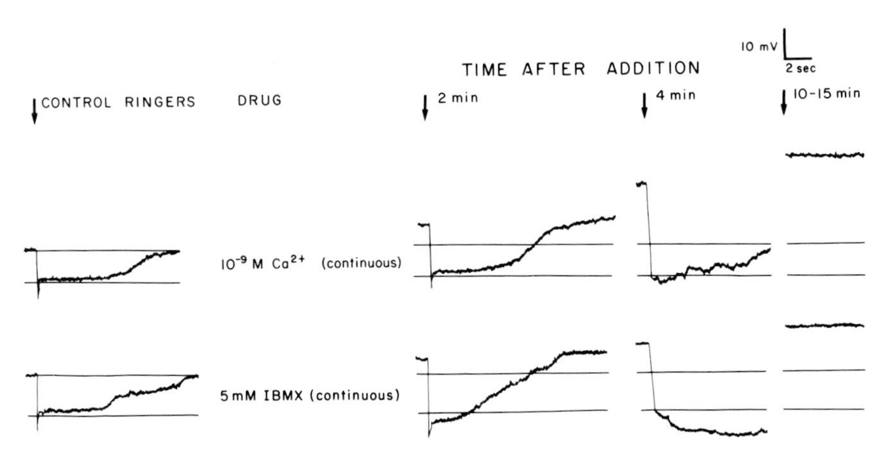

FIG. 1. Effects of Ringer's solution containing low Ca^{2+} (10^{-9} *M*) or 5 m*M* IBMX, superfused continuously, on intracellular responses of single dark-adapted toad rods. Each horizontal row shows responses from the same cell. The horizontal lines superimposed on each response indicate the original resting potential and plateau potential. Responses to one intensity (30,000 photons absorbed per rod-flash) are shown. Both test solutions caused qualitatively similar effects on the rod responses. For example, after 2 min of superfusion both test Ringer's solutions depolarized the rods and increased the response amplitudes. By 4 min, the rods had further depolarized and the amplitude of the light response had increased even more. In addition the waveforms of the light responses were altered in a characteristic way; the initial transient seen in the control responses had disappeared, and the time to peak and duration of the response had lengthened (quantitatively more with IBMX than low Ca^{2+}). By 15 min of perfusion, the rods were highly depolarized, and light-evoked responses had disappeared.

amplitude, these manipulations of intracellular messenger levels had little effect on the sensitivity of the receptor in either light or dark.[1]

Between 6 and 15 min of perfusion in these agents we observed that the rods depolarized somewhat further but that they also began to lose responsiveness in a very characteristic fashion (Fig. 3). Response amplitudes decreased with time but, more interestingly, σ (the light intensity necessary to evoke a half-saturating response) shifted laterally along the intensity scale. This effect is very different from that ordinarily observed in an injured or dying cell, in which response amplitudes decrease but σ does not change significantly, suggesting that we were observing a very specific and similar effect on the photoreceptor response caused by the two test solutions. A shift of the voltage–intensity curve along the intensity axis, coupled with a reduction in the maximum response amplitude, occurs also when a rod is exposed to continuous illumination (Kleinschmidt and Dowling, 1975). We suggested that prolonged exposure of the rods to these test solutions might mimick light adaptation of the receptors by, for example, depleting an intradisk store of Ca^{2+}.

By 15–40 min of superfusion, the cells had completely lost light responsiveness, and they remained highly depolarized (Fig. 1). The effects described above were largely reversible when control Ringer's solution was reintroduced into the chamber within 6–8 min. When the test Ringer's solution was perfused for more than 8–9 min, recovery was incomplete in the penetrated cell after an additional 15–20 min in control Ringer's solution. However, other cells in the retina penetrated at this time were normal in terms of membrane potential, response amplitude, waveform, and sensitivity.

We also studied responses during dark adaptation. We observed similar effects of increasing cyclic GMP or lowering Ca^{2+} levels on the recovery of the rod response following exposure of the retina to a bright step of light which bleached a negligible fraction of the visual pigment. On incubation of a cell in either low Ca^{2+} (10^{-9} *M*) or 5 m*M* IBMX for about 5 min, the membrane potential and sensitivity of the rod returned to dark levels significantly *faster* than they did in the same rod when it was tested initially in control Ringer's solution (Fig. 4). This was a consistent finding, readily reversible on return of the rod to control Ringer's solution.

[1]More recent work by Flaming and Brown (1979) and Bastian and Fain (1979) has shown that small shifts in σ can be observed when extracellular Ca^{2+} is altered sufficiently. For example, Flaming and Brown showed that, when external Ca^{2+} was raised from 1.8 to 6 m*M*, σ shifted to higher light intensities by about 0.26 log unit. Small shifts of this magnitude were close to our error range ($\pm$0.2 log unit) and were not detected. However, it is well known that even moderate light adaptation causes large shifts of σ in rods, i.e., 2–3 log units; and it was for this reason that we concluded that altering the cytosol levels of Ca^{2+} and/or cyclic GMP had relatively little effect on the light sensitivity of the cell (for further discussion, see Lipton *et al.*, 1979).

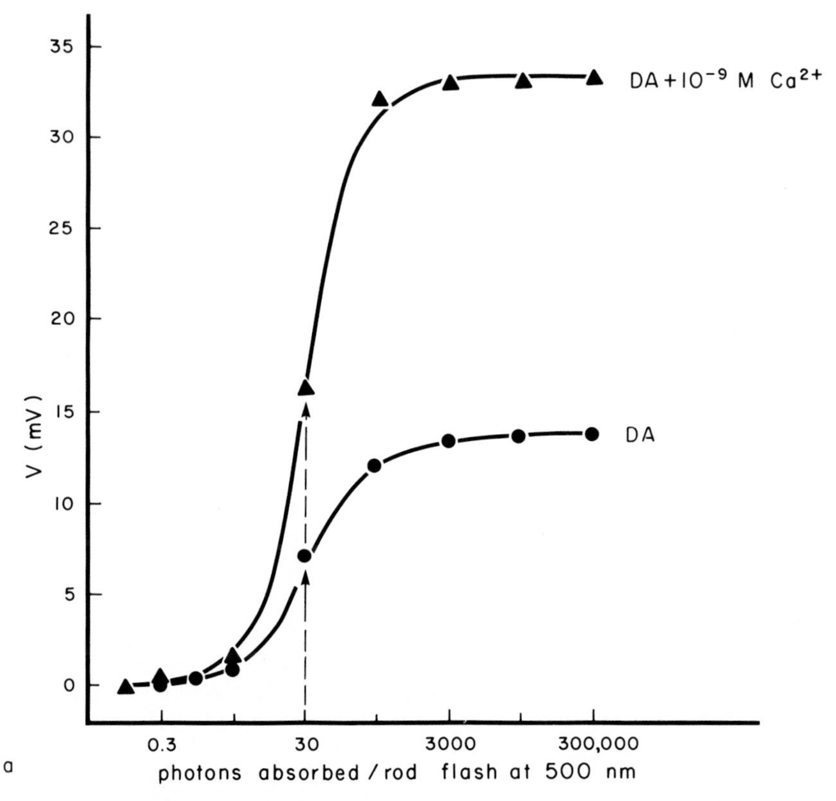

FIG. 2a. Plot of response amplitude versus flash intensity for a single dark-adapted (DA) toad rod. The duration of each flash was 200 msec, and its intensity is expressed in photons absorbed per rod–flash. The lower curve was compiled during superfusion in control Ringer's solution, and the upper curve during 4–6 min of superfusion in 10^{-9} *M* Ca^{2+}. Note that low Ca^{2+} affected the amplitude of response compared to that of the control but did not shift the curve along the abscissa, as reflected by little or no change in σ (indicated by arrows).

II. ALTERNATIVE MODELS

Additional experiments, testing the effects of substances believed to increase intracellular cyclic-GMP levels such as cyclic GMP itself, dibutryl cyclic GMP, and prostaglandin $PGF_{2\alpha}$, also showed effects similar to those obtained by lowering Ca^{2+}. Thus all our evidence suggested a close but inverse relation between the levels of Ca^{2+} and cyclic GMP in the functioning of the rod. But what might this be? There is evidence that light alters the levels of both Ca^{2+} and cyclic GMP (see, for example, Gold and Korenbrot, 1980; Woodruff and Bownds, 1979), and our evidence and that of others indicate that altering

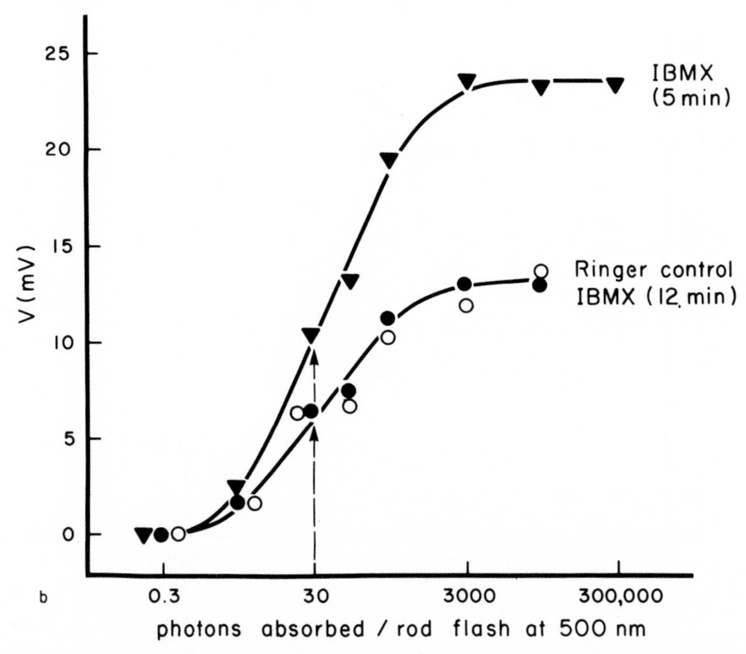

FIG. 2b. Voltage-intensity relationships before (solid circles), and after (triangles and open circles) a 10-sec application of 5 m*M* IBMX to a dark-adapted rod. Five minutes after IBMX addition (triangles) the amplitude of response had increased at all stimulus intensities, but the curve had not significantly shifted laterally on the intensity scale; σ (indicated by arrows) remained virtually the same. Response amplitudes returned to control levels by 12 min after drug addition (open circles).

the cytosol concentrations of Ca^{2+} and cyclic GMP causes substantial changes in both membrane potential level and response amplitudes (Bastian and Fain, 1979; Miller and Nicol, 1979). Thus three basic alternatives can be suggested. First, cyclic GMP and Ca^{2+} might both serve as intracellular messengers in the rod and independently regulate membrane potential. Thus light could cause a parallel increase in Ca^{2+} and a fall in cyclic GMP in the cytosol, and both substances could contribute to hyperpolarization of the membrane potential, i.e., the decrease in G_{Na}. This may be diagrammed in the following way.

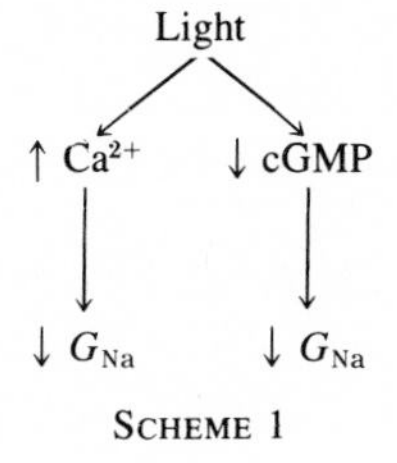

SCHEME 1

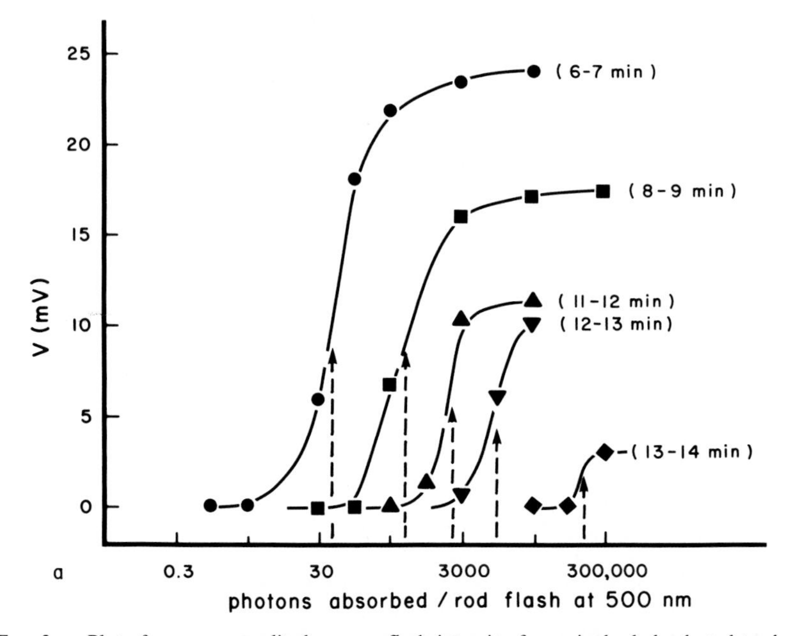

FIG. 3a. Plot of response amplitude versus flash intensity for a single dark-adapted toad rod during superfusion in long-term (6–15 min) 10^{-9} *M* Ca^{2+} Ringer's solution. Responses were to 200-msec flashes of 500-nm light; the intensity is represented on the abscissa. The numbers in parentheses beside each curve give the time intervals of continuous addition of 10^{-9} *M* Ca^{2+} during which the data were obtained. With increasing time of low Ca^{2+} superfusion of the dark-adapted rod, voltage-intensity curves shifted to the right on the intensity scale.

A second possibility is that Ca^{2+} and cyclic GMP serve as dual sequential messengers in the cell. That is, light causes an alteration in one substance or the other, and this in turn causes a change in the cytosol level of the other substance which regulates G_{Na}. These possibilities may be diagrammed as follows:

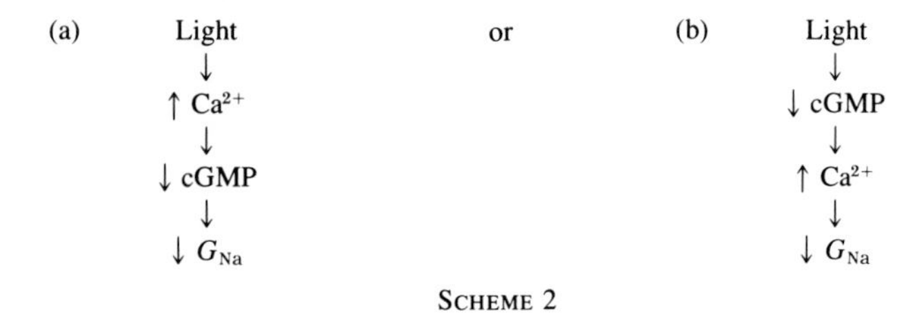

SCHEME 2

Finally, it may be that only one of these substances serves as the intracellular messenger and the other serves to modulate the intracellular concentration of the messenger substance. This may be diagrammed as follows:

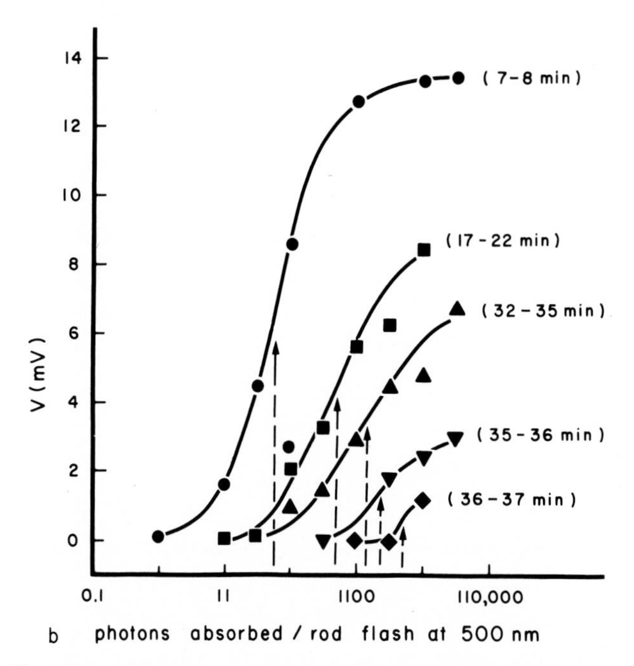

FIG. 3b. The effects of long-term superfusion of 5 m*M* IBMX on voltage-intensity curves recorded from a single, dark-adapted toad rod at the times indicated. The conditions of the experiment were identical to those for the experiment shown in Fig. 3a. Note that Ringer's solution containing 5 m*M* IBMX, like low Ca^{2+} Ringer's solution, caused the voltage-intensity curves to shift to the right on the intensity scale with time. We found often, as here, that the shifts of the V-log I curves on the intensity axis occurred over a much longer period when rods were exposed to Ringer's solution containing 5 m*M* IBMX than when they were exposed to Ringer's solution containing 10^{-9} *M* Ca^{2+} (Fig. 3a). Nevertheless, both Ringer's solutions caused qualitatively similar effects on the V-log I curves over time.

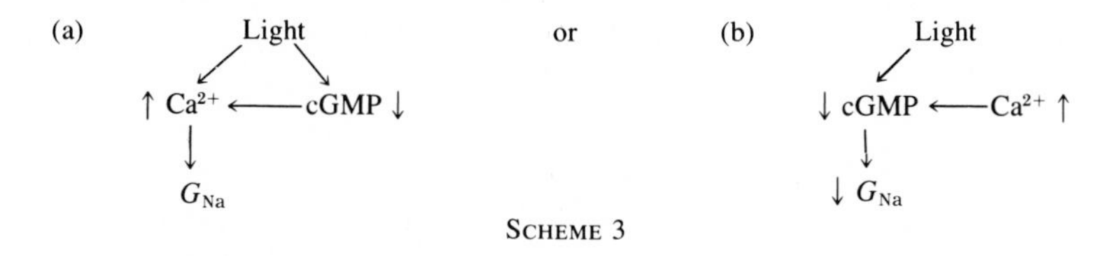

SCHEME 3

Our experiments did not distinguish among these various possibilities, although some preliminary physiological results suggested that, in the rod, raised cyclic-GMP levels may lead to lowered Ca^{2+} levels which then produce an altered membrane potential (see Lipton *et al.*, 1977b, p. 787). These preliminary results favored Scheme 2b or 3a as more likely.

More recent experiments, on the other hand, appear to distinguish more deci-

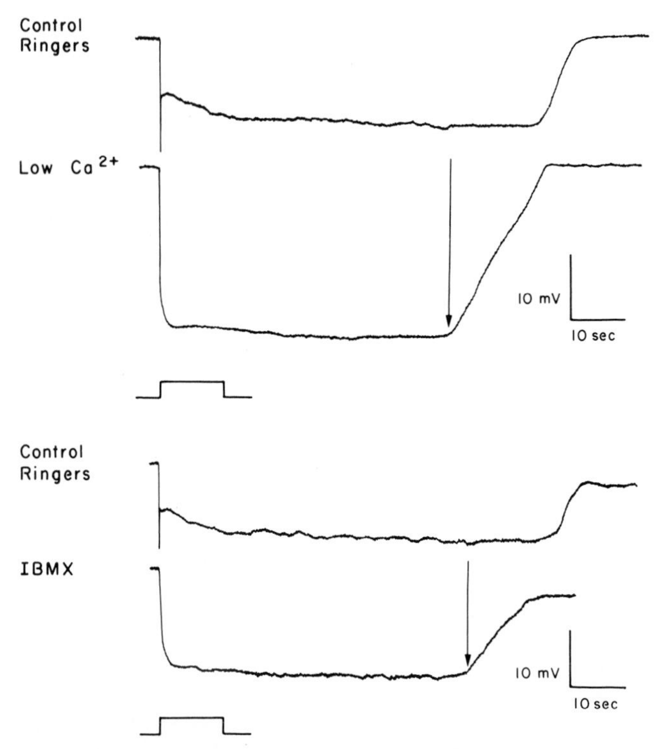

FIG. 4. Effects of Ringer's solution containing low Ca^{2+} or IBMX on recovery of membrane potential following exposure of toad rods to a bright adapting flash. The adapting procedure consisted of a 12-sec yellow (Corning cutoff filter 3484) unattenuated light with a retinal irradiance of 1.1 mW/cm^2, which bleached $\leq 3\%$ of the pigment. Each pair of traces was from a single rod. The concentration and length of addition of the test agents were the same as in Fig. 2. Both test solutions significantly accelerated the onset of membrane recovery (arrows) after the adapting light step (represented below each set of traces). In both experiments receptor sensitivity was also monitored, and a close correspondence between recovery of membrane potential and recovery of rod sensitivity was found. Thus both test solutions accelerated the return of rod sensitivity following a bright, nonbleaching test flash.

sively among the various alternatives. For example, the finding of Cohen *et al.* (1978) that lowering extracellular Ca^{2+} caused a dramatic increase in cytosol levels of cyclic GMP indicates that these substances interrelate with one another in the rod cell and that Scheme 1, which suggests that Ca^{2+} and cyclic GMP are independent messengers, is not likely to be correct. The subsequent findings of Woodruff and Bownds (1979) and Kilbride (1980) that raising Ca^{2+} could cause a decrease in cytosol cyclic-GMP levels in the rod further strengthens the view that Ca^{2+} and cyclic GMP are likely to regulate the levels of one another in the rod, and thus these biochemical results lend support to Scheme 2 or 3.

The experiments of Kilbride and Ebrey (1979), appear to distinguish between Schemes 2 and 3. These workers found that by quick-freezing retinas after light exposure, degradation of cyclic GMP within the retina was stopped within 50–100 msec. This procedure allowed a more accurate measure of the rate of fall of cyclic GMP on light excitation than previously could be achieved. Their results using this method have shown that there is no significant change in cyclic-GMP concentration in the rod after 1 sec of bright illumination. Since rods respond physiologically after illumination within a few hundred milliseconds at most, these experiments appear to rule out any direct role of cyclic GMP in the initial excitation process of the rod; i.e., these results are clearly incompatible with Scheme 2a, 2b, or 3b.

More recently, Kilbride (1980) has found that cyclic-GMP levels fall on illumination about 15-fold faster in retinas incubated in low Ca^{2+} Ringer's solution than in retinas incubated in normal Ringer's solution, indicating that under some conditions it is possible for the rod to degrade cyclic GMP fast enough for this process to be involved in rod excitation. It is interesting to note that the earlier measurements by Woodruff and Bownds (1979), which suggested a fast rate of cyclic-GMP degradation in the isolated rod outer segment on light exposure, employed a low Ca^{2+} Ringer's solution for the isolation and maintenance of the outer segments.

Recently, one of us (S.A.L), in collaboration with Geoffrey Gold, has found that the release of Ca^{2+} from the distal retina that occurs on illumination (Gold and Korenbrot, 1980) is enormously enhanced when the retina is bathed in 1–5 m*M* IBMX. This result suggests strongly that cyclic GMP exerts a profound effect on the exchange of Ca^{2+} across the rod outer segment membrane by, for example, enhancing Ca^{2+}-ATPase pumps known to be present there (see below) or perhaps by increasing a Na^{+}-Ca^{2+} exchange. In either case, this result favors Scheme 3a as the one most likely to account for the known results concerning the function of cyclic GMP in the rod cell. However, since altering Ca^{2+} also affects cyclic-GMP levels in the rod (Cohen *et al.*, 1978; Woodruff and Bownds, 1979), it appears that Scheme 3a should be altered to include the possibility that Ca^{2+} affects cyclic-GMP levels in a reciprocal, feed-forward fashion.

III. CONCLUSIONS

Figure 5 summarizes our present thinking concerning the interrelationships of Ca^{2+} and cyclic GMP within the rod outer segment. The basic idea of this scheme is that cyclic GMP regulates cytosol Ca^{2+} levels by stimulating Ca^{2+}-ATPase pumps. It has been reported that Ca^{2+}-ATPase pumps are located in both the plasma and disk membranes (Bownds *et al.*, 1971; Hagins and Yoshikami, 1974; Mason *et al.*, 1974; Ostwald and Heller, 1972); thus, when cyclic-GMP levels in the rod outer segment are artificially increased (i.e., Fig. 1), the activity of the

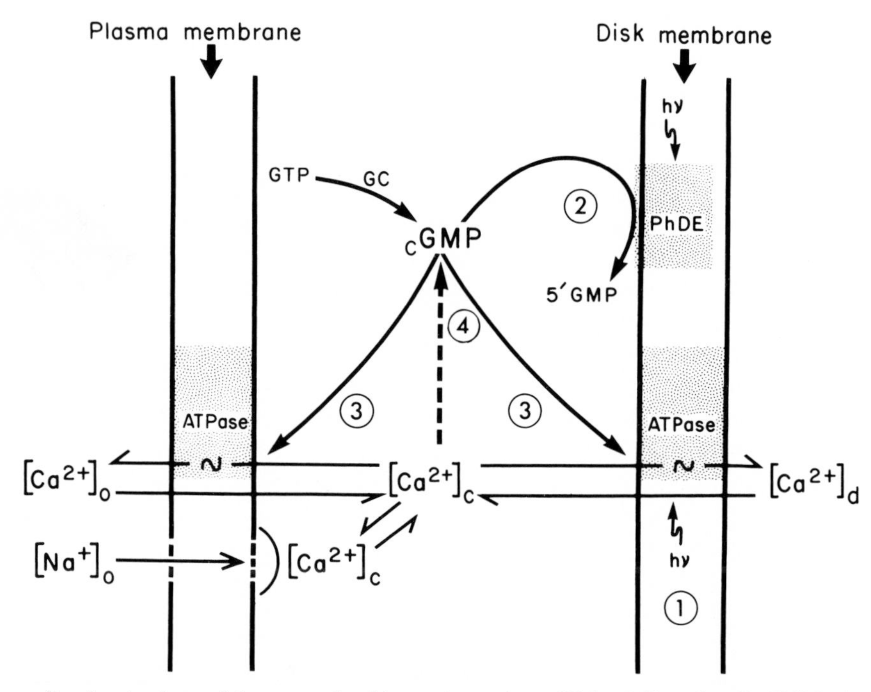

FIG. 5. A scheme of the proposed and known interactions of light, Ca^{2+}, and cyclic GMP in the rod outer segment. The numbers in the model refer to the following pathways. (1) Light putatively triggers the release of Ca^{2+} from the disks which then blocks Na^+ channels in the plasma membrane (Yoshikami and Hagins, 1971, 1973). (2) Light stimulates phosphodiesterase, resulting in the breakdown of cyclic GMP (Miki *et al.*, 1973). [We are not speculating on the mechanism of light activation of the phosphodiesterase but assume that it follows photoactivation of the visual pigment molecule (Keirns *et al.*, 1975).] (3) We propose that cyclic GMP lowers cytoplasmic Ca^{2+} concentration via Ca^{2+}-ATPase pumps. (4) Very low Ca^{2+} concentrations lead to increased cyclic-GMP levels (Cohen *et al.*, 1978; Woodruff and Bownds, 1979), while high Ca^{2+} lowers cyclic-GMP levels (Woodruff and Bownds, 1979; Kilbride, 1980). See text for discussion. GTP, guanosine 5′ triphosphate; GC, guanylate cyclase; $[Ca^{2+}]_o$, extracellular $[Ca^{2+}]$; $[Ca^{2+}]_c$, rod outer segment cytoplasmic $[Ca^{2+}]$; $[Ca^{2+}]_d$, intradisk $[Ca^{2+}]$; ATPase, Ca^{2+}-ATPase pump; PhDE, phosphodiesterase; $h\nu$, light.

Ca^{2+} pumps is enhanced and cytosol Ca^{2+} levels are lowered. With lower intracellular Ca^{2+} levels, more Na^+ channels are opened and available for blockage; the receptor depolarizes and response amplitudes increase (see also Miller and Nicol, 1979). Also, after a bright adapting flash under the same conditions (i.e., Fig. 4), cytosol Ca^{2+} levels are reduced faster, resulting in a shortened period of receptor saturation; hence the onset of membrane recovery during dark adaptation is accelerated.

The sequence of events that might occur in the rod on illumination may be summarized as follows. In response to light, rhodopsin is bleached and initiates the release of Ca^{2+} into the cytosol. Phosphodiesterase is also activated on rhodopsin bleaching, and cytosol cyclic-GMP levels subsequently fall. With a decrease in cytosol cyclic-GMP levels, the Ca^{2+} pumps in disk and plasma membrane decrease activity, which leads to a further buildup in cytosol Ca^{2+} and perhaps an amplification of the effect of bleaching each rhodopsin molecule.

With time, cytosol cyclic-GMP levels presumably recover (see, for example, Kilbride and Ebrey, 1979), which results in an acceleration of pump activity. This in turn leads to a reduction in cytosol Ca^{2+} concentrations and a return of the membrane potential toward dark levels. With lower Ca^{2+}, the cyclic-GMP levels in the rod increase (Cohen *et al.*, 1978), resulting in a feed-forward cascade that may serve to enhance the rate of membrane potential recovery. As yet we have little detailed information concerning the changes in cyclic-GMP levels that occur in rods in the intact, functioning retina under known physiological conditions during light and dark adaptation. If, however, the principal role of cyclic GMP in the rod is to regulate cytosol Ca^{2+} levels, as we suggest, such information is of fundamental importance. Further experiments along these lines are clearly needed and should provide a test of this hypothesis.

REFERENCES

Bastian, B. L., and Fain, G. L. (1979). Light adaptation in toad rods: Requirement for an internal messenger which is not calcium. *J. Physiol. (London)* **297,** 493–520.

Bownds, D., Gordon-Walker, A., Gaide-Huguenin, A. D., and Robinson, W. E. (1971). Characterization and analysis of frog photoreceptor membranes. *J. Gen. Physiol.* **58,** 225–237.

Cohen, A. I., Hall, I. A., and Ferrendelli, J. A. (1978). Calcium and cyclic nucleotide regulation in incubated mouse retinas. *J. Gen. Physiol.* **71,** 595–612.

Flaming, D. G., and Brown, K. T. (1979). Effects of calcium on the intensity-response curve of toad rods. *Nature (London)* **278,** 852–853.

Gold, G. H., and Korenbrot, J. I. (1980). Light-induced calcium release by intact retinal rods. *Proc. Natl. Acad. Sci. U.S.A.* **77,** 5557–5561.

Hagins, W., and Yoshikami, S. (1974). A role for Ca^{2+} in excitation of retinal rods and cones. *Exp. Eye Res.* **18,** 299–305.

Keirns, J. J., Miki, N., Bitensky, M. W., and Keirns, M. (1975). A link between rhodopsin and disc membrane cyclic nucleotide phosphodiesterase: Action spectrum and sensitivity to illumination. *Biochemistry* **14,** 2760–2766.

Kilbride, P. (1980). Calcium effects on frog retina cyclic guanosine 3′,5′-monophosphate levels and their light-initiated rate of decay. *J. Gen. Physiol.* **75,** 457–465.

Kilbride, P., and Ebrey, T. G. (1979). Light-initiated changes of cyclic guanosine monophosphate levels in the frog retina measured with quick-freezing techniques. *J. Gen. Physiol.* **74,** 415–426.

Kleinschmidt, J., and Dowling, J. E. (1975). Intracellular recordings from gecko photoreceptors during light and dark adaptation. *J. Gen. Physiol.* **66,** 617–648.

Lipton, S. A., Ostroy, S. F., and Dowling, J. E. (1977a). Electrical and adaptive properties of rod photoreceptors in *Bufo marinus*. I. Effects of altered extracellular Ca^{2+} levels. *J. Gen. Physiol.* **70,** 747–770.

Lipton, S. A., Rasmussen, H., and Dowling, J. E. (1977b). Electrical and adaptive properties of rod photoreceptors in *Bufo marinus*. II. Effects of cyclic nucleotides and prostaglandins. *J. Gen. Physiol.* **70,** 771-791.

Lipton, S. A., Ostroy, S. E., and Dowling, J. E. (1979). Ca^{2+} and photoreceptor adaptation. *Nature (London)* **281,** 407-408.

Mason, W. T., Fager, R. S., and Abrahamson, E. W. (1974). Ion fluxes in disk membranes of retinal rod outer segments. *Nature (London)* **247,** 562-563.

Miki, N., Keirns, J. J., Marcus, F. R., Freeman, J., and Bitensky, M. W. (1973). Regulation of cyclic nucleotide concentrations in photoreceptors: An ATP-dependent stimulation of cyclic nucleotide phosphodiesterase by light. *Proc. Natl. Acad. Sci. U.S.A.* **70,** 3820-3825.

Miller, W. H., and Nicol, G. D. (1979). Evidence that cylic GMP regulates membrane potential in rod photoreceptors. *Nature (London)* **280,** 64-66.

Ostwald, T. J., and Heller, J. (1972). Properties of magnesium- or calcium-dependent adenosine triphosphatase from frog rod photoreceptor outer segment disks and its inhibition by illumination. *Biochemistry* **11,** 4679-4686.

Woodruff, M. L., and Bownds, M. D. (1979). Amplitude, kinetics, and reversibility of a light-induced decrease in guanosine 3′,5′-cyclic monophosphate in frog photoreceptor membranes. *J. Gen. Physiol.* **73,** 629-653.

Yoshikami, S., and Hagins, W. A. (1971). Light, calcium, and the photocurrent of rods and cones. *Biophys. Soc.* **11**(2, Pt. 2), 47a (abstr.).

Yoshikami, S., and Hagins, W. A. (1973). Control of the dark current in vertebrate rods and cones. *In* "Biochemistry and Physiology of Visual Pigments" (H. Langer, ed.), pp. 245-255. Springer-Verlag, Berlin and New York.

Chapter 22

Limits on the Role of Rhodopsin and cGMP in the Functioning of the Vertebrate Photoreceptor

SANFORD E. OSTROY, EDWARD P. MEYERTHOLEN, PETER J. STEIN,[1] *ROBERTA A. SVOBODA, AND MEEGAN J. WILSON*

Department of Biological Sciences
Purdue University
West Lafayette, Indiana

I. INTRODUCTION

A number of chemical changes and reactions are initiated by light in the vertebrate rod photoreceptor. Some of these include multiple conformation changes in rhodopsin, decreases in the cyclic guanosine 3′,5′-monophosphate (cGMP) concentration of the photoreceptor, phosphorylation of rhodopsin and dephosphorylation of other protein components of the cell, and possible changes in the internal calcium concentration of the cell. Our current studies have focused on providing further understanding of the role of these reactions. To consider the functional role of these processes one must delineate the many physiological processes occurring in the vertebrate photoreceptor. As we have considered them, the processes are as follows: the primary reduction in sodium conductance

[1]Present address: Department of Pathology, Yale University Medical School, New Haven, Connecticut.

ISBN 0-12-153315-8

of the rod plasma membrane that occurs after illumination (phototransduction); the temporary decreases in electrophysiological sensitivity of the photoreceptor cell that occur subsequent to the changes in sodium conductance and which are isolated at negligible bleaching levels (rapid dark adaptation); the changes in cell sensitivity that result from the photolysis of rhodopsin and which may be observed at higher levels of bleaching (slow dark adaptation); the changes in cell response and sensitivity that occur in the presence of continuous illumination (light adaptation); and the maintenance processes of the cell including the various pumps required to maintain ion gradients and to provide cell nutrients, the processes associated with rhodopsin and opsin synthesis and regeneration, the processes associated with disk formation, sustenance, and phagocytosis, and the processes associated with neurotransmitter release and synthesis. Except for the known role of rhodopsin as the first step in the visual process, and the proven role of calcium in the process of adaptation in the invertebrate photoreceptor, there is no widespread agreement on the physiological role of any of the listed chemical changes and reactions. Our data appear to indicate certain limits on the possible functional role of the conformations of rhodopsin and the concentration changes in cGMP. The simplest interpretation of the data on rhodopsin conformation changes is that neither the Meta I_{478} conformation nor later intermediate products or reactions are directly responsible for the phototransduction or rapid dark-adaptation processes of the photoreceptor or, if these conformations are involved, a rhodopsin-inactivating reaction must be present. The simplest interpretation of the data we have obtained on the light-induced concentration changes in cGMP is that these changes are not directly involved in any of the electrophysiological processes of the photoreceptor but are more likely to be involved in some of the maintenance processes.

II. ALTERING THE cGMP CONCENTRATION OF THE VERTEBRATE ROD PHOTORECEPTOR

In studies on cGMP we were interested in finding a chemical treatment that would mimic the effect of light in reducing the cGMP concentration of the vertebrate rod photoreceptor. Since light activates a phosphodiesterase that preferentially hydrolyzes cGMP, the studies were directed toward affecting this enzyme. Miki *et al.* (1975) had shown that this enzyme exhibited maximal activity at pH 8.0, with 50% activity at pH 6.4 and 25% activity at pH 6.0. Since bicarbonate-CO_2 had been shown to reduce the internal pH of the snail neuron (Thomas, 1974) and the barnacle photoreceptor (Brown and Meech, 1979) and had been used with toad photoreceptors (Pinto and Ostroy, 1978), we tested the effect of this buffer on the cGMP concentration of toad photoreceptors. Although we do not yet know if the elimination of bicarbonate-CO_2 from the perfusate mimicked the effect of light by increasing the activity of the phosphodiesterase normally stimulated by light, the results obtained thus far have been consistent with such a mechanism. Some of the data are shown in Fig. 1 and indicate that, in the dark, elimination of bicarbonate-CO_2 from the perfusate reduced the cGMP concentration of the photoreceptor to the levels normally observed on illumination and also reduced the effect of light on the changes in cGMP concentration (to 25% of the concentration changes observed in the presence of bicarbonate-CO_2; also see Meyertholen *et al.*, 1980). The reversibility of the effect is illustrated in Fig. 2. Preliminary data suggest that the mechanism of this

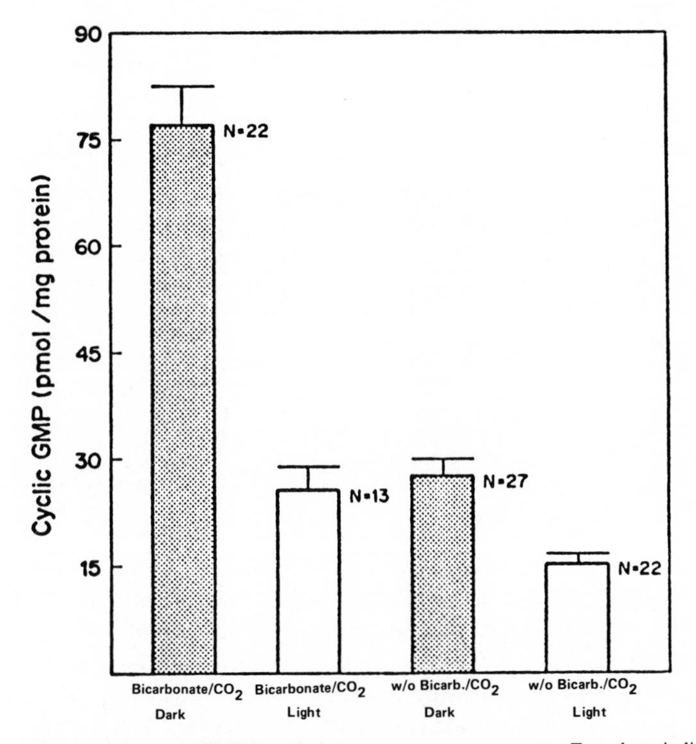

FIG. 1. Concentrations of cGMP in rod photoreceptor outer segments. Error bars indicate standard error of the mean. Bar graphs present effects of various solutions with differing light–dark conditions. Marine toads (*B. marinus*) were dark-adapted for 24 hr before all experiments. The composition of the bicarbonate solution was: 108 m*M* Na^+, 2.5 m*M* K^+, 1.0 m*M* Ca^{2+}, 1.6 m*M* Mg^{2+}, 90.5 m*M* Cl^-, 0.6 m*M* SO_4^{2-}, 24 m*M* HCO_3^-, 5.6 m*M* glucose, and 3 m*M* *N*-2-hydroxyethylpiperazine-*N*′-2-ethanesulfonic acid (HEPES). The solution was equilibrated with a gas mixture of 95% 0_2—5% CO_2, and the pH was adjusted to 7.8. The solution without bicarbonate-CO_2 was identical to the control solution except that 24 m*M* NaCl was substituted for 24 m*M* $NaHCO_3$ and the solution was equilibrated with 100% 0_2. The small difference in free Ca^{2+} concentration between the two solutions due to bicarbonate binding (1.0 versus 0.9 m*M*) was found to have minimal effects on the measurements. See text or Meyertholen *et al.* (1980) for further details.

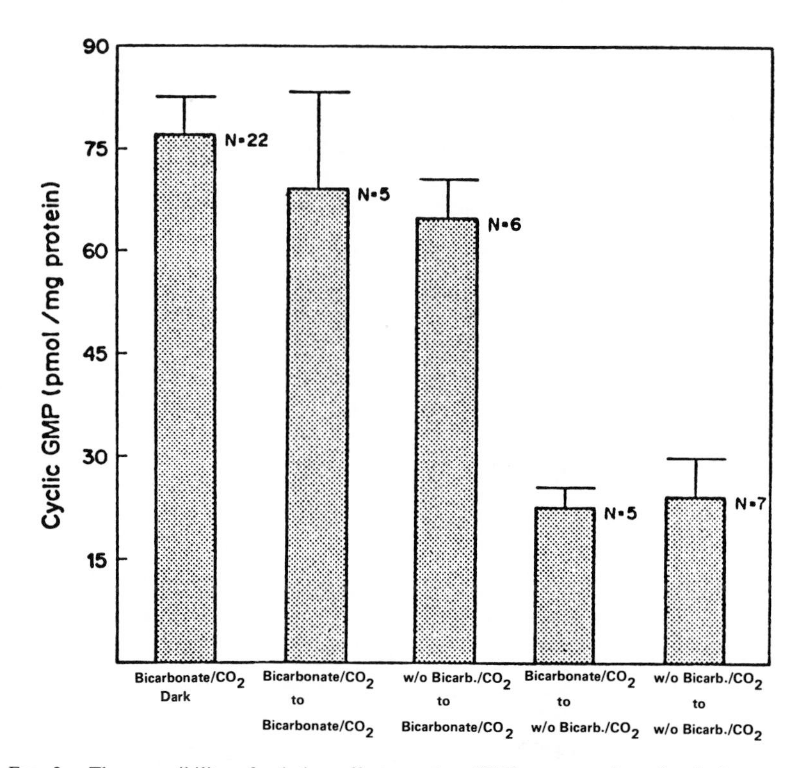

FIG. 2. The reversibility of solution effects on the cGMP concentration of rod photoreceptor outer segments. Effect of incubating dark-adapted retinas in one solution for 5 min and then transferring them to a second solution for 5 min. See Fig. 1 legend, the text, or Meyertholen *et al.* (1980) for further details.

effect is related to the concentration of bicarbonate buffer in the incubation media. Lowering the bicarbonate concentration from 24 m*M* to 5 m*M* reduced the cGMP concentration by 25%. A further lowering of the buffer concentration to 2 m*M* resulted in a 65% drop in cGMP levels, similar to that seen in the absence of bicarbonate. The increase in the cGMP levels does not appear to be the result of changes in the dissolved CO_2 concentration. Varying the CO_2 level of a 1 m*M* HCO_3 solution from 0.007 m*M* to 0.77 m*M* (the same concentration that is present in 24 m*M* bicarbonate/CO_2) did not have major effects on the cGMP levels [$28 \pm 4_{\mathrm{S.E.}}$ ($n = 11$) vs 19 ± 4 ($n = 9$) pmoles cGMP/mg protein]. In addition, use of acetate (20 m*M*) and of increased levels of HEPES buffer (24 m*M*) gave cGMP concentrations in the dark which were similar to those observed in the presence of 24 m*M* bicarbonate [81 ± 12 ($n = 7$) and 72 ± 11 ($n = 5$) pmoles cGMP/mg protein, respectively].

Some of the electrophysiological effects of eliminating bicarbonate-CO_2 from the perfusate are presented in Fig. 3. A reduction in signal amplitude was ob-

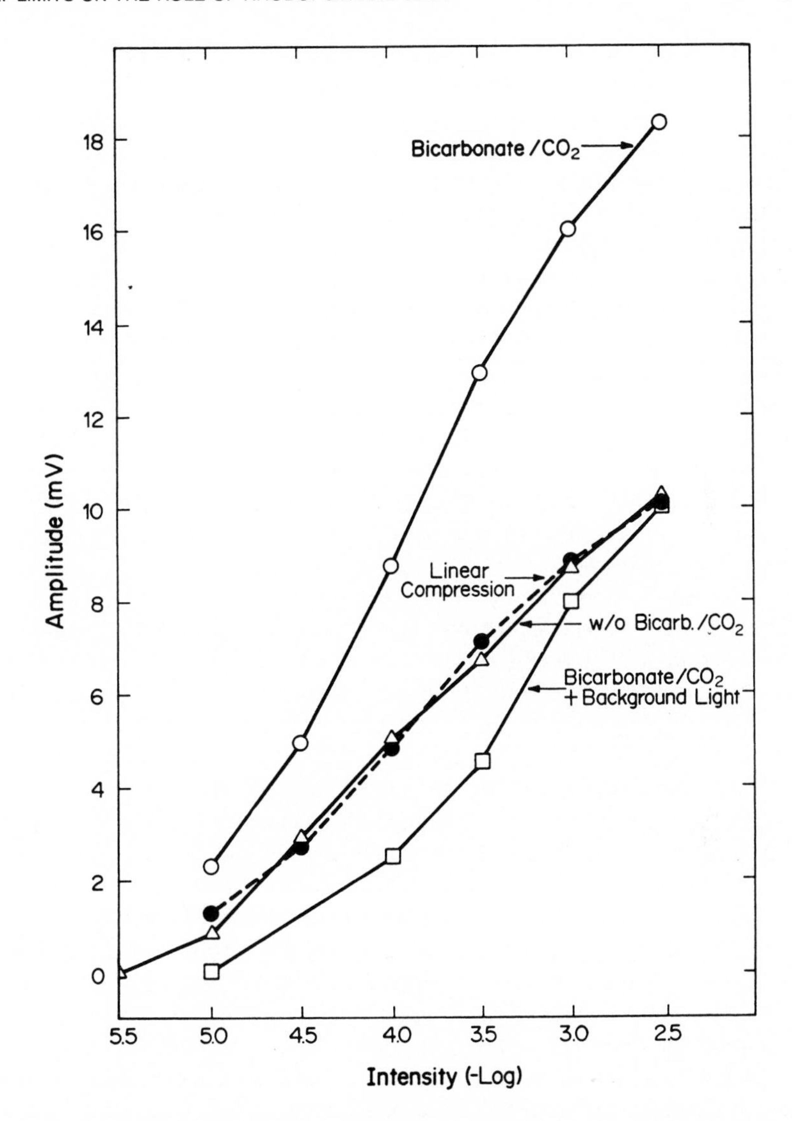

FIG. 3. Representative intracellular experiment comparing the effect of bicarbonate-CO_2 and light on the photoreceptor responses of the toad retina (*B. marinus*). Light stimuli were 200 msec in duration with 15 sec between stimuli. The unattenuated stimulus light intensity (500 nm) was 2.3×10^{-6} W/cm^2, the background light intensity (500 nm) was 1.5×10^{-10} W/cm^2, and the perfusion rate was approximately 4 ml/min. After a minimum of 5 min in the initial perfusion solution, bicarbonate-CO_2, a V-log I curve was determined. The V-log I curve without bicarbonate-CO_2 was started 9.5 min after changing solutions. The linear compression curve illustrates a 55% decrease in the bicarbonate-CO_2 curve. This experiment is representative of three individual cells that were studied completely and seven cells that were studied under various conditions and whose data were averaged. The only observed difference between the averaged cells and the data presented here was that some of the cells in solution without bicarbonate-CO_2 showed somewhat greater sensitivity.

served at all light intensities. Under the conditions of the experiment, the reduction was approximately 45%. In addition to these amplitude changes, elimination of the bicarbonate–CO_2 caused depolarization of the cells (in 8 of 11 perfusate changes, with 8 cells), while changing to perfusate *with* bicarbonate–CO_2 caused cellular hyperpolarization (in 14 of 18 changes, with 15 cells). The changes in membrane potential and signal amplitude are consistent with the report of Pinto and Ostroy (1978). The observed lack of effect of altered cGMP concentrations on the sensitivity of the cell (Fig. 3) are consistent with the results of Lipton *et al.* (1977). Thus far, the only other electrophysiological effects we have observed on elimination of bicarbonate–CO_2 are decreases in the latency of the responses. In preliminary data decreases in latency of 30–110 msec were observed for responses of comparable amplitudes with initial latencies of 120–250 msec ($n = 3$). The latency changes are in the same direction as those observed by Nicol and Miller (1978; also Miller and Nicol, 1979), although of a smaller magnitude. In preliminary experiments the elimination of bicarbonate–CO_2 caused no clear effect on the time to peak (2 of 3 comparisons) or the total time course of the responses (2 of 3 comparisons).

In many respects reducing the cGMP concentration of the cell did not mimic the effect of light on the electrophysiological properties of the photoreceptor. If the concentration changes in cGMP were directly responsible for the light-induced decreases in the sodium conductance of the rod photoreceptor (as suggested by Liebman and Pugh, 1979; Lolley *et al.*, 1977), then on decreasing the concentration of cGMP to the levels normally observed on complete bleaching of the rhodopsin, one would expect electrophysiological changes consistent with the application of a strong bleaching light. Thus one would expect cellular hyperpolarization (however, depolarization was observed) and the elimination of light-induced photoreceptor responses (photoresponses were still observed with amplitudes that were 55% of normal). If the concentration of cGMP were responsible for the effect of light on the adaptation properties of the photoreceptor cell (i.e., light adaptation, or rapid or slow dark adaptation), one would expect decreases in cell sensitivity (none were observed; see Fig. 3), decreases in the time to peak and the overall time course of the responses (none were observed), and decreases in the latency of the responses (only these were observed). The data may indicate that the overall concentrations of cGMP in the photoreceptor are not a critical factor in these electrophysiological processes. However, other interpretations are also possible. Thus one explanation for these data is that the proposed electrophysiological effects of cGMP were offset by secondary metabolic (Winkler and Riley, 1977) or pH (Brown and Meech, 1979) effects from the changes in bicarbonate or CO_2 concentration. Another possible interpretation is that the overall concentrations of cGMP are not a critical factor in these electrophysiological processes, but that the light-induced changes in cGMP (which still remain, although their amplitudes are reduced) reflect, or are responsible for, some critical process.

III. CONFORMATIONS OF RHODOPSIN AND THE ELECTROPHYSIOLOGICAL PROPERTIES OF THE PHOTORECEPTOR

The spectral intermediates of vertebrate rhodopsin are a sequential and kinetically reproducible indication of the chromophore–protein interactions of rhodopsin. They also provide defined landmarks which are indicative of the conformation changes in the protein, although all these conformation changes may not be detected spectrally. We investigated the relationship between the Meta I_{478} and Meta II_{380} conformations of rhodopsin and the adaptational and transduction states of the photoreceptor. In these experiments the concentrations of Meta I_{478} and Meta II_{380} were compared to the time of return of membrane potential and photoreceptor cell sensitivity following bleaching. The data for a 5% bleach are presented in Fig. 4 (also see Stein and Ostroy, 1978). At the time of return of membrane potential or sensitivity after bleaching, approximately 6.0×10^6 molecules of Meta I_{478} and 12×10^6 molecules of Meta II_{380} were present in each photoreceptor. With a 0.00035% bleach (data not shown) approximately 950 molecules of Meta I_{478} and 1900 molecules of Meta II_{380} were present at the time of return of membrane potential, and approximately 900 molecules of Meta I_{478} and 1800 molecules of Meta II_{380} were present at the time of return of cell sensitivity. [In experiments involving a 0.00035% bleach approximately 3150 molecules of bleached rhodopsin molecules were produced per photoreceptor, and the concentrations of Meta II_{380} were calculated using the rate constant for Meta II_{380} decay after a 5% bleach (see Donner and Hemila, 1975, for possible complications).] In conjunction with these studies we also measured the effect of temperature on the time of return of membrane potential and cell sensitivity following a 0.00035% bleach. At 10°C the time of return of membrane potential was 57 ± 1 (SD) sec and the time of return of cell sensitivity was 90 ± 6 sec; at 15°C the time of return of membrane potential was 53 ± 3 sec and the time of return of cell sensitivity was 77 ± 8 sec. These changes are 7 and 14%, respectively, whereas the rate of the Meta I_{478}-to-Meta II_{380} reaction would change by approximately 65% over this temperature range (assuming a Q_{10} of 8; Abrahamson *et al.*, 1960) and the rate of thermal decay of Meta II_{380} would change by approximately 38% over this temperature range (assuming a Q_{10} of 2.6; Stein and Ostroy, 1978). The data do not show any apparent correlation between the concentration of Meta I_{478} or Meta II_{380}, or the Meta I_{478}-to-Meta II_{380} reaction, and the time of return of membrane potential or sensitivity of the photoreceptor following bleaching. If the Meta I_{478} or Meta II_{380} conformation or the Meta I_{478}-to-Meta II_{380} reaction were directly responsible for the light-induced reductions in the sodium conductance of the photoreceptor or the rapid dark-adaptation process of the photoreceptor, one would not expect 900–6,000,000 of these molecules to be present or to continue to react after the membrane potential and sensitivity have returned to their normal dark values—particularly when one considers that the bleaching of one rhodopsin molecule is thought to be sufficient to affect the sodium conductance of the photoreceptor (Hecht *et al.*, 1942; Fain, 1975). These data may therefore be interpreted as indicating that the concentrations of Meta I_{478}, Meta II_{380}, and later photoproducts in the rhodopsin photolysis cycle, and the Meta I_{478}-to-Meta II_{380} reaction and later reactions in the photolysis cycle, are not critical factors in the process of phototransduction or rapid dark adaptation. Similar conclusions with regard to Meta II_{380} and later photoproducts were reached in other studies using a variety of techniques and animals (Frank and Dowling, 1968; Frank, 1971; Brin and Ripps, 1977). The uniqueness of our study is the measurements of Meta I_{478} and the use of both intracellular recording and physiological bleaching levels. However, other interpretations of the data are also possible. Farber *et al.* (1978) and Liebman and Pugh (1979) suggest that the phosphorylation of rhodopsin terminates its transducing function. Based on their view, phosphorylation of the photoproducts of rhodopsin would occur soon after illumination, and the photoproducts observed in our study, although spectrally indistinguishable from the photoproducts produced initially, would then be in an inactive, phosphorylated state. In fact, any spectrally undetectable inactivating reaction can be utilized to explain the data presented and still retain the view that these photoproducts or their interchanges are directly responsible for the process of phototransduction or dark adaptation. In addition, there is always the possibility that there is a special population of rhodopsin molecules that exhibits kinetic behavior somewhat different from that of the overall population that is typically monitored. A number of studies have suggested that the Meta I_{478}-to-Meta II_{380} reaction is of importance. Abrahamson and Ostroy (1967) originally suggested that the Meta I_{478}-to-Meta II_{380} reaction was likely to be responsible for the phototransduction process of the photoreceptor because it is the longest-lasting rapid reaction of rhodopsin involving both major conformation changes and interactions with the external environment. Recent studies have linked the Meta I_{478}-to-Meta II_{380} reaction with activation of the cGMP-phosphodiesterase of the photoreceptor (Parkes *et al.*, 1979) and with the release of calcium from the disk membrane (Kaupp *et al.*, 1980). Also, some studies on cellular physiology and rhodopsin intermediates have suggested a correlation between certain of these later photoproducts and the process of rapid dark adaptation in the photoreceptor (Donner and Reuter, 1967; Ernst and Kemp, 1972; Donner and Hemila, 1979). As discussed, if the Meta I_{478}-to-Meta II_{380} rhodopsin reaction or later photoproducts are involved in the physiological process of phototransduction or rapid dark adaptation, some additional factors will have to be invoked to explain our data.

IV. SUMMARY AND CONCLUSIONS

At the present stage of our understanding of the events that mediate between the absorption of light by rhodopsin and the phototransduction and adaptation

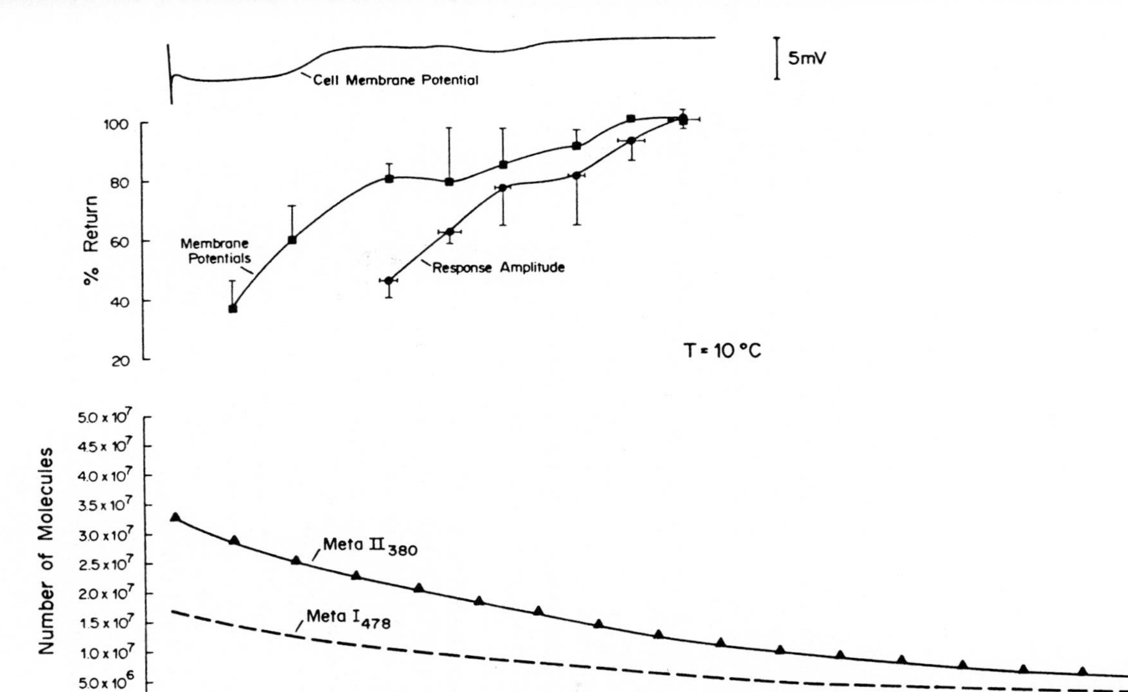

FIG. 4. The relationship between the Meta II_{380} concentration and rapid dark adaptation after a 5% bleach. To monitor the decay of Meta II_{380}, dissected retinas were placed in a chamber containing oxygenated Ringer's solution consisting of 107 m*M* Na^+, 2.5 m*M* K^+, 1.1 m*M* Ca^{2+}, 1.2 m*M* Mg^{2+}, 107 m*M* Cl^-, 0.6 m*M* SO_4^{2-}, 3.0 m*M* PO_4^{2-}, 5.6 m*M* glucose, and 3.0 m*M* HEPES at pH 7.8. The chamber was positioned in the beam of a Cary 14 recording spectrophotometer modified for microspectrophotometry. The absorbance changes at 380 nm were recorded following bleaches (500-nm light) of 5% of the rhodopsin. The Meta I_{478} concentration is based on an equilibrium constant of 2 determined for the Meta I_{478}-Meta II_{380} equilibrium at 10°C. This value compares with the value of 2.76 obtained by Baumann (1978) for this equilibrium at this temperature in the frog retina. In parallel experiments, the return of membrane potential and cell sensitivity were monitored using intracellular recording (error bars denote the standard deviation). Sensitivity was observed by monitoring the amplitude of the voltage response to a criterion test flash. The return of sensitivity was defined as the time required to return the response amplitude to a constant level.

processes of the vertebrate photoreceptor, it is premature to interpret our data in any singular way. However, singular speculative interpretations have the advantage of providing testable working hypotheses and forcing some data and literature interpretations. One such hypothesis is that the concentration changes in cGMP that occur on illumination are not responsible for the light-induced decreases in sodium conductance (the simplest interpretation of our cGMP data). In this hypothesis one could also assume that the Meta I_{478}-to-Meta II_{380} reaction is responsible for activation of the phosphodiesterase (Parkes *et al.*, 1979), perhaps via pH effects (Wong and Ostroy, 1973), and one could also interpret the rhodopsin data from our laboratory and others in the simplest manner as showing that Meta I_{478}, Meta II_{380}, the Meta I_{478}-to-Meta II_{380} reaction, and later rhodopsin intermediates and reactions are not responsible for phototransduction. In fact, if one uses the simplest interpretations, one should also suggest that cGMP changes are not involved in light adaptation (Fig. 3) or rapid dark adaptation (Fig. 4). However, one is wary of doing this because of a basic belief that such a complicated system as light activation of the phosphodiesterase would not exist without a purpose and that these electrophysiological processes could have many components. If, however, the cGMP system were involved in some of the maintenance processes of the photoreceptor (as implied but not explicitly stated in studies on the degenerative diseases of certain retinas; Lolley *et al.*, 1977; Aguirre *et al.*, 1978; Liu *et al.*, 1979), it is possible that the cGMP changes would not be directly involved in any of the electrophysiological processes we have investigated. It is also possible that the cGMP changes could be involved in the activation of a calcium or sodium pump without showing any obvious direct correlation with phototransduction or adaptation. If one assumes that cGMP concentration changes are not involved in the electrophysiological events of the vertebrate photoreceptor and proceeds to accept the basic tenets of the calcium hypothesis of visual transduction (the other major hypothesis of photoreceptor function; Yoshikami and Hagins, 1973), there are still some problems with simple data interpretations. The difficulty arises because of the recent data of Kaupp *et al.* (1980), which suggest that calcium release from the disks occurs during the Meta I_{478}-to-Meta II_{380} reaction. If this is the primary phototransduction step that releases the calcium and reduces the sodium conductance of the photoreceptor cell, the Meta I_{478}–Meta II_{380} data we have presented (Fig. 4) could again not be interpreted simplistically. Under these conditions (as with the hypothesis that cGMP changes are responsible for the light-induced decreases in sodium conductance) one would again have to invoke the notion of a reaction that terminates the phototransduction properties of rhodopsin. That some kinetically distinct rhodopsin molecules are involved in such a process does not seem as likely, since Kaupp *et al.* (1980) correlate the release of calcium with the main Meta I_{478}-to-Meta II_{380} reaction.

ACKNOWLEDGMENTS

Supported in part by grants from the U.S. Public Health Service, nos. EY00413, EY07008 (P.S.), and GM07211 (E.M.).

REFERENCES

Abrahamson, E. W., and Ostroy, S. E. (1967). The photochemical and macromolecular aspects of vision. *Prog. Biophys. Mol. Biol.* **17**, 179–215.

Abrahamson, E. W., Marquisee, J., Gavuzzi, P., and Roubie, J. (1960). Flash photolysis of the visual pigments. *Z. Elektrochem.* **64**, 177–180.

Aguirre, G., Farber, D., Lolley, R., Fletcher, R. T., and Chader, G. J. (1978). Rod-cone dysplasia in Irish setters: A defect in cyclic GMP metabolism in visual cells. *Science* **201**, 1133–1134.

Baumann, C. (1978). The equilibrium between metarhodopsin I and metarhodopsin II in the isolated frog retina. *J. Physiol. (London)* **279**, 71–80.

Brin, K. P., and Ripps, H. (1977). Rhodopsin photoproducts and rod sensitivity in the skate retina. *J. Gen. Physiol.* **69**, 97–120.

Brown, H. M., and Meech, R. W. (1979). Light induced changes of internal pH in a barnacle photoreceptor and the effect of internal pH on the receptor potential. *J. Physiol. (London)* **297**, 73–93.

Donner, K. O., and Hemila, S. (1975). Kinetics of long-lived rhodopsin photoproducts in the frog retina as a function of the amount bleached. *Vision Res.* **15**, 985–995.

Donner, K. O., and Hemila, S. O. (1979). Dark-adaptation of the aspartate-isolated rod receptor potential of the frog retina: Threshold measurements. *J. Physiol. (London)* **287**, 93–106.

Donner, K. O., and Reuter, T. (1967). Dark-adaptation processes in the rhodopsin rods of the frog's retina. *Vision Res.* **7**, 17–41.

Ernst, W., and Kemp, C. M. (1972). The effects of rhodopsin decomposition on PIII responses of isolated rat retinae. *Vision Res.* **12**, 1937–1946.

Fain, G. L. (1975). Quantum sensitivity of rods in the toad retina. *Science* **187**, 838–841.

Farber, D. B., Brown, B. M., and Lolley, R. N. (1978). Cyclic GMP: Proposed role in visual cell function. *Vision Res.* **18**, 497–499.

Frank, R. N. (1971). Properties of "neural" adaptation in components of the frog electroretinogram. *Vision Res.* **11**, 1113–1123.

Frank, R. N., and Dowling, J. E. (1968). Rhodopsin photoproducts: Effects on electroretinogram sensitivity in isolated perfused rat retina. *Science* **161**, 487–489.

Hecht, S., Shlaer, S., and Pirenne, M. H. (1942). Energy, quanta and vision. *J. Gen. Physiol.* **25**, 819–840.

Kaupp, U. B., Schnetkamp, P. P. M., and Junge, W. (1980). Metarhodopsin I/metarhodopsin II transition triggers light-induced change in calcium binding at rod disk membranes. *Nature (London)* **286**, 638–640.

Liebman, P. A., and Pugh, E. N., Jr. (1979). The control of phosphodiesterase in rod disk membranes: Kinetics, possible mechanisms and significance for vision. *Vision Res.* **19**, 375–380.

Lipton, S. A., Rasmussen, H., and Dowling, J. E. (1977). Electrical and adaptive properties of rod photoreceptors in *Bufo marinus*. II. Effects of cyclic nucleotides and prostaglandins. *J. Gen. Physiol.* **70**, 771–791.

Liu, Y. P., Krishna, G., Aguirre, G., and Chader, G. J. (1979). Involvement of cyclic GMP phosphodiesterase activator in an hereditary retinal degeneration. *Nature (London)* **280**, 62–64.

Lolley, R. N., Farber, D. B., Rayborn, M. E., and Hollyfield, J. G. (1977). Cyclic GMP accumulation causes degeneration of photoreceptor cells: Simulation of an inherited disease. *Science* **196**, 664–666.

Meyertholen, E. P., Wilson, M. J., and Ostroy, S. E. (1980). Removing bicarbonate/CO_2 reduces cGMP concentration of the vertebrate photoreceptor to the levels normally observed on illumination. *Biochem. Biophys. Res. Commun.* **96**, 785–792.

Miki, N., Baraban, J. M., Keirns, J. J., Boyce, J. J., and Bitensky, M. W. (1975). Purification and properties of the light-activated cyclic nucleotide phosphodiesterase of rod outer segments. *J. Biol. Chem.* **250**, 6320–6327.

Miller, W. H., and Nicol, G. D. (1979). Evidence that cyclic GMP regulates membrane potential in rod photoreceptors. *Nature (London)* **280**, 64–66.

Nicol, G. D., and Miller, W. H. (1978). Cyclic GMP injected into retinal rod outer segments increases latency and amplitude of response to illumination. *Proc. Natl. Acad. Sci. U.S.A.* **75**, 5217–5220.

Parkes, J. H., Liebman, P. A., and Pugh, E. N., Jr. (1979). Comparison of delay in hydrolysis of cGMP in ROS suspensions with rate of formation of metarhodopsin II. *Invest. Ophthalmol. Visual Sci.* **18**, Suppl. 22.

Pinto, L. H., and Ostroy, S. E. (1978). Ionizable groups and conductances of the rod photoreceptor membrane. *J. Gen. Physiol.* **71**, 329–345.

Stein, P. J., and Ostroy, S. E. (1978). Metarhodopsin II_{380} does not control rapid dark adaptation in vertebrate photoreceptors. *Biophys. J.* **21**, 172a.

Thomas, R. C. (1974). Intracellular pH of snail neurones measured with a new pH-sensitive glass microelectrode. *J. Physiol. (London)* **238**, 159–180.

Winkler, B. S., and Riley, M. V. (1977). Na^+-K^+ and HCO_3^- ATPase activity in retina: Dependence on calcium and sodium. *Invest. Ophthalmol. Visual Sci.* **16**, 1151–1154.

Wong, J. K., and Ostroy, S. E. (1973). Hydrogen ion changes of rhodopsin I. Proton uptake during the metarhodopsin I_{478}-metarhodopsin II_{380} reaction. *Arch. Biochem. Biophys.* **154**, 1–7.

Yoshikami, S., and Hagins, W. A. (1973). Control of the dark current in vertebrate rods and cones. *In* "Biochemistry and Physiology of Visual Pigments" (H. Langer, ed), pp. 245–255. Springer-Verlag, Berlin and New York.

Chapter 23

$[Ca^{2+}]_i$ Modulation of Membrane Sodium Conductance in Rod Outer Segments

BURKS OAKLEY II[1] *AND LAWRENCE H. PINTO*

Department of Biological Sciences
Purdue University
West Lafayette, Indiana

I. Na^+ DARK CURRENT: DECREASED BY LIGHT

When a vertebrate retinal rod absorbs light, a change in voltage occurs across the plasma membrane of its outer segment. The transduction process, by which the absorption of light eventually leads to the change in membrane voltage, is not fully understood. In this chapter, we will summarize the data that are relevant to one hypothesis for this process, and we will report the results of an experiment designed to test this hypothesis.

Several properties of the transduction process have been deduced by measuring the membrane current of rods. In the dark, there is a large current which flows out of the proximal portion of the rod and into the rod outer segment; light reduces the magnitude of this current (Penn and Hagins, 1969). Manipulations of the extracellular ionic concentration have shown that under normal conditions

[1]Present address: Department of Electrical Engineering, University of Illinois at Urbana-Champaign, Urbana, Illinois.

ISBN 0-12-153315-8

this current is carried by Na^+ (Sillman *et al.*, 1969; Hagins and Yoshikami, 1975). The reduction in Na^+ current entering the rod outer segment is thought to be associated with a decrease in Na^+ conductance (Toyoda *et al.*, 1969; Hagins, 1972; Fain *et al.*, 1978), as if light were blocking Na^+ channels in the rod outer segment that were open in the dark. This decrease in Na^+ conductance results in a hyperpolarizing receptor potential (Toyoda *et al.*, 1969; Brown and Pinto, 1974).

Measurements of the membrane current associated with the rod outer segment (Penn and Hagins, 1972; Baylor *et al.*, 1979) have shown that an intense stimulus is able to suppress completely the dark current, most likely by blocking all the Na^+ channels in the outer segment. In addition, a stimulus that causes an average of only one quantum of light to be absorbed per outer segment is able to suppress as much as 5% of the dark current. From this result, it has been calculated that more than 10^6 Na^+ ions fail to enter a rod when it absorbs a single quantum of light (Hagins *et al.*, 1970). Thus there is a very large amplification associated with the response to dim stimuli.

II. IS Ca^{2+} RELEASED FROM DISKS?

Nearly all the photopigment in rods is located in the membranes of intracellular disks contained within the rod outer segment (Hall *et al.*, 1969). Since these disks are physically separate from the plasma membrane (Cohen, 1968, 1970), several models for the transduction process have postulated the existence of an intracellular transmitter for excitation (Baylor and Fuortes, 1970; Yoshikami and Hagins, 1971). According to these models, this transmitter is released by photoexcited disks and diffuses to the plasma membrane; on arrival at the plasma membrane, the transmitter binds to Na^+ channels and blocks them. A transmitter that blocks Na^+ channels could account for several key features of the transduction process, such as saturation and amplification. In 1971, Yoshikami and Hagins suggested that calcium ion was the intracellular transmitter, and during the past decade much effort has been expended in testing this "calcium hypothesis."

Biochemical experiments have provided support for the calcium hypothesis. The intracellular disks contain sufficient Ca^{2+} (2–10 m*M*) for it to serve as the excitatory transmitter (Hendriks *et al.*, 1974; Liebman, 1974; Weller *et al.*, 1975; Hagins and Yoshikami, 1975; Szuts and Cone, 1977), and isolated rod outer segments retain the ability to sequester Ca^{2+} in the dark (Hemminki, 1975b; Schnetkamp *et al.*, 1977; Daemen *et al.*, 1977). There are, however, conflicting reports regarding the ability of disks or vesicles prepared from disk membranes to release Ca^{2+} on illumination (Hendriks *et al.*, 1974; Liebman, 1974; Mason *et al.*, 1974; Hemminki, 1975a; Shevchenko, 1976; Smith *et al.*, 1977; Szuts and

Cone, 1977; Kaupp and Junge, 1977; Kaupp *et al.*, 1979). These experiments have used various conditions in preparing the disks, and there have been no independent measures of the viability of the transduction mechanism in these preparations. The presence (or absence) of a light-evoked increase in intracellular calcium in intact rod outer segments remains to be demonstrated.

III. PREVIOUS TESTS OF THE CALCIUM HYPOTHESIS

Experiments in which the intracellular free calcium concentration, $[Ca^{2+}]_i$, has been manipulated also have provided support for the calcium hypothesis. In some of these experiments, the extracellular calcium concentration, $[Ca^{2+}]_o$, has been changed in an effort to produce changes in $[Ca^{2+}]_i$ of like sign. For example, Yoshikami and Hagins (1971) found that increasing $[Ca^{2+}]_o$ (which presumably increased $[Ca^{2+}]_i$) suppressed the dark current; this effect was similar to that caused by light. Brown and Pinto (1974) and Lipton *et al.* (1977) found that an increase in $[Ca^{2+}]_o$ caused the rod membrane to hyperpolarize. Again, presumed increases in $[Ca^{2+}]_i$ (produced by increases in $[Ca^{2+}]_o$) were seen to produce effects similar to those produced by light. Hagins and Yoshikami (1974) used the divalent cation ionophore X537a to increase the permeability of the plasma membrane to Ca^{2+} and showed that the level of $[Ca^{2+}]_o$ required to suppress the dark current was reduced 2000-fold in the presence of the ionophore. They concluded that Ca^{2+} had an intracellular site of action and that the dark current was affected in a similar way by an increase in $[Ca^{2+}]_i$ and by light. Changes in $[Ca^{2+}]_o$ also have been shown to affect the Na^+ permeability of isolated rod outer segments, most likely by changing $[Ca^{2+}]_i$. Using an osmotic technique, Korenbrot and Cone (1972) found that the Na^+ permeability of the outer segment was markedly reduced when $[Ca^{2+}]_o$ was raised from 2 to 10 m*M*. However, Wormington and Cone (1978) found that, when the ionophore X537a was present in the extracellular solution, raising $[Ca^{2+}]_o$ from 0.1 to only 10 μM was sufficient to suppress the Na^+ permeability. Since this ionophore allows Ca^{2+} to equilibrate across the plasma membrane, a given increase in $[Ca^{2+}]_i$ appears to be much more effective than a given increase in $[Ca^{2+}]_o$ in reducing the Na^+ permeability of rod outer segments.

One difficulty in interpreting the results of experiments that have used changes in $[Ca^{2+}]_o$ to produce changes in $[Ca^{2+}]_i$ concerns the direct effects on the rod membrane of $[Ca^{2+}]_o$ itself. In other systems, such as squid giant axon (Frankenhaeuser and Hodgkin, 1957), changes in $[Ca^{2+}]_o$ directly affect membrane conductance. It is possible therefore that changes in $[Ca^{2+}]_o$ have a direct effect on the rod membrane in addition to an indirect effect mediated by changes in $[Ca^{2+}]_i$. Thus any experiment in which retinal $[Ca^{2+}]_o$ is altered cannot be expected to produce effects on the rod membrane that are exactly the same as those of the hypothetical light-evoked change in $[Ca^{2+}]_i$ (see Brown and Pinto, 1974).

A more direct method of increasing $[Ca^{2+}]_i$ is to inject Ca^{2+} iontophoretically into a rod outer segment. Brown *et al.* (1977) impaled toad rods with double-barrel micropipets; one barrel was filled with 4.0 *M* potassium acetate and the other with 4.0 *M* potassium acetate plus 0.1 *M* calcium acetate. Ca^{2+} could be injected into a rod by passing current between the two barrels. When this technique was used, the rod membrane potential was not altered directly by transmembrane current. After the iontophoretic injection of Ca^{2+}, hyperpolarization occurred that recovered with a time course similar to that of the response to light. Again, increasing $[Ca^{2+}]_i$ produced an effect on the membrane potential that was similar to that of light.

The calcium hypothesis has also been tested by experiments in which $[Ca^{2+}]_i$ was buffered to a low value. Brown *et al.* (1977) iontophoretically injected ethylene glycol bis-β-aminoethyl ether *N,N,N′,N′*-tetraacetic acid (EGTA), a chelator with a high specificity for Ca^{2+} as compared to Mg^{2+}, into the outer segments of toad rods. They observed the effects of intracellular EGTA both on the resting membrane potential and on the receptor potential evoked by a small spot of light; EGTA produced membrane depolarization and attenuation of the receptor potential. These results are difficult to interpret, however, since any depolarization of the rod reduces the driving force on Na^+ and will therefore, of itself, attenuate the receptor potential. In control experiments, Brown *et al.* (1977) found that attenuation of the receptor potential was greater during the membrane depolarization produced by EGTA than during an equivalent depolarization produced by current. They concluded that lowering $[Ca^{2+}]_i$ with EGTA produced a change in rod membrane potential opposite that produced by light, and that sequestering cytoplasmic Ca^{2+} attenuated the receptor potential.

Hagins and Yoshikami (1977) used a vesicle-fusion technique to introduce EGTA into rods. After the rods were loaded with EGTA, the receptor potential (recorded extracellularly) was attenuated in response to dim stimuli but was of normal amplitude in response to intense stimuli. This result is consistent with the hypothesis that Ca^{2+} is the excitatory transmitter, if one assumes that the low intracellular EGTA concentrations ($<15\ \mu M$) produced using this technique could attenuate a (presumably) small increase in $[Ca^{2+}]_i$ in response to dim stimuli but could not affect a (presumably) much larger increase in $[Ca^{2+}]_i$ in response to an intense stimulus.

The above experiments, in which EGTA was used to buffer $[Ca^{2+}]_i$ to a low value, were potentially very important tests of the calcium hypothesis. There were, however, several difficulties with these experiments that prevented stronger conclusions from being reached. In the experiments with EGTA-filled vesicles (Hagins and Yoshikami, 1977), a reduction in $[Ca^{2+}]_o$ by more than three orders of magnitude was required in order to observe the desensitizing effects of intracellular EGTA. When the retina was bathed in normal Ringer's solution, intracellular EGTA was without effect, presumably because of the low intracellular concentrations of EGTA that could be attained with the technique. In the experiments in which EGTA was injected iontophoretically into rod outer segments (Brown *et al.*, 1977), EGTA not only attenuated the voltage response to a small spot stimulus but also depolarized the rod. Such a depolarization could have been due to an increase in sodium conductance (g_{Na}) but also could have been produced by a change in another ionic conductance. For example, a decrease in the potassium conductance (g_K) would have depolarized the membrane. The change in membrane conductance associated with the EGTA-evoked depolarization was not measured by Brown *et al.* (1977). In order to conclude that EGTA does affect g_{Na}, as predicted by the calcium hypothesis, it is necessary to assess the specific membrane conductance affected by EGTA.

IV. Ca^{2+} MODULATION OF Na^+ CONDUCTANCE

This section describes the results of experiments in which the effects of intracellular EGTA were examined in greater detail. EGTA was injected into rod outer segments under conditions such that (1) the membrane potential could be held constant, and (2) the specific membrane conductance altered by EGTA could be measured (Oakley and Pinto, 1980). The results of these experiments demonstrated that $[Ca^{2+}]_i$ modulated primarily the sodium conductance of the plasma membrane of rods. A complete report of these results is in preparation.

In the following experiments, EGTA was pressure-injected into outer segments of rods (most likely the larger and more numerous red rods) in an isolated retina preparation of the toad *Bufo marinus*. All data were obtained from cells that responded to light with hyperpolarizing plateaus greater than 10 mV. The retina was mounted receptor side up in a chamber having a transparent bottom. Oxygenated Ringer's solution (Brown *et al.*, 1977) flowed continuously over the retina. The chamber was placed on the stage of an inverted compound microscope and viewed using infrared illumination and an infrared-to-visible image converter. A rod outer segment was impaled with a double-barrel micropipet (as shown schematically in Fig. 1) which was made from "theta" tubing and had a coupling resistance of less than 3 MΩ. One of the electrode barrels was filled with 1.0 *M* potassium acetate; the other was filled with 0.5 *M* EGTA, 1.0 *M* potassium acetate, and 10 m*M* morpholinopropanesulfonic acid (MOPS, a pH buffer with a pK_a of 7.2) titrated to pH 7.1 with KOH. The potassium acetate barrel was used to measure the membrane voltage V_m. In order to prevent EGTA-induced changes in membrane voltage, the cell was "point-clamped" to its resting membrane voltage in the dark. With the use of a conventional voltage-clamp circuit, the membrane voltage was compared to a holding voltage, and the difference between these two voltages was used to drive a current source

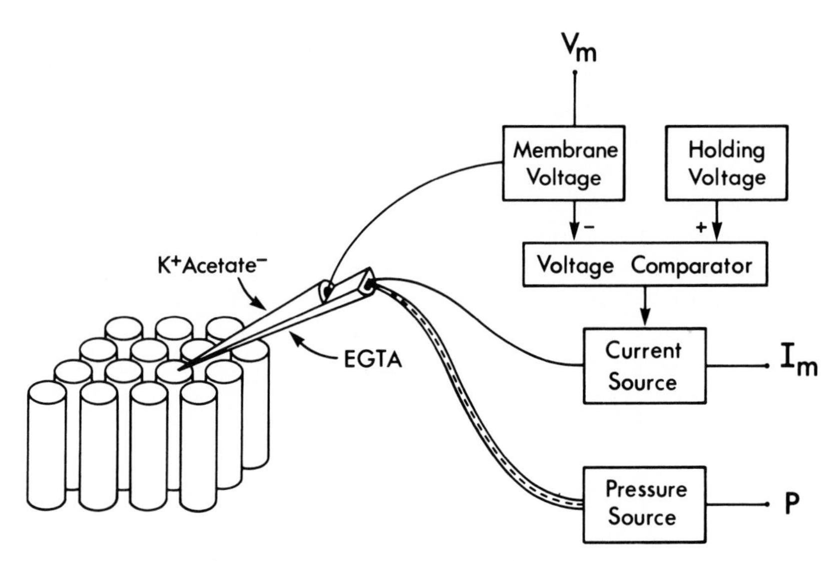

FIG. 1. A schematic diagram of the experimental technique used to pressure-inject EGTA into a rod outer segment during the point clamp. See text for additional details.

which injected current into the rod through the EGTA barrel. Since the holding voltage was always made equal to the cell's resting membrane voltage, no net current was passed into the cell prior to the injection of EGTA. The EGTA barrel was also connected to a pressure source, and the EGTA solution could be injected into the outer segment by applying pressure to the electrode (Meech, 1972). Although the rods were electrically coupled in this preparation (Fain *et al.*, 1976; Leeper *et al.*, 1978; Gold, 1979; Griff and Pinto, 1981), it was valid to use this point-clamp technique to examine the responses to injected EGTA because (1) none of the rods surrounding the impaled rod were altered in any way, (2) the impaled rod itself always remained clamped to its resting voltage, and (3) EGTA was injected into the impaled rod near the origin of the clamp current.

The effects of intracellular EGTA were observed by point-clamping the rod outer segment to its resting voltage and recording the membrane current while pressure-injecting EGTA (Fig. 2). Every 7 sec, a small (<10 mV), hyperpolarizing command pulse, 1 sec in duration, was given to the voltage clamp, and the input conductance of the cell was calculated by measuring the incremental membrane current required to hyperpolarize the membrane (Brown and Brown, 1973). In order to inject the EGTA solution, a pressure of about 4 bars was applied to the EGTA barrel. (One bar is approximately 1 atm.) During the time when the cell was point-clamped, the injection of EGTA evoked an inward membrane current I_m which reached a maximum of about 0.22 nA (Fig. 2). When the pressure was reduced to zero, the membrane current slowly returned to zero. The inward membrane current was associated with an increase in input conductance, as evidenced by the increase in the incremental membrane current required to hyperpolarize the membrane by 8.0 mV. For the cell illustrated in Fig. 2, the input conductance was initially 11.5 nS and was increased by 43% to 16.4 nS during the injection of EGTA. In other experiments, when the electrode was known to be able to eject EGTA (that is, when applying pressure to the electrode

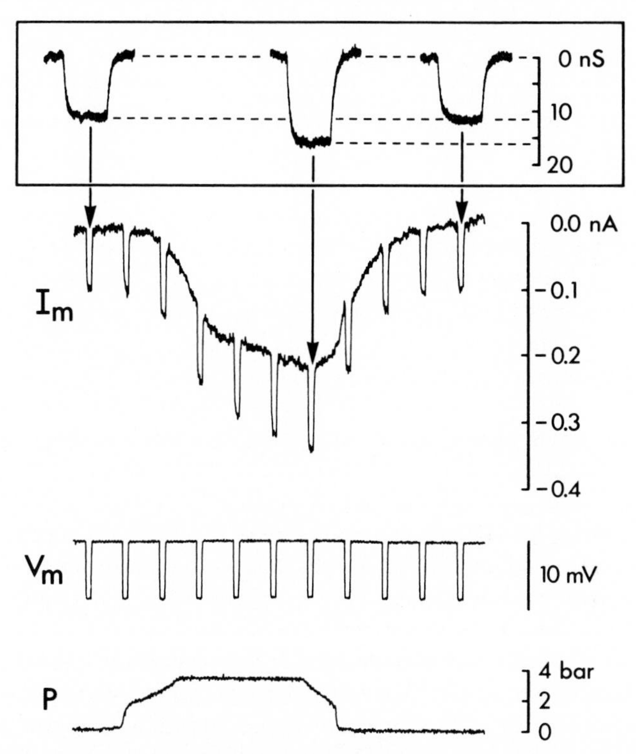

FIG. 2. Responses to the injection of EGTA. The rod outer segment was point-clamped to its dark resting voltage of −34 mV, as indicated by the initial value of the trace marked V_m. Every 7 sec, an 8.0-mV hyperpolarizing command pulse (1 sec in duration) was given to the voltage clamp, resulting in negative deflections of V_m. The EGTA solution was injected into the rod by applying pressure (from a source of compressed nitrogen gas) to the solution in the electrode. A pressure transducer was used to measure the applied pressure; its output is shown as the trace marked P. The membrane current required to point-clamp the outer segment is shown in the trace marked I_m. The incremental membrane current required to hyperpolarize the membrane by 8.0 mV was a direct measure of the input conductance of the cell. Three of these incremental membrane currents are displayed on an expanded time base in the inset at the top of the figure, and a scale representing conductance has been drawn next to them. The input conductance was initially 11.5 nS (equivalent to the reciprocal of 87 MΩ) and was increased by 43% to 16.4 nS (equivalent to the reciprocal of 61 MΩ) during the injection of EGTA. The effect of the injection of EGTA was reversible; after the injection, the input conductance returned to its control value.

caused a decrease in electrode resistance; see Bader *et al.*, 1976), similar effects of EGTA were observed in every injected outer segment. In a total of 21 cells from 13 retinas, the input conductance averaged 8.7 nS before the injection of EGTA and increased by 38% during the injection of EGTA. This result is consistent with an EGTA-evoked increase in g_{Na}.

To show that this result was specific to EGTA and not an artifact due to pressure and/or volume changes within the cell, a control experiment was performed. EGTA was omitted from the solution filling the electrode, so that the electrode contained only potassium acetate and MOPS at pH 7.1. In five cells, injection of this potassium acetate solution did not affect membrane current or input conductance. Thus the effects of injected EGTA (Fig. 2) were specific to EGTA itself.

Since the rods are electrically coupled, the point-clamp technique does not provide a true space clamp. It is possible therefore that some patches of membrane could have depolarized during the injection of EGTA. The rod membrane contains several voltage-sensitive conductances (Fain *et al.*, 1977, 1978) which lead to nonlinearities in the membrane's current–voltage curve. Thus membrane depolarization during the injection of EGTA could cause an increase in input conductance, and this effect would be attributed incorrectly to a direct effect of EGTA. To test this possibility, the retina was bathed in a solution of pharmacological agents chosen to abolish voltage-sensitive conductances and thus linearize the membrane's current–voltage curve. This solution was similar to the one used by Bader *et al.* (1979), and contained 10 m*M* TEA^+, 10 m*M* Cs^+, 0.5 m*M* Co^{2+}, and 1.0 m*M* 4-aminopyridine. In the presence of these pharmacological agents, the injection of EGTA still evoked an inward membrane current and an increase in input conductance. Thus the effects of EGTA are not likely to be artifacts due to limitations of the point-clamp technique.

According to the calcium hypothesis, it is g_{Na} specifically that is increased by lowering $[Ca^{2+}]_i$ with EGTA. The following experiment was designed to test for such specificity. A rod outer segment was impaled and, as a control, EGTA was injected during the point clamp. This injection produced an inward membrane current and an increase in input conductance similar to those shown in Fig. 2. The voltage clamp was then turned off, and the solution bathing the retina was switched to one in which Na^+ was replaced on an equimolar basis by choline$^+$. The rod hyperpolarized, and its response to light was abolished (Brown and Pinto, 1974). The membrane was then point-clamped to its new hyperpolarized resting value. The EGTA-evoked changes in membrane current and input conductance were greatly attenuated in the low-Na^+ solution. For a total of 11 cells in low-Na^+ solution, the EGTA-evoked changes in input conductance were reduced below the limit of detectability, which was about a 10% change in conductance.

A recent study by Fain and Bastian (1980) found that bathing the retina in a low-Na^+ solution seemed to produce an increase in $[Ca^{2+}]_i$, presumably because of a lack of Na^+-Ca^{2+} exchange (Baker *et al.*, 1969). If $[Ca^{2+}]_i$ were increased appreciably by the low-Na^+ solution, then the pressure-injection of EGTA might not chelate sufficient Ca^{2+} to decrease $[Ca^{2+}]_i$ by a significant amount. For this reason, a series of experiments were conducted in which the low-Na^+ solution did not contain any added Ca^{2+} or in which Ca^{2+} was buffered to a low value (10^{-8} *M*) with EGTA. In four cells from retinas bathed in low-Na^+, low-Ca^{2+} solutions, the EGTA-evoked changes in input conductance were again reduced below the limit of detectability (less than a 10% change in conductance). Since the injection of EGTA evoked increases in input conductance of nearly 40% when the retina was in a control solution, we concluded that the pressure-injection of EGTA into rod outer segments increased primarily the sodium conductance of the plasma membrane.

Our experiments are a valid test of the calcium hypothesis only if the effects of injected EGTA are due to the specific sequestration of Ca^{2+} rather than the sequestration of other cytoplasmic cations. EGTA chelates Ca^{2+} with an affinity approximately six orders of magnitude greater than its affinity for Mg^{2+} (Sillén and Martell, 1971), so in the rod outer segment EGTA chelates Ca^{2+} specifically in the presence of Mg^{2+}. It is unlikely therefore that the observed effects of intracellular EGTA are due to the chelation of Mg^{2+}. Although EGTA chelates Ca^{2+} with a high affinity, it actually has a greater affinity for certain other divalent cations, such as Pb^{2+}, Zn^{2+}, Cd^{2+}, and Ni^{2+} (Sillén and Martell, 1971). It is possible therefore that the effects we observed on injection of EGTA occurred as a result of the chelation of a trace metal ion for which EGTA had high affinity. To test this possibility, a solution containing approximately equimolar EGTA and Ca^{2+} was injected into a rod outer segment. Since the EGTA in this solution was already nearly saturated with Ca^{2+}, it would not have been able to chelate a significant fraction of the intracellular free calcium. On the other hand, if EGTA produced its effects by chelating a trace metal ion for which its affinity was much greater than its affinity for Ca^{2+}, EGTA would still chelate this trace ion in the presence of a saturating amount of Ca^{2+}. In four cells, there were no detectable changes in membrane current or input conductance when the Ca^{2+}-EGTA solution was injected during the point clamp. Thus the effects of injected EGTA did not result from the chelation of a trace metal for which EGTA had a high affinity. We concluded that these effects most likely resulted from the lowering of $[Ca^{2+}]_i$.

V. CONCLUSIONS

In summary, when a rod outer segment is point-clamped to its resting voltage, the pressure-injection of EGTA evokes an inward membrane current and an increase in input conductance. The amplitudes of both these effects are attenuated when the extracellular sodium concentration is reduced. Thus the injection of EGTA increases primarily g_{Na}. Since the injection of a solution containing EGTA to which is bound a saturating amount of calcium is without effect, the relevant intracellular ionic concentration that is altered is most likely that of calcium. Thus lowering $[Ca^{2+}]_i$ in rod outer segments increases primarily the sodium conductance of the plasma membrane. This finding is consistent with the calcium hypothesis of visual transduction.

From the above experiments, we cannot say with certainty that lowered $[Ca^{2+}]_i$ itself directly increases the sodium conductance of the plasma membrane, or if lowered $[Ca^{2+}]_i$ alters the intracellular concentrations of other substances thought to regulate the sodium conductance, such as cyclic nucleotides (see Miller and Nichol, 1979, and this volume Chapters 24 and 25). Additional experiments must now be done to investigate the specific conductance changes associated with the injection of cyclic nucleotides, as well as the specific conductance changes that result from increased $[Ca^{2+}]_i$ (Brown *et al.*, 1977). Finally, we await the results of experiments in which $[Ca^{2+}]_i$ is measured during the illumination of intact rods known to be capable of performing the transduction process.

ACKNOWLEDGMENT

This work was supported by NIH grants EY01221 and EY07008.

REFERENCES

Bader, C. R., Baumann, F., and Bertrand, D. (1976). Role of intracellular calcium and sodium in light adaptation in the retina of the honey bee drone (*Apis mellifera, L.*) *J. Gen. Physiol.* **67**, 475–491.

Bader, C. R., MacLeish, P. R., and Schwartz, E. A. (1979). A voltage-clamp study of the light response in solitary rods of the tiger salamander. *J. Physiol. (London)* **296**, 1–26.

Baker, P. F., Blaustein, M. P., Hodgkin, A. L., and Steinhardt, R. A. (1969). The influence of calcium on sodium efflux in squid axons. *J. Physiol. (London)* **200**, 431–458.

Baylor, D. A., and Fuortes, M. G. F. (1970). Electrical responses of single cones in the retina of the turtle. *J. Physiol. (London)* **207**, 77–92.

Baylor, D. A., Lamb, T. D., and Yau, K.-W. (1979). The membrane current of single rod outer segments. *J. Physiol. (London)* **288**, 589–611.

Brown, A. M., and Brown, H. M. (1973). Light response of a giant *Aplysia* neuron. *J. Gen. Physiol.* **62**, 239–254.

Brown, J. E., and Pinto, L. H. (1974). Ionic mechanism for the photoreceptor potential of the retina of *Bufo marinus*. *J. Physiol. (London)* **236**, 575–591.

Brown, J. E., Coles, J. A., and Pinto, L. H. (1977). Effects of injections of calcium and EGTA into the outer segments of retinal rods of *Bufo marinus*. *J. Physiol. (London)* **269**, 707–722.

Cohen, A. I. (1968). New evidence supporting the linkage to extracellular space of outer segment saccules of frog cones but not rods. *J. Cell Biol.* **37**, 424–444.

Cohen, A. I. (1970). Further studies on the question of the patency of saccules in outer segments of vertebrate photoreceptors. *Vision Res.* **10**, 445–453.

Daemen, F. J. M., Schnetkamp, P. P. M., Hendriks, T., and Bonting, S. L. (1977). Calcium and rod outer segments. *In* "Vertebrate Photoreception" (H. B. Barlow and P. Fatt, eds.), pp. 29-40. Academic Press, New York.

Fain, G. L., and Bastian, B. L. (1980). The role of Ca^{2+} in light-adaptation in rods. *Invest. Ophthalmol. Visual Sci.* **19**, 192.

Fain, G. L., Gold, G. H., and Dowling, J. E. (1976). Receptor coupling in the toad retina. *Cold Spring Harbor Symp. Quant. Biol.* **40**, 547-561.

Fain, G. L., Quandt, F. N., and Gerschenfeld, H. M. (1977). Calcium-dependent regenerative responses in rods. *Nature (London)* **269**, 707-710.

Fain, G. L., Quandt, F. N., Bastian, B. L., and Gerschenfeld, H. M. (1978). Contribution of a caesium-sensitive conductance increase to the rod photoresponse. *Nature (London)* **272**, 467-469.

Frankenhaeuser, B., and Hodgkin, A. L. (1957). The action of calcium on the electrical properties of squid axons. *J. Physiol. (London)* **137**, 218-244.

Gold, G. H. (1979). Photoreceptor coupling in retina of the toad, *Bufo marinus*. II. Physiology. *J. Neurophysiol.* **42**, 311-328.

Griff, E. R., and Pinto, L. H. (1981). Interactions among rods in the isolated retina of *Bufo marinus*. *J. Physiol. (London)* **314**, 237-254.

Hagins, W. A. (1972). The visual process: Excitatory mechanisms in the primary receptor cells. *Annu. Rev. Biophys. Bioeng.* **1**, 131-158.

Hagins, W. A., and Yoshikami, S. (1974). A role for Ca^{2+} in excitation of retinal rods and cones. *Exp. Eye Res.* **18**, 299-305.

Hagins, W. A., and Yoshikami, S. (1975). Ionic mechanisms in excitation of photoreceptors. *Ann. N. Y. Acad. Sci.* **264**, 314-325.

Hagins, W. A., and Yoshikami, S. (1977). Intracellular transmission of visual excitation in photoreceptors: Electrical effects of chelating agents introduced into rods by vesicle fusion. *In* "Vertebrate Photoreception" (H. B. Barlow and P. Fatt, eds.), pp. 97-139. Academic Press, New York.

Hagins, W. A., Penn, R. D., and Yoshikami, S. (1970). Dark current and photocurrent in retinal rods. *Biophys. J.* **10**, 380-412.

Hall, M. O., Bok, D., and Bacharach, A.D.E. (1969). Biosynthesis and assembly of the rod outer segment membrane system: Formation and fate of visual pigment in the frog retina. *J. Mol. Biol.* **45**, 397-406.

Hemminki, K. (1975a). Light-induced decrease in calcium binding to isolated bovine photoreceptors. *Vision Res.* **15**, 69-72.

Hemminki, K. (1975b). Accumulation of calcium by retinal outer segments. *Acta Physiol. Scand.* **95**, 117-125.

Hendriks, T., Daemen, F. J., and Bonting, S. L. (1974) Biochemical aspects of the visual process. XXV. Light-induced calcium movements in isolated frog rod outer segments. *Biochim. Biophys. Acta* **345**, 468-473.

Kaupp, U. B., and Junge, W. (1977). Rapid calcium release by passively loaded retinal discs on photoexcitation. *FEBS Lett.* **81**, 229-232.

Kaupp, U. B., Schnetkamp, P.P.M., and Junge, W. (1979). Light-induced calcium release in isolated intact cattle rod outer segments upon photoexcitation of rhodopsin. *Biochim. Biophys. Acta* **552**, 390-403.

Korenbrot, J. I., and Cone, R. A. (1972). Dark ionic flux and the effects of light in isolated rod outer segments. *J. Gen. Physiol.* **60**, 20-45.

Leeper, H. F., Normann, R. A., and Copenhagen, D. R. (1978). Red rods of the toad retina: Evidence for passive electrotonic interactions. *Nature (London)* **275**, 234-236.

Liebman, P. A. (1974). Light-dependent Ca^{++} content of rod outer segment disc membranes. *Invest. Ophthalmol.* **13**, 700-701.

Lipton, S. A., Ostroy, S. E., and Dowling, J. E. (1977). Electrical and adaptive properties of rod photoreceptors in *Bufo marinus*. I. Effects of altered extracellular Ca^{2+} levels. *J. Gen. Physiol.* **70**, 747-770.

Mason, W. T., Fager, R. S., and Abrahamson, E. W. (1974). Ion fluxes in disk membranes of retinal rod outer segments. *Nature (London)* **247**, 562-563.

Meech, R. W. (1972). Intracellular calcium injection causes increased potassium conductance in *Aplysia* nerve cells. *Comp. Biochem. Physiol. A* **42A**, 493-499.

Miller, W. H., and Nichol, G. D. (1979). Evidence that cyclic GMP regulates membrane potential in rod photoreceptors. *Nature (London)* **280**, 64-66.

Oakley, B., II, and Pinto, L. H. (1980). $[Ca^{2+}]_i$ modulates membrane sodium conductance in rod outer segments. *Invest. Ophthalmol. Visual Sci.* **19**, Suppl., 102.

Penn, R. D., and Hagins, W. A. (1969). Signal transmission along retinal rods and the origin of the electroretinographic a-wave. *Nature (London)* **223**, 201-205.

Penn, R. D., and Hagins, W. A. (1972). Kinetics of the photocurrent of retinal rods. *Biophys. J.* **12**, 1073-1094.

Schnetkamp, P.P.M., Daemen, F.J.M., and Bonting, S. L. (1977). Biochemical aspects of the visual process. XXXVI. Calcium accumulation in cattle rod outer segments: Evidence for a calcium-sodium exchange carrier in the rod sac membrane. *Biochim. Biophys. Acta* **468,** 259-270.

Shevchenko, T. F. (1976). Change in the activity of calcium ions on illumination of a suspension of the fragments of the external segments of visual cells. *Biophysics* **21,** 327-330.

Sillén, L., and Martell, A. (1971). "Stability Constants Supplement," Vol. 1, pp. 733-734. Chemical Society, London.

Sillman, A. J., Ito, H., and Tomita, T. (1969). Studies on the mass receptor potential of the isolated frog retina. II. On the basis of the ionic mechanism. *Vision Res.* **9**, 1443-1451.

Smith, H. G., Fager, R. S., and Litman, B. J. (1977). Light-activated calcium release from sonicated bovine retinal rod outer segment discs. *Biochemistry* **16**, 1399-1405.

Szuts, E., and Cone, R. A. (1977). Calcium content of frog rod outer segment and discs. *Biochim. Biophys. Acta* **468**, 194-208.

Toyoda, J.-I., Nosaki, H., and Tomita, T. (1969). Light-induced resistance changes in single photoreceptors of *Necturus* and *Gekko*. *Vision Res.* **9**, 453-463.

Weller, M., Virmaux, N., and Mandel, P. (1975). Role of light and rhodopsin phosphorylation in control of permeability of retinal rod outer segment disks to Ca^{2+}. *Nature (London)* **256,** 68-70.

Wormington, C. M., and Cone, R. A. (1978). Ionic blockage of the light-regulated sodium channels in isolated rod outer segments. *J. Gen. Physiol.* **71**, 657-681.

Yoshikami, S., and Hagins, W. A. (1971). Light, calcium, and the photocurrent of rods and cones. *Biophys. J.* **11**, 47a.

Chapter 24

Cyclic-GMP-Induced Depolarization and Increased Response Latency of Rods: Antagonism by Light

WILLIAM H. MILLER AND GRANT D. NICOL[1]

Yale University School of Medicine
New Haven, Connecticut

I. INTRODUCTION

A. Sensitization and Latent Periods

"The mechanism of photoreception is not a single process. Corresponding with the division of the reaction time into an exposure or sensitization period and a latent period, there is a fundamental division of the underlying machinery into

[1]Present address: Department of Electrical Engineering and Computer Sciences, University of California, Berkeley, California.

ISBN 0-12-153315-8

an initial photochemical reaction and a consequent ordinary chemical reaction'' wrote Selig Hecht (1919–1920). The underlying machinery of the photochemical reaction manifests itself in (1) the close correspondence between the absorption properties of rhodopsin and the visibility curve for scoptopic vision (Hecht and Williams, 1922–1923); (2) the reciprocal relation between intensity and duration for short flashes reflecting the photochemical law of Bunsen and Roscoe (Hartline, 1934); and (3) the insensitivity to temperature in the physiological range of the minimum intensity needed for excitation (Hecht, 1919–1920) that corresponds to the low Q_{10} (about 1) of the photochemical reaction. With regard to this last point, ''the quantum efficiences for the [photo]isomerization of *mono-cis* isomers to all-*trans*-retinal in hexane are virtually independent of temperature,'' according to Kropf and Hubbard (1970).

After the initial effects of a short flash, which are clearly photochemically inspired, there follows a latent period. The underlying machinery of the processes in this dark interval between the absorption of light and the first sign of a response appears to be dominated by ''ordinary chemical reactions.'' The photochemical reaction, 11-*cis*- to all-*trans*-retinal, occupies picoseconds; the latent period ranges from milliseconds to seconds. In contrast to the photochemical reaction, the latent period shows the strong temperature dependence characteristic of most chemical reactions (Hecht, 1918–1919; Hartline *et al.*, 1952; Bornschein and Charif, 1959). Finally, the kinetics of the light response of single *Bufo* rod outer segments are consistent with a set of reactions involving four stages of delay which clearly are separate from the rapid photochemical reaction (Baylor *et al.*, 1979).

B. Chemical Reactions of the Latent Period and the Response

Interest in the transduction mechanism before 1971 centered principally on known dark reactions subsequent to the photochemical conversion of rhodopsin to prelumirhodopsin. The dark reactions, prelumirhodopsin to metarhodopsin I, and Meta I to Meta II, with activation energies on the order of 60 kcal/mole (Hubbard *et al.*, 1965), both showed the strong temperature dependence characteristic of the latent period. Yet neither could unequivocally be shown to lie in the direct chain of events linking the photochemical reaction with control of the rod plasma membrane sodium permeability. By 1971 two other molecular constituents known to be important in the regulation of cellular function for virtually all cell types were proposed as intracellular regulators of vertebrate photoreceptor organelle function—calcium (Yoshikami and Hagins, 1971) and cyclic adenosine 3′,5′-monophosphate (cAMP) (Bitensky *et al.*, 1971).

cAMP had been discovered to be the intracellular mediator of hormone-induced glycogenolysis by Earl Sutherland and his colleagues in 1957, and by

1970 it was believed to be present in virtually every cell type and to act as an intracellular regulator of a wide variety of cellular functions in response to a hormone signal. Although the hormone precedent was missing in the rod outer segment (ROS), cAMP had been shown to be the intracellular mediator of vasopressin-induced increased sodium permeability of the mucosal surface of the toad bladder (Orloff and Handler, 1967). It therefore seemed of interest to determine whether light controlled the concentration of this ubiquitous intracellular regulator, which in turn could control the sodium permeability of the rod plasma membrane. No other comparable intracellular regulators were known, and although cyclic guanosine 3′,5′-monophosphate (cGMP) had been shown to be the only other naturally occurring cyclic nucleotide, its function at that time was totally undefined. To test the concept [8-^{14}C] adenosine 5′-triphosphate (ATP) was incubated with homogenized ROS that had been either light- or dark-adapted, and the ROS were assayed to measure the amount of cAMP made under each condition. The result of this experiment was that after a 10-min incubation there was over a sixfold greater amount of cAMP in the dark-adapted than in the light-adapted preparation (Bitensky *et al.*, 1971). With this demonstrated control of cAMP intracellular concentration by illumination, the cyclic-nucleotide enzymatic cascade thereby became a candidate for a role in the postulated ordinary chemical reactions of the latent period and in the generation of the response to light. The idea that the regulation of intracellular cyclic-nucleotide levels by illumination constituted a link in the intermediary processes that regulated rod plasma membrane sodium permeability was not defined in detail, yet the concept later proved to be sufficiently sound to serve as a working hypothesis. Details were not long in coming. It quickly became obvious that cGMP, not cAMP, was the native ROS cyclic nucleotide and that its concentration was controlled by light activation of cyclic-nucleotide phosphodiesterase (PDE). But, as PDE hydrolyzes cAMP (though inefficiently) as well as cGMP, so the cyclase catalyzes the formation of cAMP from ATP as well as cGMP from guanosine 5′-triphosphate GTP (Fleischman, this volume, Chapter 6). This is a probable explanation for the success of the first experiment utilizing ATP (Bitensky *et al.*, 1971).

The cGMP concentration is controlled by PDE activation, and the first step in this activation is mediated by the exchange of GTP for bound guanosine 5′-diphosphate (GDP) catalyzed by photolyzed rhodopsin (Fung and Stryer, 1980). Further, PDE activation is rapid enough to mediate excitation (Yee and Liebman, 1978; Woodruff and Bownds, 1979). This evidence that PDE activation by light regulates the cGMP concentration is basic to the hypothesis that cGMP may in turn regulate the ROS membrane potential. We selected the method of intracellular injection of cGMP as a test of this hypothesis. Intracellular injection was selected because it avoids the pharmacological uncertainties of extracellular application. Nucleotides act inside the cell and penetrate membranes poorly and may give uncertain results when applied extracellularly. PDE inhibitors have complex and diverse actions independent of their effect on cyclic nucleotides (Weber and Herz, 1968; Smith *et al.*, 1979). The most widely used PDE inhibitor, isobutylmethylxanthine (IBMX), raises ROS intracellular levels of both cGMP and cAMP (Cohen, this volume, Chapter 12) and does not allow one to determine which nucleotide causes which effect. In addition, injection is the most accurate method for observation of the crucially important latency that occupies the dark period during which the chemical reactions of the cyclic-nucleotide system must begin to operate to generate the physiological response to light. Working with this method we have found physiological evidence suggesting that, corresponding to the demonstrated enzymatic control of cGMP concentration, cGMP in turn controls important aspects of the membrane potential including the latency, amplitude, rise time, and duration of the response to light. This control appears to be exerted through light-regulated hydrolysis rates of cGMP, which influence the membrane potential locally by as yet unknown mechanisms. The keystone of this edifice of cGMP control of ROS membrane potential is the observation that excess ROS cGMP increases the response latency (Nicol and Miller, 1978). This observation (Section II,C,2) shows that excess cGMP must be hydrolyzed in order to achieve a response to light and thereby connects the cyclic-nucleotide cascade with known properties of the normal response such as increasing latency with decreasing stimulus intensity.

II. EFFECTS OF cGMP INJECTED INTO ROD OUTER SEGMENTS

A. Depolarization by Excess cGMP

1. Antagonism of Transient Depolarization by Light

When cGMP is injected into a ROS in the dark-adapted retina of the toad *Bufo marinus,* it produces a transient depolarization. Details of the injection method are given in Nicol and Miller (1978) and Miller and Nicol (1979). The basis of the method is that the injection is made through the recording pipet which contains only the potassium salt of cGMP. The rationale of this method is that negative current should tend electrophoretically to force the cGMP anion intracellularly. Under favorable circumstances a ROS can be held for 1 hr or more and yield stable, repeatable injection results.

Injection of $\sim 1 \times 10^{-9}$ A ($\sim$ 1 nA) cGMP in the dark for 1, 5, and 10 msec causes transient graded depolarizations, the amplitude and duration of which are directly proportional to the injection time (Fig. 1). We hypothesize that the transient depolarization caused by increased intracellular cGMP results from increased ROS plasma membrane permeability to Na^+. The repolarization can be

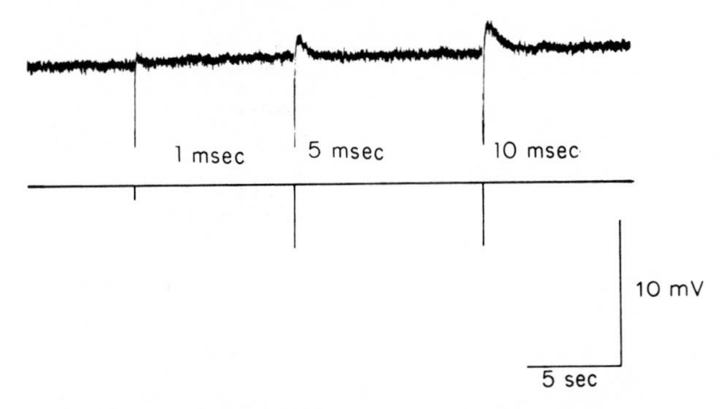

FIG. 1. Transient depolarizations of the ROS plasma membrane in the dark, graded in amplitude and duration, which are caused by intracellular injections of excess cGMP. The downward deflections of the bottom trace indicate negative current injections of about 1 nA for the specified durations. Similar injections outside the cell or with NaCl and KCl pipets produce only 1-, 5- or 10-msec transients and no depolarization. (From Nicol and Miller, 1981).

explained by classical Michaelis–Menten enzyme kinetics: As the PDE substrate (cGMP) concentration increases, so does the hydrolysis rate of cGMP. As the increased hydrolysis destroys the excess cGMP, its concentration, hence the Na^+ permeability and membrane potential, approach the resting condition which is determined by the balance of steady-state cyclase and PDE activity.

The depolarization shown in Fig. 1 must be the result of introducing cGMP intracellularly. The only other ion in the pipet is K^+. The repolarization must be the result of decreased cGMP concentration. The mechanism of decreased cGMP concentration may be enzymatic, as explained in the previous paragraph. On the other hand, the repolarization may simply be the result of cGMP diffusing away. A test of the mechanism of repolarization would be to increase the velocity of hydrolysis of cGMP to determine if this action independently suppresses the depolarization. The velocity of hydrolysis of cGMP can be increased by light, which activates PDE according to numerous biochemical studies. In our model, PDE would be expected not only to be activated by light but also to become inactivated with the time constant of dark adaptation. Prior light adaptation does in fact antagonize the depolarization that would have been produced by injection of cGMP in the dark as shown in Fig. 2. Following the control injection in the dark in Fig. 2A is a −1.6-log unit 0.1-sec flash. After the flash, cGMP is injected. The negative current hyperpolarizes the membrane for 10 msec and causes downward deflection of the trace, which is also seen just preceding the control injection. Note that the injection fails to produce depolarization and that further test injections show that the depolarization returns as dark adaptation. When the light is maintained, depolarization by such test injections is suppressed as long as the light is on. The degree of sup-

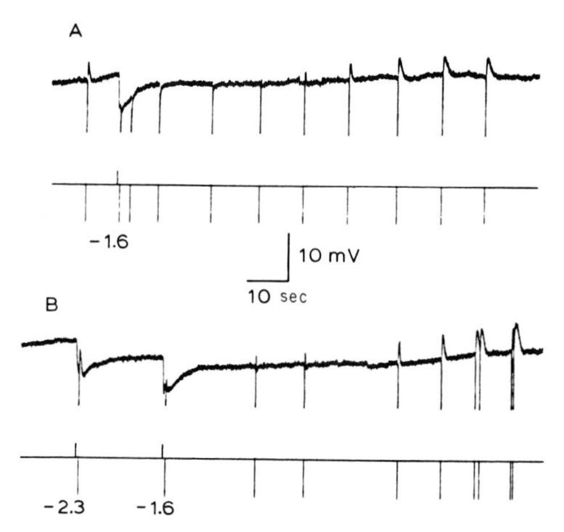

FIG. 2. Depolarization produced by excess injected cGMP, which is antagonized by illumination. (A) Injection (10-msec duration) of cGMP produces transient depolarization, and a −1.6-log unit flash (upward deflection in the signal trace) suppresses the cGMP-induced depolarizations that would have been produced by subsequent injections. Depolarization returns as dark adaptation. (B) A continuation of the trace in (A). Depolarization produced by the 10-msec injection of cGMP immediately following the −2.3-log unit (first) flash shows a recovery more rapid than that of the control and depolarization immediately after the −1.6-log unit (second) flash is smaller and recovery proceeds as in the trace in (A). (From Nicol and Miller, 1981.)

pression is directly proportional to the light intensity. Figure 2B is a continuation of Fig. 2A. The first flash is −2.3 log units. This weaker flash fails to suppress depolarization, but it both decreases the time of depolarization and increases the rate of repolarization consistent with the hypothesis that depolarization results from excess cGMP and that the effects of excess intracellular cGMP in the ROS are antagonized by increasing the activity of PDE. The next flash is −1.6 log units. It prevents depolarization. Further test injections after the −1.6 log unit flash in Fig. 2B show the time course of the decrease in PDE activation according to our interpretation. Our interpretation is that (1) the concentration of cGMP determines the membrane potential, and (2) under physiological conditions, the number of bleached rhodopsin molecules (Rh*) determine the number of activated PDE molecules. The injection of excess cGMP transiently increases the velocity of active PDE molecules but does not affect the number of PDE molecules that are active. The test injections are interpreted as giving a measure of the number of active PDE molecules.

Repeating the test injections at closer and closer intervals toward the end of the record in Fig. 2B results not only in depolarizations of increasing amplitude but also of increasing duration as the number of active PDE molecules decreases with dark adaptation. When the injection time is very long (seconds), as in Fig. 3, the membrane potential approaches zero or becomes slightly positive and

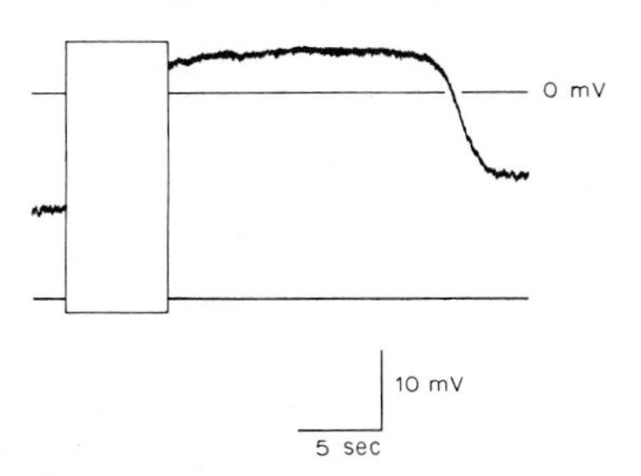

FIG. 3. Prolonged depolarization in the dark caused by a 5-sec injection of cGMP. The level of depolarization approaches the sodium equilibrium potential, calculated to be about +5 mV. The box on the voltage trace represents the period of time that the membrane voltage cannot be recorded because of current passage, about −1 nA. [Reprinted by permission from *Nature*. Miller and Nicol (1979). Copyright © 1979, Macmillan Journals Limited.]

stays at this saturated, stable level for periods ranging up to a minute. In a relatively dark-adapted preparation, the small number of activated PDE molecules are evidently unable to hydrolyze the excess cGMP quickly, even with the increased velocity caused by the high cGMP concentration. Thus injection of cGMP in darkness transiently depolarizes the ROS membrane and, after a delay proportional to the amount injected, spontaneously repolarizes to approach the resting membrane potential.

2. Evidence that All Na^+ Channels Are Not Open in the Dark

The exact amount of cGMP that is injected is unknown, but an upper bound may be estimated by Eq. (1), from Tsien (1973):

$$\text{No. of molecules injected} = T \cdot Q \cdot N/F \quad (1)$$

where T is the transfer number, Q is the charge in coulombs, N is Avagadro's number, and F is Faraday's constant. If we assume a transfer number of 1 and a ROS volume of 3×10^{-12} liter, the ROS concentration of cGMP will be nominally increased by 5 mM/sec for a current of 1.5 nA. Maximally, the 1-msec, 1-nA injection in Fig. 1 would therefore transiently increase the cGMP concentration by less than 5 μM. The saturating injections which depolarize to about 0 mV imply that excess cGMP causes the ROS permeability to approach the sodium equilibrium potential. This deduction is based on internal sodium concentrations estimated by Normann and Perlman (1978) and by Bader *et al.* (1978). If the depolarization in fact results from increased sodium permeability, the normal dark resting potential of about −30 mV indicates that, even in the fully dark-adapted ROS which is estimated to contain 70 μM cGMP (Woodruff *et al.*, 1977), not all the potentially available sodium channels are open.

B. Decrease in Resistance Caused by Excess cGMP

This interpretation depends on the assumption that cGMP is a normally occurring intracellular substance, that its concentration is controlled by light and darkness and, importantly, that its concentration determines the plasma membrane Na^+ permeability. The first two assumptions are based on solid biochemical findings reviewed in Part II of this volume. The third has yet to be proven, but preliminary evidence that the PDE inhibitor IBMX decreases ROS membrane resistance supports it. A 20-sec pulse of 5 m*M* IBMX in the superfusate caused the abrupt depolarization accompanied by a decrease in membrane resistance depicted in Fig. 4. IBMX is known to increase intracellular levels of cyclic nucleotides. Because IBMX causes both depolarization and decreased resistance, we tentatively conclude that depolarization is more likely caused by increased Na^+ permeability than by decreased K^+ permeability. This conclusion is tainted by the well-known ambiguities associated with IBMX, and further evidence on this point would be welcome.

C. Increase in Latency and Amplitude of Response Caused by Excess cGMP

1. Increased Latency and Amplitude

The recovery phase following the injection of cGMP in darkness is in a hyperpolarizing direction. We hypothesized in Section II,A that this hyper-

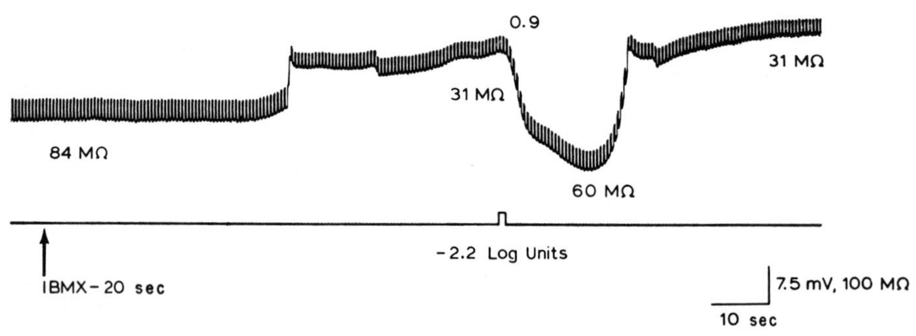

Fig. 4. The effects of superfusion with 5 m*M* IBMX–Ringer's solution on the membrane resistance of the rod photoreceptor. IBMX–Ringer's solution was applied as a 20-sec pulse to the superfusion flow while hyperpolarizing current pulses were applied through the recording electrode. The membrane resistance in the dark was about 84 MΩ. Application of IBMX produced a depolarization of about 15 mV, which was associated with a decrease in membrane resistance to about 31 MΩ. A −2.2-log unit 1-sec flash produced a large light response which had a latency of 0.9 sec as opposed to a pre-IBMX value of 0.180 sec. The light response was also accompanied by a large increase in membrane resistance to about 60 MΩ. The membrane potential and resistance recovered to the pre-IBMX level (not shown). (From Nicol and Miller, 1981.)

polarizing recovery phase is controlled by the rate of hydrolysis of excess injected cGMP. On illumination the ROS membrane potential responds with hyperpolarization as well. It is therefore tempting to look for evidence that the light response may be controlled by the rate of cGMP hydrolysis.

The strongest evidence that the rate of cGMP hydrolysis controls the response to light comes from observation of the latent period when excess cGMP is injected into the ROS (Nicol and Miller, 1978). When large amounts of cGMP are injected, the membrane potential saturates at a strongly depolarized level (Fig. 5). If a light flash is delivered while the response is saturated, the latency, as measured from the beginning of the light flash to the first detectable hyperpolarization, is increased in a remarkable manner. The control latency in Fig. 5 is 0.336 sec. However, after the injection of cGMP, the ROS is depolarized and the latency is increased to 5.72 sec. After the next cGMP injection, which is a control injection in the dark, the ROS stays depolarized for a much longer time and spontaneously repolarizes toward the resting potential at a slow rate. This is in contrast to the light response following the cGMP injection, which more rapidly repolarizes and actually hyperpolarizes in relation to the resting potential. The increase in latency from 0.336 to 5.72 sec suggests that the excess injected cGMP had to have been hydrolyzed in order to produce the light response. This

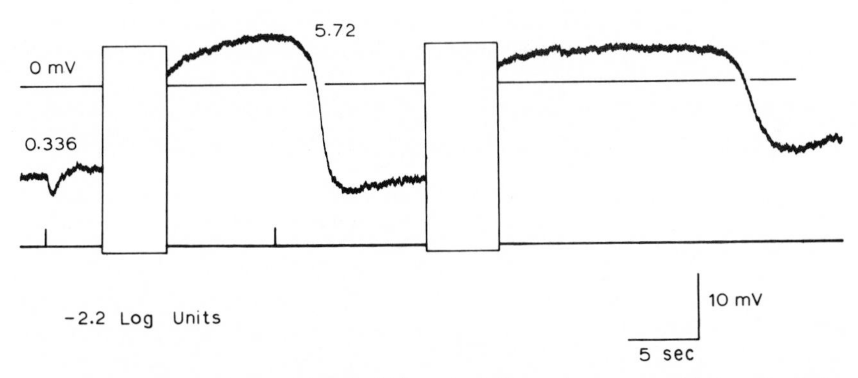

FIG. 5. The effects of excess injected cGMP on the ROS membrane potential followed by light or darkness. A control −2.2-log unit 0.1-sec flash produces a small response with a latency of about 0.336 sec. The injection of cGMP represented by the box (≃1 nA) produces a large maintained depolarization. The −2.2 log unit flash during the depolarization produces a response with a latency of 5.72 sec, much larger in amplitude than that of the control. After recovery of the membrane potential to the baseline level, cGMP is injected and again produces a depolarization to about +5 mV. The cell is left in the dark, and the depolarization is maintained for many seconds. After some time the potential spontaneously repolarizes back toward the original baseline level. Note that the hyperpolarization of the membrane by light is much more rapid than the spontaneous hyperpolarization in the dark. [Reprinted by permission from *Nature*. Miller and Nicol (1979). Copyright © 1979, Macmillan Journals Limited.]

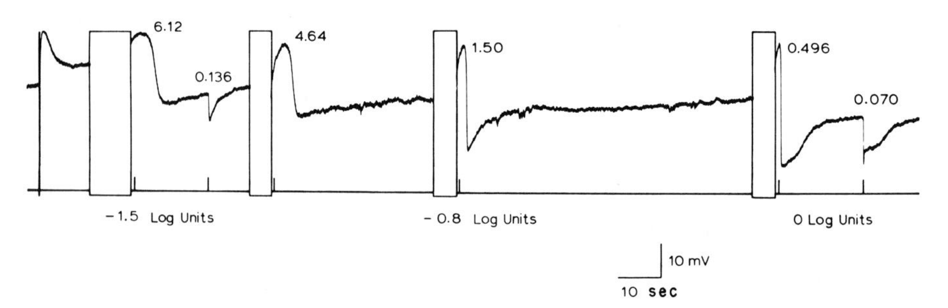

FIG. 6. Intensity series showing that the increase in response latency produced by excess cGMP is inversely proportional to the stimulus light intensity. The amount of current passed was about −1 nA.

increase in the latent period is inversely proportional to the intensity of the light flash (Fig. 6), as would be expected if light controls PDE activation.

The control of PDE activation and the subsequent PDE-mediated cGMP hydrolysis by light would constitute one of Hecht's "ordinary chemical reactions" of the latent period. An effect of excess cGMP, whether caused by IBMX (Lipton *et al.*, 1977) or cGMP directly injected in the ROS (Miller and Nicol, 1979; Waloga and Brown, 1979) is to depolarize the ROS. The effect of this depolarization on the light response would be to reduce the driving force, hence the amplitude of the response to light. However, the amplitude of the response is not decreased—it is *increased* by higher cGMP levels (Lipton *et al.*, 1977; Nicol and Miller, 1978; Waloga and Brown, 1979). The simpest inference from this observation is that light reduces the concentration of cGMP and the Na^+ conductance, rapidly increasing the driving force and receptor potential as $[Na]_i$ decreases.

2. Importance of Increased Latency

Although the inference that light-activated PDE reduces the concentration of cGMP is suggested by the depolarization and increased response amplitude caused by excess cGMP, the key and most important finding is that excess cGMP increases the latency of the response (Nicol and Miller, 1978). The latency observation alone uniquely links the cGMP concentration to its hydrolysis rate, because the light response cannot occur until the excess cGMP is hydrolyzed (Nicol and Miller, 1978). The latency of the response to light is increased in proportion to the amount of cGMP injected (Nicol and Miller, 1978; Miller and Nicol, 1979) and, further, the response latency in the depolarized condition caused by excess cGMP is inversely proportional to the intensity of the stimulating light (Nicol and Miller, 1978; Miller and Nicol, 1979). Our conclusion is that the latency is proportional to the hydrolysis rate of cGMP, as suggested by the *in vitro* biochemical observations relating light intensity to cGMP hydrolysis rates. There is thus an indirect correlation between cGMP levels and the light response, indirect because cGMP levels have not been directly measured simultaneously with membrane current or voltage. In the absence of such direct proof, one can conclude that the best current evidence suggests that the membrane potential in both darkness and light is regulated by the local internal ROS cGMP level which is in turn controlled by the synthesis and hydrolysis rates of cGMP.

The normal hyperpolarizing response to a 0.1-sec light flash of high intensity lasts several seconds. If the response is controlled by a light-activated PDE that locally depletes cGMP, the long length of the normal response may be explained by assuming that the light-activated PDE is not inactivated immediately but has a finite active phase. Further physiological evidence of the finite light activation time of PDE is seen in Fig. 7. A change in noise level during the injection indicated the possibility that the pipet tip was broken by the current and that large amounts of cGMP may have spontaneously leaked into the ROS. In other words, it appears possible that so much cGMP was in the ROS that the light activation of PDE by a 0.1-sec flash was insufficient to deplete enough cGMP to unsaturate the depolarizing response to excess cGMP. This could explain why the first flash following the injection in Fig. 7 did not produce a response. When a response occurred after the next light flash, the activation time was long enough and the cGMP depleted enough to carry the membrane potential to a level that would have been determined by the light-activated hydrolysis rate. The amplitude of the response approximately equals the control amplitude plus the depolarization (Figs. 5-7). On the other hand, the hyperpolarization resulting from injections without light never repolarized beyond the previous resting potential (Figs. 1-3). The large amplitude of the light response following cGMP injections resulted

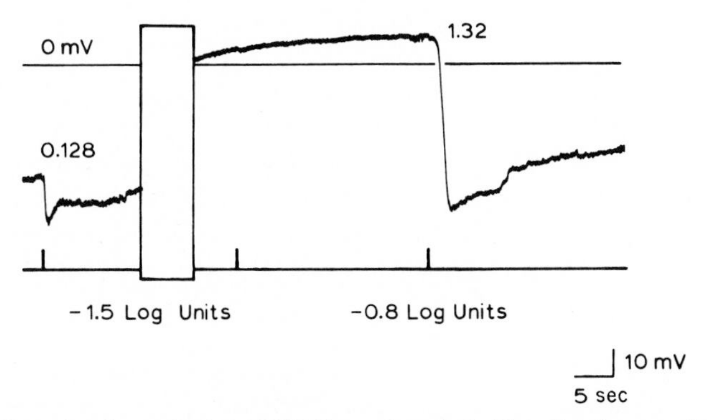

FIG. 7. Short-duration activation of PDE by a short flash. Note that the second flash fails to hyperpolarize the membrane. See text. [Reprinted by permission from *Nature*. Miller and Nicol (1979). Copyright © 1979, Macmillan Journals Limited.]

from increased PDE activity stimulated by two factors: an abnormally high concentration of the PDE substrate artificially caused by the injection of cGMP combined with the effect of light activation of the PDE.

2. Is PDE Fast Enough?

We conclude from the effects of intracellular injections of cGMP that cGMP controls the membrane potential and that the means for this control is the light-regulated PDE. It would therefore be reassuring to know that the rates of cGMP hydrolysis by light-activated PDE *in vivo* are consistent with these physiological findings. Yee and Liebman (1978) have found that light-activated PDE is capable of hydrolyzing 4×10^5 cGMP molecules/sec per Rh* molecule. The 0.1-sec flashes on the traces in Fig. 8 bleach 1.3×10^5 rhodopsin molecules. Such a

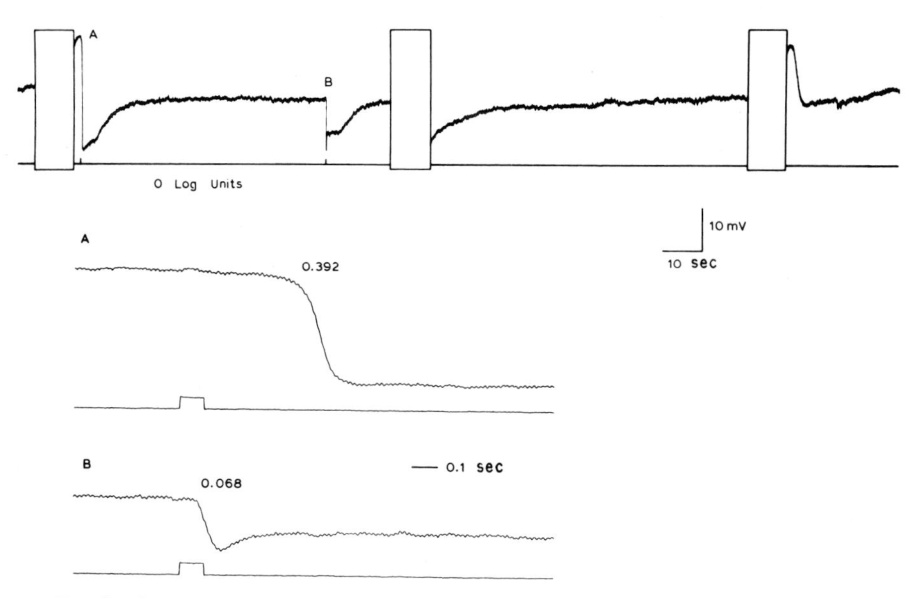

FIG. 8. Suppression, by previous light adaptation, of effects of excess cGMP on the membrane potential. The top trace shows the effects of excess injected cGMP on the membrane potential, characterized by depolarization, and on the light response (A) to a 0-log unit flash, characterized by an increase in response latency and an increase in response amplitude. After the potential has returned to the baseline level, another 0-log unit flash produces a normal response (B). The following injection of cGMP fails to produce depolarization. The cell is allowed to recover in the dark for about 1 min, and the injection of cGMP again produces depolarization. Trace A (bottom) is an expansion of response A in the top trace after the injection of cGMP, while trace B (bottom) is an expansion of response B in the top trace. Note that in trace A, after cGMP injection, the response latency is greatly increased over the normal condition as in trace B. The rates of hyperpolarization for both responses are nearly identical. [Reprinted by permission from *Nature*. Miller and Nicol (1979). Copyright © 1979, Macmillan Journals Limited.]

flash should be capable of hydrolyzing 5.2×10^{10} cGMP molecules/sec. The shortest latency we have observed after the injection of cGMP is the 0.392-sec latency shown in Fig. 8. In 0.392 sec, 2×10^{10} cGMP molecules could be hydrolyzed based on Yee and Liebman's (1978) data. According to Eq. (1) 10^{10} cGMP molecules/sec were injected, assuming a transfer number of 1. It would be useful to know how long it would have taken to repolarize after the injection in the absence of light. The control in Fig. 8 indicates that the answer is well under a second, even allowing for differences in adaptational states. Clearly the PDE is fast enough to explain both the 0.392-sec latency and the normal latency of 0.070 sec in Fig. 8.

3. Ca^{2+}

If cGMP is to control the ROS membrane potential, obviously its action must be unique. No other naturally occurring molecular species should exert the same regulatory function. It is therefore important to know the effects of Ca^{2+} compared to those of cGMP. The retina in the experiment shown in Fig. 9 was superfused with $^{1}/_{10}$ the normal Ringer's solution concentration of Ca^{2+}, giving rise to both depolarization and increased amplitude of the response to light as described in Lipton *et al.* (1977). When the traces are examined in detail (Fig. 9, inset), it is apparent that there is a slight increase in latency at the greatest levels of depolarization. According to our hypothesis there should not only be an increase in latency when the system is saturated with cGMP, as for example in Fig. 5, but also for nonsaturating cGMP injections and for naturally higher levels that may occur during dark adaptation. Our reasoning is that it takes time to

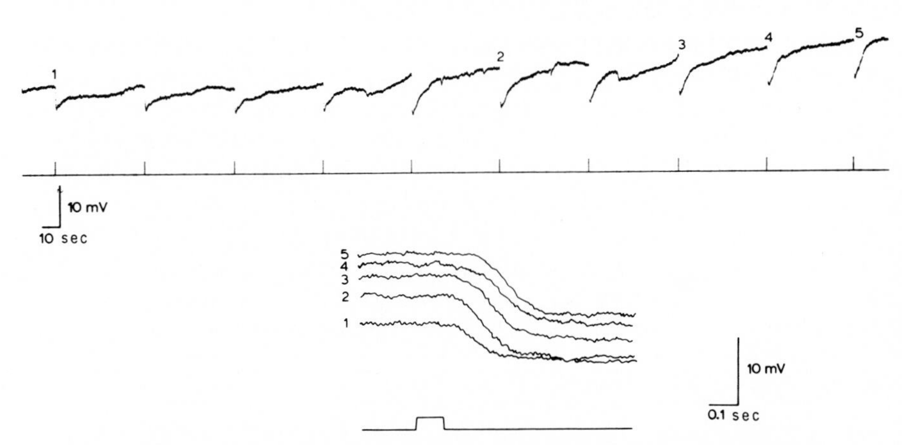

FIG. 9. Effects of 0.16 mM Ca^{2+}-Ringer's solution on the membrane potential and light response as recorded from the toad rod. The inset shows an expansion of each corresponding light response observed in the top trace. See text. (From Nicol and Miller, 1981.)

deplete the cloud of cGMP locally, and the higher the cGMP concentration, the more time it takes, all other things being equal. Then, as the local region is depleted, the light-activated hydrolysis rate at the channel becomes rate-limiting and is reflected in the rising phase of the receptor potential. The effect of decreased Ca^{2+} levels is consistent with the fact that low Ca^{2+} in the superfusate raises intracellular cGMP levels (Cohen *et al.*, 1978; Cohen, this volume, Chapter 12). The injection of EGTA, which sequesters Ca^{2+}, both depolarizes and blocks the response to light. However, in contrast to the effects of excess cGMP, when the response returns, it is of approximately normal latency and at first smaller in amplitude than the control (Brown *et al.*, 1977, cf. their text—Fig. 6).

4. cAMP

There are complex interactions between Ca^{2+} and cGMP (Cohen, this volume, Chapter 12), and it is not possible to verify if one can change the concentration of Ca^{2+} independently of cGMP; however, Ca^{2+}-mediated latency changes are small compared with those caused by cGMP. cAMP is another excellent control, naturally occurring and in fact an analogue of cGMP. cAMP can serve as a control to determine whether or not the action of cGMP in the ROS is really unique. Both cGMP and cAMP are well known to be affected by ROS enzyme systems. ROS guanylate cyclase also catalyzes the formation of cAMP from ATP (Fleischman, this volume, Chapter 6; Bitensky *et al.*, 1971), and ROS light-activated PDE hydrolyzes both cAMP and cGMP *in vitro*. A naturally occurring PDE inhibitor is said to cause the enzyme to favor cGMP as a substrate (Dumler and Etingof, 1976). Finally, both cAMP and cGMP are present in the ROS (Cohen, this volume, Chapter 12), and for this reason alone it is crucial to test the effects of increased intracellular cAMP as a control.

Excess cAMP injected into the ROS depolarizes, as does cGMP, guanosine 5′-monophosphate (GMP) (Nicol and Miller, 1978), and EGTA (Brown *et al.*, 1977). However, there is no change in latency, and the response to light is smaller than in the control (Fig. 10). Without speculating on the mechanisms by which cAMP produces these effects, it is clear that light-activated PDE *in vivo* distinguishes very well between cAMP and cGMP. After the response to light, the membrane again depolarizes to its previous level determined by cAMP, and the hydrolysis of cAMP is seen to proceed independently as the potential returns to the control level. An identical result is obtained with the injection of either the Na^+ or the K^+ salt of cAMP. The response of the ROS to excess injected cAMP establishes the uniqueness of the light-regulated action of cGMP on the membrane potential.

5. Coupling

Injections of cGMP and cAMP are administered to single ROS, but their effects may be influenced by the fact that toad rods are electrically coupled to one

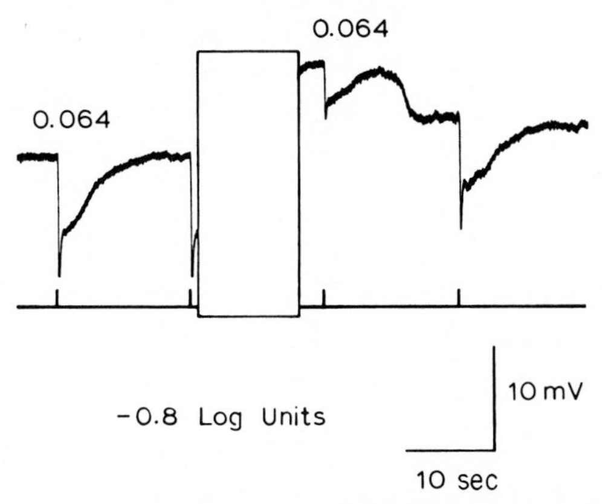

FIG. 10. Effects of excess injected cAMP on the membrane potential and light response of the rod photoreceptor. Note that the response latency is the same before and after cyclic adenosine 3′,5′-monophosphate (cAMP) injection. cAMP produces a small depolarization of the membrane by an unknown mechanism. This depolarization does not appear to be affected by illumination. (From Nicol and Miller, 1981.)

another at the inner segment level with a space constant of about 25 μm (Gold, this volume, Chapter 4). Two problems have to be considered: whether the injected nucleotide diffuses into neighboring receptors and whether the altered responses of the injected receptor are seriously distorted by the coupling.

The diffusion coefficient for cAMP is 4.4×10^{-6} cm^2/sec (Dworkin and Keller, 1977). With a similar value for cGMP, the leading edge of the cGMP would diffuse 1 μm from the pipet tip in the first millisecond, 3 μm in 10 msec, 9 μm in 0.1 sec, 30 μm in 1 sec, and 94 μm in 10 sec. Anatomical barriers to diffusion and interacting flows resulting from other charged particles are likely to make these figures smaller. Molecules only slightly larger than cGMP, such as Procion yellow, do not diffuse from cell to cell. For these reasons it is unlikely that injected nucleotides diffuse to neighboring cells. This conclusion is supported by our finding that all the effects described for cGMP injections have been observed to occur in under a second using millisecond cGMP injection regimens.

For single-photon events the longitudinal spread of excitation and desensitization along the toad rod outer segment has a diffusion coefficient for both excitation and desensitization of approximately 10^{-7} cm^2/sec (McNaughton *et al.*, 1980; Yau *et al.*, this volume, Chapter 2). Because of the barrier to diffusion presented by the disks, these processes must be mediated by a molecule with a diffusion coefficient 100 times greater or $\simeq 10^{-5}$ cm^2/sec (Yau *et al.*, this volume, Chapter 2), which is within a factor of 2 of the diffusion coefficient for both cGMP and Ca^{2+}.

The coupling is a dual-edged sword. On the one hand, the shunting caused by coupling reduces the *IR* drop across the ROS membrane caused by the current injection and therefore aids in stabilizing the preparation. On the other hand, the shunting reduces the amplitude of effects such as de- and hyperpolarization and theoretically makes them more difficult to detect. The fact that these effects are so dramatic indicates that they are strong local effects easily recorded by the techniques being used.

D. Potentiation of Light Aftereffects by Excess cGMP

When small amounts of cGMP are injected into ROS as in Figs. 1 and 2, they cause transient depolarization, the amplitude and duration of which become greater with increasing dark adaptation. We explained this result by assuming that the test injections monitored only the number of active PDE molecules, i.e., that the velocity of the PDE was transiently increased by the increased substrate without affecting the underlying activity. This simplifying assumption holds for injections of cGMP in the dark but must be modified for the case when a light stimulus is delivered during the time the cell is depolarized by excess cGMP. Such a flash, for example, the second flash in Fig. 11, produces a hyperpolarizing response that implies that the cell was super dark-adapted; the amplitude of the response approximately equals the depolarization plus the control flash amplitude. The absolute sensitivity of the cell in millivolts per photon is increased as if the cell were more dark-adapted than it is. In the normal ROS the cell depolarizes with dark adaptation, and the latency and amplitude of the response increase. Artificially introducing cGMP produces the same results, but

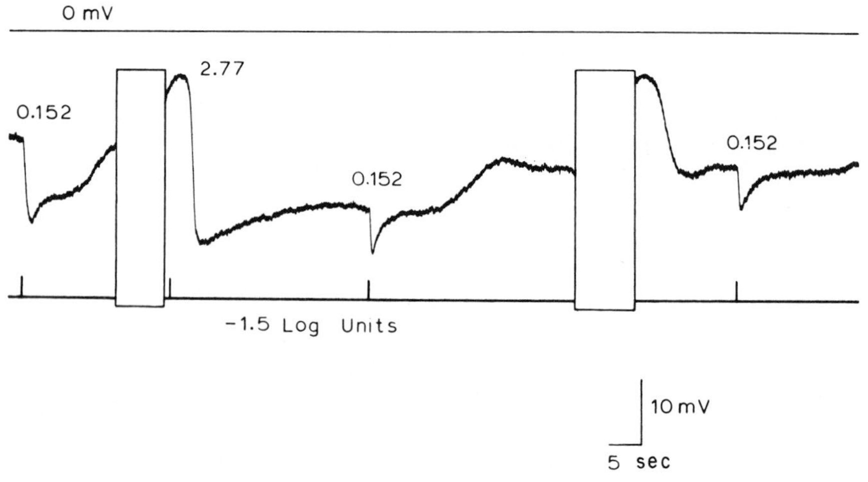

FIG. 11. Excess cGMP-accentuated plateau. See text. [Reprinted by permission from *Nature*. Miller and Nicol (1979). Copyright © 1979, Macmillan Journals Limited.]

not only are there immediate effects of increased dark adaptation not warranted by light conditions, but also other effects resembling dark adaptation are prolonged after the initial transient as if the cyclic-nucleotide system "remembered" that it was super dark-adapted. The signs of this "memory" of super dark adaptation are a prolonged plateau and decreased sensitivity to light following the light flash delivered during the depolarization caused by excess cGMP (Fig. 11). The ROS acts as if it were more dark-adapted than it is, and for a considerable period. This behavior cannot be explained by the hypothesis that the excess injected cGMP simply monitors PDE activity. Excess cGMP plus light activation of PDE acts as if more PDE were activated for a longer period of time than it would have been in its absence. The properties of the cyclic-nucleotide system that could explain this behavior are unknown. However, it is known that the kinetics of the enzymes may be changed simply by changing the substrate concentration, since Robinson *et al.* (1980) have shown that the Michaelis constant K_m of ROS PDE increases with increasing substrate concentration. Such kinetic studies plus increased knowledge of the control mechanisms of cyclic-nucleotide enzyme systems are now needed to understand the physiological data.

E. Antagonism of cGMP Effects by Light

Just as the injection of excess cGMP has long-lasting effects on what we interpret as PDE hydrolysis rates, so does light. Light of course causes light adaptation. Light also antagonizes the depolarization that would have been caused by cGMP injection (Figs. 2 and 8). Observations that excess cGMP potentiates long-term light effects (adaptation) and that light antagonizes long-term effects of cGMP injections raise the possibility that both the long- and short-term effects of light are mediated through the control of intracellular cGMP levels. Such long-lasting adaptational effects therefore provide additional evidence that cGMP is capable of regulating the transduction of light into the membrane potential signal, both for excitation and adaptation, and an eventual goal will be to understand and model these effects.

III. A MODEL FOR cGMP CONTROL OF MEMBRANE POTENTIAL

We propose a model based on the Michaelis–Menten kinetic behavior of light-regulated PDE as a preliminary effort in interpreting our data indicating that the local concentration of cGMP controls the ROS membrane potential. The mechanism by which this control is exerted is unknown. The model assumes that in the steady-state dark-adapted ROS, cyclase and PDE rates are in a steady state. The Michaelis–Menten relation is

$$V = V_{max}[S]/([S] + K_m) \qquad (2)$$

where V is the rate at which PDE hydrolyzes cGMP, V_{max} is the maximal rate, [S] is the concentration of the substrate, cGMP, and K_m is the cGMP concentration at which the hydrolysis rate is half the maximal value.

This model is depicted in Fig. 12A and B. Let us assume that, in the dark, [S] = 70, $V_{max} = 2$, and $K_m = 70$. Then, according to Eq. (2) and the curve labeled $V_{max} = 2$ in Fig. 12A, the PDE velocity in the dark is 1. Illumination raises the PDE V_{max} to 2×10^4, which results in $V = 676$, which is shown on the ordinate in Fig. 12A as a solid circle. The lower trace in Fig. 12B shows V_{max} and V rising after the light flash. At the same time, the increased PDE velocity causes a decrease in [S] and the membrane potential (two top traces in Fig. 12B). The decrease in [S] and a time-dependent decrease in the light activation of PDE take the PDE velocity along a hypothetical curve starting at the circle on the ordinate in Fig. 12A. As the velocity decreases, [S] increases (Fig. 12B). The membrane potential does not accurately follow [S]. The peak and plateau are generated by voltage- and time-dependent conductance changes (Werblin, 1979; Owen and Torre, this volume, Chapter 3).

The high concentration of cGMP in the ROS should provide considerable buffering capacity. After a bright flash that reduces the local cGMP concentration to almost zero, the local concentration may rise rapidly even though there is still a considerable amount of light-activated PDE. Thus the membrane potential may, within a few millivolts, approach the dark-adapted value, as after the second flash of the top trace in Fig. 8. Even though the membrane potential has almost recovered, by this argument there is a large potential for increased PDE velocity indicated by the ◊'s in Fig. 12A. This potential is unmasked by failure of the injection of increased cGMP to elicit depolarization except after prolonged dark adaptation, as shown in Fig. 8. Such models are necessarily sketchy and inexact and can only improve as our knowledge of complex controls in the cyclic-nucleotide cascade increases.

IV. INCOMPLETE KNOWLEDGE OF cGMP EFFECTS

Our experiments suggest that light- and dark-regulated hydrolysis rates of cGMP control aspects of both excitation and adaptation. The latency observations provide the strongest argument linking cGMP to transduction. Of all the known diverse ROS molecular species, only cGMP has hydrolysis rates that regulate latency. Therefore the light-activated hydrolysis rate of cGMP appears to be the rate-limiting step in excitation. That latency is inversely related to stimulus intensity fits this concept. However, it is not clear whether or not the increased latency observed with light adaptation (Grabowski and Pak, 1975) is consistent with this hypothesis. We need more knowledge about the effects of light on interactions with other components, such as the cyclase and Ca^{2+}, before these more complex effects can possibly be explained.

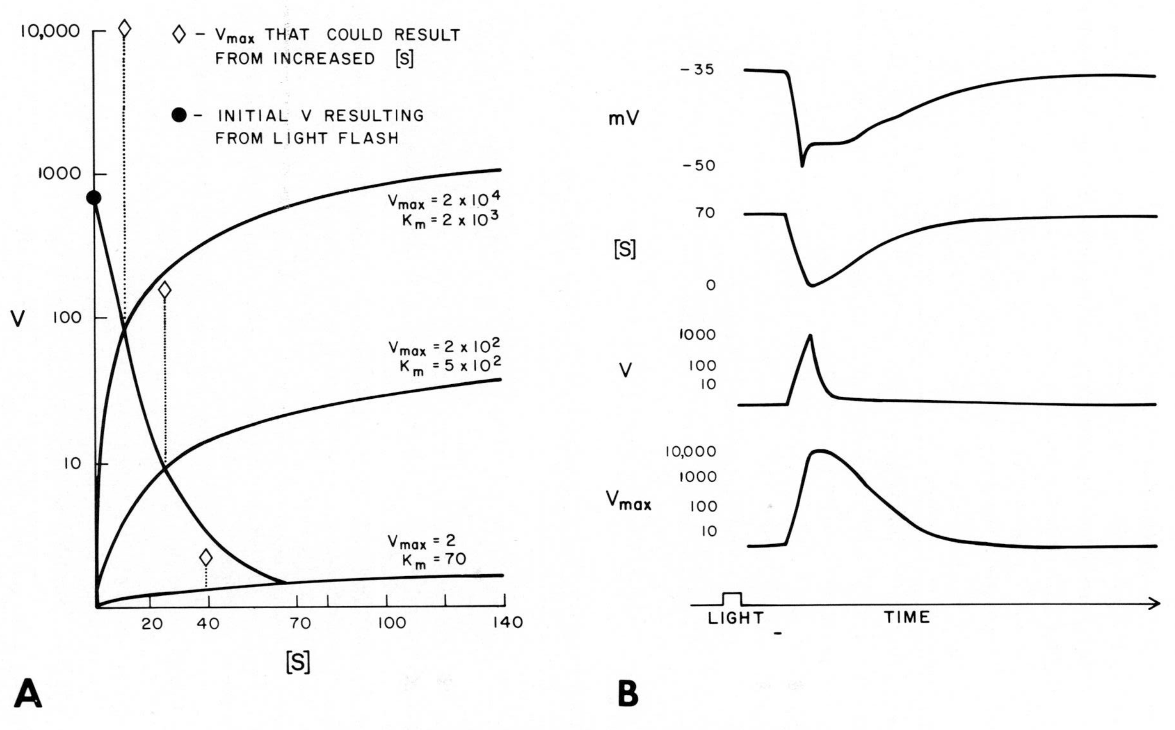

Fig. 12. Tentative model of cGMP effects on rod membrane potential.

V. CONCLUSION

The results of the injection of excess cGMP suggest that hydrolysis rates of cGMP in the ROS function in mediating transduction by regulating the Na^+ permeability for some properties of both excitation and adaptation. This interpretation is supported by observations that excess intracellular cGMP in darkness depolarizes the ROS in a graded and transient manner, decreases membrane resistance, and uniquely regulates the latency, amplitude, and duration of the response to light. The effects of cGMP are antagonized by light in a dose-dependent manner. These physiological experiments suggest that PDE activation is the rate-limiting step in excitation, that the latency and rising phase of the response may be triggered by cGMP hydrolysis, and that adaptation may reflect residual effects on either PDE activity or its balance with other regulatory components superimposed on inherent membrane time and voltage conductances.

ACKNOWLEDGMENT

This research was supported mainly by United States Public Health Service NIH grant EY3196.

REFERENCES

Bader, C. R., MacLeish, P. R., and Schwartz, E. A. (1978). Responses to light of solitary rod photoreceptors isolated from the tiger salamander retina. *Proc. Nat. Acad. Sci. U.S.A.* **75**, 3507-3511.

Baylor, D. A., Lamb, T. D., and Yau, K.-W. (1979). The membrane current of single rod outer segments. *J. Physiol. (London)* **288**, 589-611.

Bitensky, M. W., Gorman, R. E., and Miller, W. H. (1971). Adenyl cyclase as a link between photon capture and changes in membrane permeability of frog photoreceptors. *Proc. Natl. Acad. Sci. U.S.A.* **68**, 561-562.

Bornschein, H., and Charif, S. E. (1959). Die Bedeutung konstanter Temperaturbedingungen für ERG-Untersuchungen bei Kleinsäugern. *Experientia* **15**, 296-300.

Brown, J. E., Coles, J. A., and Pinto, L. H. (1977). Effects of injections of calcium and EGTA into the outer segments of retinal rods of *Bufo marinus*. *J. Physiol.* **269**, 707-722.

Cohen, A. I., Hall, I. A., and Ferrendelli, J. A. (1978). Calcium and cyclic nucleotides regulation in incubated mouse retinas. *J. Gen. Physiol.* **71**, 595-612.

Dumler, I. L., and Etingof, R. N. (1976). Protein inhibitor of cyclic adenosine 3′:5′-monophosphate phosphodiesterase in retina. *Biochim. Biophys. Acta* **429**, 474-484.

Dworkin, M., and Keller, K. H. (1977). Solubility and diffusion coefficient of adenosine 3′:5′-monophosphate. *J. Biol. Chem.* **252**, 864-865.

Fung, B.K.-K., and Stryer, L. (1980). Photolyzed rhodopsin catalyzes the exchange of GTP for bound GDP in retinal rod outer segments. *Proc. Natl. Acad. Sci. U.S.A.* **77**, 2500-2504.

Grabowski, S. R., and Pak, W. L. (1975). Intracellular recordings of rod responses during dark-adaptation. *J. Physiol. (London)* **247**, 363-391.

Hartline, H. K. (1934). Intensity and duration in the excitation of single photoreceptors. *J. Cell. Comp. Physiol.* **5**, 229-247.

Hartline, H. K., Wagner, H. G., and MacNichol, E. F., Jr. (1952). The peripheral origin of nervous activity in the visual system. *Cold Spring Harbor Symp. Quant. Biol.* **17**, 125-141.

Hecht, S. (1918–1919). The effect of temperature on the latent period in the photic response of *Mya arenaria*. *J. Gen. Physiol.* **1**, 667–685.

Hecht, S. (1919–1920). The photochemical nature of the photosensory process. *J. Gen. Physiol.* **2**, 229–246.

Hecht, S., and Williams, R. E. (1922–1923). The visibility of monochromatic radiation and the absorption spectrum of visual purple. *J. Gen. Physiol.* **5**, 1–33.

Hubbard, R., Bownds, D., and Yoshizawa, T. (1965). The chemistry of visual photoreception. *Cold Spring Harbor Symp. Quant. Biol.* **30**, 301–315.

Kropf, A., and Hubbard, R. (1970). The photoisomerization of retinal. *Photochem. Photobiol.* **12**, 249–260.

Lipton, S. A., Rasmussen, H., and Dowling, J. E. (1977). Electrical and adaptative properties of rod photoreceptors in *Bufo marinus*. II. Effects of cyclic nucleotides and prostaglandins. *J. Gen. Physiol.* **70**, 771–791.

McNaughton, P. A., Yau, K.-W., and Lamb, T. D. (1980). Spread of activation and desensitization in rod outer segments. *Nature (London)* **283**, 85–87.

Miller, W. H., and Nicol, G. D. (1979). Evidence that cyclic GMP regulates membrane potential in rod photoreceptors. *Nature (London)* **280**, 64–66.

Nicol, G. D., and Miller, W. H. (1978). Cyclic GMP injected into retinal rod outer segments increases latency and amplitude of response to illumination. *Proc. Natl. Acad. Sci. U.S.A.* **75**, 5217–5220.

Nicol, G. D., and Miller, W. H. (1981). In preparation.

Normann, R. A., and Perlman, I. (1978). Cytoplasmic ion composition and the plasma membrane permeabilities of rod photoreceptors in the light and dark. ARVO abstr., *Suppl. to Invest. Ophthalmol. Visual Sci.* p. 219.

Orloff, J., and Handler, J. (1967). The role of adenosine 3′,5′-phosphate in the action of antidiuretic hormone. *Am. J. Med.* **42**, 757–768.

Robinson, P. R., Kawamura, S., and Bownds, M. D. (1980). Control of cyclic-GMP phosphodiesterase of frog photoreceptor membranes by light and calcium ions. *Fed. Proc., Fed. Am. Soc. Exp. Biol.* **39**, 2138.

Smith, P. A., Weight, F. F., and Lehne, R. A. (1979). Potentiation of Ca^{++}-dependent K^{+} activation by theophylline is independent of cyclic nucleotide elevation. *Nature (London)* **280**, 400–402.

Tsien, R. W. (1973). Adrenaline-like effects of intracellular iontophoresis of cyclic AMP in cardiac Purkinje fibers. *Nature (London), New Biol.* **245**, 120–122.

Waloga, G., and Brown, J. E. (1979). Effects of cyclic nucleotides and calcium ions on *Bufo* rods. ARVO abstr. *Suppl. to Invest. Ophthalmol. Visual Sci.* p. 5.

Weber, A. (1968). The mechanism of the action of caffeine on sarcoplasmic reticulum. *J. Gen. Physiol.* **52**, 760–772.

Werblin, F. S. (1979). Time- and voltage-dependent ionic components of the rod response. *J. Physiol. (London)* **294**, 613–626.

Woodruff, M. L., and Bownds, M. D. (1979). Amplitude, kinetics, and reversibility of a light-induced decrease in guanosine 3′,5′-monophosphate in frog photoreceptor membranes. *J. Gen. Physiol.* **73**, 629–653.

Woodruff, M. L., Bownds, D., Green, S. H., Morrisey, J. L., and Shedlovsky, A. (1977). Guanosine 3′,5′-monophosphate and the *in vitro* physiology of frog photoreceptor membranes. *J. Gen. Physiol.* **69**, 667–679.

Yee, R., and Liebman, P. A. (1978). Light-activated phosphodiesterase of the rod outer segment. *J. Biol. Chem.* **253**, 8902–8909.

Yoshikami, S., and Hagins, W. A. (1971). Light, calcium, and the photocurrent of rods and cones. *Biophys. J.* **11**, 47a.

Part IV
An Editorial Overview

Both calcium and cyclic GMP are involved in transduction

Sisyphus, a legendary king of Corinth, was consigned to Hades unceasingly to "heave the stone against the rising mount" (Dryden). Is cyclic GMP a molecular Sisyphus, ceaselessly controlling the calcium stone? (Athenian black figure vase painter Bucci, sixth century B.C.; Munich Antikensammlungen, with permission)

CURRENT TOPICS IN MEMBRANES AND TRANSPORT, VOLUME 15

Chapter 25

Ca^{2+} and cGMP[1]

WILLIAM H. MILLER

I. EXPERIMENTAL FACTS

The transduction process is unlikely to be mediated by any one simple reaction. Superimposed on the basic membrane properties of voltage- and time-dependent conductances (*1*) influenced by coupling with neighboring receptors (*2*), the light response can be modeled as if it were generated by a series of four sequential first-order delay processes (*3*). This complexity leaves room for enzymatic processes, including components of both the cyclic-nucleotide and calcium systems. But these physiological data are incapable of ruling out any particular component; for example, the diffusion coefficient for the ROS longitudinal spread of both excitation and adaptation is within a factor of 2 of that for both Ca^{2+} and cGMP (*4*). Nevertheless, based on the following generally agreed on facts, both Ca^{2+} and the cyclic-nucleotide cascade may now safely be included along with rhodopsin in the phototransduction mechanism by which the absorption of light causes decreased Na^+ permeability of the ROS plasma membrane:

1. One Rh*, probably metarhodopsin II (*5*),catalyzes the exchange of about 500 GTPs for GDP molecules by a like number of guanyl-nucleotide-binding protein molecules, transducin; the transducin–GTP complex activates

[1]Abbreviations used: ROS, rod outer segment; cGMP, cyclic guanosine 3′,5′-monophosphate; Rh*, photolyzed rhodopsin; PDE, phosphodiesterase; GDP, guanosine 5′-diphosphate; GTP, guanosine 5′-triphosphate; GTPase, guanosinetriphosphatase; ATP, adenosine 5′-triphosphate; IBMX, isobutylmethylxanthine.

ISBN 0-12-153315-8

PDE (*6*). That Rh* communicates directly with transducin and not PDE is suggested by light-induced binding of only the former to disk membranes (*7*), though both transducin and PDE are located on the outside of the disk membrane (*8*). In fact, when purified rhodopsin is photylyzed in reconstituted systems, transducin, but not PDE, is activated (*6, 8, 9*). The activation of transducin by Rh* and the activation of PDE by transducin–GTP are sequential first-order processes (*5*).

2. The kinetics of these sequential first-order processes initiated by Rh* are consistent with what would be required for the mediation of transduction. When activated, PDE hydrolyzes $>10^5$ cGMP molecules/sec/Rh* molecule with a latency comparable to that of excitation in both isolated ROS (*10*) and rod disk preparations (*5*). In the absence of PDE activation the dark ROS cGMP concentration is about 50 μM, an order of magnitude higher than the usual cellular cAMP concentration in other cells.

3. Increased cGMP concentration correlates with increased dark permeability of the ROS plasma membrane to Na^+ (*10*) and with depolarization of the ROS and increased amplitude of the ROS membrane voltage response to light (*11, 12, 13*).

4. Excess cGMP injected into the ROS delays the response to a light flash, increasing the response latency in a dose-dependent manner as if the excess were hydrolyzed by light-activated PDE in order to produce the physiological response to light (*12*).

5. PDE is inactivated by GTPase activity of transducin (*6*), which is conferred by a "helper" protein (*9*). PDE inactivation is aided by ATP (*5*) which may act through phosphorylation of Rh* by rhodopsin kinase (*8*), as well as directly on PDE (*10*). Rhodopsin kinase, like transducin, undergoes light-induced binding to disk membranes (*7*) and is located on the outer surface of the disk membrane (*8*).

6. Ca^{2+} decreases the ROS Na^+ dark current (*14, 15*). Thus the actions of Ca^{2+} and cGMP are opposed to one another (*11*).

7. There is no evidence for the active accumulation of Ca^{2+} by disks or for its release from disks by light in physiologically important quantities (*16*), but modulation of ROS Ca^{2+} by light could operate on cytosol stores (*16*).

8. Light causes the release of Ca^{2+} from ROS of the intact retina into the extracellular space both with a sufficiently short latency and in sufficient quantity to mediate transduction (*17, 18*).

9. Cyclic GMP decreases Ca^{2+} binding to disks by 25% (*19*). The implication is that light increases Ca^{2+} binding to disks and perhaps to plasma membranes (*19*). This finding is therefore consistent with the above observations on Ca^{2+} fluxes, assuming that Ca^{2+} binds to the inside of the Na^+ pore on illumination and is passed through the pore to the extracellular space.

II. POSTULATED SEQUENCE OF EVENTS

Although the above facts seem sufficient to place rhodopsin, cGMP, and Ca^{2+} in the transduction mechanism, the roles of all of these substances are the subject of controversy. For example, ROS PDE, with $k_{cat} \simeq 10^3$ (*20*), has $k_{cat}/K_M > 10^7$ M^{-1} sec^{-1}, which makes it one of the most efficient enzymes; its second-order rate constant approaches the diffusion-controlled encounter of enzyme and substrate. Activation of PDE by a single Rh* causes this ratio to act as if it were raised by seemingly impossible orders of magnitude (*5*); this activation results in an amplification which is interpreted as the consequence of activation of hundreds of PDE molecules per Rh* molecule (*5*). The power of this system is evident from experiments on isolated ROS (*10*) and intact ROS (*12*) as well. Though it would be illogical and wasteful for such a system to serve ROS functions slower than transduction and less intimately tied to illumination, the matter is controversial because of the failure to detect rapid and dramatic changes in ROS cGMP levels caused by illumination in the intact retina (*21*). This failure could stem from both the expected very localized nature of the PDE activity and the presence of guanylate cyclase located in the axonemes (*22*), which may respond via feedback mechanisms almost as rapidly as PDE responds to light. The purified disk (*5*) and isolated ROS (*10*) preparations, as open-loop situations lacking cyclase, may provide better indications of the potential for actual PDE effects. Only further work can resolve this controversy.

In addition to the controversies surrounding data interpretation, there are gaps in our knowledge of the mechanisms of action of Ca^{2+} and cGMP and of the relations between these two agents. Lowered extracellular Ca^{2+} increases the ROS cGMP concentration (*23*). This effect may be exerted through control of enzymes, because Ca^{2+} inhibits cyclase (*22*), activates a calmodulin-dependent PDE in the immature rod (*24*), and enhances light activation of PDE in the adult ROS (*10*). Ca^{2+} also affects the phosphorylation of small ROS proteins of unknown function (*10*). With this degree of complexity, it is not surprising that no one molecular species exactly mimics the effects of light and/or darkness (*25, 26*).

There is a similar gap in our knowledge of how the effects of cGMP are mediated. Efforts to prove involvement of the traditional effectors of cyclic-nucleotide actions, phosphorylating kinases, have not been definitive and no phosphorylase has been identified. The possibility that Ca^{2+} acts directly at the Na^+ pore should be considered, because cGMP at physiological concentrations decreases Ca^{2+} binding to disks (*19*). Ca^{2+} also decreases the dark current within 90 msec of being applied to ROS either by itself or in the presence of IBMX, which delays the response peak (*27*). Perhaps none of these experiments proves a direct action of Ca^{2+}. The 90-msec time resolution may not be sufficient to rule out a primary effect through enzymes, and IBMX may alter the system so that excess Ca^{2+} acts via enzymes (e.g., cyclase inhibition; *22*) to cause a rapid physiological response.

While these controversies and gaps in our knowledge demonstrate that there is no proof of the sequence of reactions of transduction, there is a reasonable logic that can serve as a working hypothesis. The weight of evidence suggests that the chain of command in the control of transduction is initiated by Rh* and devolves in sequence first to cGMP and then to Ca^{2+}. There is no evidence of a physiologically important link between Rh* and Ca^{2+} (*28*). On the contrary, Rh* communicates directly with disk enzymes (*6, 5, 10*) rapidly and causes the hydrolysis of cGMP, an excess of which delays transduction (*12*). That cGMP acts directly, rather than either in parallel with Ca^{2+}, or in a restorative fashion (e.g., by activating a Ca^{2+} pump) (*11*) is suggested by the fact that a strong light flash does not cause a physiological response of normal latency in the presence of excess cGMP (*12*). Yet such a restorative process, if regulated by a PDE of sufficient power (*5*) and with sufficient range and control (*6, 29*), could be capable of mediating both short- and long-term physiological effects of transduction.

In this hypothesis, the chain is initiated by Rh* and is linked to cGMP by transducin and PDE; cGMP in turn regulates Ca^{2+}. Thus the biochemical and physiological evidence of the sequence of events suggests that the cyclic-nucleotide cascade may provide the basic machinery for the control of transduction and that this control may be exercised through regulation of Ca^{2+}. Determining whether even this sketchy perception of the sequence of molecular events is correct must await the assembly of further knowledge of intermediate controls, as well as characterization of the Na^+ channel and its molecular regulators.

REFERENCES

1. Owen, G., and Torre, V., this volume, Chapter 3.
2. Gold, G. H., this volume, Chapter 4.
3. Matthews, G., and Baylor, D. A., this volume, Chapter 1.
4. Yau, K. W., Lamb, T. D., and McNaughton, P. A., this volume, Chapter 2.
5. Liebman, P. A., and Pugh, E. N., Jr., this volume, Chapter 9.
6. Stryer, L., Hurley, J. B., and Fung, B. K.-K., this volume Chapter 5.
7. Kühn, H., this volume, Chapter 10.
8. Shichi, H., this volume, Chapter 15.
9. Bitensky, M. W., Wheeler, G. L., Yamazaki, A., Rasenick, M. M., and Stein, P., this volume, Chapter 14.
10. Bownds, M. D., this volume, Chapter 11.
11. Lipton, S. A., and Dowling, J. E., this volume, Chapter 21.
12. Miller, W. H., and Nicol, G. D., this volume, Chapter 24.
13. Brown, J. E., and Waloga, G., this volume, Chapter 20.
14. Yoshikami, S., and Hagins, W. A. (1971). Light, calcium and the photocurrent of rods and cones. *Biophys. J.* **11,** 47a.
15. Oakley II, B., and Pinto, L. H., this volume, Chapter 23.
16. Szuts, E. Z., this volume, Chapter 16.
17. Yoshikami, S., George, J. S., and Hagins, W. A. (1980). Light-induced calcium fluxes from outer segment layer of vertebrate retinas. *Nature (London)* **286,** 395–398.

18. Gold, G. H., and Korenbrot, J., this volume, Chapter 17.
19. Sorbi, R. T., this volume, Chapter 18.
20. Baehr, W., Devlin, M. J., and Applebury, M. L. (1979). Isolation and characterization of cGMP PDE from bovine ROS. *J. Biol. Chem.* **254,** 11669–11677.
21. Ebrey, T. G., Kilbride, P., Hurley, J. B., Calhoon, R., and Tsuda, M., this volume, Chapter 7.
22. Fleischman, D., this volume, Chapter 6.
23. Cohen, A. I., this volume, Chapter 12.
24. Chader, G. J., Liu, Y. P., Fletcher, R. T., Aguirre, Santos-Anderson, R., and T'so, M., this volume, Chapter 8.
25. Bastian, B. L., and Fain, G. L., this volume, Chapter 19.
26. Ostroy, S. E., Meyertholen, E. P., Stein, P. J., Svoboda, R. A., and Wilson, M. J., this volume, Chapter 22.
27. Yoshikami, S., and Hagins, W. A. (1980). Kinetics of control of the dark current of retinal rods by Ca^{++} and by light. *Fed. Proc., Fred. Am. Soc. Exp. Biol.* **39,** 1814.
28. Kaup, U. B., Schnetkamp, P.P.M., and Junge, W. (1980). Metarhodopsin I/metarhodopsin II transition triggers light-induced change in calcium binding at rod disk membranes. *Nature (London)* **286,** 638–640.
29. Hurley, J. B. (1982). Isolation and assay of a phosphodiesterase inhibitor from retinal rod outer segments. *Meth. Enzymol.* **81** (in press).

Index

R